Development and Organization of the Retina

From Molecules to Function

NATO ASI Series

Advanced Science Institutes Series

A series presenting the results of activities sponsored by the NATO Science Committee, which aims at the dissemination of advanced scientific and technological knowledge, with a view to strengthening links between scientific communities.

The series is published by an international board of publishers in conjunction with the NATO Scientific Affairs Division

A	Life Sciences	Plenum Publishing Corporation
B	Physics	New York and London
C	Mathematical and Physical Sciences	Kluwer Academic Publishers
D	Behavioral and Social Sciences	Dordrecht, Boston, and London
E	Applied Sciences	
F	Computer and Systems Sciences	Springer-Verlag
G	Ecological Sciences	Berlin, Heidelberg, New York, London,
H	Cell Biology	Paris, Tokyo, Hong Kong, and Barcelona
I	Global Environmental Change	

PARTNERSHIP SUB-SERIES

1. Disarmament Technologies	Kluwer Academic Publishers
2. Environment	Springer-Verlag
3. High Technology	Kluwer Academic Publishers
4. Science and Technology Policy	Kluwer Academic Publishers
5. Computer Networking	Kluwer Academic Publishers

The Partnership Sub-Series incorporates activities undertaken in collaboration with NATO's Cooperation Partners, the countries of the CIS and Central and Eastern Europe, in Priority Areas of concern to those countries.

Recent Volumes in this Series:

Series A: Life Sciences

Development and Organization of the Retina

From Molecules to Function

Edited by

Leo M. Chalupa

University of California at Davis
Davis, California

and

Barbara L. Finlay

Cornell University
Ithaca, New York

Springer Science+Business Media, LLC

Proceedings of a NATO Advanced Study Institute on
Development and Organization of the Retina: From Molecules to Function,
held June 18 – 28, 1997,
in Crete, Greece

Library of Congress Cataloging-in-Publication Data

Development and organization of the retina : from molecules to
   function / edited by Leo M. Chalupa and Barbara L. Finlay.
       p.   cm. -- (NATO ASI series. Series A, Life sciences ; v.
   299)
     "Proceedings of a NATO Advanced Study Institute on Development and
   Organization of the Retina: From Molecules to Function, held June
   18-28, 1997, in Crete, Greece"--T.p. verso.
     Includes bibliographical references and index.
     ISBN 978-0-306-45906-1   ISBN 978-1-4615-5333-5 (eBook)
     DOI 10.1007/978-1-4615-5333-5
     1. Retina--Growth--Congresses. 2. Retina--Differentiation-
   -Congresses.   I. Chalupa, Leo M. II. Finlay, Barbara L. (Barbara
   Laverne), 1950-   . III. NATO Advanced Study Institute on
   Development and Organization of the Retina: From Molecules to
   Function (1997 : Crete, Greece) IV. Series.
     [DNLM: 1. Retina--physiology congresses. 2. Retina--cytology
   congresses. 3. Vision--physiology congresses.   WW 270D475 1998]
   QP479.D47 1998
   612.8'43--dc21
   DNLM/DLC
   for Library of Congress   98-21939
                                                            CIP

ISBN 978-0-306-45906-1

© 1998 Springer Science+Business Media New York
Originally published by Plenum Press, New York in 1998

http://www.plenum.com

10 9 8 7 6 5 4 3 2 1

PREFACE

Viewed under the microscope, the organization of the vertebrate retina appears rather simple and utterly beautiful. Comprised of no more than a dozen different cell types, arranged into three distinct cell and two plexiform layers, the job of the neural retina is to transform photons captured by the eye into the electrical and chemical signals giving rise to vision. How this process actually transpires remains a daunting challenge for researchers, but through the use of modern neuroanatomical, immunocytochemical, and neurophysiological methods substantial progress has been achieved on this front in recent years. The retina and the long range projections of ganglion cells, the output neurons by which all information is conveyed from the eyes to the visual centers of the brain, has also been exploited by researchers who have been primarily interested in extra-visual issues such as developmental neurobiology. Indeed, for much of this century, retinal projections have been a favorite model for those seeking to understand the mechanisms underlying the formation of specific connection patterns. Work on this front has also progressed substantially with the advent of modern molecular and cellular techniques.

In 1997, we organized a NATO Advanced Study Institute, held in Crete, whose major goal was to introduce students to the diversity of current research issues and methodologies dealing with the retina and retinal projections. Our intention was to bring together two groups of colleagues, those concerned primarily with the organization of the mature retina and those working mainly on issues dealing with development. In particular, we wanted to relate to the students (some of whom were senior colleagues in other fields) the broad range of topics addressed by modern retinal researchers as well as the powerful arsenal of techniques being used to tackle these problems. The chapters in this volume provide an account of the material covered by the lecturers. We are grateful to our colleagues for their participation and for providing the written materials in a timely fashion. We are also appreciative of the inputs from the students, many of whom exhibited informative posters depicting their own work. Virtually all the students took part in animated discussions before, during, and after the more or less formal lectures. Finally, we thank NATO as well as the National Science Foundation (which funded some of the USA students) for financial support and Plenum Press for publishing this volume.

CONTENTS

Development and Organization of the Retina
From Molecules to Function

DEVELOPMENT OF VISION AND THE PRE-VISUAL SYSTEM

S. S. Easter, Jr.,[1] G. N. Nicola,[1,2] and J. D. Burrill[1,3]

[1]Department of Biology
University of Michigan
Ann Arbor, Michigan
[2]Case Western Reserve Medical School
Cleveland, Ohio
[3]Molecular Neurobiology Laboratory
Salk Institute
San Diego, California

1. ABSTRACT

Two topics are covered, both of which are relevant to all vertebrates, but all the results here were obtained in the zebrafish embryo and larva. The development of a system of tracts in the forebrain, the "pre-visual system," is described. It includes the tract of the postoptic commissure, the dorsoventral diencephalic tract, and the tract of the posterior commissure, and homologs have been found in the embryos of several classes of vertebrates. All three tracts are intimately associated with the retinal axons that appear later. The onset of vision was studied behaviorally (examining the startle and the optokinetic responses) and related to the anatomical development of the eye and extraocular muscles.

2. PREFACE

This paper is written in both the first person singular (I, meaning SE) and first person plural (we, meaning my co-authors). The co-authors contributed most of the work, but the paper is drawn from two talks given by myself. In keeping with the nature of this volume as a summary of talks at a meeting, it is written in the first person, singular and plural, as a talk is given. When opinions are offered, they will be mine, but results will be ours. (Admittedly, in some cases it would be more appropriate to use third person, plural. As the saying goes, "When I say, 'I,' I mean 'we,' and when I say, 'we,' I mean, 'they.'")

Development and Organization of the Retina, edited by Chalupa and Finlay.
Plenum Press, New York, 1998.

3. THE DEVELOPING NERVOUS SYSTEM

We are attracted to a particular field of study for a variety of reasons, among which is its inherent structure, the rigor of the investigative methods, and the aesthetics of the product. All of these attracted me to development, but I suspect the main reason was that I realized that I would never understand the mature nervous system, and so I might just as well try to understand an immature one, which was probably simpler. I have been moving backwards in ontogenetic time ever since, and have now settled in at the embryonic stage when the first tracts begin to form in the vertebrate brain, just after the neural tube. I find it comforting working on a little organism in which there may be a half dozen tracts, half that many commissures, and probably no synapses; I may be able to understand the workings of such a nervous system, at least rudimentarily. So part of this chapter is a summary of what we have been able to find out about these early stages of the development of the nervous system. The other part is about a later event, the development of the retina and of vision. We will attempt to link the two because, surprisingly, the central part, the tracts in the brain, are in place before the more peripheral part, the retina, so when the retina begins to function it has a ready made network of tracts already in place in the brain.

4. THE ZEBRAFISH

The study of embryonic development requires embryos, and one of the main advantages of studying the zebrafish is that it produces embryos with quite astonishing regularity. A standard aquarium of breeding fish can produce about 50 fertile eggs per day. They are kept in a room with a controlled light/dark cycle, and when the lights go on in the morning, the zebrafish begin to breed. The eggs are fertilized outside of the mother and immediately sink to the floor of the aquarium. The parents will normally eat the eggs, but if the bottom is structured to provide nooks and crannies for them, then the parents can't reach them and the experimenter gets them. The eggs are small, but the egg shell (chorion) is quite transparent, and with the help of a dissecting microscope the number of cells can be counted. From this number one can infer, with an uncertainty of less than 20 minutes, the time of fertilization, and if the embryos are kept at a constant temperature (we use 28.5°C) then development proceeds along a fairly predictable schedule. So the ease of acquisition of embryos is a major advantage of the zebrafish. What else?

They develop very rapidly. From fertilization to hatching is only about three days. Given that conception takes place at sunup on one morning, that means that the fish's life outside the egg will begin around the morning of the fourth day post fertilization (dpf). Usually this is an uneventful day, as the embryo has been spending the first three days turning yolk into fish, and at the time it hatches, a good bit of yolk still remains, so the hatchling stays on the bottom and completes the yolk transformation. Then, on the fifth day it is up in the water column feeding on small plankton, a behavior that looks like a vision-based predation. Rapid development is useful to the experimenter, in that it shortens the turnaround time between the initiation and the completion of a study. But it has its downside, too, which is that everything happens in a hurry.

The adults and the embryo are small. This is an advantage because of the modest needs for space, and therefore of money, to house the animals. The adults are happy in standard aquaria. The embryos are happy in small petri dishes inside benchtop water incubators. When one manipulates them, and especially when one sections them for histology, the small size is a help because it is possible to cut a section that includes a cross-section of the entire organism, and so the context of whatever one studies is not necessarily lost.

They are optically clear. This is a consequence of the fact that they keep their yolk separate from the other cells; it has a different index of refraction than the aqueous cytoplasm, and so if it were dispersed inside the cytoplasm, as it is in frogs, the cells would be milky. The small size coupled with the optical clarity of the animal contributed to the feasibility of some of the visual studies that will be described below. The optical clarity has been exploited most cleverly in the ingenious studies of Kimmel and his collaborators [12] who have injected blastomeres with non-toxic fluorescent dyes and followed the progeny in living embryos. As a result of these studies, we know more about cellular movements, particularly during gastrulation, in zebrafish than in any other vertebrate.

They normally develop outside of the uterus (in contrast to mammals) and when pressed, they develop normally outside their chorions (in contrast to chicks). The ease with which one can modify their environment makes experiments on the embryos feasible, but it must be admitted that they are fragile (relative to *Xenopus* and chick) when subjected to surgery.

They are genetically accessible, as a large number of developmental mutations have recently been produced that should prove useful for the analysis of development. But the genetic tricks that have made the study of mice so successful (such as knocking out genes by homologous recombination) are not available in fish, nor are most of the ones (e.g., transposable elements) that make *Drosophila* development the trendsetter in almost every aspect of development.

4.1. The Pre-Visual System

I am probably the only person to use this terminology, and it reflects my origins as a visual scientist, even a visual chauvinist. What I mean by the "pre-visual system" is that part of the CNS that develops before the retinal axons, and may interact with and possibly influence their growth and termination. The several vertebrate embryos that have been examined all have a similar set of tracts that appear early, soon after the formation of the neural tube[9,11,16,18]. They include the tract of the postoptic commissure (TPOC), the dorsoventral diencephalic tract (DVDT) and the tract of the posterior commissure (TPC), and all are shown in Figure 1.

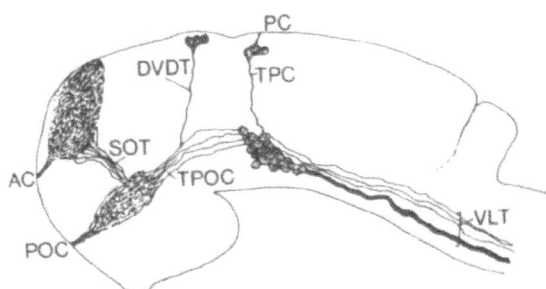

Figure 1. This is a schematic diagram of the brain of a 24 hpf zebrafish embryo, seen from the left side (dorsal is up and anterior to the left). The eyes have been removed. There are three commissures (anterior, AC; posterior, PC; and postoptic, POC), and four tracts (tract of the postoptic commissure, TPOC; tract of the posterior commissure, TPC; dorsoventral diencephalic tract, DVDT; and supraoptic tract, SOT). The dorsoventral thickness of the brain is about 100 μm. From reference 14 with permission.

The most important is the TPOC, which is pioneered by a few axons that issue from neurons (the only ones in the forebrain at this time) at the base of the optic stalks. Our knowledge of this tract comes from several species, and all aspects have not been seen on all species, so I will emphasize what we know about zebrafish. The first axons appear at 18 hours post fertilization (hpf) in zebrafish, are the only neurons in the forebrain, and their axons form the first longitudinal tract there[14]. They extend into the midbrain and hindbrain, but we do not know where they terminate or indeed if they terminate in the classical sense, with presynaptic endings. It is possible that they are transient axons, as the neurons may not persist in the adults. Around 24 hpf, some of these neurons send axons across the midline, thus forming the postoptic commissure (POC), so named because of its position behind the optic stalks[18]. About 32 hpf the first retinal axons reach the midline, where they pass very close to the dorsal boundary of the POC (thus forming the optic chiasm), grow alongside the TPOC through most of the diencephalon, and then diverge from it dorsally to enter the presumptive tectum by about 42 hpf[2,3]. As we will see below, this all occurs long before visual behavior can be evoked.

The other two tracts in the pre-visual system are oriented dorsoventrally; both join the TPOC and are close to the optic axons. The DVDT originates from neurons at the base of the epiphysis, the presumptive pineal body. This tract is pioneered by one or two axons within an hour or so after the TPOC appears, and they grow ventrally to meet the dorsal-most fascicles of the TPOC within a few hours. They invariably turn rostrally at this point, fasciculate with the TPOC axons, and cross the midline to terminate in the contralateral hypothalamus. As with the TPOC, no adult tract with this origin and termination is known to exist, therefore the DVDT axons may be transient as well. The pineal contains photoreceptors a day after the first DVDT axons appear, and they are closely associated with the somata that produced the axons[13], so we infer that it serves a photoreceptive function, but the details have not been examined. Given the prominent role of the pineal gland in circadian rhythms, it seems likely that the precocity of the fish pineal may serve to get the fish in phase with the diurnal cycle, but this has not been investigated. The remaining previsual tract, the TPC, also originates from a small cluster of neurons in the alar plate; and its axons also grow ventrally, meet the TPOC and fasciculate with it, but in contrast to the DVDT axons, they invariably turn caudally. Later, a small population of optic axons diverge from the main bundle and grow dorsally in association with the TPC to ramify in a dorsal pretectal area[2].

When Jeremy Taylor and I first discovered the TPOC in *Xenopus*[10] we thought that the optic axons fasciculated with the TPOC axons and were guided by them. Since then, two other findings have led me to doubt that the TPOC axons play an essential role in retinal axon guidance. First an experimental study in *Xenopus* showed that retinal axons grew successfully to the tectum in the absence of a TPOC[5]. (In a similar experiment in zebrafish, the TPOC was removed surgically and the TPC axons were tested to see if they would still grow caudally at the level where they normally meet the TPOC[4]. Some did, supporting the idea that the TPOC axons are not essential to the navigation by late comers.) Second, we found that the retinal axons in zebrafish never actually contacted the TPOC; they were always near, but always separate from it, and did not fasciculate with the TPOC axons[3]. The most economical explanation for these results is that the TPOC and optic tract axons all follow the same guidance signals, which essentially point caudally along the longitudinal axis. (I must note that Dr. Taylor does not share this view.) No comparable evaluations of the DVDT or the TPC have been made.

In summary, the first retinal axons arrive in a brain that already has several tracts in place, and while the retinal axons grow in close association with three of them, the functional role of the pre-existing tracts, the "pre-visual system," is apparently not too important. The retinal axons probably follow the same cues in the neuroepithelium that the axons of the pre-existing axons followed.

4.2. The Development of Zebrafish Vision

Although the first retinal axons are in place early on the second day of life, the retina is not structurally mature until many hours later. Synapses are evident electron microscopically beginning on the second day. The first photoreceptors are born at 48 hpf[1], but do not have outer segments until about 60 hpf, which must therefore be the earliest age at which one could anticipate photoreception to be possible.

We have investigated the onset of vision behaviorally[7,8]. A description of the eye and nervous system may lead us to infer that an animal can see, and electrophysiological signs such as the electroretinogram give further objective evidence of photoreceptivity, but the only way to be sure that an animal sees is by doing a behavioral test. If the animal responds to a visual signal, then it can see. We have used two unlearned behavioral responses: the startle response and the optokinetic response. The startle response is a sudden twitch, a contraction of the longitudinal muscles. It can be evoked by a variety of stimuli; we used a sudden decrease in light intensity. The utility of this response to the animal is unclear, but it is easy to detect and almost certainly depends only on the animal's ability to detect a change in light intensity, and not on any pattern vision. The optokinetic response, in contrast, requires pattern vision. The animal is put inside a cylinder with high contrast vertical stripes lining the inside, and the drum is rotated about the vertical axis. All animals, including humans, do the same thing when put in such an environment: they move their eyes or head or both in the direction of rotation. The biological utility of this behavior is clear; it stabilizes the retinal image of the outside world, and a stable image is believed to be easier to analyze than a moving one. If the eyes are not able to resolve the stripes, then the eyes will not move because a defocussed set of vertical stripes is indistinguishable from a uniform gray surface.

We studied the onset of vision by exposing 20 animals of the same age to identical tests and recording their individual responses. Every group of 20 was naive; having been through the tests at one age, they were not used again.

The startle response was evoked both mechanically and visually; a light touch to the flank of an embryo will evoke a twitch, and this served as a useful indicator that the fish could twitch. The touch-evoked startle was evident by 48 hpf in all fish. The shadow-evoked startle appeared much later. A single twitch to just one of 5 presentations was scored as a response; such a permissive criterion was acceptable because of the virtual absence of twitches in the absence of a visual stimulus. As Figure 2 shows, none of the fish responded before 68 hpf, and the fraction of fish that did respond was initially quite small. Over time, all of the fish responded, and those that did so responded more frequently. Eyeless fish did not respond, so the behavior was not mediated through some light sense outside the eyes. Fish did not respond to sham stimuli (an activated shutter without the light on) so the response was to a decrease in light. Thus, by the criterion of the startle response, vision developed over about 10 hours, beginning 8 hours after the first photoreceptor outer segments appeared, and about the same time that the full set of 10 retinorecipient targets in the brain were receiving arborizations of retinal axons[2]. This development would normally be occurring on the fourth day post fertilization, the first day outside of the egg.

The optokinetic response is a more subtle response than the axial twitch. In the adult, it has two components; one a drift in the direction of stripe movement with a veloc-

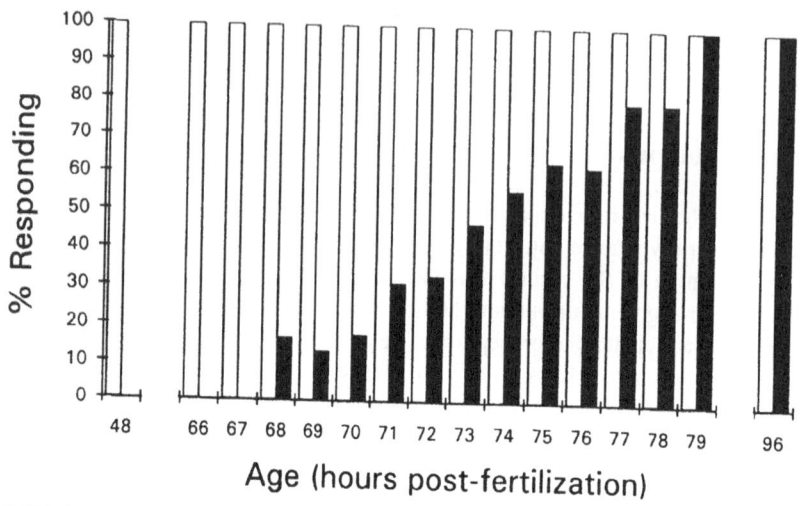

Figure 2. This bar graph shows the time course of the development of the two startle responses, to touch (light bars) and to shadow (dark bars). The vertical axis indicates the percentage of fish at each age that responded to at least one of the five repeated stimuli. All subjects received both kinds of stimuli, and no subjects were used at more than one age (horizontal axis). From reference 7 with permission.

ity graded according to how fast the drum rotates, and the other a fast reset movement in the opposite direction. The need for two movements is obvious, because the stripes move around 360°, but the eyes can not, so what they do normally is to track the stripes for a few degrees and then quickly spring back to a position closer to where they started and begin tracking again. This "optokinetic nystagmus" is an oscillatory movement of which only the slow phase is visually driven, and it was the slow phase that we sought in our embryos and larvae. Slow eye movements are not easily detected, but by videotaping the fish and then playing the tape back at high speed, they were spotted. Figure 3 shows the

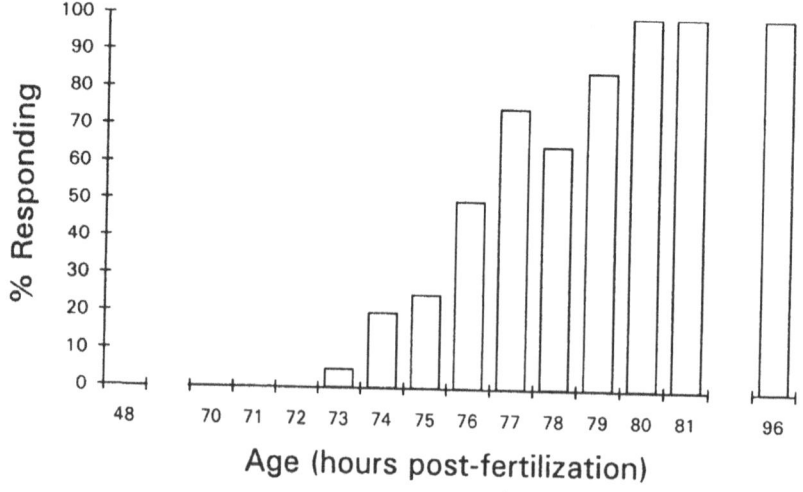

Figure 3. This bar graph shows the time course of the onset of the optokinetic response. As with Figure 2, all subjects were used at only one age. The vertical axis indicates the percentage that followed the rotating striped drum at some time, however briefly, during its rotation. From reference 7 with permission.

results. The optokinetic response began slightly later than the startle response, and took about the same period of time to develop.

The initial responses were different from the adult response because of the absence of reset movements (the fast phase), but the fast movements appeared shortly after the slow phase[8]. Like the tracking movements, the fast ones never occurred spontaneously, and when they appeared, it was always in association with the slow phase.

4.2.1. Behavioral Variability. The variability of both the startle and the optokinetic responses was unexpectedly large. Our expectations had two bases, one our intuition and the other our previous knowledge of zebrafish morphological variability.

To consider the latter first: some variation was expected, as animals of the same clutch, known to be developmentally synchronized soon after fertilization, often develop at different rates later on, and the same morphological state may be achieved by embryos whose ages differ by as much as 2–3 hours[2]. The data of Figures 2 and 3 suggest a much greater variability than that, because every age between 68 and 79 hpf produced both responsive and unresponsive subjects. Thus the most retarded larva became responsive 11 hours after the most precocious, a behavioral variability of 11 hours.

Our intuition led us to anticipate that the development of responsiveness was less an all-or-none phenomenon (non-responsive abruptly changing to responsive) than a gradual shift. We anticipated that individual fish would progress from non-responsive to sluggishly responsive to vigorously responsive, and we supposed that the number of responses to each set of stimuli would indicate the maturation of this vigor. Thus, the same larva would be expected to respond to 1/5 at one age, 2/5 a few hours later, 3/5 still later, etc. If that were the case (and remember, we never tested any fish at more than one age), then development of an individual fish could be described by a graph in which the fraction of stimuli that evoked a response is plotted on the vertical axis vs. age on the horizontal, and we would expect to see a sigmoid relation. Different individuals would be expected to have their curves located at different places along the x-axis, and the population response would then be a sum of out of phase sigmoids. The data of Figures 2 and 3 look like that, with an unexpectedly high variability (see previous paragraph), but when each age is examined in detail, we find that the model of individual development is inconsistent with the data. That conclusion is not evident from the data as plotted here, because our criterion for responsiveness (1 response per 5 presentations) masks a considerable degree of complexity in the data. If we consider the data not in terms of individual animals, but in terms of the total number of responses to the stimuli, at 68 hpf, 100 stimuli were given and only 1 evoked a response. By 75 hpf (to choose just one age), most of the animals responded, and most of those that responded did so to more than 1 of the stimuli presented, but a substantial number continued to be unresponsive to any. All right, you say, that just reflects the greater than expected variability, and you could be right. But by the age of 78 hpf, all the fish that responded, responded vigorously (to 3 or more of the 5 stimuli), and none responded to 1 or 2, but several still remained unresponsive. This void of sluggish responders at the older ages is puzzling, because if individuals were progressing from nonresponsive to sluggishly responsive to vigorously responsive, then we would expect that some of those that were non responsive at 74 hpf would have responded sluggishly to the stimuli when they were an hour older. But they didn't; at the older ages, every individual was either vigorously responsive or non-responsive. Instead of a continuous development of this "visual maturation" in all individuals, there seems to be a non-linearity at work such that some of the fish remain mired in non-responsiveness as their cohorts are approaching adult levels of behavior, and when the non-responsive ones become responsive, they leap into a vigorous mode, never passing through the sluggish one. This non-linearity is surprising, and we have not done any experiments to investigate its origin. We speculate that it may reflect a set of steps in series (to oversimplify: first, detection of the stimulus; second, transmission of the message reporting detection; and third, activation of the muscle contraction), and the later steps (message transmission and muscle contraction) may mature even when the first (detection in the eye) is not functional. Thus, non-responsive fish could be unable to detect the stimulus even though their transmission and activation abilities were advanced, so when the detection becomes functional they would immediately behave in a relatively mature mode. While this accounts for the non-linearity, the greater temporal variability remains unexplained, as the scheme that we suggest would require a 10 hour variability in the development of the detection operation.

4.2.2. Light Detection vs. Form Vision. The delay between the onset of the visual startle response and the optokinetic response, about 3 hours, is probably attributed to two limiting steps: the development of functional extraocular muscles and the formation of the retinal image [7]. The extraocular muscles are essential for moving the eyes, and as we have noted, eye movements are an integral part of the fish's visual analysis of the world. We have examined the extraocular muscles using light and electron microscopy and immuno-cytochemistry, and have shown that they are very rudimentary at 66 hpf but quite mature by 72 hpf, about the time that the eyes first start to follow the drums. The formation of the retinal image was examined directly, using the optics of the compound microscope (Fig. 4). To be of any use in form vision, an eye must be able to form an image of an object at optical infinity (a fancy term for objects that lie many focal lengths away). An eye that can do this is "emmetropic;" one that forms an image that falls behind the photoreceptors is "hyperopic" or far sighted; and if the image falls inside the eyeball in front of the photoreceptors, then the eye is "myopic" or near-sighted. We showed that the lens could image a grating at optical infinity quite early, but the image lay well behind the eye. Thus the eye was initially hyperopic. The focal length of the eye gradually decreased as it matured, and by about 72 hpf, when the larva would normally begin its first day outside the egg shell, the eye was emmetropic. All this dioptric maturation occurred in the lens, which is the only optically active element in the eyes of aquatic animals. Paradoxically (because smaller lenses have shorter focal lengths in adult fish) the shortening of the focal length was associated with an increase in the lens diameter, but, more importantly, with the development of a dense refractile core in the lens center (Fig. 5). This active adjustment of the focal length continues through life for the fish, as their eyes enlarge continually and remain emmetropic[6].

The life history suggests that some of the visual maturation that we have described might be experience-dependent. We have tested that hypothesis and rejected it. We reared the embryos in total darkness from 24 hpf (before the birth of any retinal neurons) to 5 dpf, and then evaluated the optics and the optokinetic response immediately on bringing them into the light. We found that both were normal; the eyes were emmetropic and both slow and fast movements of the optokinetic response were appropriately directed.

The conclusion that the optokinetic response was independent of experience is consistent with decades of research on the retinotectal projection in fishes and frogs. A wide variety of experimental procedures, including eye rotation, optic nerve deviation, and others, have led to the conclusion that the polarity of the projection from anamniotes' eyes is hard-wired from the outset, and that this polarity rules visuomotor behavior[15]. But recent work, mostly on chick and monkey, has shown that the emmetropization of the eye is an active process that is dependent on the visual experience of the eye[17]. We anticipate that the same is true for later stages of the development of fishes' eyes, too, and we suggest that the apparent difference is attributable to the difference in stages of development that the two bodies of work have covered. The experimental chicks and monkeys were living independently (as hatchlings and newborns) in contrast to the larval fish, which had only begun to fend for themselves hours earlier. We suggest that the very earliest stage of dioptric development in all animals is experience-independent; indeed, the eyes of chicks and monkeys *in ovo* or *in utero* have no visual experience, so it must be.

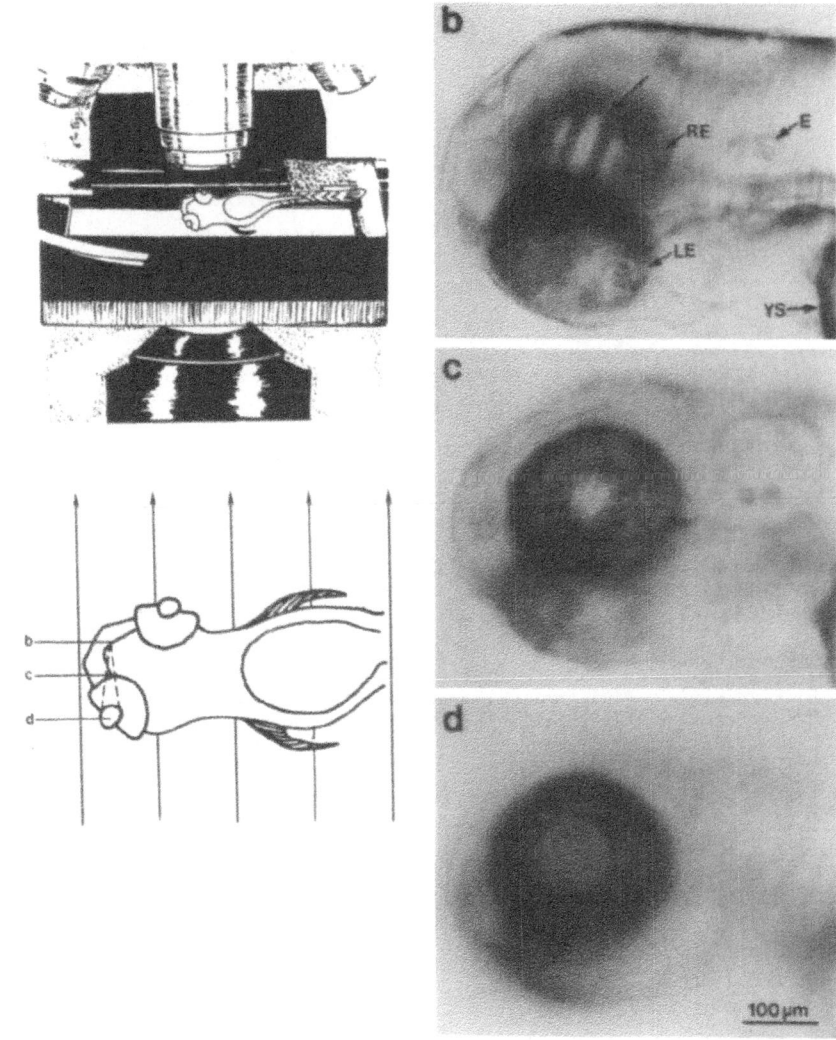

Figure 4. This shows the optical arrangement used to assess the refractile state of the eye. **a.** The zebrafish lay on its side in an optically flat water-filled chamber, and a grating was presented at optical infinity to the downward looking eye. **b–d.** The experimenter directly examined the back focal plane of this eye, and the three photomicrographs on the right illustrate the images at each of the three focal planes indicated by the letters, b, c, and d on the left. This 66 hpf embryo was hyperopic (the image was formed behind the eye). From reference 7 with permission.

5. CONCLUSIONS

The development of the visual system in zebrafish, and probably in all vertebrates, is preceded by the formation of a small set of tracts in the forebrain of which some will be intimately associated with the retinal axons. Although the role of these tracts in the guidance of the retinal axons appears to be minor, the full story is probably not yet told, and what appears to be a non-essential role may yet be shown to have some positive features.

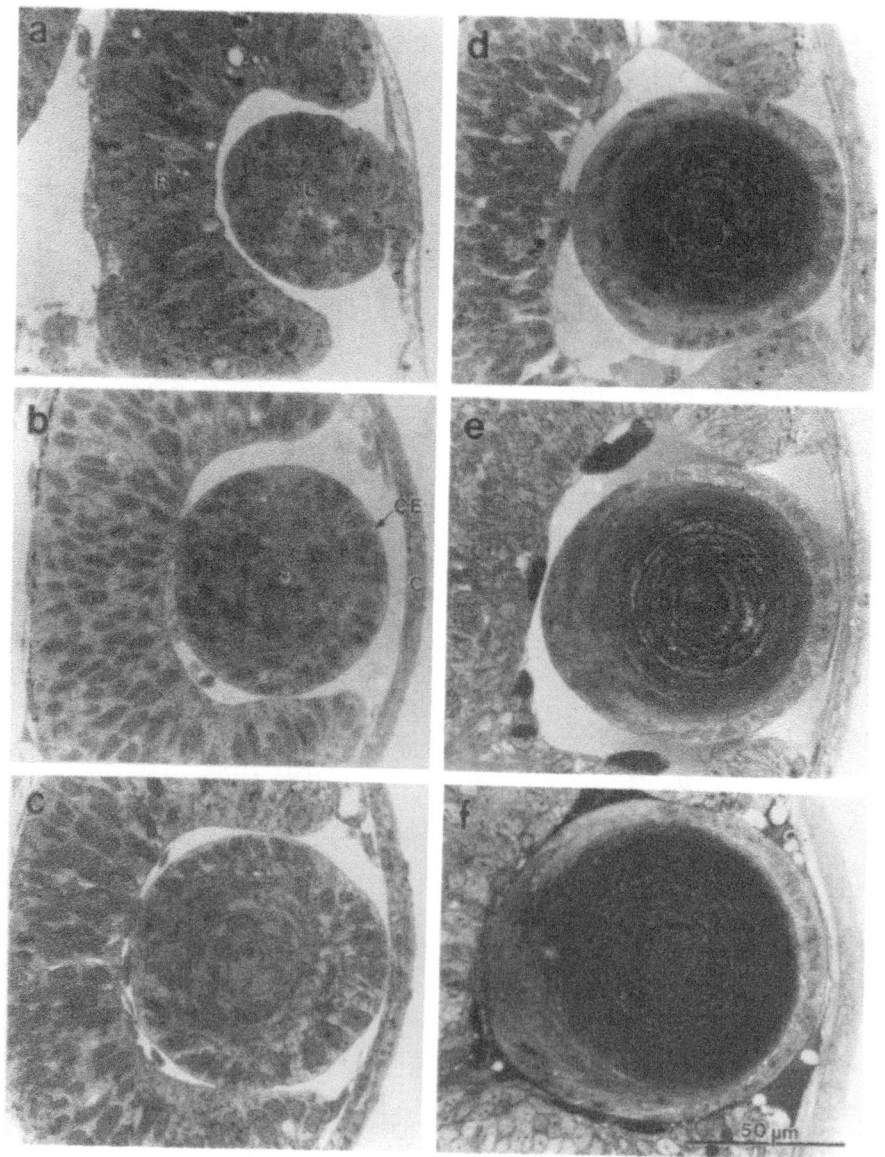

Figure 5. This set of semithin sections illustrates the size and structure of the lens in progressively older fish. The ages (all in hpf) were: a. 24, b. 30, c. 36, d. 48, e. 60, f. 72. From reference 7 with permission.

The later development of form vision and the optokinetic response resembles the "just in time" manufacturing strategy that has become so widespread in recent years. Although the assembly of the sensorimotor circuit is underway throughout the second and third days, the finishing touches that permit image formation and eye movement are added only at the last minute, when the larva begins to live on its own, beginning what will be a lifetime of ocular growth and modification.

ACKNOWLEDGMENT

The work described here was all supported by a research grant from NIH, R01-EY-00168 to SSE.

REFERENCES

1. Branchek, T., and BreMiller. R. (1984). The development of photoreceptors in the zebrafish, *Brachydanio rerio*. I. Structure. J. Comp. Neurol. *224*: 107–115.
2. Burrill, J. D., and S. S. Easter, Jr. (1994). The development of the retinofugal projections in the embryonic and larval zebrafish, (*Brachydanio rerio*). J. Comp. Neurol. *346*: 583–600.
3. Burrill, J. D., and S. S. Easter, Jr. (1995). The first retinal axons and their microenvironment in zebrafish: cryptic pioneers and the pre-tract. J. Neurosci. *15*: 2935–2947.
4. Chitnis, A. B., and J. Y. Kuwada (1991). Elimination of a brain tract increases errors in pathfinding by follower growth cones in the zebrafish embryo. Neuron *7*: 277–285.
5. Cornel, E., and C. E. Holt (1992). Precocious pathfinding: retinal axons can navigate in an axonless brain. Neuron *9*: 1001–1011.
6. Easter, S. S., Jr., P. R. Johns, and L. R. Baumann (1977). Growth of the adult goldfish eye-I: Optics. Vision Res. *17*: 469–477.
7. Easter, S. S., Jr., and G. N. Nicola (1996). The development of vision in the zebrafish (*Danio rerio*). Dev. Biol. *180*: 646–663.
8. Easter, S. S., Jr., and G. N. Nicola (1997). The development of eye movements in the zebrafish (*Danio rerio*). Dev. Psychobiol. (In press.)
9. Easter, S. S., Jr., L. S. Ross, and A. Frankfurter (1993). Initial tract formation in the mouse brain. J. Neurosci. *13*: 285–299.
10. Easter, S. S., Jr. and J. S. H. Taylor (1989). The development of the *Xenopus* retinofugal pathway: optic fibres join a pre-existing tract. Development *107*: 553–573.
11. Easter, S. S., Jr., J. D. Burrill, R. C. Marcus, L. S. Ross, J. S. H. Taylor, and S. W. Wilson (1994). Initial tract formation in the vertebrate brain. In: *The Self-Organizing Brain: From Growth Cones to Functional Networks*. (Edited by J. van Pelt, M. A. Corner, H. B. M. Uylings, and F. H. Lopes da Silva). Prog. Brain Res. *102*: 79–94.
12. Kimmel, C. B., and R. Warga (1986). Tissue-specific cell lineages originate in the gastrula of the zebrafish. Science *231*: 365–368.
13. Masai, I, C. P. Heisenberg, K. A. Barth, R. Macdonald, S. Adamek, S. W. Wilson (1997). *Floating head* and *masterblind* regulate neuronal patterning in the roof of the forebrain. Neuron *18*: 43–57.
14. Ross, L. S., T. Parrett, and S. S. Easter, Jr. (1992). Axonogenesis and morphogenesis in the embryonic zebrafish brain. J. Neurosci. *12*: 467–482.
15. Sperry, R.W. (1943). Effect of 180 degree rotation of the retinal field on visuomotor coordination. J. Exp. Zool., *92*: 263- 279.
16. Taylor, J. S. H. (1991). The early deveopment of the frog retinotectal projection. Development *Supp. 2*: 95–104.
17. Wallman, J. (1993). Retinal control of eye growth and refraction. Prog. Retinal Res. *12*: 133–153.
18. Wilson, S., L. Ross, T. Parrett, and S. S. Easter, Jr. (1990). The development of a simple scaffold of axon tracts in the brain of the embryonic zebrafish, *Brachydanio rerio*. Development *108*: 121–147.

2

CELLULAR AND MOLECULAR ASPECTS OF PHOTORECEPTOR DIFFERENTIATION

Ruben Adler

Wilmer Ophthalmological Institute
Johns Hopkins University School of Medicine
Baltimore, Maryland

INTRODUCTION

The investigation of genetic programs and microenvironmental signals that regulate retinal cell differentiation in vertebrates is a very active field that has attracted the attention of many investigators. A comprehensive review of the relevant literature is beyond the scope of this presentation, the goal of which is to summarize studies from this laboratory focusing on cellular and molecular mechanisms regulating the differentiation of chick retinal cells in general, and photoreceptor cells in particular; the reader is referred to recent reviews of the field (e.g., Adler, 1993; Hausman, 1993; Layer and Willbold, 1993; Reichenbach, 1993; Lillien, 1994; Cepko, 1996; Goldowitz et al., 1996; Hicks, 1996; Rapaport, 1996; Szel et al., 1996).

By the end of its embryonic development, the mature chick neural retina consists of a variety of cell types that are already highly differentiated, are postmitotic, and are segregated in a characteristic pattern of layers, with each cell type occupying a defined laminar position. At early stages of embryonic development, however, the future neural retina is a very simple neuroepithelium, in which all cells are mitotically active (Fujita and Horii, 1963). Each mature retinal cell, therefore, is generated through a series of mitotic divisions; a neuron is considered "born" after it undergoes its last division. After this, terminal mitosis cells migrate to one of the retinal layers, differentiate into a cell type that corresponds to their laminar position within the retina, and establish synaptic connections with specific pre- and/or postsynaptic partners. Lineage-tracing studies have shown, in the chick and in other species, that proliferating precursors can give rise to two or more types of mature derivatives, indicating that retinal development does not involve a deterministic lineage mechanism (Turner and Cepko, 1987; Holt et al., 1988; Wetts and Fraser, 1988; Turner et al., 1990). While studies of this type suggest that precursor cells remain multipotential while they are mitotically active, they do not show whether the commitment of each cell to a specific phenotypic fate occurs at the time of, or sometime after, its terminal

Development and Organization of the Retina, edited by Chalupa and Finlay.
Plenum Press, New York, 1998.

mitosis. The elucidation of the timing of these events is potentially very important, since it could provide useful clues for the investigation of the underlying cellular and molecular mechanisms, and is the subject of some of the experiments described below.

Cell differentiation involves the coordinated expression of a large number of molecular, structural and functional properties. How is such coordination established during the transition of retinal precursor cells, from their initial undifferentiated phenotype, to their mature differentiated state? This question has also been addressed by our studies, which have tested predictions from two different models. One of these scenarios would propose that each new property expressed by the cell is induced by a separate inductive signal, and that the final phenotype is determined by the nature and sequence of those signals. An alternative (but not completely exclusive) model postulates the existence of "developmental master programs" controlling the coordinated expression of differentiated properties. The relative contributions of each of these two hypothetical mechanisms to the differentiation of retinal cells in general, and photoreceptor cells in particular, is not well understood, but some of the studies summarized below have suggested that master regulatory programs do play a protagonistic role in the development of chick photoreceptor cells.

CELL BIRTH ANALYSIS DURING CHICK RETINAL DEVELOPMENT USING THE "WINDOW-LABELING" TECHNIQUE

The initial analysis of questions regarding the role of terminal mitosis in cell differentiation was based on the investigation of cell birth *in vivo* using either ^{3}H-thymidine (^{3}HT) autoradiography or bromodeoxyuridine (BrDU) immunocytochemistry; these two precursor molecules are incorporated into the DNA of dividing cells during the S phase of the mitotic cycle. Many different studies have shown that, while different retinal cell types are generated in a predictable sequence, there is also considerable overlap in their time of birth, as well as significant interspecific differences (Fujita, 1963; Fujita and Horii, 1963; Hollyfield, 1968; Sidman, 1970; Morris, 1973; Kahn, 1974; Young, 1985; Stone, 1988, Spence and Robson, 1989; Snow and Robson, 1994). Much of this information was obtained using either a "cumulative" labeling paradigm (in which ^{3}HT or BrDU are kept constantly available after their initial administration), or a "pulse-labeling" method, in which the DNA precursor is only made available to the cells for a limited period of time. Cumulative labeling allows analysis of cells born *before* the initial administration of the label, whereas pulse-labeling is usually used to analyze "heavily-labeled" cells, that is to say, cells that stop dividing within a relatively short time after the labeling pulse, and therefore do not undergo dilution of the label through further rounds of division. Although these methods have provided much useful information, they have some limitations, including lower temporal resolution than what is needed for detailed investigation of possible correlations between terminal mitosis and cell differentiation. To overcome these limitations, we have developed a technique that allows determining the time of terminal mitosis with a resolution of hours, rather than days (Repka and Adler, 1992b; Belecky-Adams et al., 1996). As shown in Fig. 1, the method is based on an initial injection of ^{3}HT, followed 3–5 hours later by an injection of BrDU; the latter is repeated as needed to keep this precursor constantly available to the cells. Concomitant analysis by ^{3}HT autoradiography and BrDU immunocytochemistry makes it possible to distinguish three populations of cells in preparations labeled in this manner (Fig. 1): unlabeled cells, born *before* thymidine injection, BrDU(+) cells, born *after* BrDU administration, and ^{3}HT(+)/BrDU(−), "window-labeled"

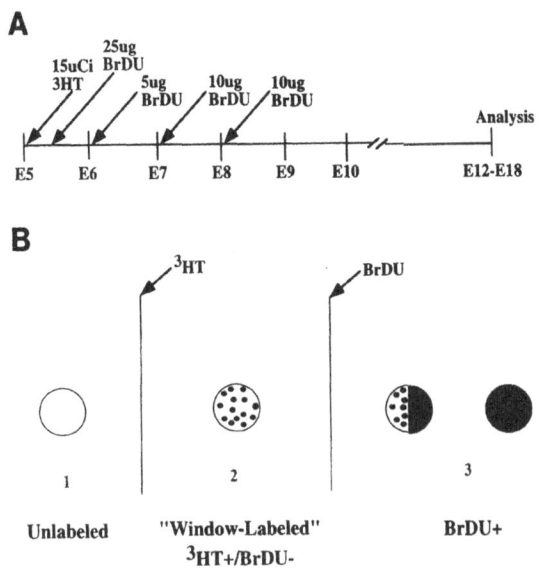

Figure 1. (A) Diagrammatic representation of a "window-label" protocol used to identify cells undergoing terminal mitosis during a 5 hr period on ED 5. Fifteen μCi of ^{3}HT was added through an opening made in the eggshell, followed 5 hours later by 25 μg of BrDU. BrDU administration was repeated thereafter until ED 8, to keep it constantly available to cells that continue to divide. The embryos were fixed and processed for BrDU immunocytochemistry and ^{3}HT autoradiography. (B) Diagrammatic representation of the three populations of cells observed in a window-label procedure (adapted from Repka and Adler, 1992b). Cells that undergo terminal mitosis (are "born") prior to the addition of ^{3}HT are unlabeled (1), while cells that continue to divide following addition of BrDU are labeled either with BrDU alone or with BrDU and ^{3}HT (3). The only cells labeled with ^{3}HT, but not with BrDU, are those undergoing their last round of DNA duplication after ^{3}HT autoradiography but before BrDU administration ["window-labeled" cells (2)]. From Belecky-Adams et al., 1996, with permission from Academic Press.

cells, which are those undergoing terminal mitosis during the interval separating the administration of both precursors. It is important to note that, in addition to providing higher temporal resolution than what is achievable with pulse- or cumulative labeling with individual precursors, the window-labeling technique makes it possible to identify cohorts of contemporary cells not only within their normal microenvironment, but also when they are explanted or transplanted, thus allowing comparisons of their behavior under different experimental conditions.

The window-labeling technique was recently used to reinvestigate the kinetics of cell generations during normal development of the chick embryo retina *in vivo* (Belecky-Adams et al., 1996); groups of embryos were "window-labeled" for 5 hours on different embryonic days (ED) and allowed to develop until embryonic day 18, a stage at which retinal cell differentiation resembles closely that observed in the adult. Quantitative analysis of ED 18 eyes, processed for autoradiography and BrDU immunocytochemistry, showed a pattern of laminar distribution of window-labeled cells that, despite some differences in detail, was generally consistent with observations reported on the basis of pulse- or cumulative-labeling approaches (Fujita and Horii, 1963; Morris, 1973; Kahn, 1974; Dutting et al., 1983; Spence and Robson, 1989; Prada et al., 1991; Snow and Robson, 1994); the results are summarized in Figs. 2 and 3. Several general conclusions can be reached based on this analysis. First, the studies confirmed that there is extensive overlap

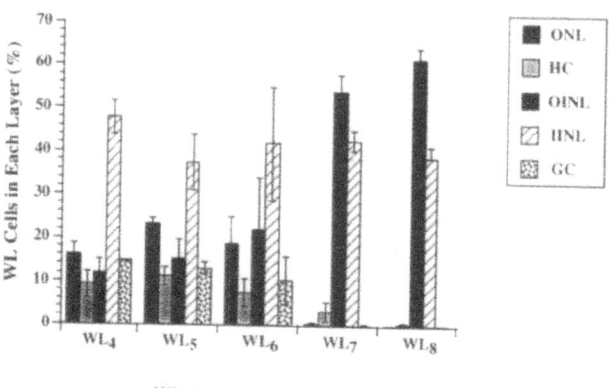

Figure 2. Quantitative analysis of the number of cells born during 5-hr periods on ED 4–8. Embryos were window-labeled on ED 4, ED 5, ED 6, ED 7, or ED 8, allowed to survive until ED 18, and processed for BrDU immunocytochemistry and autoradiography. The number of window-labeled cells in each layer/sublayer was counted in an area adjacent to the choroid fissure. For each WL period, the number of cells in each retinal sublayer was expressed as a percentage of the total number of WL cells in the sampled area. From Belecky-Adams et al., 1996, with permission from Academic Press.

in the generation of different cell types within individual 5-hr periods, suggesting that the time at which precursor cells undergo terminal mitosis is not, by itself, a cell fate determinant. Second, the developmental potential of proliferating precursor cells seems to become progressively restricted during normal development; thus, precursor cells undergoing their last mitotic division during the window-labeling period on ED 4 or 5 contribute to all retinal cell layers, whereas progenitors window-labeled on ED 7 give rise only to inner

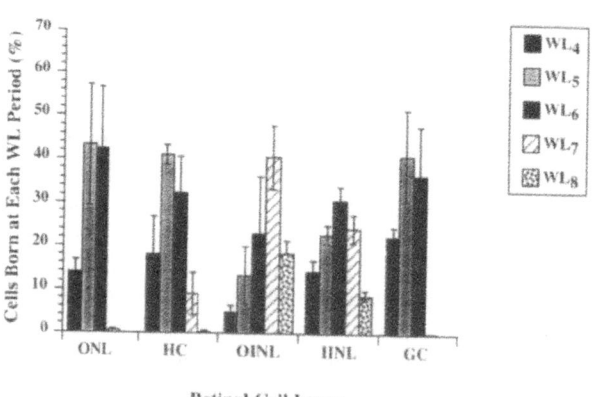

Figure 3. Time of birth of cells in each retinal layer. Embryos were window-labeled for 5 hr on ED 4, ED 5, ED 6, ED 7, or ED 8, allowed to survive until ED 18, and processed for BrDU immunocytochemistry and autoradiography. The number of WL cells in each retinal layer/sublayer was counted. For each retinal layer, the number of cells contributed by each WL period is expressed as a percentage of all the WL cells observed in that layer in all the samples. Note that the generation of cells in the two external regions (ONL and GC) is completed before that of the central (OINL and IINL) regions. From Belecky-Adams et al., 1996, with permission from Academic Press.

nuclear layer cells. Another conclusion is that there is no obvious inside-out or outside-in pattern of cell generation within the retina; cells generated at earlier embryonic stages contribute to all retinal layers, and addition of new cells ceases more or less concomitantly in the innermost and the outermost layers of the retina (prospective ganglion cell and photoreceptor cell layers, respectively). Finally, cell generation is largely completed in the fundal region of the eye by ED 8, but retinal ganglion cells are the only ones showing overt signs of structural differentiation at this stage (Coulombre, 1955; Meller, 1984; Grun, 1982).

DEVELOPMENTAL DISTRIBUTION OF HOMEOBOX GENE EXPRESSION IN THE CHICK EMBRYO RETINA

The results described in the preceding section indicate that the precursors of retinal photoreceptors and non-photoreceptor neurons are already postmitotic by ED 8 but, with the exception of ganglion cells, are still morphologically undifferentiated. This, however, does not preclude the possibility that the differentiation of the remaining cells may have also started at that time, albeit at a more subtle, molecular level, perhaps involving the differential expression of DNA-binding transcriptional regulators. Such possibility is suggested by many recent studies of genetic mechanisms regulating cell differentiation, which have been particularly successful in the case of *Drosophila* in general and its compound eye in particular, in which many DNA-binding transcription factors have been shown to control the coordinated expression of networks of cell-specific genes (e.g., Banerjee and Zipursky, 1990; Moses, 1991; Greenwald and Rubin, 1992; Reh and Cagan, 1994). Genetic regulation of retinal differentiation in vertebrates is much less well understood, but a variety of candidate genes have been found to be expressed in the undifferentiated neuroepithelium (e.g., Stadler and Solursh, 1994; Dorsky et al., 1995; Guillemot and Joyner, 1993), whereas others show a laminar pattern of expression that correlates with the differentiation of retinal cell types (Jasoni et al., 1994; Carriere et al., 1993; Liu et al., 1994; Levine et al., 1994; Xiang et al., 1993, 1995; Turner et al., 1994; Hatini et al., 1994; Deitcher et al., 1994; Nornes, 1990; among others). The functional significance of some of these candidate genes has been demonstrated through their deletion by homologous recombination (e.g., Gan et al., 1996).

Our laboratory is investigating three candidate genes, Pax-6, Prox 1, and Chx10. Pax-6 is the vertebrate homologue of *Drosophila* eyeless, which triggers development of normal eyes when expressed ectopically in flies (Halder et al., 1995); Pax-6 mutations cause congenital abnormalities such as aniridia in humans and the small eye phenotype in mice (Glaser et al., 1994). The chicken Prox 1 gene (Tomarev et al., 1997) is homologous to the mouse Prox 1 (Oliver et al., 1993) and to the *Drosophila prospero* genes (Doe et al., 1991), which are important for neuroblastic differentiation in the fly. Chx10 was initially cloned in the mouse (Liu et al., 1994), and it was subsequently shown that its mutations cause the ocular retardation phenotype in mice (Burmeister et al., 1996). Our study (Belecky-Adams et al., 1997) used Northern blot analysis, *in situ* hybridization and immunocytochemistry to investigate the expression of these genes in chick embryo retinas at ED 5 (when most retinal precursor cells are still proliferating), ED 8 (when most precursor cells are already postmitotic, but still morphologically undifferentiated), ED 15 (when extensive overt cell differentiation is underway), and ED 18 (when the organization of the retina closely resembles that present in the adult). Northern blot analysis showed that the three genes were expressed at all the stages of development studied (although Prox 1 could only be detected in overexposed blots on ED 5). The relative abundance of Chx10 remained fairly constant throughout development. Pax-6 levels appeared to increase between ED 5 and ED 8, when a second band also became visible which likely represents alternatively spliced products; this pattern remained unchanged thereafter. As already mentioned, Prox 1 was barely detectable on ED 5, but was readily observed at later stages, when it appeared as one major band and several small bands that are likely to correspond to alternatively spliced products (see also Tomarev et al., 1997). Much more dramatic changes, however, were observed by *in situ* hybridization. Most cells in the proliferating, undifferentiated neuroepithelium showed low but detectable levels of expression of all three genes at early embryonic stages (ED 4–5), but this diffuse pattern of distribution was replaced by a topographically restricted, laminar pattern of expression by ED 7–8 (summarized in Table 1), which coincides with and/or predicts the positions occupied by individual cell types in the mature retina. For example, Pax-6 mRNA is abundant in the ganglion cell layer, the amacrine cell sublayer of the INL, and a subpopulation of putative horizontal cells, but is not detectable in putative bipolar and Müller cells. On the other hand, Prox 1 becomes concentrated in putative horizontal cells, decreased towards the vitreal side of the retina and is undetectable in ganglion cells (as well as in putative photoreceptors), whereas Chx10 mRNA becomes concentrated in the bipolar-Müller region of the inner nuclear layer, decreases gradually towards both the horizontal and ganglion cells, and is negative in the photoreceptor layer. This pattern of expression is qualitatively maintained, but tends to decrease in intensity, at later stages of development. Immunocytochemical analysis showed similar distribution of the corresponding proteins which, in addition, appeared to be localized to the cell nucleus.

In summary, the studies indicate that by ED 8, when most retinal cells are still morphologically undifferentiated, they already display characteristic patterns of Pax-6, Prox 1 and Chx10 expression (Table 1), with putative photoreceptors being negative for all three genes, putative horizontal cells being negative for Chx10 but strongly positive for Prox 1 and, in some cases, weakly positive for Pax-6, cells in the Müller-bipolar region being strongly positive for Chx10 and weakly for Prox 1 (and, in an irregular pattern, for Pax-6 as well), putative amacrine cells being strongly positive for Pax-6 and, depending on their position within the inner nuclear layer, showing also variable levels of Chx10 expression, and cells in the ganglion cell layer being rich in Pax-6 but devoid of the other two factors. This distribution is consistent with the possibility that the differentiated fate of each retinal precursor cell may be influenced by the presence of the products of these genes, acting either individually or in a combinatorial manner, with photoreceptor differentiation perhaps requiring the simultaneous absence of the products of all three genes (Belecky-Adams et al., 1997). This possibility is currently under investigation in this laboratory using a cell culture system (see below) in which these genes are also expressed in a cell-specific manner.

Table 1.

Stage	Cell type	PAX-6	PROX-1	CHX10
ED 4–5	Neuroepithelial	+	+	+
ED 8	Photoreceptors	–	–	–
	Horizontal	+/–	++++	–
	Bipolar/Muller	–	+/–	++++
	Amacrine	++++	–	+
	Ganglion	++++	–	–

DIFFERENTIATION OF ISOLATED RETINAL PRECURSOR CELLS *IN VITRO*

The presence of characteristic combinations of transcription factors in the precursors of each retinal cell type could indicate that the cells are already committed to specific developmental pathways by ED 8. This would in turn predict that each precursor cell would follow a predetermined pathway of differentiation not only during normal development within the retina *in vivo*, but also when exposed to different microenvironmental conditions through transplantation or *in vitro* experiments. We have investigated this possibility using cells dissociated from ED 8 chick embryo retina, which are cultured at low density on polyornithine-coated substrata. The cells attach to this substratum as individual units, and are therefore devoid of contacts with other cells at culture onset, when they also show a morphologically undifferentiated appearance. Under these conditions, it becomes possible to investigate: 1) whether these isolated cells can differentiate at all; 2) if they can, whether all the cells express the same pattern of differentiation or, rather, whether they generate different cell types in defined proportions; and 3) the extent to which the isolated cells can reach a level of molecular, structural and functional differentiation resembling those of their *in vivo* counterparts. These questions have been investigated in our laboratory during the past several years using a multidisciplinary battery of analytical techniques; a brief summary of these studies will be presented below.

Some of the cells present in the cultures differentiate as photoreceptors, while others become non-photoreceptor, multipolar neurons (glial cell development does not occur under these culture conditions); interestingly, neighboring cells frequently follow divergent patterns of differentiation. The process of morphological differentiation of the precursor cells begins shortly after culture onset and can be recognized through changes in cell shape. The isolated retinal precursor cells that differentiate as photoreceptors, for example, undergo a transition from the initial round configuration to a highly elongated, polarized and compact neutralized organization, with a single short neurite (Madreperla and Adler, 1989). Neighboring cells that differentiate as multipolar neurons can be seen to produce several long, branched neurites, while their cell body grows larger and acquires a circular or polyhedrical appearance. Quantitative analysis shows that the proportion of precursor cells that differentiate as either photoreceptors or non-photoreceptor neurons in culture is predictable for different stages of development. In cultures of cells isolated on ED 8, for example, approximately 80% of the differentiating cells show a non-photoreceptor phenotype, while the remaining 20% of differentiated cells become photoreceptors; surprisingly, the frequency of photoreceptor cells is reproducibly *higher* in cultures of cells isolated at earlier developmental stages (Adler and Hatlee, 1989). These observations appear to provide answers to the first two questions outlined above: isolated precursor cells *can* differentiate *in vitro*, and they *do* follow divergent pathways of development.

The studies showed that the differentiated features developed by cultured cells are very complex, involving not only the expression of cell-specific genes, but also mimicking their *in vivo* counterparts in their pattern of structural organization, and even in the capacity to perform cell-specific functional behaviors. The cells that become photoreceptors in culture develop an elongated, compartmentalized phenotype, with a short axon, a cell body occupied almost exclusively by the nucleus, an inner segment that contains the metabolic machinery of the cell, and a small outer segment connected to the inner segment by a cilium (Adler et al., 1984; Adler, 1986; Madreperla and Adler, 1989; Saga et al., 1996). Sequential photographic analysis showed that photoreceptor precursors undergo a predictable series of morphological transitions from their initial process-free, circular outline to their elongated and polarized phenotype (Madreperla and Adler, 1989); this suggested the active involvement of intracellular forces, since these phenotypic transformations occur in cells devoid of contacts with other cells. Results from experiments with cytoskeletal inhibitors, such as nocodazole and cytochalasin D, were consistent with this possibility and showed that the development and maintenance of the structural polarity of photoreceptors results from the balance between microtubule-dependent forces (which tend to elongate the cells) and actin-dependent forces (which tend to shorten them). It is noteworthy that, although photoreceptors undergo extensive structural changes in response to these inhibitors, they recover their normal pattern of organization when the drugs are removed (Madreperla and Adler, 1989). The polarity of cultured photoreceptors can be recognized at the molecular level as well; as is the case *in vivo*, visual pigments are concentrated in their outer segment-like process (Adler, 1986; Saga et al., 1996), while the enzyme Na^+-K^+-ATPase is localized predominantly to plasma membrane of the inner segment of the cultured photoreceptors (Madreperla et al., 1989). ATPase polarity appears to depend on interactions between enzyme molecules and cytoskeletal elements (Madreperla et al., 1989).

Photoreceptor length is regulated by light in the retinas of non-mammalian vertebrates; light causes rod elongation and cone contraction, and the opposite changes occur in darkness (rev: Burnside and Dearry, 1986; Kunz, 1990). These photomechanical responses are also influenced by the neuromodulators dopamine and melatonin, which act as reciprocal antagonists and mimic the effects of light and darkness, respectively (Burnside and Dearry, 1986; Dowling and Ehinger, 1986; Besharse et al., 1988; Cahill and Besharse, 1991). To evaluate the responsiveness of cultured photoreceptors to light, we investigated their capacity for photomechanical responses by growing them under cycles of 12 hours light/12 hours darkness. Measurements of photoreceptor length at various time points confirmed that many of the cells that differentiate *in vitro* as photoreceptors do acquire the capacity of responding to light with photomechanical movements, with most of the responsive photoreceptors elongating in light and contracting in darkness (Stenkamp and Adler, 1993). A series of biochemical, pharmacological and autoradiographic studies suggested that dopamine and melatonin are involved in the regulation of these responses *in vitro*, resembling the *in vivo* situation (Stenkamp et al., 1994). The evidence for dopamine's involvement was extensive. Photoreceptor responses to light could be blocked with dopamine D_2 receptor antagonists, and they were attenuated by dopamine synthesis inhibitors. The possible existence of an endogenous source of dopamine in the cultures was suggested by the immunological detection of the dopamine-synthesizing enzyme, tyrosine hydroxylase, by the autoradiographic detection of a Na^+-dependent uptake mechanism for dopamine, associated with non-photoreceptor cells, and by the demonstration that ^3H-dopamine release occurred in the cultures both through a Ca^{++}-dependent mechanism and through reverse function of a nomifensin-sensitive transporter (Stenkamp et al., 1994). The finding that both dopamine release mechanisms could be regulated by light and by melatonin suggested the development in the cultures of complex networks of neuromodulatory mechanisms; consistent with this possibility is the finding that retinal cells also contain serotonin N-acetyltransferase, a key enzyme in melatonin synthesis, which appears to be localized to the photoreceptor cells (Iuvone et al., 1990).

Vertebrate visual pigments are based on a protein (opsin) covalently bound to the vitamin A derivative 11-cis retinaldehyde, without which they are not light-sensitive; the response of a photoreceptor to light is initiated by the transformation of 11-cis retinaldehyde into all-trans retinol; the latter is transported to the retinal pigment epithelium, where it is re-converted into 11-cis retinaldehyde and transported back to the photoreceptors

(Wald, 1968; Hubbard and Kropf, 1958; Saari, 1990; Bok, 1990; Rando et al., 1991). The finding that cultured photoreceptors were light-sensitive *in vitro* raised challenging questions about the chromophores involved in these responses, because the medium used in our experiments does not contain 11-cis retinaldehyde, and retinal pigment epithelial cells are not present in the cultures. Since other retinoids are present in the medium, however, we investigated by high pressure liquid chromatography (HPLC) whether the cultured cells could metabolize these vitamin A derivatives (Stenkamp and Adler, 1994). Relevant to the issue under consideration here was the finding that the cultured neuronal elements could generate retinaldehyde from retinol or retinyl esters. This is consistent with (but does not prove) the possible use by the cells of 11-cis retinaldehyde as a chromophore for phototransduction, because the studies did not detect the 11-cis isomer of retinaldehyde. This could be due to limitations in the sensitivity of the HPLC method used, because the minute amounts that would be sufficient to generate a response to light could be generated by photoisomerization of the retinaldehyde produced *in vitro*, and were likely to escape detection by HPLC (Stenkamp and Adler, 1994).

Taken together, the studies summarized above demonstrate that morphologically undifferentiated retinal precursor cells can differentiate as either photoreceptor or non-photoreceptor multipolar neurons while developing within the same microenvironment, and in the absence of contact-mediated cell interactions. Moreover, the divergent differentiation of precursor cells involves the expression of very complex phenotypes that resemble those expressed by their *in vivo* counterparts. Such behavior suggests that, at the time of their isolation, the undifferentiated precursor cells already have complex "master programs" of development, which they can express to a considerable degree, in a cell autonomous manner. Are those programs determined at the time of, or sometime after terminal mitosis of the cells?

COMMITMENT VS. PLASTICITY OF POSTMITOTIC RETINAL CELLS

The precise laminar distribution of particular cell types in the mature retina suggests that, if cell determination were to occur precisely *at* the time of terminal mitosis, it would be necessary for each precursor cell to simultaneously acquire not only the commitment to develop as a particular cell type, but also the information necessary for migrating to, and homing in, the retinal layer (or sublayer) corresponding to that cell type. This model would predict, moreover, that the microenvironment to which the cells are exposed after terminal mitosis would not be able to "switch" precursors from one differentiated fate to another. An alternative scenario would propose that postmitotic precursor cells remain "plastic" (i.e., uncommitted to specific fates) for some time after terminal mitosis, with their differentiated fate being determined by position-dependent inductive signals to which they become exposed during their intraretinal migration. In addition to providing an explanation for the correlation between the phenotype and the laminar position of each cell, this model would predict that changes in the composition of the microenvironment surrounding the post- mitotic precursor cells could "switch" them from one to another differentiated fate.

We have tested predictions from these two models using several complementary approaches. The first indication that retinal precursor cells do remain plastic after terminal mitosis was provided by studies in which "cumulative" ^{3}H-thymidine labeling was started on ED 5, making it possible to identify unlabeled cells, born before that day, from cells born after the time of initial ³HT administration (Adler and Hatlee, 1989). *In vitro* studies showed that the cells born before ED 5 gave rise predominantly to non-photoreceptor neurons when they were allowed to develop *in vivo* until embryonic day 8, prior to their isolation for culture. The same cell population, however, gave rise predominantly to photoreceptor cells when they were isolated two days earlier, on ED 6. While the experiments suggested that the differentiated fate of postmitotic cells was changed by their exposure to the retinal microenvironment, this interpretation was somewhat complicated by the temporal heterogeneity of the population under investigation, and by the fact that the *in vitro* fate of the cell population of interest could not be compared with its normal development within the retina. These limitations were overcome more recently using the window-labeling technique which, as mentioned above, makes it possible to identify cells born during narrowly defined time intervals both *in situ* and under various experimental conditions (Belecky-Adams et al., 1996). For these experiments the embryos received an injection of ³H-thymidine (³HT) on ED 5, followed 5 hours later by the administration of bromodeoxyuridine (BrDU), which was constantly available to the cells thereafter. As shown in Fig. 4, when the *in vivo* fate of these cells was investigated in ED 18 retinas, we found that approximately 80% of the cells gave rise to non-photoreceptor neurons. A similar result was observed when the window-labeled cells spent at least 72 hrs within the retina (i.e., until ED 8) before being isolated for culture; under these conditions the cultured cells mimicked their *in vivo* fate, giving rise predominantly (near 80%) to non-photoreceptor neurons. However, a completely different behavior was observed when the cells, window-labeled on embryonic day 5, were removed for culture on embryonic day 6, that is to say, after a much shorter exposure to the retinal microenvironment (Fig. 4). In this case,

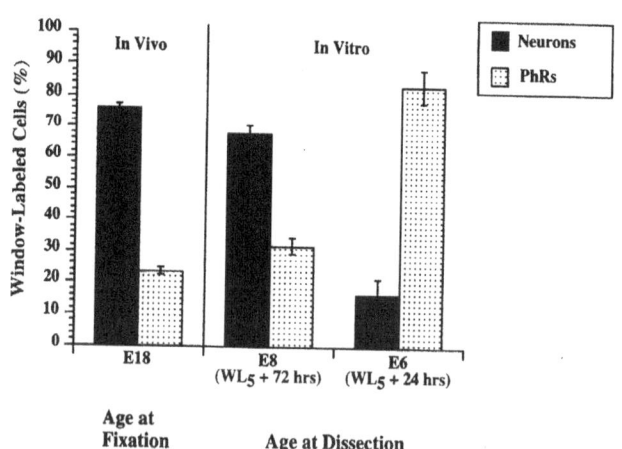

Figure 4. *In vitro* fate of a cohort of contemporary cells, as a function of their exposure to the retinal microenvironment prior to their isolation for culture. Embryos were window-labeled for 5 hr on ED 5, and either allowed to develop until ED 18, or processed for dissociated retinal cultures 24 hr or 72 hr after the end of the WL period (WL + 24 and WL + 72 hr, respectively). Cell cultures fixed after 4 days *in vitro*, and histological sections were processed for BrDU immunocytochemistry and ³HT autoradiography. In ED 18 embryos, approximately three times as many WL₅ cells were found in the INL and GC layers as in the ONL. In cell culture, a similar 3:1 ration between WL non-photoreceptor neurons with respect to photoreceptors was observed when the cells were exposed to the *in vivo* microenvironment for 72 hours before isolation for culture. On the other hand, WL₅ cells that were exposed to the *in vivo* microenvironment for only 24 hr gave rise predominantly (84%) to photoreceptors, and became non-photoreceptor neurons in only 16% of the cases. From Belecky-Adams et al., 1996, with permission from Academic Press.

the window-labeled cells gave rise predominantly to photoreceptors, with only some 20% of them differentiating as non-photoreceptor neurons (Belecky-Adams et al., 1996). Since additional analysis showed that differential cell death could not explain these results, the experiments are consistent with the hypothesis that many postmitotic precursor cells remain plastic for some time, with their differentiated fate being influenced by factors from the microenvironment to which they were exposed after terminal mitosis. Whether this applies to *all* precursor cells is questionable, however, because it has been reported that some cells express ganglion cell markers immediately after undergoing their last mitotic division (Waid and McLoon, 1995).

These experiments, together with those of Adler and Hatlee (1989), suggest in addition that photoreceptor differentiation may be the "default pathway" followed by many retinal precursor cells when they are prevented from interacting with the retinal microenvironment. This suggestion is supported by the observation that practically 100% of the isolated cells that undergo their terminal mitosis *in vitro* differentiate predominantly as photoreceptors (Repka and Adler, 1992b). It must be noted, however, that Austin et al. (1995) reported predominant expression of ganglion cell markers when retinal cells are isolated from younger chick embryos and grown *in vitro*. This discrepancy could be due to differences in culture media and substrata, which could influence cell differentiation (Belecky-Adams et al., 1996). It is also noteworthy that, while cultured photoreceptors have been characterized through the analysis of many molecular, structural and functional properties (see above), cultured ganglion cells have been characterized almost exclusively by immunocytochemical markers (Austin et al., 1995). These markers could conceivably be transiently expressed, either *in vivo* or *in vitro* by cells that are already differentiating but are not yet committed to specific fates (Larison and BreMiller, 1990).

CONCLUDING COMMENTS

Our ongoing investigations of the mechanisms of retinal cell differentiation in the chick embryo are aimed at testing predictions of a working hypothesis based on the following elements: 1) retinal precursor cells remain plastic for some time after terminal mitosis; 2) as they migrate within the retina toward a particular laminar position, postmitotic cells are exposed to position-dependent inductive signal; 3) their exposure to these micro environmental signals determines the expression of particular DNA-binding transcription factors, which in turn activate the expression of networks of cell-specific properties; and 4) these developmental "master programs" can be expressed by the cells autonomously, even when they are removed from their normal micro environment and grown in the absence of contact-mediated cell interactions. The dissociated culture system described above is well suited for the experimental analysis of predictions derived from this hypothetical model, since it is not only possible to control the composition of their micro environment, but the cells are also amenable to genetic manipulation (Werner et al., 1990; Kumar et al., 1996).

ACKNOWLEDGMENTS

Work in the author's laboratory was supported by NIH grants EY 04859 and 05404; and core grant EY 01765. R.A. is a senior investigator of the Research to Prevent Blindness, Inc. The author is grateful to Ms. Elizabeth Bandell for secretarial assistance.

REFERENCES

Adler, R. (1986) Developmental predetermination of the structural and molecular polarization of photoreceptor cells. Dev. Biol. 117:520–527.

Adler, R. (1993) Plasticity and differentiation of retinal precursor cells. In: *International Review of Cytology*, (Friedlander, M., ed.) Academic Press, San Diego, Vol. 146, pp. 145–190.

Adler, R., Lindsey, J. D. and Elsner, C. L. (1984) Expression of cone-like properties by chick embryo neural retina cells in glial-free monolayer cultures. J. Cell Biol., 99:1173–1178.

Adler, R. and Hatlee, M. (1989) Plasticity and differentiation of embryonic retinal cells after terminal mitosis. Science, 243:391–393.

Austin, C.P., Feldman, D.E., Ida, J.A., and Cepko, C.L. (1995) Vertebrate retinal ganglion cells are selected from competent progenitors by the action of Notch. Development 121:3637–3650.

Banerjee, U. and Zipursky, S. L. (1990) The role of cell-cell interaction in the development of the *Drosophila* visual system. Neuron 4:177–187.

Belecky-Adams, T., Cook, B., and Adler, R. (1996) Correlations between terminal mitosis and differentiated fate of retinal precursor cells in vivo and in vitro: analysis with the "window-label" technique.

Belecky-Adams, T., Tomarev, S., Li, H-S., Ploder, L., McInnes, R.R., Sundin, O., and Adler, R. (1997) Prox 1, Pax-6 and chx10 homeobox gene expression correlate with phenotypic fate of retinal precursor cells. Invest. Ophthalmol. Vis. Sci. 38:1293–1303.

Besharse, J. C., Iuvone, P. M. and Pierce, M. E. (1988) Regulation of rhythmic photoreceptor metabolism: A role for post-receptoral neurons. Prog. Ret. Res. 721–61.

Bok, D. (1990) Processing and transport of retinoids by the retinal pigment epithelium. Eye 4:376–332.

Burmeister, M., Novak, T., Liang, M.Y., Basu, S., Ploder, L., Hawes, N.L., Vidgen, D., Hoover, F., Goldman, D., Kalnins, V.I., Roderick, T.H., Taylor, B.A., Hankin, M.H., and McInnes, R.R. (1996) Ocular retardation of the mouse caused by chx10 homeobox null allele—impaired retinal progenitor proliferation and bipolar cell differentiation. Nature Gen. 12:376–384.

Burnside, B. and Dearry, A. (1986) Cell motility in the retina. In: The Retina: A Model for Cell Biology Studies (R. Adler and D. Farber, eds.). Academic Press, Orlando, pp. 152–206.

Cahill, G. M. and Besharse, J. C. (1991) Resetting the circadian clock in cultured Xenopus eyecups—regulation of retinal melatonin rhythms by light and D2 dopamine receptors. J. Neurosci. 11:2959–2971.

Carriere, C., Plaza, S., Martin, P., Quantannens, B., Bailly, M., Stehelin, D., and Saule, S. (1993) Characterization of quail Pax-6 (Pax-QNR) proteins expressed in the neuroretina. Mol. Cell. Biol. 13:7157–7166.

Cepko, C. L. (1996) The patterning and onset of opsin expression in vertebrate retinae [Review]. Curr. Opin. Neurobiol. 6:542–546.

Coulombre, A. J. (1955) Correlations of structural and biochemical changes in the developing retina of the chick. Am. J. Anat. 96:153–189.

Deitcher, D.L., Fekete, D.M., Cepko, C.L. (1994) Asymmetric expression of a novel homeobox gene in vertebrate sensory organs. J. Neurosci. 14:486–498.

Doe, C.Q., Chu-LaGraff, Q., Wright, D.M., and Scott, M.P. (1991) The prospero gene specified cell fates in the Drosophila central nervous system. Cell 65:451–464.

Dorsky, R.I., Rapaport, D.H., and Harris, W.A. (1995) Xotch inhibits cell differentiation in the Xenopus retina. Neuron 14:487–496.

Dowling, J.E., and Ehinger, G. (1986) Dopamine: a retinal neuromodulator. Trends Neurosci. 9:236–266.

Dutting, D., Gierer, A., and Hansmann, G. (1983) Self renewal of stem cells and differentiation of nerve cells in the developing chick retina. Dev. Brain Res. 10:21–32.

Fujita, S. (1963) The matrix cell and histogenesis on the developing central nervous system. J. Comp. Neurol. 120:37–42.

Fujita, S. and Horii, S. (1963) Analysis of cytogenesis in the chick retina by tritiated thymidine autoradiography. Archumhistol. Japan 23:295–366.

Gan, L., Xiang, M.Q., Zhou, L.J., Wagner, D.S., Klein, W.H., and Nathans, J. (1996) POU domain factor Brn-3B is required for the development of a large set of retinal ganglion cells. Proc. Natl. Acad. Sci. USA 93:3920–3925.

Glaser, T., Jepeal, L., Edwards, J.G., Young, S.R., Favor, J., and Maas, R.L. (1994) Pax-6 gene dosage effect in a family with congenital cataracts, aniridia, anophthalmia and central nervous system defects. Nature Genetics 8:203.

Goldowitz, D., Rice, D. S., and Williams, R. W. (1996) Clonal architecture of the mouse retina [Review]. Prog. Brain Res. 108:3–15.

Greenwald, I. and Rubin, G. M. (1992) Making a difference - the role of cell-cell interactions in establishing separate identities for equivalent cells. Cell 68(2):271–281.

Grun, G. (1982) The development of the vertebrate retina: a comparative study. Adv. Anat. Embryol. Cell. Biol. 78:7–85.

Guillemot, F. and Joyner, A. L. (1993) Dynamic expression of the murine achaete-scute homologue mash-1 in the developing nervous system. Mech. Develop. 42(3):171–185.

Halder, G., Callaerts, P., and Gehring, W.J. (1995) Induction of ectopic eyes by targeted expression of the eyeless gene in Drosophila. Science 267:1788–1792.

Hatini, V., Tao, W., and Lai, E. (1994) Expression of winged helix genes, BF-1 and BF-2, define adjacent domains within the developing forebrain and retina. J. Neurobiol. 25:1293–1309.

Hausman, R. E., Rao, A.S.M.K., Ren, Y., Sagar, G.D.V. and Shah, B. H. (1993) Retina-cognin, cell signaling, and neuronal differentiation in the developing retina [Review]. Develop. Dynam. 196(4):263–266.

Hicks, D. (1996) Characterization and possible roles of fibroblast growth factors in retinal photoreceptor cells [Review]. Keio J. Med. 45:140–154.

Hollyfield, J. G. (1968) Differential addition of cells to the retina in Rana pipiens tadpoles. Dev. Biol. 18, 163–179.

Holt, C. E., Bertsch, T. W., Ellis, H. M. and Harris, W. A. (1988) Cellular determination in the Xenopus retina is independent of lineage and birth date. Neuron 1:15–26.

Hubbard, R. and Kropf, A. (1958) The action of light on rhodopsin. Proc. Natl. Acad. Sci. 44:130–139.

Iuvone, P. M. (1990) Development of melatonin synthesis in chicken retina - regulation of serotonin N-acetyltransferase activity by light, circadian oscillators, and cyclic AMP. J. Neurochem. 54(5):1562–1568.

Jasoni, C.L., Walker, M.B., Morris, M.D., Reh, T.A. (1994) A chicken achaete-scute homolog (cash-1) is expressed in a temporally and spatially discrete manner in the developing nervous system. Development 120(4):769–783.

Kahn, A. J. (1974) An autoradiographic analysis of the time of appearance of neurons in the developing chick neural retina. Dev. Biol. 38:30–40.

Kumar, R., Scheurer, D., Duh, E., Regemtulla, A., Swaroop, A., Adler, R., and Zack, D.J. (1996) The bZIP transcription factor Nrl stimulates rhodopsin promoter activity in primary retinal cell cultures. J. Biol. Chem. 271:29612–29618.

Kunz, Y. W. (1990) Ontogeny of retinal pigment epithelium-photoreceptor complex and development of rhythmic metabolism under ambient light conditions. Prog. Reg. Res. 9:135–196.

Larison, K. D. and BreMiller, R. (1990) Early onset of phenotype and cell patterning in the embryonic zebra fish retina. Development 109:567–576.

Layer, P. G., and Willbold, E. (1993) Histogenesis of the avian retina in reaggregation culture: from dissociated cells to laminar neuronal networks [Review]. Int. Rev. Cytol. 146:1–47.

Levine, E.M., Hitchcock, P.F., Glasgow, E., and Schechter, N. (1994) Restricted expression of a new paired-class homeobox gene in normal and regenerating adult goldfish retina. J. Comp. Neurol. 348:596–606.

Lillien, L. (1994) Neurogenesis in the vertebrate retina [Review]. Perspectives Dev. Neurobiol. 2:175–182.

Liu, I.S.C., Chen, J.D., Ploder, L., Vidgen, D., Vanderkooy, D., Kalnins, V.I., McInnes, R.R. (1994) Developmental expression of a novel murine homeobox gene (CHX10)—evidence for roles in determination of the neuroretina and inner nuclear layer. Neuron 13:377–393.

Madreperla, S. A. and Adler, R. (1989) Opposing microtubule-and actin-dependent forces in the development and maintenance of structural polarity in retinal photoreceptors. Dev. Biol. 131:149–160.

Madreperla, S. A., Edidin, M. and Adler, R. (1989) Na^+,K^+-Adenosine triphosphatase polarity in retinal photoreceptors: a role for cytoskeletal attachments. J. Cell Biol. 109:1483–1493.

Meller, K. (1984) Morphological studies on the development of the retina. Prog. Ret. Res. 3:1–18.

Morris, V. B. (1973) Time differences in the formation of the receptor types in the developing chick retina. J. Comp. Neurol. 151:323–330.

Moses, K. (1991) The role of transcription factors in the developing Drosophila eye. Trends Genet. 7(8):250–255.

Nornes, H.O., Dressler, G.R., Knapik, E.W., Deutsch, U., and Gruss, P. (1990) Spatially and temporally restricted expression of Pax-2 during murine neurogenesis. Development 109:797–809.

Oliver, G., Sosa-Pineda, B., Geisendorf, S., Spana, E.P., Doe, C.Q., and Gruss, P. (1993) Prox 1, a prospero-related homeobox gene expressed during mouse development. Mech. Dev. 1993;44:3–16.

Prada, C., Puga, J., Perez-Mendez, L., Lopez, R., and Ramirez, G. (1991) Spatial and temporal patterns of neurogenesis in the chick retina. Eur. J. Neurosci. 3:559–569.

Rando, R. R., Bernstein, T. S. and Barry, R. J. (1991) New insights into the visual cycle. Prog. Ret. Res. 10:161–178.

Rapaport, D.H., Rakic, P., and LaVail, M.M. (1996) Spatiotemporal gradients of cell genesis in the primate retina [Review]. Perspect. Dev. Neurobiol. 3:147–159.

Reh, T.A., and Cagan, R.L. (1994) Intrinsic and extrinsic signals in the developing vertebrate and fly eyes—viewing vertebrate and invertebrate eyes in the same light. Perspectives Dev. Neurobiol. 2:183–190.

Reichenbach, A. (1993) Two types of neuronal precursor cells in the mammalian retina—a short review. J. Hirnforsche. 34:335–341.

Repka, A. M. and Adler, R. (1992a) Differentiation of retinal precursor cells born in vitro. Dev. Biol. 153:242–249.

Repka, A. M. and Adler, R. (1992b) Accurate determination of the time of cell birth using a sequential labeling technique with <H-3>-thymidine and bromodeoxyuridine (window labeling). J. Histochem. Cytochem. 40(7):947–953.

Saari, J. C. (1990) Enzymes and proteins of the mammalian visual cycle. Prog. Ret. Res. 9:363–381.

Saga, T., Scheurer, D., and Adler, R. (1996) Development and maintenance of outer segments by isolated chick embryo photoreceptor cells in culture. Invest. Ophthalmol. Vis. Sci. 37:561–573.

Sidman, R. L. (1970) Autoradiographic methods and principles for study of the nervous system, with thymidine-H3. IN: Contemporary Research in Neuroanatomy, (E. Nauta, ed.). Springer Verlog, New York, pp. 252–274.

Snow, R.L., and Robson, J.A. (1994) Ganglion cell neurogenesis, migration and early differentiation in the chick retina. Neuroscience 58(2):399–409.

Spence, S. G. and Robson, J. A. (1989) An autoradiographic analysis of neurogenesis in the chick retina in vivo and in vitro. Neurosci. 32:801–812.

Stadler, H.S., and Solursh, M. (1994) Characterization of the homeobox-containing gene GH6 identifies novel regions of homeobox gene expression in the developing chick retina. Dev. Biol. 161:251–262.

Stenkamp, D. and Adler, R. (1993) Photoreceptor differentiation of isolated retinal precursor cells includes the capacity for photomechanical responses. Proc. Natl. Acad. Sci. USA, 90:1982–1986.

Stenkamp, D.L., and Adler, R. (1994) Cell-type- and developmental-stage-specific metabolism and storage of retinoids by embryonic chick retinal cells in culture. Exp. Eye Res. 58:675–687.

Stenkamp, D.L., Iuvone, P.M., Adler, R. (1994) Photomechanical movements of cultured embryonic photoreceptors: Regulation by exogenous neuromodulators and by a regulable source of endogenous dopamine. J. Neurosci. 14:3083–3096.

Stone, J. (1988) The origins of the cells of vertebrate retina. Proc. Ret. Res. 7:1–19.

Szel, A., Rohlich, P., Caffe, A. R., and van Veen, T. (1996) Distribution of cone photoreceptors in the mammalian retina [Review]. Microsc. Res. Tech. 35:445–462.

Tomarev, S.I., Sundin, O., Banerjee-Basu, S., Duncan, M.K., Yang, J-M., and Piatigorsky, J. (1997) A chicken homeobox gene Prox 1, related to Drosophila prospero, is expressed in the developing lens and retina. Dev. Dynamics 206:354–367.

Turner, D.L., and Weintraub, H. (1994) Expression of achaete-scute homolog 3 in Xenopus embryos converts ectodermal cells to a neural fate. Genes & Development 8:1434–1447.

Turner, D. L., Snyder, E. Y. and Cepko, C. L. (1990) Lineage-independent determination of cell type in the embryonic mouse retina. Neuron 4:833–845.

Turner, D. L. and Cepko, C. L. (1987) A common progenitor for neurons and glia persists in rat retinas late in development. Nature (London) 328:131–136.

Waid, D.K., and McLoon, S.C. (1995) Immediate differentiation of ganglion cells following mitosis in the developing retina. Neuron 14:117–124.

Wald, G. (1968) The molecular basis of visual excitation. Nature 219:800–807.

Werner, M., Madreperla, S., Lieberman, P. and Adler, R. (1990) Expression of transfected genes by differentiated, postmitotic neurons and photoreceptors in primary cell cultures. J. Neurosci. Res. 25:50–57.

Wetts, R. and Fraser, S. E. (1988) Multipotent precursors can give rise to all major cell types of the frog retina. Science 239:1142–1145.

Xiang, M. Q., Zhou, L. J., Peng, Y. W., Eddy, R. L., Shows, T. B. and Nathans, J. (1993) Brn-3B—a POU domain gene expressed in a subset of retinal ganglion cells. Neuron 11:689–701.

Xiang, M. Q., Zhou, L. J., Macke, J. P., Yoshioka, T., Hendry, S. H. C., Eddy, R. L., Shows, T. B., and Nathans, J. (1995) THE Brn-3 family of POU-domain factors—primary structure, binding specificity, and expression in subsets of retinal ganglion cells and somatosensory neurons. J. Neurosci. 15(7 Part 1):4762–4785.

Young, R. W. (1985) Cell differentiation in the retina of the mouse. Anat. Record 212, 199–205.

3

EMBRYONIC PATTERNING OF CONE SUBTYPES IN THE MAMMALIAN RETINA

K. C. Wikler* and D. L. Stull

Section of Neurobiology
Yale School of Medicine
New Haven, Connecticut

Vertebrate color vision is mediated by different subtypes of cones whose visual pigments or opsins are maximally sensitive to long (LWS cones), middle (MWS cones), or short (SWS cones) wavelengths of light (Dartnall et al., 1983). Quantitative assessment of the distribution of these wavelength-sensitive cones in several mammalian species has revealed dramatically different arrangements and combinations of cone subtypes across the retinal sheet that coincide with differences in color vision and photopic acuity. For example, in the retina of both rhesus monkey and man, opsin-specific cone subtypes are arranged into reiterative patterns, in which each SWS cone is surrounded by approximately ten L/MWS cones (Szel et al., 1988; Curcio et al., 1991; Wikler and Rakic, 1990). In contrast, in the mouse cone subtypes are topographically segregated with ventral retina occupied exclusively by SWS cones and dorsal retina dominated by MWS cones (Szel et al., 1992; Rohlich et al., 1994; Calderone and Jacobs, 1995). Thus, the different adult cone arrangements in murine and primate retina suggest that color vision may be tightly linked to the specification of both the position and the relative ratios of wavelength-sensitive cone subtypes.

Clinical findings in humans demonstrate a correlation between specific visual deficits and developmental anomalies in the position or ratio of cone subtypes, offering additional support for this scenario. For example, visual deficits associated with rod monochromacy are believed to result from abnormalities in the positioning of cones in the retinal mosaic rather than a failure of the retina to generate cones (Falls et al., 1965; Glickstein and Heath, 1975). Additionally, an inherited autosomal dominant color vision defect is characterized by mild macular dystrophy and a reduction in visual acuity coincident with selective disruption of the SWS cone system (Bresnick et al., 1989). Finally, "enhanced S cone syndrome", with symptoms such as night blindness, is characterized by a decreased response of rods and

* Send correspondence to: Dr. Kenneth C. Wikler, Yale School of Medicine, Section of Neurobiology, SHM C323, 333 Cedar St., New Haven, CT 06510. Phone: (203) 737-2190; fax: (203) 785-5263; e-mail: Ken_Wikler@Yale.edu

Development and Organization of the Retina, edited by Chalupa and Finlay.
Plenum Press, New York, 1998.

L/MWS cones and a selective hypersensitivity of the SWS cone system attributed to a selective increase in the number of SWS cones (Hood et al., 1995).

Despite the critical roles of the generation and distribution of wavelength-sensitive cone subtypes for visual acuity and perception in the mammalian retina, few studies have investigated underlying mechanisms. We have observed a subset of periodically-positioned, early-differentiating cones at mid-gestation in the fetal monkey retina that are identified by antibodies specific to the L/MWS opsin, a synaptic vesicle protein, or a photoreceptor-specific membrane bound epitope (Wikler and Rakic, 1991; Wikler and Rakic, 1994). These arrays of precocious cones, which are surrounded by nascent, postmitotic cones, raised the hypothesis that lateral interactions between neighboring cells are critical to the emergence of the periodic spacing and opsin phenotypes of cone subtypes in the mature retinal mosaic (Wikler and Rakic, 1991; 1994; 1996a). Two possibilities were proposed for the timing of these interactions between neighboring cones; these interactions could occur either during the period close to the time of cone genesis or at a later stage of development coincident with the onset of opsin expression. We initiated a series of studies in embryonic monkey and mouse retina to examine these scenarios and identify the cellular and/ or molecular bases for these interactions.

To determine if cones are positioned before their opsin phenotypes are specified we needed to label immature cones during the time separating cone genesis from opsin protein expression. Because opsin-specific antibodies identify cones weeks after their final mitotic division, they are not useful for examining early events in cone patterning. Therefore, we used a cone and primate-specific monoclonal antibody, 7G6, to determine if immature cones are arranged periodically before they express an opsin (Wikler et al., 1997). Although 7G6 does not distinguish separate cone subtypes, it does label generic cones earlier than all other cone markers which makes it useful for examining the embryonic formation of the photoreceptor mosaic. Our analysis revealed that the onset of 7G6 immunoreactivity precedes the immunocytochemical detection of the L/MWS and SWS opsins by two weeks and the formation of synaptic contacts in the outer plexiform layer by at least two months. Early-differentiating 7G6-positive cones are organized into a regular array in immature regions of the E65 fetal monkey retina. This early onset and spatial distribution of 7G6 immunoreactivity suggests that retinal cells commit to a generic cone phenotype and are positioned into an array soon after their final mitotic division and *prior* to the formation of synapses between cones and horizontal or bipolar cells.

Although 7G6 labeling reveals an early cone array, it does not indicate whether these cones have committed to a specific wavelength phenotype. In order to test the plasticity of cone opsin phenotypes, we began a series of studies to manipulate the local cellular environment of postmitotic cones in the postnatal mouse retina using an organotypic culture system. We tested the plasticity of cone opsin phenotypes in co-culture experiments by exposing a phenotypically homogeneous population of immature SWS cones (ventral retina), prior to the expression of their class-specific opsin, to mature or immature MWS cones restricted to dorsal retina. Initial experiments included two culture paradigms that were run in parallel. Identified dorsal (90% MWS cones) and ventral (100% cones) retinal regions were either dissociated, reaggregated, and then incubated as retinal pellets, or were incubated as explants. These approaches were used to assay for the importance of diffusable factors (retinal explants) and contact-mediated mechanisms (reaggregation pellets) in the development of opsin phenotypes.

The results of these studies revealed fundamental differences in the development of the opsin phenotype of the MWS and SWS cone subtypes (Wikler et al., 1996b). Analyzing both types of retinal cultures harvested from newborn mice revealed that the SWS opsin emerged in ventral retina after five days *in vitro*, consistent with the temporal and spatial pattern of opsin expression *in vivo*. In addition, the topographic separation of SWS cones into distinct dorsal and ventral fields in these preparations was obvious, indicating that the development of opsin phenotype and patterning of SWS cones in organotypic and reaggregation cultures paralleled that seen *in vivo*. The MWS opsin was not expressed in these cultures unless retinae were harvested from postnatal day (P) 3 or older pups. In both explant and reaggregation preparations, however, MWS cones were always restricted to dorsal retina, as seen *in vivo*. No increase in the numbers of SWS cones in dorsal retinal explants was found despite the absence of MWS opsin expression in newborn retinal cultures. These results argue against the hypothesis that postnatal cones can be induced to switch their opsin.

To confirm this idea, we co-cultured dorsal and ventral retinal explants at various stages of maturation and assayed for the ectopic regional expression of either the MWS or

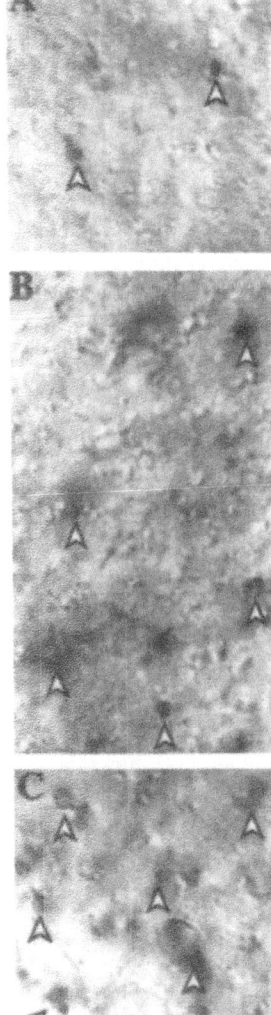

Figure 1. Digital image of SWS cones immunoreactive to the JH455 antibody found distributed in a regional gradient in P0 retinal explants, despite being co-cultured with explants from P12 dorsal retina. Arrowheads indicate examples of immunolabeled SWS cones. There is a four-fold increase in the density of SWS cones when moving from dorsal (A) through the transitional zone (B) to ventral retina (C). Scale bar = 10 μm.

SWS opsin. Neither the emergence of the cone fields nor the difference in the regional and temporal development of the opsins were affected in these experiments (Figure 1, Tables 1 and 2). These findings indicate that the opsin phenotype of postnatal cones is not altered by exposure to putative diffusable factors. Moreover, exposure to factors known to influence photoreceptor differentiation embryonically, including RA and tri-iodo-thyronine (T3), failed to modify the opsin phenotype of postnatal cones (Table 3; Wikler et al., 1996b). Although contact-mediated mechanisms may still prove to be important for the postnatal development of cone subtypes, these results suggest that the determination of both positional information in the retina and the opsin identity of individual cones occurs at embryonic ages.

These results from our culture studies, combined with the early array of 7G6-labeled cones in the fetal monkey retina suggested that cones might be specified to an opsin phenotype earlier than detected with opsin-specific antibodies. To examine this possibility we performed RT-PCR amplification of the L/MWS and SWS opsin mRNAs to maximize our chances of detecting opsin mRNAs that may be in low abundance in immature tissue. We looked for opsin mRNA expression as evidence for the early commitment of a cell to a cone subtype. Retinae were taken from rhesus monkeys sacrificed at embryonic (E) day E48, E65, E70, E72, and E90, prepared as wholemounts, and divided into three samples: the first sample containing the fovea and optic disc, the second the parafoveal eccentricities, and the third the peripheral margins of the retina. Primers for PCR amplification of the L/MWS and SWS opsins were selected based on the coding sequences of the human opsins. We found that by E48 *both* opsins were expressed throughout the retina, even in the sample taken from the retinal margin. The generation of cones in the peripheral margins of the retina begins at E45 (LaVail et al., 1991), three days prior to our earliest detection of opsin mRNAs. Since opsin expression may begin prior to our earliest sample point (E48), these data sets suggest that cones commit to a specific subtype within hours of their final mitotic division.

Table 1. Cultured retinal explants

AGE	10 DIV	14 DIV	21 DIV	28 DIV
P0- P1				
Wholemount	SWS (20)	SWS (14)	SWS (8)	SWS (8)
Dorsal explant	SWS (10)	SWS (10)	SWS (8)	SWS (8)
Ventral explant	SWS (10)	SWS (10)	SWS (8)	SWS (8)
P2- P3				
Wholemount	SWS (24)	MWS, SWS (10)	MWS, SWS (6)	
Dorsal explant	SWS (12)	MWS, SWS (10)	MWS, SWS (6)	
Ventral explant	SWS (12)	SWS (10)	SWS (6)	
P4- P5				
Wholemount	MWS, SWS (10)	MWS, SWS (6)		
Dorsal explant	MWS, SWS (8)	MWS, SWS (6)		
Ventral explant	SWS (8)	SWS (6)		
P6- P7				
Wholemount	MWS, SWS (6)	MWS, SWS (6)		
Dorsal explant	MWS, SWS (6)	MWS, SWS (6)		
Ventral explant	SWS (6)	SWS (6)		
P8- P9				
Wholemount	MWS, SWS (6)	MWS, SWS (6)		
Dorsal explant	MWS, SWS (4)	MWS, SWS (4)		
Ventral explant	SWS (4)	SWS (4)		
P10- P12				
Wholemount	MWS, SWS (8)			
Dorsal explant	MWS, SWS (6)			
Ventral explant	SWS (6)			
P16- P18				
Wholemount	MWS, SWS (8)			
Dorsal explant	MWS, SWS (6)			
Ventral explant	SWS (6)			

Summary of opsin expression in retinal explants harvested from pups ranging in age from the day of birth (P0) to P18 and incubated for 10, 14, 21, or 28 days *in vitro* (DIV). One half of the explants were incubated with attached retinal pigment epithelium (RPE) and the RPE was dissected free from remaining explants. No differences were seen between these groups. Number of retinae examined is indicated in parentheses.

Table 2. Co-culture "sandwiches" of dorsal and ventral retinal explants

AGE	*AGE OF CO-CULTURE RETINAE			
P0	**P0**	**P8**	**P14**	**P21**
Dorsal/Dorsal*	SWS/SWS (14)	SWS/SWS (10)	SWS/SWS (12)	SWS/SWS (6)
Dorsal/Ventral	SWS/SWS (14)	SWS/SWS (10)	SWS/SWS (12)	SWS/SWS (6)
Ventral/Dorsal	SWS/SWS (14)	SWS/SWS (10)	SWS/SWS (12)	SWS/SWS (6)
Ventral/Ventral	SWS/SWS (14)	SWS/SWS (10)	SWS/SWS (12)	SWS/SWS (6)
P3	**P3**	**P8**	**P14**	**P21**
Dorsal/Dorsal	MWS,SWS/MWS,SWS (14)	MWS,SWS/MWS,SWS (12)	MWS,SWS/MWS,SWS (10)	MWS,SWS/MWS,SWS (6)
Dorsal/Ventral	MWS,SWS/SWS (14)	MWS,SWS/SWS (12)	MWS,SWS/SWS (10)	MWS,SWS/SWS (6)
Ventral/Dorsal	SWS/MWS,SWS (14)	SWS/MWS,SWS (12)	SWS/MWS,SWS (10)	SWS/MWS,SWS (6)
Ventral/Ventral	SWS/SWS (14)	SWS/SWS (12)	SWS/SWS (10)	SWS/SWS (6)

Summary of opsin expression in dorsal or ventral retinal explants harvested from pups at either P0 or P3 and co-cultured with dorsal or ventral explants harvested at either P0, P8, P14, or P21. *Sandwiched explants are indicated by X/X with the region of either P0 or P3 tissue listed first followed by the region of co-cultured retinae. Explants were incubated for 10 days in vitro. One half of the co-cultures were incubated with the two photoreceptor layers facing each other ("belly to belly") and the others with the photoreceptor layer of one explant adjacent to the ganglion cell layer of the other ("belly to back"). No differences were seen between these groups. Number of retinae examined is indicated in parentheses.

Table 3. Retinal explants cultured with exogenous factors

AGE	Defined Medium	T3	RA	RA & T3
P0				
Wholemount	SWS (6)	SWS (10)	SWS (14)	SWS (10)
Dorsal explant	SWS (6)	SWS (10)	SWS (14)	SWS (10)
Ventral explant	SWS (6)	SWS (10)	SWS (14)	SWS (10)

Summary of opsin expression in retinal explants incubated for 10 DIV in either defined medium or defined medium supplemented with tri-iodo-thyronine (T3), all-*trans* retinoic acid (RA) or both. Number of retinae examined is indicated in parentheses.

The findings outlined above strengthen our conviction that the specification of the opsin phenotype and the positioning of cones into species-specific mosaics are not mediated by late-occurring interactions between neighboring cones or cones and bipolar cells. Rather, cone patterning results from a series of early inductive events in the embryonic photoreceptor layer. To pursue this thesis, we are currently examining the expression patterns of candidate molecules involved in the emergence of the dorsoventral segregation of cone subtypes in the murine retina. To date, this work has provided evidence implicating endogenous retinoids in the early patterning of the cone mosaic.

Retinoids have been shown to participate in the patterning of cells in regions as diverse as the developing limb bud, hindbrain, and inner ear (Tabin, 1991; Kelley et al., 1993; Capecchi, 1994), and are thought to regulate pattern formation by mediating the transcriptional activity of target genes through two families of nuclear receptors, the retinoic acid (RARs) and the retinoid X (RXRs) receptors (reviewed in Leid et al., 1992a; Linney, 1993; Kastner et al., 1993). Each family consists of three subtypes: α, β, and γ, and through alternative promoter usage and alternative splicing each subtype gene can produce numerous mRNA isoforms that differ only in their N-terminal region (reviewed in Leid et al., 1992a, b). RAR α, β, and γ are preferentially activated by all-trans RA, though all can also be activated by 9-cis RA (Allenby et al., 1993; and references within). RXR α, β, and γ, however, can only be activated by 9-cis RA (Mangelsdorf et al., 1990, 1992; Heyman et al., 1992; Levin et al., 1992; Hamada et al., 1989; Leid et al., 1992b; Allenby et al., 1993; Heery et al., 1993; for review Leid et al., 1992a; Linney, 1993; Kastner et al., 1993 and references within). Additionally, RXRs can form heterodimer complexes with RARs which, in vitro, bind with RA to RA responsive elements (RAREs) more effectively than either RAR or RXR monomers or homodimers (Yu et al., 1991; Leid et al., 1992b; Zhang et al., 1992), indicating that RAR-RXR heterodimers direct retinoid-mediated gene transcription in vivo (Durand et al., 1992; Husmann et al., 1992; Heery et al., 1993; Nagpal et al., 1993; Dey et al., 1994).

In addition to the two families of nuclear receptors, cellular RA binding proteins (CRABP I and II) as well as cellular retinol binding proteins (CRBP I and II) have been characterized. The CRABPs are small cytoplasmic proteins that bind RA, but do not act like transcription factors (Ong et al., 1994). They are believed to modulate RA availability to the RARs or the metabolic enzymes (Boylan and Gudas, 1991, 1992; Fiorella and Napoli, 1991, 1994; Dekker et al., 1994). For example, overexpressing the Xenopus homologue xCRABPI during embryogenesis resulted in abnormalities similar to those observed after exposure to excess RA. These experiments suggest that CRABPI directs RA to the nuclear receptors, thus raising intracellular RA concentrations (Means and Gudas, 1997).

Two lines of research suggest that endogenous retinoids are candidate molecules for patterning the MWS and SWS cone subtypes into dorsal and ventral compartments in the embryonic murine retina. First, previous studies have shown that RA participates in establishing the dorsoventral retinal axis in both mouse and zebrafish. McCaffery and Drager have determined that in the E13 murine embryo, ventral retina is more effective than dorsal retina at inducing retinoid-dependent transgene activation when co-cultured with a retinoid-dependent reporter cell line (McCaffery et al., 1992), suggesting that RA levels may be elevated in ventral retina at this time. Furthermore, the enzymes that synthesize RA from retinaldehyde are restricted to either the dorsal or ventral retinal compartment during this period of development (McCaffery et al., 1992; 1993). The acidic dehydrogenases located in ventral retina are more efficient at synthesizing RA than aldehyde dehydrogenase class-1 isoform (AHD-2) which is restricted to dorsal retina at E13. Thus, differential RA availability may regulate dorsoventral axial development in the embryonic retina.

Dowling and colleagues have shown that manipulating RA levels in the embryonic zebrafish produces both retinal malformations and the ectopic expression of transcription factors normally restricted to either the embryonic dorsal (msh[c]) or ventral (pax2) retina (Hyatt et al., 1992; 1996; Marsh-Armstrong, 1994). Augmenting RA levels induced the duplication of ventral retina and caused larger optic discs and thicker optic stalks (Hyatt et al., 1992), while depleting levels with citral reduced the size of ventral retina, creating a "half-retina" (Marsh-Armstrong et al., 1994). Local application of RA to dorsal retina resulted in the ectopic expression of pax2 and appearance of an additional optic stalk as well as the decreased or abolished expression of msh[c] and aldehyde dehydrogenase in dorsal retina.

Finally, studies on mice that are null mutants for specific receptor combinations indicate that RA (RAR) and retinoid X (RXR) receptors are critical for the development of the dorsoventral axis. For example, mouse mutants lacking RARα and RARβ2 (Lohnes et al., 1994), RARα and RARγ (Mendelsohn et al., 1994), RARγ and RARβ2 (Kastner et al., 1994), RARβ2 and RARγ2 (Grondona et al., 1996), or RXRα (Kastner et al., 1994) together display all of the ocular abnormalities seen in fetal vitamin A deficiency (VAD) syndrome, such as shortened ventral retina, retinal dysplasia, and anopthalmia (Warkany and Schraffenberger, 1946; Wilson et al., 1953).

The second set of findings that support the hypothesis that retinoids are involved in the patterning of cones indicate that RA directly influences the commitment of an immature retinal cell to a photoreceptor fate. For example, *in vitro* studies demonstrate that retinoids, including retinol, 11-cis retinaldehyde, and all-trans retinoic acid (RA) regulate cell fate decisions and photoreceptor differentiation in both the chick and rat. Exposing chick or rat retinal neuroblasts to vitamin A metabolites increased the proportion of cells that differentiated as rods without affecting cell proliferation (Stenkamp et al., 1993; Stenkamp and Adler, 1994; Kelley et al., 1994). Rod development in embryonic zebrafish was also potentiated *in vivo* after the administration of all-trans RA without influencing cell proliferation (Hyatt et al., 1996). These results suggest that endogenous retinoids act on postmitotic cells to regulate the development of photoreceptor fates in the vertebrate retina.

These studies suggest that RA and retinoid receptors are involved in compartmentalizing the MWS and SWS cones in the murine retina into segregated dorsal and ventral regions. RA could be influencing photoreceptor differentiation by regulating the development of positional information along the dorsoventral axis or alternatively, regulate the differentiation of the cone subtypes directly. To begin our investigation of retinoids and cone patterning we examined the spatial and temporal pattern of retinoid activated gene transcription in the embryonic murine retina prior to and during cone genesis. We analyzed a line of indicator mice that possess a retinoid-dependent transgene. This transgene has been successfully used to map the location of retinoid-responsive neuroblasts in the developing olfactory system and spinal cord (Balkan et al., 1992; Colbert et al., 1993, 1995; Anchan et al., 1997) in two separate lines of transgenic mice, indicating that the expression pattern of this transgene is reproducible and not dependent on insertion site (Balkan et al., 1992). In addition, this RARE-tk transgene has been shown to be 100 times more sensitive to all-trans RA than to other retinoid isomers (Balkan et al., 1992). Thus, transgene expression in these mice represents RARE activation by a receptor bound to a retinoid ligand, which is presumably all-trans RA.

We found that early in retinogenesis (E11.5) retinoid-dependent transgene activation is restricted to neuroblasts located in the dorsal region of the retina. This early asymmetric response to RA is consistent with the restriction of AHD-2 and CRABPI to dorsal retina during early retinogenesis (McCaffery et al., 1992; 1993). This co-localization of AHD-2 and CRABP I to dorsal retina could result in higher levels of RA in this region, since it has been suggested that CRABP increases intracellular RA levels (Means and Gudas, 1997). Although no studies have directly measured *in vivo* levels of retinoids in the embryonic murine retina, these results indicate that all-trans RA levels are elevated in the dorsal E12 retina, thus inducing retinoid-dependent gene activation in this restricted region.

We observed that retinoid-activated transgene expression shifts dramatically from exclusively dorsal retina at E11.5 to predominately ventral retina at E14.5 (Figure 2). This change in the location of activated neuroblasts suggests that all trans RA levels are now elevated in ventral retina at this later stage of development, coincident with the peak in cone genesis. Consistent with this hypothesis is the report that at E13 ventral retina is more effective than dorsal retina at inducing retinoid-dependent transgene activation when co-cultured with a reporter cell line (McCaffery et al., 1992). Together these results support the hypothesis that RA levels are elevated in ventral retina by E13, perhaps due to the differential activity of retinoid-synthesizing enzymes (McCaffery et al., 1992; 1993).

Although the dorsal-to-ventral shift we observe in retinoid responsiveness suggests that RA levels are higher in dorsal retina at the onset of cone genesis and in ventral retina later in development, it remains unclear if this shift is due to regional changes in the local availability of retinoids or alterations in the capacity of neuroblasts to respond to retinoids. To evaluate the importance of RA levels in the shifting gradient of transgene activation we gavage-fed timed-pregnant mice all-trans RA (80 mg/kg) to augment RA levels or citral (3 mg/kg), a drug that inhibits retinoid synthesis by competing preferentially with acidic dehydrogenases in ventral retina, to reduce retinoid levels.

The hypothesis that local availability of RA is critical for determining which neuroblasts respond to endogenous retinoids is supported by our finding that manipulating RA levels altered transgene activity. Early RA treatments induced transgene activation

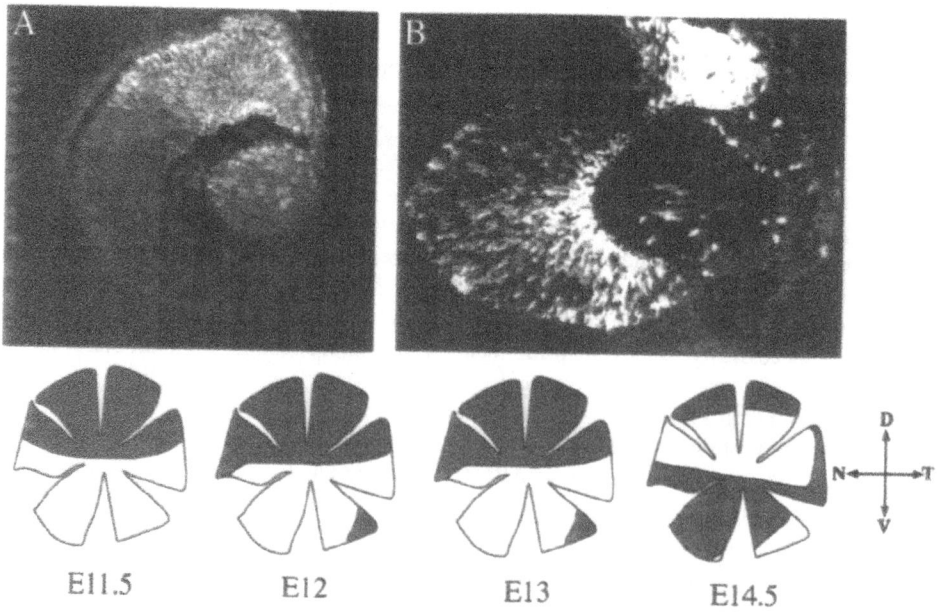

Figure 2. Distribution of retinoid-responsive cells in the embryonic mouse retina. b-gal labeled neuroblasts in RARE-tk-*lacZ* indicator mice are restricted to dorsal retina at E11 (A) and shift to predominantly ventral retina by E14 (B). Schematic of retinoid responsive neuroblasts (represented by shaded areas) from E11.5 (b-gal labeling restricted to dorsal retina) to E14.5 (b-gal labeling located primarily in ventral retina).

throughout the E11.5 retina, while citral administration reduced retinoid-dependent gene transcription in dorsal retina and potentiated a response in ventral retina. Therefore, neuroblasts in the E11.5 retina, regardless of their location, have the capacity to respond to endogenous retinoids; their responsiveness is in part limited by the availability of endogenous RA. However, RA availability is a limiting factor in determining which neuroblasts respond to endogenous retinoids only during an early developmental window. When we expose older embryos to either exogenous RA or citral, we see no changes in the regional localization of RA induced transgene activation; responsive neuroblasts are consistently located at the dorsal margins and ventral retina regardless of experimental treatment. These results demonstrate that between E11.5 and E14.5 neuroblasts in a region dorsal to the optic disc lose their capacity to respond to RA even when exogenous RA is provided. Although it remains unclear why neuroblasts lose their ability to respond to available RA this result suggests that other factors, perhaps intrinsic to the cell, are critical for regulating the capacity of neuroblasts to respond to retinoids.

Because transgene activation requires the presence of both ligand and receptor, we next examined the expression patterns of nuclear retinoid receptors, as possible intrinsic factors that might be important for modulating retinoid responsiveness along the dorsoventral axis, . Although surveys of receptor localization in mouse embryos suggested that retinoid receptor subtypes are uniformly expressed in the retina (Dolle et al., 1990; 1994; Ruberte et al., 1990; 1991; 1993; Mangelsdorf et al., 1992; Morriss-Kay, 1992 and references within), regional malformations observed in the embryonic retinae of mice that are null mutants for specific receptor combinations indicated that these receptor subtypes are critical for the development of the dorsoventral axis of the embryonic eye (Kastner et al., 1994; 1997; Lohnes et al., 1994; Mendelsohn et al., 1994; Grondona et al., 1996). Using nonradioactive wholemount *in situ* hybridization, we found that RARβ mRNA is asymmetrically expressed along the dorsoventral axis; RARβ mRNA is restricted to dorsal retina at E12 and then shifts to ventral retina by E14. Thus, RARβ expression parallels retinoid-induced transgene activation during this period. For example, at E14, RARβ mRNA was observed in both ventral retina as well as in the dorsal retinal margins, as was transgene activity, demonstrating that nonresponsive neuroblasts located immediately dorsal to the optic disc also fail to express RARβ mRNA. RARβ expression, therefore, may participate in regulating retinoid responsiveness in conjunction with another receptor family member, as retinoid-mediated gene transcription often requires a RAR-RAR or RAR-RXR heterodimer (Berrodin et al., 1992; Marks et al., 1992; Giguere et al., 1994; Zhang et al., 1992; Zhang and Pfahl, 1993; Mandelsdorf and Evans, 1995; Meyer et al., 1996 and references within).

We also found that administering either RA or citral early in retinogenesis dramatically altered the size and shape of the retina and of the optic disc. In particular, RA treatment resulted in a smaller retina with a larger optic disc that was positioned more ventrally than the near midpoint position seen in control E11.5 retina. Citral treatment shortened retinal length along the dorsoventral axis and enlarged the optic disc, but did not affect the length of the retina along the nasotemporal axis. Unique to this treatment we found that administering citral resulted in additional ventral regions, separated from the primary ventral retina piece by pigment epithelial cells in the murine embryo. Administering RA or citral later in development had only minimal effects on retinal morphometry.

These results are consistent with the finding from zebrafish studies that changes in retinal morphology can be induced by RA perturbations only during a restricted developmental window (Hyatt et al., 1992; 1996; Marsh-Armstrong et al., 1994). In the mouse this period appears to end by E14.5. However, studies in zebrafish have shown that RA-induced retinal morphological abnormalities are accompanied by the ectopic expression of region-specific transcription factors such as pax2 and msh[c], markers of embryonic ventral and dorsal retina respectively, leading to the hypothesis that RA is critical for the development of ventral retina (Hyatt et al., 1996).

Our results in the mouse, however, do not support this hypothesis. Although we observed significant morphological changes, as well as regional shifts in retinoid-activated gene transcription after manipulating RA levels, we found that giving exogenous RA shortened ventral retina and enlarged dorsal retina. Furthermore, RA treatment did not induce the ectopic expression of region-specific markers of the embryonic murine retina. For example, Pax2 expression remained restricted to ventral retina and AHD-2 expression to dorsal retina after retinoid manipulations. These differences in the effects of RA manipulations on the morphogenesis of the zebrafish and murine retina may be due to species differences in early eye development or to the regulation of retinoid-activated gene transcription. Alternatively, the longer gestation period in the mouse may allow for the temporal resolution of retinoid-induced perturbations of retinal development, not observed in zebrafish. Our results suggest that RA may independently regulate the formation of the optic stalk, the dorsal/ventral growth of the retina, and the regional restriction of transcription factors in the embryonic murine retina.

RA and citral given prior to the onset of cone genesis altered the regional expression of the RARE-tk-lacZ transgene, but not of transcriptional markers of dorsal and ventral retina, suggesting that the pattern of retinoid responsiveness is not directly correlated with the establishment of the dorsoventral axis. We next asked whether experimentally-induced shifts in retinoid-dependent gene transcription are associated with corresponding changes in the patterning of cone subtypes in the adult photoreceptor mosaic. Our results indicate that retinoid manipulation results in region- and stage-specific changes in cone patterning that we believe are due to changes in the determination of the phenotypic fate of these cells. This hypothesis is supported by experimental evidence from zebrafish, chick, and rat that indicates that retinoids act upon postmitotic rather than dividing cells; RA induced changes do not include an increased number of dividing neuroblasts (Stenkamp et al., 1993; Stenkamp and Adler, 1994; Kelley et al., 1994; Hyatt et al., 1996). Although studies have reported that retinoid treatment enhances the survival of photoreceptors in low density cultures of dissociated chick retinae (Stenkamp et al., 1993), our analysis of cone patterning in the mouse reveals that manipulating embryonic retinoid levels changes the *proportion* of cones expressing a wavelength-sensitive opsin, but not overall cone number. These findings indicate that retinoid manipulations regulate the *differentiation* of embryonic cells into specific cone subtypes rather than cone genesis or survival.

For example, citral treatment at E8 *decreased* expression of the MWS opsin in dorsal retina and *increased* expression of the MWS opsin in ventral retina. This region-specific change in MWS opsin expression altered the composition of the adult cone mosaic; we observed a decrease in the density of MWS cones in dorsal retina (peak MWS cone density: controls: 13.7×10^3 cones/ mm^2, treated; 5.6×10^3 cones/ mm^2) and the ectopic appearance of both MWS and M/SWS cones ventral to the optic disc. These results suggest that retinoids influence the development of MWS cones, perhaps by regulating MWS opsin expression directly, as suggested by the presence of RAREs in the promoter region of the human L/MWS opsin genes (see Hyatt et al., 1996). Early citral treatment, however, did not only affect MWS cones. We also observed region-specific changes in SWS opsin expression in these citral-treated retinae such as the ectopic appearance of M/SWS cones in dorsal retina and a decreased density of SWS cones in ventral retina (peak SWS cone density: controls: 12.8×10^3 cones/ mm^2, treated; 4.6×10^3 cones/ mm^2). These reciprocal effects on opsin expression after retinoid manipulation suggest that enhancing the development of one cone subtype inhibits the development of the other.

Manipulating retinoid levels at later embryonic ages (E14) induced SWS opsin expression in dorsal retina (Figure 3). For example, RA-treatment at E14 resulted in a 7-fold increase in the density of SWS cones in dorsal retina (peak SWS cone density: controls; 1.2×10^3 cones/mm^2, treated; 9.7×10^3 cones/mm^2). Importantly, peak SWS cone densities in ventral retina of both control and treated mice were similar (peak SWS cone density: controls; 12.8×10^3 cones/mm^2, treated; 13.5×10^3 cones/mm^2). Although peak SWS cone density increased in dorsal retina of treated animals, we saw no evidence for a reciprocal decrease in the density of MWS cones (peak MWS cone density: controls; 13.7×10^3 cones/mm^2, treated; 14.3×10^3 cones/mm^2). The stability of the ventral cone population indicates that RA treatment selectively induces SWS opsin expression in dorsal retina. Cone number estimates indicate that the number of dorsal cones remains stable despite these dramatic changes in the proportion of cone subtypes. This result suggests that exposing MWS cones to exogenous RA induces them to co-express the SWS opsin, similar to the effect of early citral treatment.

CONCLUSIONS

By correlating regional changes in the response of embryonic neuroblasts to RA with regional changes in the composition of the cone mosaic after retinoid manipulation we postulate two phases in the determination of the wavelength-sensitive cone subtypes.

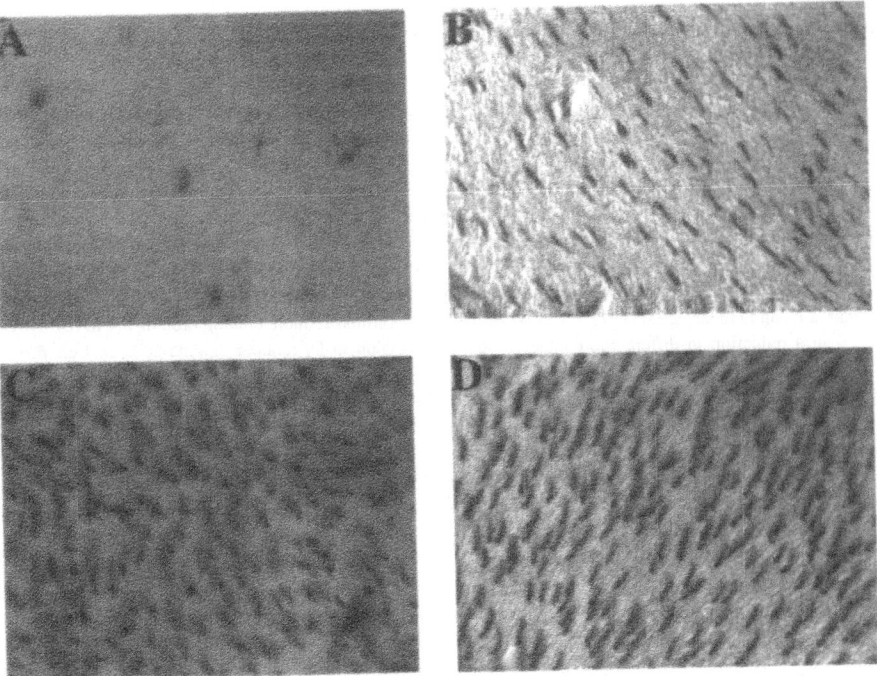

Figure 3. Distribution of SWS cones before (A, C) and after RA (B, D) treatment at E14. RA treatment selectively increases dorsal SWS cone density (compare A and B).

At early stages of development, retinoid-activated gene transcription is associated with enhanced MWS opsin expression and decreased SWS opsin expression. For example, early citral treatment induces retinoid-activated gene transcription in ventral retina and the expression of the MWS opsin by ventral cones. In the dorsal retina of these animals, however, retinoid responsivenss is attenuated, and a corresponding decrease in the expression of the MWS opsin as well as an increase in SWS opsin expression is observed in dorsal retina. Later in cone genesis, however, retinoid-activated gene transcription is associated with SWS opsin expression. For example, late RA treatment increases the magnitude of the response of neuroblasts to retinoids in dorsal retina and the expression of the SWS opsin by dorsal cones.

Taken all together, these results suggest that disrupting the normal regional segregation of cone subtypes in the mouse retina by manipulating retinoid levels does not involve disrupting this retinal axis; instead, wavelength-sensitive cone subtypes appear to be specified coincident with the period of cone genesis and this occurs after the murine retina is parceled into dorsal and ventral compartments. We hypothesize that dividing neuroblasts are equivalent in dorsal and ventral regions of the retina prior to cone genesis and that regional differences in the development of cone phenotypes emerge later, at the onset of cone differentiation. This specification step appears to be regulated by gene transcription under the control of endogenous retinoids. The mechanism(s) underlying retinoid-mediated changes in cone differentiation and opsin expression are the focus of our current studies.

REFERENCES

Allenby, G., Bocquel, M.T., Saunders, M., Kazmer, S., Speck, J., Rosenberger, M., Lovey, A., Kastner, P., Grippo, J.F., Chambon, P., and Levin, A.A. (1993). Retinoic acid receptors and retinoid X receptors: interactions with endogenous retinoic acids. *Proc. Natl. Acad. Sci. USA* **90**, 30–34.

Anchan, R.M., Drake, D.P., Haines, C.F., Gerwe, E.A., and LaMantia, A.-S. (1997). Disruption of local retinoid-mediated gene expression accompanies abnormal development in the mammalian olfactory pathway. *J. Comp. Neurol.* **379**, 171–184.

Balkan, W., Colbert, M., Bock, C., and Linney, E. (1992). Transgenic indicator mice for studying activated retinoic acid receptors during development. *Proc. Natl. Acad. Sci. USA* **89**, 3347–3351.

Berrodin, T.J., Mark, M.S., Ozato, K., Linney, E., and Lazar, M.A. (1992). Heterodimerization among thyroid hormone receptor, retinoic acid receptor, retinoid X receptor, chicken ovalbumin upstream promoter transcription factor, and an endogenous liver protein. *Mol. Endo.* **6**, 1468–1478.

Boylan, J.F., and Gudas, L.J. (1991). Overexpression of the cellular retinoic binding protein-I (CRABP-I) results in a reduction in differentiation-specific gene expression in F9 teratocarcinoma cells. *J. Cell Bio.* **122**, 965–979.

Boylan, J.F. and Gudas, L.J. (1992). The level of CRABP-I expression influences the amounts and types of all-trans retinoic acid metabolism in F9 teratocarcinoma stem cells. *J. Biol. Chem.* **267**, 21486–21491.

Bresnick, G.H., V.C. Smith, and J. Pokorny (1989). Autosomal dominantly inherited macular dystrophy with preferential short-wavelength sensitive cone involvement. *Am.J.Ophthalmol.* *108*:265–276.

Calderone, J.B. and G.H. Jacobs (1995). Regional variations in the relative sensitivity to UV light in the mouse retina. Visual Neurosci. *12*:463–468.

Capecchi, M.R. (1994). Targeted gene replacement. *Sci.Amer.* **270**, 52–59.

Clagett-Dame, M., Verhalen, T. J., Biedler, J. L., and Repa, J. J. (1993). Identification and characterization of all-trans retinoic acid receptor transcripts and receptor protein in human neuroblastoma cells. *Arch. Biochem. Biophys.* **300**, 684–693.

Colbert, M.C., Rubin, W.W., Linney, E., and LaMantia, A.-S. (1995). Retinoid signaling and the generation of regional and cellular diversity in the embryonic mouse spinal cord. *Dev. Dyn.* **204**, 1–12.

Colbert, M.C., Linney, E., and LaMantia, A.-S. (1993). Local sources of retinoic acid coincide with retinoid-mediated transgene activity during embryonic development. *Proc. Natl. Acad. Sci. USA* **90**, 6572–6576.

Crowe, D. L., Hu, L., Gudas, L. J., and Rheinwald, J. G. (1991). Variable expression of retinoic acid receptor beta (RARβ) mRNA in human oral and epidermal keratinocytes; relation to keratin 19 expression and keratinization potential. *Differentiation* **48**, 199–203.

Curcio, C.A, Allen, K.A., Sloan, K.R., Lerea, C.L., Hurley, J.B., Klock, I.B., and Milam, A.H. (1991) Distribution and morphology of human cone photoreceptors stained with anti-blue opsin. J.Comp.Neurol. *312*:610–624.

Dartnall, H.J.A., Bowmaker, J.K., and Mollon, J.D. (1983). Human visual pigments: microspectrophotometric results from the eyes of seven persons. Proc.R.Soc.Lond.B. *220*:115–130.

Dekker, E.-J., Vaessen, M.-J., van den Berg, C., Timmermans, A., Godsave, S., Holling, T., Nieukwoop, P., Kessel, A. G., and Durston, A. (1994). Overexpression of a cellular retinoic acid binding protein (xCRABP) causes anteroposterior defects in developing Xenopus embryos. *Development* **120**, 973–985.

de The, H., Marchio, A., Tiollais, A., and Dejean, A. (1989). Differential expression and ligand regulation of the retinoic acid receptor alpha and beta genes. *EMBO J.* **8**, 429–433.

Dolle, P., Ruberte, E., Kastner, P., Petkovich, M., Stoner, C. M., Gudas, L. J., and Chambon, P. (1989). Differential expression of genes encoding alpha, beta, and gamma retinoic acid receptors and CRABP in the developing limbs of the mouse. *Nature* **342**, 702–705.

Dolle, P., Ruberte, E., Leroy, P., Morriss-Kay, G., and Chambon, P. (1990). Retinoic acid receptors and cellular retinoid binding proteins. I. A systemic study of their differential pattern of transcription during mouse organogenesis. *Development* **110**, 1133–1151.

Dolle, P., Fraulob, V., Kastner, P., and Chambon, P. (1994). Developmental expression of murine retinoid X receptor (RXR) genes. *Mech. Dev.* **51**, 91–104.

Durand, B., Saunders, M., Leroy, P. Leid, M., and Chambon, P. (1992). All-trans and 9-cis retinoic acid induction of CRABP II transcription is mediated by RAR-RXR heterodimers bound to DR1 and DR2 repeated motifs. *Cell* **71**, 73–85.

Falls, H.F., Wolter, J.R., and Alpern, M. (1965) Typical total monochromacy—a histological and psychophysical study. Arch.Ophthal. *74*:610–616.

Fiorella, P.D., and Napoli, J.L. (1991). Expression of cellular retinoic acid binding protein (CRABP) in Escherichia coli. Characterization and evidence that holo-CRABP is a substrate in retinoic acid metabolism. *J. Biol. Chem.* **266**, 16572–16759.

Giguere, V. (1994). Retinoic acid receptors and cellular retinoid binding proteins: complex interplay in retinoid signaling. *Endo. Rev.* **15**, 61–79.

Glickstein, M., and Heath, R. (1975). Receptors in the monochromat eye. Vis.Res. *15*:633–636.

Grondona, J.M., Kastner, P., Gansmuller, A., Decimo, D., Chambon, P., and Mark, M. (1996). Retinal dysplasia and degeneration in RARβ2/RARγ2 compound mutant mice. *Development* **122**, 2173- 2188.

Hamada, K., Gleason, S.L., Levi, B.Z., Hirschfield, S., Appella, E., and Ozato, K. (1989). H-2RIIBP, a member of the nuclear receptor superfamily that binds to both the regulatory element of major histocompatibility class I genes and the estrogen response elements. *Proc. Natl. Acad. Sci. USA* **86**, 8289–8293.

Heery, D.M., Zacarewski, T. Pierrat, B., Gronemeyer, H., Chambon, P., and Losson, R. (1993). Efficient transactivation by retinoic acid receptors in yeast requires retinoid X receptors. *Proc. Natl. Acad. Sci. USA* **90**, 4281–4285.

Heyman, R.A., Mangelsdorf, D.J., Dyck, J.A., Stein, R.B., Eichele, G., Evans, R.M., and Thaller, C. (1992). 9-cis retinoic acid is a high affinity ligand for retinoid X receptor. *Cell* **66**, 397–406.

Hood, D.C., Cideciyan, A.V., Roman, A.J., and Jacobson, S.G. (1995). Enhanced S cone syndrome: evidence for an abnormally large number of S cones. Vision Res. **35**:1473–1481.

Hyatt, G.A., Schmitt, E.A., Marsh-Armstrong, N.R., and Dowling, J.E. (1992). Retinoic acid induced duplication of the zebrafish retina. *Proc. Natl. Acad. Sci. USA* **89**, 8293–8297.

Hyatt, G.A., Schmitt, E.A., Marsh-Armstrong, N., McCaffery P., Drager, U.C., and Dowling, J.E. (1996). Retinoic acid establishes ventral characteristics. *Development* **122**, 195–204.

Hyatt G. A. and Dowling, J. E. (1997). Retinoic acid: A key molecule for eye and photoreceptor development. *Invest. Ophth. Vis. Sci.* **38**, 1471–1475.

Kastner, P., Grondona, J.M., Mark, M., Gansmuller, A., LeMeur M., Decimo, D., Vonesch, J.-L., Dolle, P., and Chambon, P. (1994). Genetic analysis of RXRα developmental function: Convergence of RXR and RAR signaling pathways in heart and eye morphogenesis. *Cell* **78**, 987–1003.

Kastner, P., Mark, M., Ghyselinch, N., Krezel, W., Dupe, V., Grondona, J.M., and Chambon, P. (1997). Genetic evidence that the retinoid signal is transduced by heterodimeric RXR/RAR functional units during mouse development. *Development* **124**, 313–326.

Kato, S., Mano, H., Kumazawa, T., Yoshizawa, Y., Kojima, R., and Masushige, S. (1992). Effect of retinoid status on alpha, beta, and gamma retinoic acid mRNA levels in various rat tissues. *Biochem. J.* **286**, 755–60.

Kelley, M.W., Xu, R.A., Wagner, M.A., Warchol, M.E., and Corwin, J.T. (1993). The developing organ of Corti contains retinoic acid and forms supernumerary hair cells in response to exogenous retinoic acid in culture. *Development* 119, 1041–1053.

Kelley, M.W., Turner, J.K., and Reh, T.A. (1994). Retinoic acid promotes differentiation of photoreceptors in vivo. *Development* 120, 2091–2102.

Leid, M., Kastner, P., and Chambon, P. (1992a). Multiplicity generates diversity in the retinoic acid signaling pathways. *Trends Biochem. Sci.* 17, 427–433.

Leid, M., Kastner, P., Lyons, R., Nakshatri, H., Saunders, M., Zacharewski, T. Chen, J.-Y., Staub, A., Garnier, J.-M., Mader, S., and Chambon, P. (1992b). Purification, cloning, and RXR identity of the HeLa cell factor with which RAR or TR heterodimerizes to bind target sequences efficiently. *Cell* 68, 377–395.

Levin, A.A., Sturzenbecker, L.J., Kazmer, S., Bosakowski, T., Huselton, C., Allenby, G., Speck, J., Kratzeisen, C., Rosenberger, M., Lovay, A. (1992). 9-cis retinoic acid stereoisomer binds and activates the nuclear receptor RXR alpha. *Nature* 355, 359–361.

Linney, E. (1992). Retinoic acids: transcription factors modulating gene regulation, development, and differentiation. *Curr. Top. Dev. Bio.* 27, 309–350.

Lohnes, D., Mark, K., Mendelsohn, C., Dolle, P., Dierich, A., Gorry, P., Gansmuller, A., and Chambon, P. (1994). Function of the retinoic acid receptors (RARs) during development. I. Craniofacial and skeletal abnormalities in RAR double mutants. *Development* 120, 2723–2748.

Lovat, P.E., Pearson, A.D., Malcolm, A., and Redfern, C.P. (1993). Retinoic acid receptor expression during the in vitro differentiation of human neuroblastomas. *Neurosci. Letters* 162, 109–13.

Mangelsdorf, D.J., Ong, E.S., Dyck, J.A., and Evans, R.M. (1990). Nuclear receptor that identifies a novel retinoic acid response pathway. *Nature* 345, 224–229.

Mangelsdorf, D.S., Borgmeyer, U., Heyman, R.A., Zhou, J.Y., Ong, E.S., Oro, A.E., Kakizuka, A., and Evans, R.M. (1992). Characterization of 3 RXR genes that mediate the action of 9-cis retinoic acid. *Genes Dev.* 6, 329–344.

Mangelsdorf, D.S., and Evans, R.M. (1995). The RXR heterodimers and orphan receptors. *Cell* 83, 841–850.

Mark, M.S., Hallenbeck, P.L., Nagata, T., Segars, J. H., Appella, E., Nikodem U. M., and Ozato, K. (1992). H-2RIIBP (RXR β) Heterodimerization provides a mechanism for combinatorial diversity in the regulation of retinoic acid and thyroid hormone responsive genes. *EMBO J.* 11, 1419–35.

Marsh-Armstrong, N., McCaffery, P., Gilbert, W., Dowling, J. E., and Drager, U. C. (1994). Retinoic acid is necessary for development of the ventral retina in zebrafish. *Proc. Natl. Acad. Sci. USA* 91, 7286–7290.

McCaffery, P., Lee, M.-O., Wagner, M.A., Sladek, N.E., and Drager, U.C. (1992). Asymmetrical retinoic acid synthesis in the dorsoventral axis of the retina. *Development* 115, 371–382.

McCaffery, P., Posch, K.C., Napoli, J.L., Gudas, L., and Drager, U.C. (1993). Changing patterns of the retinoic acid system in the developing retina. *Dev. Bio.* 158, 390–399.

McMenamy K.R., and Zachman, R.D. (1993). Effect of gestational age and retinol (vitamin A) deficiency on fetal rat lung nuclear retinoic acid receptors. *Pediatric Res.* 33, 251–255.

Means A.L. and Gudas, L.J. (1994). The roles of retinoids in vertebrate development. *Ann. Rev. Biochem.* 64, 201–233.

Means, A.L., and Gudas, L.J. (1997). The CRABP I gene contains two separable, redundant regulatory regions active in neural tissues in transgenic mouse embryos. *Dev. Dyn.* 209, 59–69.

Mendelsohn, C., Lohnes, D., Decimo, D., Lufkin, T., LeMeur, M., Chambon, P., and Mark, M. (1994). Function of the retinoic acid (RARs) during development. II. Multiple abnormalities at various stages of organogenesis in RAR double mutants. *Development* 120, 2749–2771.

Meyer, M., Sonntag-Buck, V., Keaveney, M., and Stunnenberg, H. G. (1996). Retinoid dependent transcription: the RAR/RXR-TBP-EIA/EIA-LA connection [review]. *Biochem.* 62, 97–109.

Morriss-Kay, G. (1992). Retinoic acid receptors in normal growth and development [review]. *Cancer Surveys* 14, 181–193.

Nagpal, S., Friant, S., Nakshatri, H., and Chambon, P. (1993). RARs and RXRs: Evidence for two autonomous transactivation functions (AF-1 and AF-2) and heterodimerization in vivo. *EMBO J.* 12, 2349–2360.

Rohlich, P., Th. van Veen, and A. Szel (1994). Two different visual pigments in one retinal cone cell. *Neuron* 13:1159–1166.

Ruberte, E., Dolle, P., Krust, A., Zelent, A., Morriss-Kay, G., and Chambon, P. (1990). Specific spatial and temporal distribution of retinoid acid receptor gamma transcripts during mouse embryogenesis. *Development* 108, 213–222.

Ong, D.E., Newcomer, M.E., and Chytil, F. (1994). Cellular retinoid binding proteins. in "The Retinoids: Biology, Chemistry, and Medicine," 2nd ed., Sporn, M B., Roberts, A. B., Goodman, D. S. (eds). New York: Raven Press, pp. 283–317.

Ruberte, E., Dolle, P., Krust, A., Zelent, A., Morriss-Kay, G., and Chambon, P. (1990). Specific spatial and temporal distribution of retinoid acid receptor gamma transcripts during mouse embryogenesis. *Development* **108**, 213–222.

Ruberte, E., Dolle, P., Chambon, P., and Morriss-Kay, G. (1991). Retinoic acid receptors and cellular binding proteins. II. Their differential pattern of transcription during early morphogenesis in mouse embryos. *Development* **111**, 45–60.

Ruberte, E., Friederich, V., Chambon, P., and Morriss-Kay, G. (1993). Retinoic acid receptors and cellular retinoid binding proteins. III. Their differential transcript distribution during mouse nervous system development. *Development* **118**, 267–282.

Stenkamp, D.L., Gregory, J.K., and Adler, R. (1993). Retinoid effects in purified cultures of chick embryo retina neurons and photoreceptors. *Invest.Ophthal.Vis.Sci.* **34**, 2425–2436.

Stenkamp, D.L., and Adler, R. (1994). Cell-type and developmental- stage-specific metabolism and storage of retinoids by embryonic chick retinal cells in culture. *Exp.Eye Res.* **58**, 675–687.

Szel, A., Diamantstein, T., and Rohlich, P. (1988). Identification of the blue-sensitive cones in the mammalian retina by anti-visual pigment antibody. J.Comp.Neurol. *273*:593–602.

Szel, A., Rohlich, P., Caffe, A.R., Juliusson, B., Aguirre, G., and Van Veen, T. (1992). Unique topographic segregation of two spectral classes of cones in the mouse retina. J.Comp.Neurol. *325*:327–342.

Szel, A., Rohlich, P., Mieziewska, K., Aguirre, G., and Van Veen, T. (1993). Spatial and temporal differences between the expression of short- and middle-wave sensitive cone pigments in the mouse retina: a developmental study. J.Comp.Neurol. *331*:564–577.

Tabin, C.J. (1991). Retinoids, homeoboxes, and growth factors: Toward molecular models for limb development. *Cell* **36**, 199–217.

Warkany, J. and Schraffenberger, S. (1946). Congenital malformations induce in rats by maternal vitamin A deficiency. I. Defects of the eye. *Arch. Ophth.* **35**, 150–169.

Wikler, K.C. and Rakic, P. (1991). Relation of an array of early-differentiating cones to the photoreceptor mosaic in the primate retina. Nature, 351, 397–400.

Wikler, K.C. and Rakic, P. (1994). An array of early-differentiating cones in the fetal monkey retina precedes the emergence of the photoreceptor mosaic. Proc. Natl. Acad. Sci., 91, 6534–6538.

Wikler, K.C. and Rakic, P. (1996a). Development of the primate photoreceptor mosaic. Perspectives on Developmental Neurobiology, 3, 161–175.

Wikler, K.C., Szel, A., and Jacobsen, A-L. (1996b). Evidence for prenatal determination of cone opsin phenotypes and position in the mouse retina. J. Comp. Neurol., 374, 96–107.

Wikler, K.C., Rakic, P., Bhattacharyya, N., and MacLeish, P.R. (1997). Early emergence of photoreceptor mosaicism in the primate retina revealed by a novel cone-specific monoclonal antibody. J. Comp. Neurol., 377, 500–508.

Wilson, J. G., Roth, C. B., and Warkany, J. (1953). An analysis of the syndrome of malformations induced by maternal vitamin A deficiency: Effects of restoration of vitamin A at various times during gestation. *Am. J. Anat.* **92**, 189–217.

Wu, T.C., Wang, L., and Wan, Y.J. (1992). Retinoic acid regulates gene expression of retinoic acid receptors alpha, beta, and gamma in F9 mouse teratocarcinoma cells. *Differentiation* **51**, 219–224.

Yu, V.C., Delsert, C., Andersen, B., Holloway, J.M., Devary, O., Naar, A.M., Kim, S.Y., Boutin, J.M., (1991). Retinoic acid, thyroid hormone, and vitamin D receptors to their cognate response elements. *Cell* **67**, 1251–1266.

Zhang, X.K., Hoffman, B., Tran, P.B., Graupner, G., and Pfahl, M. (1992). Retinoid X receptor is an auxiliary protein for thyroid hormone and retinoic acid receptors. *Nature* **355**, 441–6.

Zhang, X.K. and Pfahl, M. (1993). Hetero- and homodimeric receptors in thyroid hormone and vitamin A action. *Receptor* **3**, 183–191.

4

DEVELOPMENT OF CONE DISTRIBUTION PATTERNS IN MAMMALS

Á. Szél,[1] B. Vígh,[1] T. van Veen,[2] and P. Röhlich[1]

[1]Semmelweis University of Medicine
Department of Human Morphology and Developmental Biology
Budapest, Tűzoltó u. 58., H-1094, Hungary
[2]University of Göteborg
Department of Zoomorphology
Göteborg, Medicinaregatan 18, S-41390, Sweden

1. ABSTRACT

Color vision is mediated by a number of photoreceptor cell classes housing different visual pigments. Immunocytochemistry provides a means to distinguish among spectrally different cone types. This method has revealed the occurrence of various photoreceptor distribution patterns in mammals. Irrespective of the functional significance of the observed cone topographies, an important question is the development of these patterns with respect to the differentiation of individual cone types. Immunocytochemistry also enabled us to follow the ontogeny of cones in a number of species. The combination of antibody labeling with various experimental paradigms has furthered our knowledge about cone development. Recent data on this subject are summarized in the present review.

2. INTRODUCTION

The spatial distribution of various photoreceptor types across the retina has been an important question ever since rods and cones were distinguished. The existence of central areas with higher cone and lower rod densities was established long ago. The highly specialized photopic visual center of primates, the fovea centralis, with the partial or total exclusion of rods, was discovered with microscopic observations (Schultze, 1866; Chievitz, 1891; Osterberg, 1935). The introduction of histochemical methods, based on selective uptake of dyes, enabled the discrimination of spectrally different cone types (Ahnelt, 1985; DeMonasterio et al., 1981; Harwerth and Sperling, 1975). The next step in studying the photoreceptor mosaic was the production of antibodies recognizing colour-specific visual pigments (Szél

Development and Organization of the Retina, edited by Chalupa and Finlay.
Plenum Press, New York, 1998.

et al., 1986; Lerea et al., 1986, 1989). Recently, *in situ* hybridization with probes identifying the mRNAs of various visual pigments furthered this approach at the level of the message (Raymond et al., 1995; Bumsted et al., 1997). A considerable body of information has been accumulated about the highly regular array of cone types in fish (Cameron and Easter, 1993; Marc and Sperling, 1976; Stenkamp et al., 1995), the dorso-ventral gradient of short and middlewave sensitive cones in certain rodents (Szél et al., 1992), the hexagonal array of blue-sensitive cones in the primate fovea (DeMonasterio et al., 1985), as well as the lack of blue-sensitive cones in the very center of the fovea (Williams et al., 1981; DeMonasterio et al., 1985; Szél et al., 1988; Wikler and Rakic, 1990; Curcio et al., 1991).

Whether the physiological significance of these patterns in colour vision is understood or not, a very interesting problem with respect to photoreceptor distribution is how the spatial arrangement of cones is generated during development. The basic questions are: 1. which cone type emerges first, and 2. how is the identity of individual cones determined during the differentiation process. We have been studying the distribution of photoreceptors in vertebrates for about a decade using visual-pigment specific antibodies that distinguish among cone subpopulations. As compared to histochemistry, and more direct physiological methods, such as microspectrophotometry or intracellular recording, the greatest advantage of immunocytochemistry is that there is no need for sophisticated instrumentation or surviving retinal preparations. Furthermore, even whole retinas can be reacted with the antibodies and studied as wholemounts, enabling the precise topographic mapping of each visual cell type. Yet another advantage of the method for developmental biology studies is that cone types can be identified even before they are actually taking part in the visual functions of the animal. The different phenotypes can be distinguished as early as the synthesis of type-specific pigments has started. To follow the developmental fate of photoreceptors, several mammalian species exhibiting various sorts of visual cell topography were selected, and retinas at different ages were compared. In addition to studies carried out on normal development, photoreceptor differentiation was also investigated under various experimental conditions such as retinal transplantation or retinal organ cultures. Comparing the results of these experiments, various scenarios can be envisaged which might take place either in combination with one another or alone in various species.

3. DISTRIBUTION OF CONES AS DETECTED WITH IMMUNOCYTOCHEMISTRY

Colour discrimination in the retina of non-primate mammals is served by two cone types, carrying middle-wave (green) and short-wave (blue or ultra-violet) sensitive visual pigments, respectively (Jacobs et al., 1991; Jacobs, 1993; Jacobs and Deegan, 1994; Chiu and Nathans, 1994). In primates, a third cone type containing a red-sensitive pigment has been evolved from the green cones (Nathans et al., 1986; Yokoyama and Yokoyama, 1989). The amino acid sequences of the red (long-wave, L) and green (middle-wave, M) cone pigments are almost identical, precluding the generation of antibodies that distinguish them. The mammalian short-wave (S) pigments, however, are markedly different from the M/L pigments, providing a means of unequivocally discriminating among them. Selective labelling of S and M/L pigments with immunocytochemistry can be carried out using pairs of monoclonal and/or polyclonal antibodies (Szél et al., 1986; Lerea et al., 1986 and 1989; Wang et al., 1992).

In our studies two monoclonal antibodies (COS-1 and OS-2) were most frequently used. Both were generated against a crude mixture of chicken photoreceptor proteins (Szél et al., 1986). The most prospective cell lines were selected on the basis of their specificity to cone outer segments. The anti-visual pigment specificity was confirmed using electrophoresis and immunoblotting. The colour-specificity was established with selective photic damage and immuno-electron microscopy (Szél et al., 1988). MAb COS-1 proved to be specific for middle-to-long wavelength sensitive cones of all mammals investigated so far, whether they contain green or red sensitive visual pigments. The other antibody (OS-2) labels the short-wave sensitive mammalian cones (Röhlich and Szél, 1993). Since the ultraviolet-sensitive cones of certain rodents were discovered (Jacobs et al., 1991), we have to face the problem of distinguishing between UV and blue-sensitive cones. Although mAb OS-2 and the other presently available antibodies cannot selectively label these two cone types, this failure does not pose any practical difficulty, since no species possessing both S cone subtypes has been found so far. Mammals either have blue-sensitive or ultraviolet-sensitive S cones, therefore, this cone population can unequivocally be distinguished from other cones. Besides OS-2, another antibody, the polyclonal 108B, was also often used, especially for double label immunocytochemistry (Lerea et al., 1986 and 1989), but this serum also fails to distinguish between UV and blue cones. Interestingly, a few species were found to exist without any S cones (Wikler et al., 1990; Szél et al., 1994a, Calderone and Jacobs, 1995; von Schantz et al., 1997).

Concerning the immunocytochemical approach it is very important to keep in mind that the actual spectral sensitivity of the cone contingent of a given species cannot be identified by antibodies alone. The predictions based on the presence of immunolabelled photoreceptors must be confirmed by physiological methods, and only after the effective absorbance spectra have been determined, can colour sensitivities be assigned to the individual elements of the photoreceptor mosaic in a given species.

Soon after the advent of the immunocytochemical method for distinguishing colour cones, it was established that the two basic cone types comprised rather constant percentages of the total retinal cone contingent. M/L cones consistently made about 90 percent, whereas S cones were found to form only about 10 percent (Fig. 1). Another observation that seemed to hold true for all mammalian species was that both cone types populate the retina homogeneously. All mammals studied with mAbs COS-1 and OS-2 were found to be populated by the two cone types in the constant 10:1 ratio all over the retina. No areas devoid of any cone type or exclusively populated by only one cone type were encountered except for the very center of the primate fovea where blue cones are missing (DeMonasterio et al., 1985; Szél et al., 1988; Wikler and Rakic, 1990; Curcio et al., 1991). The other

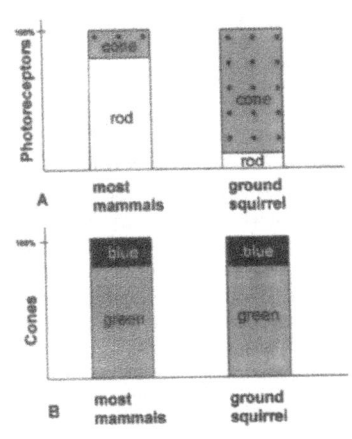

Figure 1. Ratio of photoreceptors in the mammalian retina as shown diagrammatically. Most mammals possess a rod dominant retina, which means that the majority of photoreceptors (marked with white column) are rods (A). The cones are either middle-to-longwave sensitive (grey column) or shortwave sensitive (black dots). A few mammals (ground squirrel, tree shrew) are known for their abundance in cones. When comparing the ratio of S and M/L cones (B), all mammals, whether rod- or cone-dominant, were found to be populated by about ten times more M/L than S cones.

dogma that used to dominate our view about cone patterns in mammals was that each cone outer segment necessarily contains one and only one visual pigment (Fig. 1).

4. TOPOGRAPHIC SEPARATION OF CONES

The extension of the immunocytochemical analysis on species that have not previously been investigated, however, dramatically changed our view on photoreceptor distribution. The common house mouse was the first species contradicting the mentioned dogmas. Surprisingly, an unexpected topographic separation of M and S cones was found in this species (Szél et al., 1992). Whereas the dorsal retina is populated by both cone types in the characteristic 1:10 ratio, the ventral retinal half presents an unusual territory in that its exclusive cones are of the shortwave-sensitive type. This means that although the global cone density is roughly uniform all over the mouse retina, the two cone types occupy opposite retinal halves (Fig. 2). Later, other mammals, among them the rabbit (Juliusson et al., 1993; Famiglietti and Sharpe, 1995) and the guinea pig (Röhlich et al., 1994), also proved to be heterogeneous with respect to the distribution pattern of cones; however, the actual extension of the M-cone-free area was found to be different (Fig. 3). To designate this peculiar area we have introduced the term, blue-field. When present, the blue-field always occupies the ventralmost area of the retina whether confined to a small band-like crescent as in the rabbit, or to the whole ventral retina as in the mouse. Although the biological significance of this arrangement is still unknown, a reasonable hypothesis was that having a ventral, shortwave-sensitive screen continuously scanning the sky for predators might be an advantageous feature. Likewise, the dorsal M-cone-rich retina

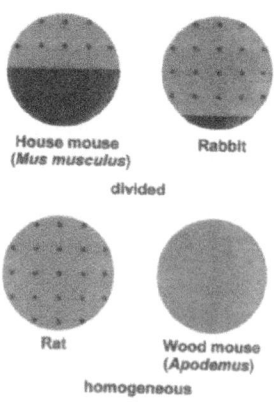

House mouse
(*Mus musculus*)   Rabbit

divided

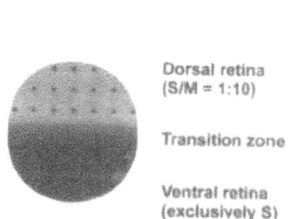

Dorsal retina
(S/M = 1:10)

Transition zone

Ventral retina
(exclusively S)

Rat   Wood mouse
(*Apodemus*)

homogeneous

Figure 2. Distribution of S and M cones in the mouse retina. In contrast to most mammals, the mouse retina stands out with its divided pattern of cone distribution. Whereas the dorsal retina (grey background with black dots) corresponds to the usual pattern (many M/L cones and a few S cones), the ventral half is exclusively populated by S cones. There is a transition zone between the two cone-fields where cones expressing both pigments are present.

Figure 3. Four distribution patterns identified so far in mammalian retinas. A and B represent the divided patterns. The lower half (house mouse) or crescent (rabbit) contains only S cones, and it is only the dorsal part of these retinas that is homogeneously populated by both cone types. The majority of the mammals studied so far, however, exhibit a homogeneous distribution all over the retina. This pattern can be composed of either both cone types (rat) or only M/L cones (wood mouse). In addition to the wood mouse there are other species (nocturnal primates, hamster, etc.) lacking M/L cones.

might be useful for viewing the green vegetation. In spite of the undoubted physical rele-vance of such a division, this hypothesis seems to be rather controversial, given the fact that rodents with the described dorsoventral gradient of colour cones share their habitat with other mouse species that are not equipped with the proposed shortwave scanner (Szél et al., 1994a). The co-existence of such species (Fig. 3) on the same habitat strongly argues against a decisive biological advantage of having blue-cone rich retinal zones. Obviously, both species enjoy the very same chances to be perceived by predators. If, however, the divided retina were to produce any advantage for survival, the other species lacking this feature would have become extinct. Therefore, we suggest that the topo-graphic separation of cones must have a different explanation.

When examining the transition zone between the upper and lower retina more closely, another striking observation has been made. Within a band of approximately 2–300 μm width, large number of cones were found to express both visual pigments at the same time (Röhlich et al., 1994). Two independent methods were used to confirm this finding. M/L-specific antibody COS-1 and S-specific antibody 108B raised in mouse and rabbit, respectively, were used to perform double label immunocytochemistry on whole-mounts of the mouse retina. The majority of cones located in the transition zone bound both antibodies, indicating the simultaneous synthesis of both visual pigments (Fig. 2). To exclude any artificial double staining originating from the cross reaction of primary or secondary antibodies, another approach was also used. Mouse retinas were embedded in araldite, and serial sections were cut on an ultramicrotome. When alternately reacting adjacent semithin sections with S-specific and M/L-specific antibodies, large numbers of cones were identified that were labelled on both consecutive sections, providing convinc-ing evidence in support of the co-existence of both pigments within the same cone cells. The same co-existence of two visual pigments within the very same cone cells was also true for the rabbit and guinea pig retina (Röhlich et al., 1994). When present, the extension of the co-expressing cone area showed considerable variation among species. Investiga-tions are underway to establish the functional properties of cells containing two visual pig-ments with different absorbance spectra. Even though the majority of mammalian species investigated so far lack this feature, the existence of a few species with divided retina and "double" cones refute the "one cone one pigment" dogma as well as generalizations about homogeneous cone distributions and the constant ratio between spectrally different cone types (Szél et al., 1996).

5. DEVELOPMENT OF COMPLEMENTARY CONE FIELDS

The co-existence of two retinal fields with spectrally different cone contingents within the same eye raises the question: how do these retinal fields come about during the development of the mouse eye? To address this issue, whole retinas derived from mice representing different developmental stages were reacted with anti-visual pigment anti-bodies against rods, S cones and M/L cones, respectively (Szél et al., 1993). The density of each component was measured across the entire retina, yielding a series of isodensity maps. These maps demonstrated the emergence and spreading of photoreceptors during postnatal development. The main findings of this study were as follows: The first immu-nopositive visual cells to appear are the rods which emerge centrally on the first-second postnatal days (P1–2) and later spread towards the periphery, retaining a centro-peripheral gradient during the entire developmental process. The first cones appear only later. Immu-nopositive S cones were shown in the lower retina already on the fourth-fifth postnatal days (P4–5). Rather than centro-peripherally, the S cones populate the retina according to a ventral-dorsal gradient. This gradient persists throughout the life and results in the afore-mentioned immunocytochemical pattern as observed in adults. The last elements of the mouse photoreceptor contingent to appear are the M/L cones. No earlier than P9–10 do the first M/L cones emerge in the dorsal retina. This cone type was never identified in the ventral retina. In summary, the sequence of emergence is rods - S cones - M/L cones. It is well known that cone cells are generated earlier than rods (Carter-Dawson and LaVail, 1979). The precedence of the rod phenotype argues that the order of opsin expression does not necessarily coincide with the sequence in which the photoreceptors are born. The same order was revealed by similar immunocytochemical studies carried out on other species (rat, gerbil, rabbit), indicating that there is not a causative relationship between the topographic separation of cones and the sequence of the emerging phenotypes (Szél et al., 1993).

6. RETINAL TRANSPLANTATION IN RABBIT

The next question we addressed was whether this sequence can be modified under experimental conditions. To this end, we employed retinal transplantations to see whether the development of transplanted embryonic retinal cells follows the same timetable as normal retinal cells. The method of preparing the retinal cell suspension from 15-day-old rabbit embryos has been described elsewhere in detail (Bergström et al., 1994; Szél et al., 1994c). The graft was inoculated in the subretinal space of adult rabbits. In general, the transplants showed a relatively normal organotypic maturation even though the integration between host and graft tissue showed minor irregularities. Rosettes, lumen-containing globules with a stratification corresponding to that of the retinal layers, were often observed. Signs of degeneration of the host retina around the graft tissue have also been observed. The transplants, however, exhibited a normal stratification with the usual arrangement of all retinal layers.

The host animals were sacrificed at different intervals of time and the obtained transplants were compared with age-matched normal rabbit retinas. The three main photoreceptor cell types were labelled with the same antibodies as in the previous studies. Due to the patchy and globular arrangement of the graft cells and the presence of a green-cone-free crescent of the normal rabbit retina, absolute cell numbers or densities would have been irrelevant in a comparison between transplants and normal retinas. Instead, we calculated the ratio of M/L and S cones, and these ratios were compared between transplants and age-matched controls. The age-matching was based on the number of postconceptional rather than postnatal days. Whereas the order of emergence of the photoreceptor types was found to be the same as in normal animals, that is rods - S-cones - M/L cones, the paucity of M/L cones as compared to S cones was a striking but consistent feature of the transplants. Despite the presence of a blue-streak in the rabbit retina from which M/L cones are totally excluded, the total estimated number of M/L cones in the whole rabbit retina exceeded that of the S cones by a factor of approximately 2. In contrast, the transplants of the same postconceptional age exhibited a ratio of only about 0.25. This finding indicates that whereas the development of S cones in the transplant seem to take place undisturbed, the differentiation of M/L-cones is slowed considerably. It is also possible that larger numbers of M/L cones fall victim to the transplantation than S cones, resulting in the dominance of the normally less frequent phenotype. Since the development of rods as well as other retinal cell types (Bergström et al., 1994) also seems to remain unaltered, we speculate that the microenvironment provided by the subretinal space of the adult rabbit is

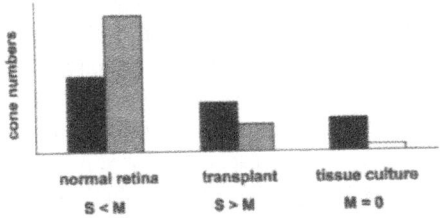

Figure 4. Ratio of the two cone types in normal and experimental conditions. The diagrams are shown to demonstrate the relative occurrence of the two basic cone types in normal, transplanted and cultured retinas. Since different species were used for both respective paradigms, the absolute values are of no importance, and only the tendencies are depicted. In the normal mammalian retina the M/L cones generally outnumber the S cones. In (rabbit) transplants, the green cones are rare components of the graft tissue as compared to the S cones. Finally, no M/L cones develop in the (mouse) culture paradigm, whereas viable S cones are present in relatively great numbers.

devoid of factors needed for the differentiation of just one photoreceptor cell line. The results of the transplantation experiments indicate that age-specific environmental cues selectively influence the normal differentiation of M/L cones, even when the species- and organ-specific factors are available in the host retina (Fig. 4).

7. RETINAL ORGAN CULTURES

In another set of experiments we have tried to omit the organ-specific environment by explanting immature retinas on nitro-cellulose paper to culture them in tissue culture medium. The retinal tissue culture method was elaborated on mouse retinas (Caffé et al., 1989 and 1993; Söderpalm, 1994). Since these experiments were performed in other species, the conclusions from the rabbit transplant and mouse explant paradigms are not readily comparable. In the basic experiments, new-born mouse retinas were cultured after the connective tissue layers of the eye were enzymatically digested away. The enucleated eyes were treated with proteinase K which removed the fibrous coat. The retinal pigmented epithelium, however, was retained. The anterior segment and the vitreous body were also removed and the remaining retina together with the retinal pigment epithelium was flatmounted on nitro-cellulose paper. We allowed these retinal pieces to grow for several weeks. Cultured material was sacrificed and fixed at various time intervals. The retinas were either cryosectioned and reacted with visual pigment antibodies or immunoreacted as wholemounts. In the latter case, the neural retina was carefully detached from the pigmented epithelium and the supporting nitro-cellulose filter, and floated in buffer so that the outer segments were exposed to the antibodies.

Similar to the transplants, the histotypic differentiation of the cultured retinas was close to normal, so that all layers of the retina as well as the major cell classes were easily identifiable. Cones and rods emerged, roughly according to the usual timetable as evidenced by lectin cytochemistry and antirhodopsin immunocytochemistry. Similar to the normal retina and the rabbit transplants, the first phenotype to emerge was the rod. The first immunopositive outer segments sprouted at P3–5, and these elements were also recognized by wheat germ agglutinin (WGA) lectin, specific for rods. As shown by PNA lectin cytochemistry, cones appeared a few days later. The first PNA-positive elements emerged at P5–7 as tiny elongate structures scleral to the outer limiting membrane. When, however, colour-specific cone types were identified with anti-cone antibodies, a surprising absence of M/L cones was found. Whereas rods and S cones emerged on time, the M/L cones were completely missing from all investigated retinal explants, even after prolonged culturing. To avoid a false negative result, we serially sectioned whole transplants, but not a single M/L cone was found in a total of about 50 individual retinas. Similarly, our trials to screen large areas of wholemounted transplants for middle-to-longwave cones were also without success. The longest survival time with the mouse-retinal culture paradigm was around two months, and this time was not enough to achieve the sprouting of a single M/L cone (Fig. 4).

In another series of experiments we have sacrificed older pups to obtain material for culturing (Wikler et al., 1996). If, before transplantation, the retinas were permitted to develop for more than three-to-four days following birth, viable M/L cones emerged. Their density, however, remained smaller in the considerably more mature retina than expected on the basis of their occurrence in normal age-matched retinas. By increasing the age of the grafts towards the time-point (9–10 days postnatal), when M/L cones appear, the development of this cone type became more and more similar to the process that takes place in the normal retinas. The comparison of these two sets of experiments allows one of the following speculations: 1. The selective backlog of the M/L cones seems to be a consistent feature of both paradigms. 2. Immature cones destined to become M/L sensitive are more prone to fall victim to one of the two experimental situations. 3. The absence of organ- and species-specific environmental conditions favors the total drop-out of this cone type, unless more mature retinas are explanted carrying cones that have reached a more advanced stage of commitment.

The striking and common feature of both developmental experiments is the asymmetry of the two lines of phenotypes during cone differentiation. It is difficult to interpret the apparently normal differentiation of photoreceptors into rods and cones, when the cone contingent is exclusively or dominantly represented by only one phenotype, namely by those that express shortwave-sensitive visual pigment. The most logical explanation would be the selective vulnerability of the other cone type, predestining its drop-out under the described experimental conditions. Several findings, however, contradict this argument. When cones in more advanced stages of development, but still before the onset of visual pigment synthesis are cultured, green cones remain normal constituents of the explants (Wikler et al., 1994). Second, blue cones rather than green cones are known for their higher susceptibility to various noxious agents (DeMonasterio et al., 1991). If the selective death of the green cones were responsible for the absence of this cone type, degenerating or dying COS-1 positive cell remnants would have been observed in the retinas, which was not the case. Moreover, the combination of the cone-specific peanut agglutinin (PNA) lectin and OS-2, the antibody specific for S cones, revealed the absence of cones that were left unstained with OS-2 (Söderpalm et al., 1994). Considering that all cones are supposed to be marked with the lectin, each and every cone must contain the short-wave pigment. Therefore, we can rule out the assumption that cones with modified pigments that are unidentifiable for any antibodies are present, and/or cones without any visual pigment develop.

The next speculation is that totally different conditions are needed for the undisturbed development of the two cone types. Those factors that are indispensable for M/L cones might be missing from our paradigms, whereas those required for the differentiation of S cones are available. Although this assumption does not seem to be reasonable, it cannot be refuted at present. Large scale experimentation is needed to address the question as to which factor(s) might have a role in directing or influencing the differentiation of two cone types.

Presumably, the manipulation coincides with events crucial for the differentiation of the green cones but not for the rods or blue cones. Although visual pigment expression is the first detectable sign of commitment to a certain phenotype, the fate of photoreceptors is obviously determined earlier. It is logical to assume that the three photoreceptor types become committed in the same order as their characteristic visual pigments emerge. In contrast, it is well known that the birthday of cones precedes that of the rods (Carter-Dawson and LaVail, 1979). One possible explanation might be that cones differentiate from other cell types very early, but, from one another they differentiate relatively late, probably after the generation of rods. In this case rods might already be committed at the time of the explantation or transplantation, and thus not disturbed by the manipulation. The separation of the two cone classes, however, might not have started yet (as in the case of the transplantation) or might have just started (as in the case of the explantation).

Logically, blue cones are either already committed at the time of the manipulation, and thus their further development goes on undisturbed, or there is a default pathway favouring the differentiation towards the synthesis of the shortwave pigment. Although the two scenarios are not mutually exclusive when various mammals are compared, our data, obtained on the rat and the gerbil, supports the latter alternative, at least in these two species (see below). Our interpretation implies that originally the two cone types do not represent two independent sets of populations. Rather, the younger M/L cones might develop from the precocious S cones with transdifferentiation. If this is true, all (or most) cones would develop so as to reach the first developmental stage where the synthesis of shortwave pigment starts. Unfavorable conditions, however, could stop this process resulting in the freezing of many (or all) cones in this stage. The possible time points when the commitment of each photoreceptor type takes place can be inferred from transgenic mouse lines. Using exogenous (human) visual pigment genes fused with appropriate reporter genes, the transcript of the gene construct can be easily followed, and the appearance of the reporter gene product signals the time of the activation of the promoter gene. Based on such transgenic mouse studies, it is probable that the time of the determination of S cones and M cones might be the 13th embryonic day (E13) and postnatal day 4 (P4), respectively (Wang et al., 1992; Chen et al., 1992). This supports our hypothesis concerning the mouse retinal cultures, whereby the manipulation takes place between these two time points. As far as the transplantation is concerned, no direct comparisons can be made, since no such data are available in rabbits. However, by extrapolation we might assume that the blue cone determination also takes place around midgestation, followed by the commitment of the other cone type much later. If this is so, the timing of the manipulation also falls between the two events, allowing for an undisturbed blue cone differentiation but perturbing the development of green cones.

8. TRANSDIFFERENTIATION OF COLOR CONES

The evidence that transdifferentiation plays a role in cone differentiation has been found in other species that were not included in our experimental paradigms. When the spreading of the sprouting cones was mapped in the developing rat retina, a surprising observation, not encountered in the mouse and rabbit, was made. The first cone type to emerge was the shortwave-sensitive population, with a density in the immature retina that was considerably higher than that observed in the adult. The rat retina stands out from most mammals with its relatively low S cone densities (Szél et al., 1993); still, the two-week-old rat retinas contain about one order of magnitude more S cones than the adults.

When following the kinetics of both cone densities as a function of time (Fig. 5), the density peak of S cones declines within a few weeks. By the end of the third postnatal week, the S cone density returns to that observed in the adult. The M/L cones that emerge somewhat later also rise in density; however, they reach a plateau without a sharp peak and remain relatively high throughout the life of this rodent. Remarkably, the sharp drop of S cones coincides with the rise in the number of M/L cones, while the total cone cell number is not subject to any considerable change in density.

This observation also raises the question as to whether S cones might indeed transdifferentiate into M/L cones. It seems logical to assume that if the total cone cell number does not change, yet the ratio of the two cone densities undergoes significant changes, there must be a transition from one cone phenotype to the other. To pursue this line, we have simultaneously used two antibodies, one against the M/L and another against the S cones. Our prediction was that if S cones switch from expressing shortwave pigment to expressing middlewave sensitive pigments, there must be cones that simultaneously contain both pigments. The reason for this assumption is that the shedding of the apical disks is considerably slower than the randomization of the newly synthesized outer segment proteins (Young, 1976; Bok, 1985).

In the rat and the gerbil (*Meriones unguiculatus*) retina we were successful in detecting the temporary coexpression of both pigment types (Fig. 6). In retinas taken from two/three-week-old rats and gerbils, a considerable amount of cones proved to be double labelled. Importantly, such "dual cones" never occurred in older animals, indicating that after the original visual pigment molecules have been removed from the dual cones, only one pigment remained (Fig. 7). In our interpretation, the default pathway of the visual pigment synthesis is the expression of the shortwave pigment. Only a small percentage of

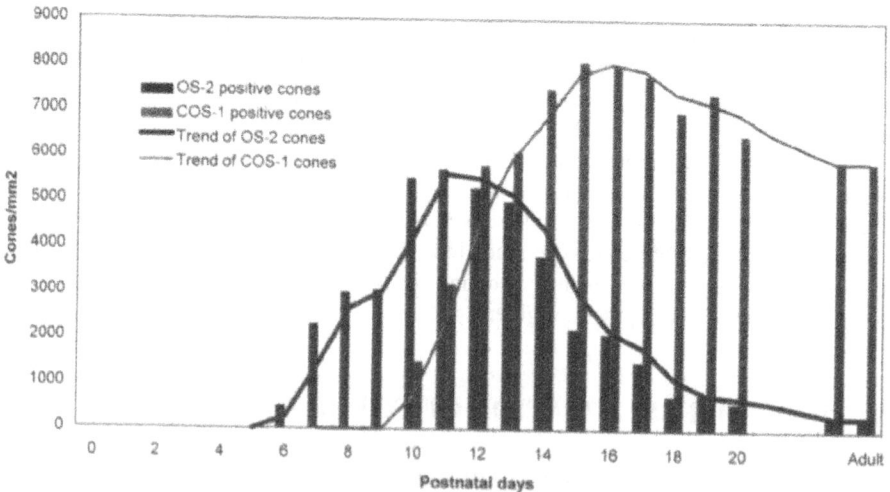

Figure 5. The density of immunopositive cones in the developing rat retina as a function of postnatal age. The samples were taken from about halfway between the optic nerve head and the ora serrata in the superior quadrant. Note that the first cones to appear are labelled by mAb OS-2. The density of this cone population rises steeply and peaks at about 6000 cone/mm^2, then decreases to about 500 cone/mm^2. The second cone type that emerges is stained with COS-1, its density also rises to reach a plateau (8000 cone/mm^2), but then does not drop to a low level, rather, after a slight decrease remains relatively high throughout the life of the animal. Note the crossing of the two trend lines around the second postnatal week.

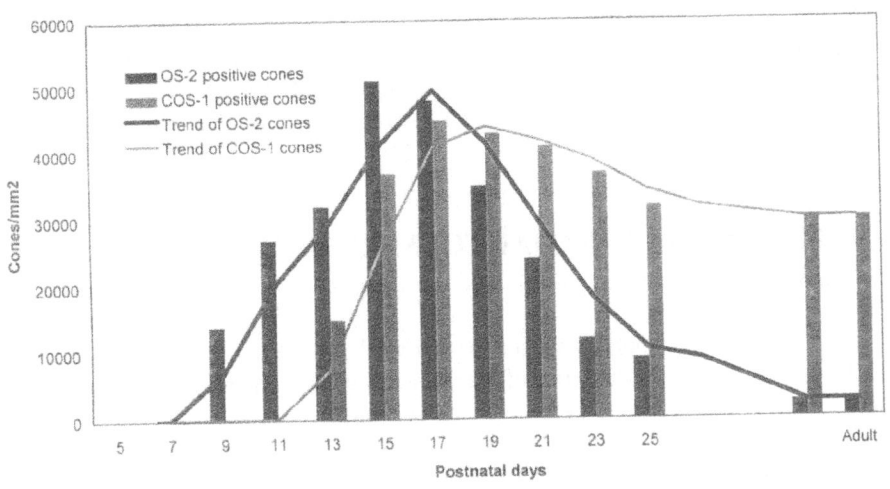

Figure 6. The kinetics of the developing cones of the gerbil retina. Although the density values and the time points of the emergence and peak of the curves are different, basically the same configuration as that obtained in the rat can be seen. After an initial rise, the S cones decrease steeply in density. Coincidentally with the drop of S cones, the density of M/L cones rises and reaches a plateau, producing a crossing of the two curves.

cones keep on expressing the shortwave pigment alone ("genuine S cones"). The majority of cones, however, start to synthesize the green pigment instead ("transforming cones"). As long as both pigments are present, these cones are identified as dual cones. Within two weeks, however, the original shortwave pigment disappears from the outer segments, and only the green pigment remains, giving rise to the definitive green cones. Those cones that do not undergo the shift give the definitive shortwave cones (Szél et al., 1994b).

The priority of S cones and their unexpected drop, coinciding with the density increase of the other cone type, can be explained by the transdifferentiation theory rather than by the selective death of one cell type. Although programmed cell death obviously takes part in forming the adult patterns, degenerating or dying cones were not a characteristic feature of the investigated material during the critical period of cone development.

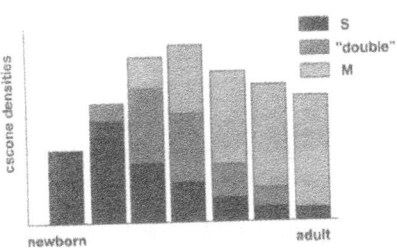

Figure 7. Diagrammatic representation of the distribution of three types of cones in the developing rat retina. Based on the immunocytochemical staining pattern, three cone types can be distinguished: S cones (stained by exclusively mAb OS-2, and marked with black columns), M cones (only labelled by mAb COS-1, and marked with light grey colour) and dual cones (binding both antibodies, and marked with dark grey columns). Note that originally all cones are of the single S type (black), later dual cones (grey) then M/L cones (light grey) appear. The dual cones, however, are only transient components of the developing rat retina.

Therefore, it seems unlikely that this mechanism accounts for the drop in S cones. The approximately 10% expansion of the retinal surface between P10 and P30, also fails to explain the decrease in S-cone density. The formation of a modified visual pigment that is undetectable for either COS-1 or OS-2 can also be ruled out, since the cone-specific PNA lectin labelled no photoreceptors other than those recognized by the mixture of COS-1 and OS-2 (Szél et al., 1994b).

9. TEMPORAL AND SPATIAL COEXPRESSION OF VISUAL PIGMENTS

It is important to emphasize that after four weeks of age, not a single cone exhibits a double immunoreaction in the retina of the rat. When comparing the spatial distribution of cones in the divided retinas with the temporal coexpression of the two visual pigments in the homogeneous retinas, a striking similarity can be observed (Fig. 8). The same cone figures appear in the developing rat retina as those that populate the retina of the mouse. The spatial arrangement of the ventral and dorsal areas of the mouse retina with a transition zone in between resembles the temporal sequence of the characteristic stages of the rat retinal development. The most ventral part of the mouse retina is populated exclusively by S cones, just as in the immature rat retina, where the only identifiable cone type is the S cone. There is a transition zone with double cones in the mouse, similar to the dual cones of the 2–3-week-old rat retina. Finally, the entire dorsal mouse retina is composed of many green cones and a few blue cones, exhibiting the same ratios as the adult rat retina, whose cones are either green or blue-sensitive without any dual elements.

The similarity can be best demonstrated by printing the diagrammatic scheme of the mouse retina and the plots of the rat cone densities below one other, so that the time axis of the latter parallels the ventral-dorsal axis of the former (Fig. 8). Using a consistent grey level coding on both diagrams for S, M/L and dual cones, respectively, the pictures imply that consecutive stages of rat retinal development are paralleled by adjacent bands of cone distribution patterns along the ventral-dorsal axis of the mouse retina. The described similarity in itself, however, does not prove a close relationship between the two phenomena,

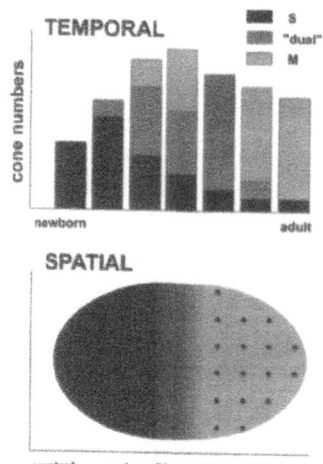

Figure 8. Comparison of two sorts of transitory visual pigment co-expression. Pictures A and B are practically the same as Figures 2 and 7, respectively. The only difference is that diagram B on Figure 8 is rotated with 90° so that the ventral-dorsal axis of the adult mouse retina lies parallel with the time axis of A. Note that the ventral part ("blue-field") of the mouse retina resembles the immature rat retina, in that S cones are the exclusive components in both. The transition zone of the mouse retina is equivalent with the 2–3-week-old rat retina, since large numbers of dual cones occur in both. Finally, the mouse dorsal retina is comparable to the adult rat retina in that dual cones are missing and all cones contain either S or M/L pigment.

and at the moment we only consider this resemblance of spatial and temporal coexpression a line to pursue in our future experiments. Our working hypothesis is that the signal that triggers the shift between blue and green cones must be determined topographically. In the animals with undivided retinas, the switch involves the entire retina; however, in those animals where there is a topographic separation, the ventral part of the retina does not undergo the shift, leaving all cones in an "immature" state where all cones synthesize the blue pigment, according to the default pathway. In this context, we consider the various patterns as parts of an evolutionary continuum, whereby the extension of the blue-field reflects the propagation of the switch. It is the retina of the common house mouse that represents the most ancestral state, insofar as the transition zone is located in the horizontal meridian of the retina. Since the rabbit retina exhibits a far more ventral transition line, we might assume that the demarcation has shifted lower in this species. In species without complementary cone fields and transition zones, the transformation of the entire retinal area has taken place, leaving no retinal territories populated with exclusively S cones.

An important assumption follows from the above reasoning: Neither the presence of dual cones in the mammalian retina nor the topographic segregation of the two basic cone types are likely to play a pivotal role in the visual functions of the species possessing these features. It is much more probable that both phenomena reflect fundamental developmental processes that underlie the differentiation of the two cone types. One difficulty in the proper interpretation of the comparison between spatial and temporal coexpression of visual pigments is that the two phenomena seem to be mutually exclusive within individual mammalian species. Whereas spatial but not temporal coexpression is a feature of the mouse and the rabbit retina, the temporary appearance of dual cones during retinal development has only been observed in the rat and gerbil retina.

The lack of coexpressing cones in the dorsal retina during the development of the rabbit eye (Fig. 9), and the absence of dual cones in the adult rat and gerbil, argue against

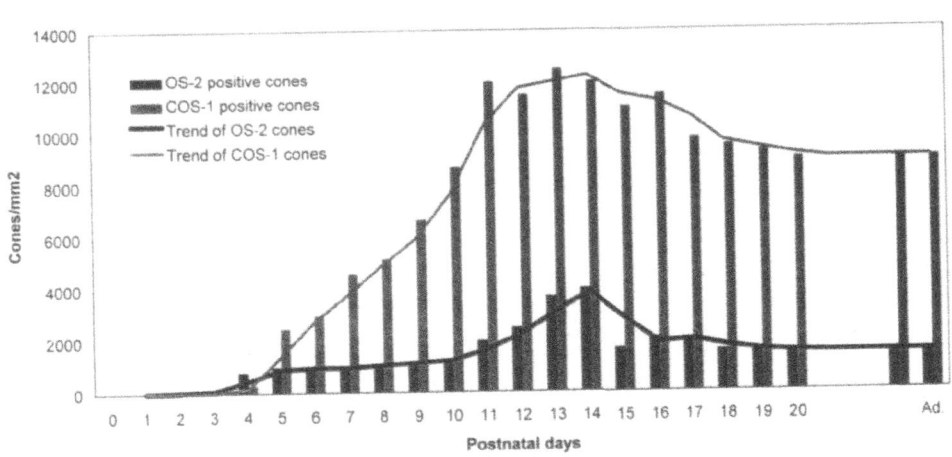

Figure 9. The density of immunopositive cones in the developing rabbit retina as a function of postnatal days. The measurements were carried out on samples from halfway between the optic nerve head and the ora serrata in the superior quadrant. Similar to the rat, the first cones to appear are labelled by OS-2. The density of this cone population however rises slowly and never reaches a high peak. The second cone type is stained with COS-1, its density rises steeply to reach the plateau (9000 cone/mm²), and after a slight decrease remains relatively high throughout the life of the animal. Note the crossing of the two trend lines.

a common mechanism taking place in both divided and non-dividing retinas. The majority of other mammalian species were not yet screened for the developmental coexpression of visual pigments; it seems likely, however, that this is not an exclusive or predominant way of cone differentiation. No dual cones have been found in developing monkey retinas (Wikler and Rakic, 1991 and 1994; Bumsted et al., 1993 and 1997), and the bovine retina also develops without the presence of cones that synthesize two pigments (Timmers and Szel, unpublished).

Two different basic mechanisms can account for the differentiation of the two cone types: with and without transdifferentiation (Figs. 10 and 11). It is not yet clear whether both processes can take place within the same retina during ontogenesis. If the two scenarios turn out to be mutually exclusive, we would still be left with the question whether species normally developing with this or the other mechanism can change their differentiation strategy under unnatural circumstances. Further studies are needed to produce an explanation that is compatible with all the aforementioned findings and at least partly contradictory pieces of information. Despite the lack of a coherent view on photoreceptor differentiation, the topographical determination of a developmental wave along the dorsal-

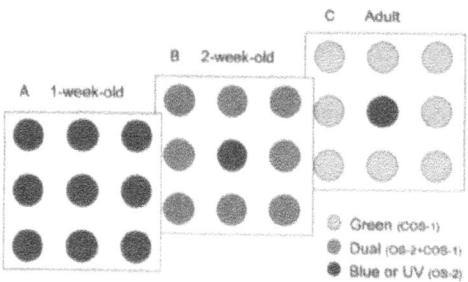

Figure 10. Cone development with transdifferentiation. The diagram shows that in the early postnatal retina, all cones express blue pigment at first (A: black circles), then most of these start to synthesize green pigment (B: dark grey circles). These transforming cones make the temporary population of dual cones stained by both COS-1 and OS-2. A smaller fraction of cones keeps producing blue pigment throughout time (B: black circles). The dual cones (B: dark grey circles) turn to definitive green cones (C: light grey circles), whereas the genuine S cones make the definitive blue cone population (C: black circles).

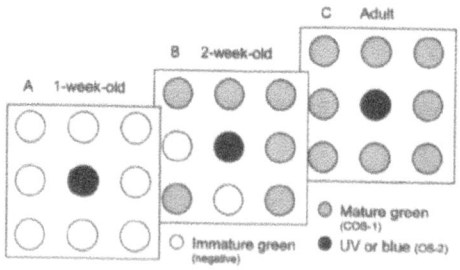

Figure 11. Cone development without transdifferentiation. Another possible scenario is that all cones are pre-destined to be either blue or green. In the early prenatal retina (A), a few cones already express blue pigment (black circles). These cones do not undergo any shift, and continuously keep producing the shortwave pigment (B and C: black circles). The majority of cones are non-expressing before the second postnatal week, when this population gradually starts to synthesize the green pigment (B and C: grey circles). There are no dual cones at any time.

ventral axis is supported by large numbers of studies, especially those dealing with the characteristic distribution of proteins in the embryonic eye (Constantine-Paton et al., 1986; McCaffery et al., 1993) and dorsal-ventral patterns of expression of exogenous opsin genes (Lem et al., 1991; Zack et al., 1991).

ACKNOWLEDGMENTS

I thank the following colleagues and friends who have contributed to this work. The names appear in alphabetical order: Aguirre, G; Bergström, A; Bonhomme, F.; Bruun, A; Bumsted, K; Caffé, A.R.; Calderone, J.B.; Chen, J.; Cooper, H.; Csorba, G.; Diamantstein, T.; Ehinger, B.; Fekete, T.; Guenet, J.L.; Hendrickson, A.; Hurley, J.; Jacobs, G.H.; Juliusson, B.; Lerea, C.; Lukáts, Á.; Mieziewska, K.; Nevo, E.; Petry, H.M.; Szél, G.; Takács, L.; von Schantz, M.; Warfvinge, K.; Wikler, K. The most important grant that supported the work is the Hungarian OTKA grant #T-017703.

REFERENCES

Bok, D., 1985, Retinal photoreceptor-pigment epithelium interactions. Invest. Ophthalmol. Vis. Sci. 26: 1659–1694.

Bumsted, K., Jasoni, C., Szél, Á., and Hendrickson, A., 1997, Spatial and temporal expression of cone opsins during monkey retinal development. J. Comp. Neurol. 378: 117–134.

Bergström, A., Ehinger, B., Wilke, K., Zucker, C.L., Adolph, A. R., and Szél, Á., 1994, Development of cell markers in subretinal rabbit retinal transplants. Exp. Eye Res. 58: 301–313.

Bumsted, K., Hendrickson, A., Erickson, A., and Szél, Á., 1993, Immunohistochemical development of cone opsins in Macaca monkey retina. Soc. Neurosci. Abstr. 19: 52.

Caffé, A. R., Visser, H., Jansen, H. B., and Sanyal, S., 1989, Histotypic differentiation of neonatal mouse retina in organ culture. Curr. Eye Res. 8: 1083–1092.

Caffé, A.R., Söderpalm, A., and van Veen T., 1993, Photoreceptor-specific protein expression of mouse retina in organ culture and retardation of rd degeneration in vitro by a combination of basic fibroblast and nerve growth factors. Curr. Eye Res. 12: 719–726.

Calderone, J. B., and Jacobs, G. H., 1994, Regional variations in sensitivity to UV light in the mouse retina. Invest. Ophthalmol. Vis. Sci. (Abstr.) 35: 2046.

Cameron, D. A. and Easter, S. S., 1993, The cone photoreceptor mosaic of the green sunfish, Lepomis cyanellus. Visual Neurosci. 10: 375–384.

Carter-Dawson, L. D., and LaVail, M.M., 1979, Rods and cones in the mouse retina. I. Structural analysis using light and electron microscopy. J. Comp. Neurol. 188: 245–262.

Chen, J., Tucker, C. L., Woodford, B., Szél, Á., Lem, J., Gianelli-Borradori, A., Simon, M. I., and Bogenmann, E., 1994, The human blue opsin promoter directs transgene expression in short-wave cones and bipolar cells in the mouse retina. Proc. Natl. Acad. Sci. USA, 91: 2611–2615.

Chievitz, J. H., 1891, Über das Vorkommen der Area centralis in den vier höheren Wirbeltierklassen. Arch. Anat. Physiol. Lpz. Anat. Abs., (Suppl.) 139: 311–334.

Chiu, M. I., and Nathans, J., 1994, Blue cones and cone bipolar cells share transcriptional specificity as determined by expression of human blue visual pigment-derived transgenes. J. Neurosci. 14: 3426–3436.

Constantine-Paton, M., Blum, A. S., Mendez-Otero, R., and Barnstable, C. J., 1986, A cell surface molecule distributed in a dordso-ventral gradient in the perinatal rat retina. Nature 324: 459–462.

Curcio, C. A., and Hendrickson, A. E., 1991, Organization and development of the primate photoreceptor mosaic. Prog. Retinal Res. 10: 89–120.

DeMonasterio, F.M., Schein, S. J., and McCrane E. P., 1981, Staining of blue-sensitive cones of the macaque retina by a fluorescent dye. Science 213: 1278–1281.

Famiglietti, E. V., Sharpe, S. J., 1995, Regional topography of rod and immunocytochemically characterized "blue" and "green" cone photoreceptors in rabbit retina. Vis. Neurosci. 12: 1151–1175.

Harwerth, R. S., and Sperling, H. G., 1975, Effects of intense visible radiation on the increment-threshold spectral sensitivity of the rhesus monkey eye. Vision Res. 15: 1193–1204.

Jacobs, G. H., Neitz, J., and Deegan, J. F. II, 1991, Retinal receptors in rodents maximally sensitive to ultraviolet light. Nature 353: 655–656.

Jacobs, G. H., 1993, The distribution and nature of colour vision among the mammals. Biol. Rev. 68: 413–471.

Jacobs, G. H., and Deegan, J. F. II, 1994, Sensitivity to ultraviolet light in the gerbil (Meriones unguiculatus): Characteristics and mechanisms. Vision Res. 34: 1433–1441.

Juliusson, B., Bergström, A., Röhlich, P., Ehinger, B., van Veen, T., and Szél, Á., 1994, Complementary cone fields of the rabbit retina. Invest. Ophthalmol. Vis. Sci. 35: 811–818.

Lem, J., Applebury, M., Falk, J. D., Flannery, J. G., and Simon, M. I., 1991, Tissue-specific and developmental regulation of rod opsin chimeric genes in transgenic mice. Neuron 6: 201–210.

Lerea, C. L, Somers, D. E., Hurley, J. B., Klock, I. B., and Bunt-Milam, A. H., 1986, Isolation of specific transducin alpha-subunits in retinal rod and cone photoreceptors. Science 234: 77–80.

Lerea, C. L., Bunt-Milam, A. H., and Hurley, J. B., 1989, Alfa-transducin is present in blue,- green-, and red-sensitive cone photoreceptors in the human retina. Neuron 3: 367–376.

Marc, R. E. and Sperling, H. G., 1976, The chromatic organization of the goldfish cone mosaic. In: Neuroanatomy of the Visual Pathways and their Development. Eds: B. Dreher and S. R. Robinson. Vol. 3 of Vision and Visual Dysfunction. Macmillan Press, Scientific & Medical, London. pp 1211–1224.

McCaffery, P., Posch, K. C., Napoli, J. L., Gudas, L., and Dräger, U. C., 1993, Changing patterns of the retinoic acid system in the developing retina. Dev. Biol. 158: 390–399.

Nathans, J., Thomas, D., and Hogness, D. S., 1986, Molecular genetics of human color vision: the genes encoding blue, green and red pigments. Science 232: 193–202.

Osterberg, G. A., 1935, Topography of the layer of rods and cones in the human retina. Acta Ophthalmol. 13: 1–97.

Raymond, P. A, Barthel, L. K., Curran, G. A., 1995, Developmental patterning of rod and cone photoreceptors in embryonic zebrafish. J. Comp. Neurol. 359: 537–550.

Röhlich, P., and Szél, Á., 1993, Binding sites of photoreceptor-specific antibodies COS-1, OS-2 and AO. Curr. Eye Res. 12: 935–944.

Röhlich, P., Ahnelt, P., Dawson, W. W., and Szél, Á., 1994a, Presence of immunoreactive blue cones in the fetal monkey fovea. Exp. Eye Res. 58: 249–252.

Röhlich, P., van Veen, T., and Szél, Á., 1994, Two different visual pigments in one retinal cone cell. Neuron 13: 1159–1166.

Söderpalm, A., Szél, Á., Caffé, A. R., and van Veen, T., 1994, Selection and development of one cone photoreceptor type in retinal organ culture. Invest. Ophthalmol. Vis. Sci. 35: 3910–3921.

Stenkamp, D. L., Hisatomi, O., Barthel, L. K., Tokunaga, F., and Raymond, P.A. Temporal expression of rod and cone opsins in embryonic goldfish retina predicts the spatial organization of the cone mosaic. Invest. Ophthalmol. Vis. Sci. 37: 363–376.

Szél, Á., Röhlich, P., Caffé, A. R., and van Veen, T., 1996, Distribution of cone photoreceptors in the mammalian retina. Microsc. Res. Techn. 35: 445–462.

Szél, Á., Takács, L., Monostori, É., Diamantstein, T., Vigh-Teichmann, I., and Röhlich, P., 1986, Monoclonal antibody recognizing cone visual pigment. Exp. Eye Res. 43: 871–883.

Szél, Á., Diamantstein, T., and Röhlich, P., 1988, Identification of the blue-sensitive cones in the mammalian retina by anti-visual pigment antibody. J. Comp. Neurol. 273: 593–602.

Szél, Á., Röhlich, P., Caffé, A. R., Juliusson, B., Aguirre, G., and van Veen, T., 1992, Unique topographic separation of two spectral classes of cones in the mouse retina. J. Comp. Neurol. 325: 327–342.

Szél, Á., Röhlich, P. and van Veen, T., 1993a, Short-wave sensitive cones in the rodent retinas. Exp. Eye Res. 57: 503–505.

Szél, Á., Röhlich, P., Mieziewska, K., Aguirre, G., and van Veen, T., 1993b, Spatial and temporal differences between the expression of short- and middle-wave sensitive cone pigments in the mouse retina: A developmental study. J. Comp. Neurol. 331: 564–577.

Szél, Á., Csorba, G., Caffé, A. R., Szél, G., Röhlich, P., and van Veen, T., 1994a, Different patterns of retinal cone topography in two genera of rodents, Mus and Apodemus. Cell Tiss. Res. 276: 143–150.

Szél, Á., van Veen, T., and Röhlich, P., 1994b, Cone differentiation in the retina. Nature, 370:336.

Szél, Á., Juliusson, B., Bergström, A., Wilke, K., Ehinger, B., and van Veen, T., 1994c, Reversed ratio of color-specific cones in rabbit retinal cell transplants. Dev. Brain Res. 81: 1–9.

Young, R. W., 1976, Visual cells and the concept of renewal. Invest. Ophthalmol. 15: 700–725.

Yokoyama, S., and Yokoyama, R., 1989, Molecular evolution of human visual pigment genes. Mol. Biol. Evol., 6:186–197.

von Schantz, M., Argamaso-Hernan, S., Szél, Á., Foster, R. G., 1997, Photopigments and circadian responses to ultraviolet light in the Syrian golden hamster. Dev. Brain Res. In press.

Wang. Y., Macke, J. P., Merbs, S. L., Zack, D. J., Klaunberg, B., Bennett, J., Gearhart, J., and Nathans, J., 1992, A locus control region adjacent to the human red and green visual pigment genes. Neuron 9: 429–440.

Wikler, K. C., and Rakic, P., 1990, Distribution of photoreceptor subtypes in the retina of diurnal and nocturnal primates. J. Neurosci. 10: 3390–3401.

Wikler, K. C., and Rakic, P., 1991, Relation of an array of early-differentiating cones to the photoreceptor mosaic in the primate retina. Nature 351: 397–400.

Wikler, K. C., and Rakic, P., 1994, An array of early-differentiating cones precedes the emergence of the photoreceptor mosaic in the fetal monkey retina. Proc. Natl. Acad. Sci. USA. 91: 6534–6558.

Wikler, K. C., Szél, Á., and Jacobsen, A. L., 1996, Positional information and opsin identity in retinal cones. J. Comp. Neurol. 374: 96–107.

Williams, D. R., MacLeod, D. I. A., and Hayhoe, M. M., 1981, Foveal tritanopia. Vision Res. 21:1341–1356.

Zack, D. J., Bennett, J., Wang, Y., Davenport, C., Klaunberg, B., Gearhart, J., and Nathans, J., 1991, Unusual topography of bovine rhodopsin promoter-lacZ fusion gene expression in transgenic mouse retinas. Neuron 6: 187–199.

5

GENESIS OF TOPOGRAPHIC AND CELLULAR DIVERSITY IN THE PRIMATE RETINA

Pasko Rakic*

Section of Neurobiology
Yale University School of Medicine
New Haven, Connecticut 06510

ABSTRACT

The retina in the developing macaque monkey is an unexcelled model system for the analysis of cellular events and the mechanisms that govern formation of the human eye. Application of the powerful methods of modern neurobiology applied to either normal embryos, or mature animals with altered visual systems, suggests that the emergence of cell classes and their segregation into primate-specific functional domains emerges as a result of multiple factors operating at each retinal layer during successive developmental stages. The available data are compatible with the hypothesis that, at early developmental stages, intrinsic mechanisms operating within the retina predominate, while at later stages reciprocal interactions with the visual centers refines the numerical cellular relationships and synaptic architecture. These studies may provide insight into the development of normal and abnormal vision in humans.

INTRODUCTION

Although the importance of studying the development of the retina directly in primates has been widely recognized by visual scientists and actively encouraged by granting agencies, the difficulty in breeding, long gestation, small number of offspring and magnitude of associated costs has caused a decline in the number of research programs devoted to this subject. This is an unfortunate trend since certain types of information, highly relevant to understanding the development and pathology of the human visual system, cannot be obtained from experimental work on non-primate species. Although in primates one cannot apply some of the molecular and genetic approaches that are now possible in flies

* Correspondence to: Pasko Rakic. M.D., Sc.D., Section of Neurobiology, Yale University School of Medicine, P.O. Box 208001, New Haven, CT 06520-8001. Tel: (203) 785-4323; Fax: (203) 785-5263.

Development and Organization of the Retina, edited by Chalupa and Finlay.
Plenum Press, New York, 1998.

or mice, investigation of the developing primate retina is not only essential but in addition offers some unique advantages.

The initial rationale behind my own decision to investigate the development of the visual system, including the retina in the macaque monkey, was the recognition that the organization, principles and mechanisms of neural development in this species is, in several major respects, remarkably similar to that of the human being. To start with, the macaque has frontally placed eyes, similar fovea, and basically the same binocular organization. Both macaque monkey and human have an almost identical number, and similar fraction, of crossed and uncrossed retinal axons, as well as a comparable number, and placement of laminae, in the lateral geniculate nucleus and primary visual cortex. Furthermore, both species have trichromatic vision with a prototypic photoreceptor mosaic and proportion of short and long wave-sensitive cones. Both species also share the exclusivity of geniculate projections to area 17 of the cerebral cortex with analogous and uniquely primate ocular dominance, laminar and columnar compartmentalization. The divergence factor (defined as the number of higher order cells to which one lower order cell projects) and the convergence factor (defined as the number of lower order cells from which a single higher order cell receives input) is comparable in rhesus monkey and human. Finally, information processing within the macaque monkey and human visual system involves two parallel pathways segregated from the retina to the cerebral cortex. Thus, on the basis of the similarity in size and organization, it is reasonable to expect that the basic principles, as well as the cellular and molecular mechanisms involved in the development of the visual system of the macaque monkey, are akin to that of human beings.

The retina in primates is large and develops very slowly so that at any given age one can distinguish, with relatively high resolution, several developmental stages. Another advantage in studying the development of the primate retina is the availability of a large amount of normative data, from the classical work that has culminated in Stephen Polyak's monumental volume (Polyak, 1957) to the new era in neurobiological research reviewed in Rodeick (1988). These quantitative and qualitative data provide an excellent background for asking relevant developmental questions. The retina in the adult macaque contains over 60 million photoreceptors (Wikler et al., 1990), which converge on about 1.2 million ganglion cells (Rakic and Riley, 1983a) interconnected radially by bipolar and laterally by horizontal and amacrine cells (Rodeick, 1988). Each cell class can be further divided into subtypes based on their molecular properties, morphology, and physiology. Although the basic cellular organization of the retina is similar in most mammals, the absolute number of cells of each class is the highest in primates, and the proportions of various cell subtypes are specific for each primate species. For example, there are significant variations in the proportion of the different wavelength-sensitive cones that make up the photoreceptor mosaic in Old and New World monkeys (Wikler and Rakic, 1990). In addition, the separation of the ganglion cell population into subtypes that project to separate midbrain and diencephalic targets is unique to primates (Livingston and Hubel, 1988; Stone, 1983). I will describe below the emergence of this diversity in macaque monkey retina based on the studies done in my laboratory over the past two decades.

TIMING AND TOPOGRAPHY OF CELL PRODUCTION

The early investigations of eye development, which relied on the detection of mitotic figures in histological preparations, provided only approximate data about timing and sequential events during retinal neurogenesis in primates. The introduction of tritiated thymidine (^{3}H-TdR) in the sixties has enabled more precise study of the timing of the origin of specific cell phenotypes in mammals (reviewed in Sidman, 1970). Over the past two decades we have carried out a series of studies on neuron origin in the macaque monkey brain using this method (e.g. Rakic, 1973, 1974, 1977a). Collection of the autoradiograms prepared from the retinas of monkeys exposed to ^{3}H-TdR at selected embryonic ages and sacrificed at various intervals were used to determine the phenotypes and location of cells that were generated at the time of injection (LaValil et al., 1991; Rapaport et al., 1992, 1995).

The first fate restriction detected using this method was divergence into cells that eventually form the retinal pigment epithelium and neural retina. The genesis in these two basic tissues begins simultaneously, between embryonic day 27 (E27) and E30 of the 165 day long gestational period (Rapaport et al., 1995). The earliest generated cells in the pigment epithelium and neural retina are found in the region of the presumptive fovea and later generated cells are gradually added to more peripheral sites. The cell production in both tissues continues throughout gestation and becomes very low by the time of birth.

All other retinal neurons, as well as the Müller glial cells, in the monkey are produced in the neural epithelium that forms the inner surface of the optic cup (LaVail. et al., 1991). As in the other vertebrates (Holt et al., 1988, Turner and Cepko, 1987, Williams and Goldowitz, 1992, Reese et al., 1995; Wets and Fraser, 1988) migration of most postmitotic cells in primates is restricted to the radial columns. This mode of migration of main projection lines from the photoreceptors *via* bipolar to ganglion cells allows tracing of the sequential genesis of each cell class within a given radial column. The cumulative frequency of the percent of ^{3}H-TdR labeled cells for each retinal cell phenotype reveals that the onset of the genesis of various retinal cell types within a given radial column proceeds in a sequential order (Fig. 1). Thus, the first cells produced are ganglion cells, followed by horizontal cells, cone photoreceptors, amacrine cells, Müller glial cells, bipolar cells, and finally rod photoreceptors. (LaVail et al., 1991). Although there is an overlap in

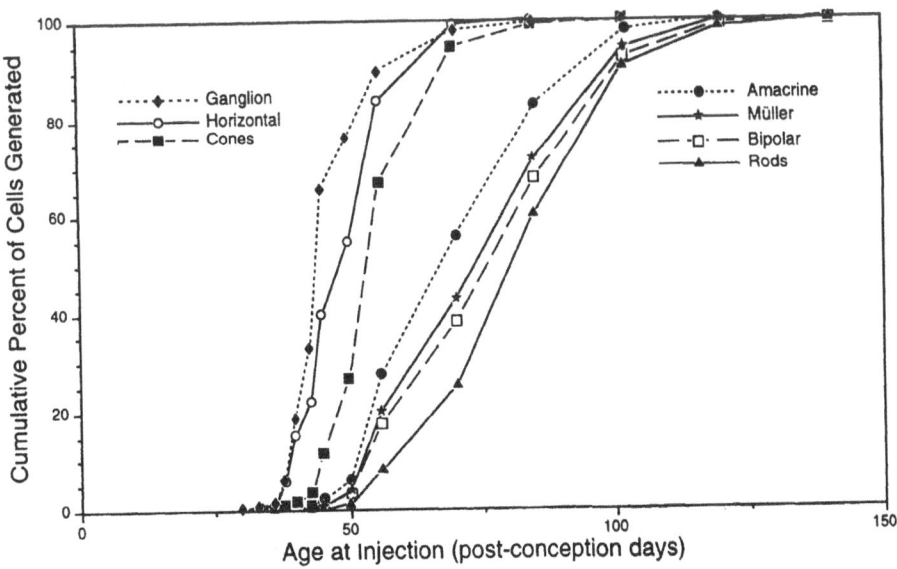

Figure 1. Graph of the cumulative percent frequency of ^{3}H-TdR labeled cells in juvenile monkey retinas after fetal exposure to this DNA-specific radioactive nucleotide. Plotting cumulatively data of the heavily radiolabeled cells, which indicate the time of their last division, allows visualization of sequences and kinetics of retinal cell genesis in primates (From Rapaport et al., 1996; based on LaVail et al., 1991).

the periods of genesis of different cell types, the sequential genesis is validated by the observation that, except at the beginning and end of their production, the plots for each cell type have a relatively constant slope (Rapaport et al., 1996). In general, the plots of individual cell types do not cross each other, the only exception being the plot of ganglion cell production which crosses the cone and horizontal cell plots as it flattens out above the 95% level (Fig. 1). This exception suggests that some other cell types in the ganglion cell layer may be generated simultaneously. Indeed, some amacrine cells are born at the end of the ganglion cell production phase. Therefore, it is likely that some displaced amacrine cells may be generated last in the ganglion cell layer (Rapaport et al., 1996).

The time separating the genesis of different cell types in the monkey retina varies from very short, such as that observed between Müller and bipolar cells, to relatively long, such as that recorded between cone photoreceptors and amacrine cells (LaVail et al., 1991). Indeed, the differences in timing of the genesis of ganglion and horizontal cells and of Müller, bipolar, and rod photoreceptor cells is so small that it can be considered negligible The longest interval between sequentially generated cell types was observed between cone photoreceptors and amacrine cells, and this interval also separates cells distinguished by several other features, the most obvious of which is the rate of cell genesis as determined by the slope of the cumulative plots (Fig. 1). Thus, retinal cells are generated in two "phases" that were initially described in marsupials (Harman and Beasley, 1989). Cytogenesis in the first phase is more rapid and follows a less pronounced spatial gradient, but has the greatest central-to-peripheral density ratios in the adult.

The number of ganglion cells produced in the macaque embryonic retina is more than twice the 1.2 million present in the adult animal (Rakic and Riley, 1983a). This overproduction is compensated for by ganglion cell death, which occurs throughout the period of cell production but is most pronounced in the second half of gestation, during the phase of binocular segregation (Fig. 2 and Rakic, 1976, 1981; Rakic and Riley, 1983b). Monocular enucleation performed before the period of maximal cell death rescues a number of ganglion cells in the remaining eye, producing a retina with a greater than normal complement of ganglion cells (Rakic and Riley, 1983b). It is not known whether, and to what extent, other retinal cell types are initially overproduced, mostly because identification and counting of immature, non-ganglion cells in the embryonic retina is difficult (Rakic, 1992).

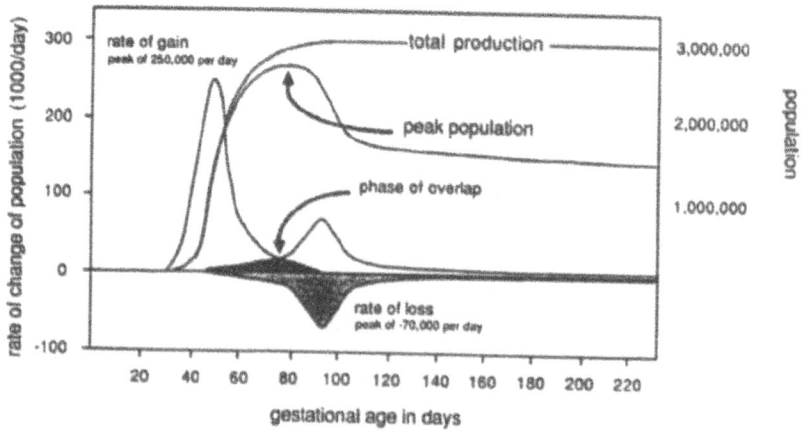

Figure 2. The assessment of the number of ganglion cells produced and eliminated during development of the macaque retina (Based on Rakic and Riley, 1983a).

The central-to-peripheral gradient of cell genesis is apparent for all cell classes in the primate retina, including the pigment epithelium (LaVail et al., 1991). For example, ganglion cells are born first, and their initial progeny in postnatal monkeys becomes located approximately midway between the center of the fovea and the optic nerve head. All subsequently generated cell types initiate their genesis in concentric waves, starting with the fovea. This is also clearly evident for the rods, which complete their genesis last. Thus, it appears that the first cells generated may not have cues as to the site of the future fixation point; rather, their genesis acts to define this site (Rapaport et al., 1996). In addition, the cell types generated in the first phase of cytogenesis (ganglion, horizontal, and cone photoreceptor cells) display a less precise spatial pattern of genesis than the cells generated in the second phase (amacrine, Muller, bipolar and rods). For the cell types produced by the second phase of cytogenesis, cell division ceases perifoveally before the onset of the genesis of cells in the periphery. It has been suggested that the relatively rapid central-peripheral expansion of early generated cells, such as cones, may be the basis for the eventual development of their mosaic distributions by producing a precocious set of post-mitotic cells across the retinal surface (see below and Wikler and Rakic, 1991, 1994).

EMERGENCE OF PHOTORECEPTOR MOSAIC

The macaque monkey retina contains about 60 million rods and 3 million cones (Wikler and Rakic, 1990) that are organized in a mosaic pattern aligned radially in register with bipolar and ganglion cells. The absolute number, size and packing density of rods and cones varies among primate species as well as according to retinal eccentricity (Wikler and Rakic, 1990). The most dramatic example of this regional variability in primates is the centrally located foveola, which in macaque contains only a few rods and in human consists exclusively of cones (Wikler et al., 1990; Curcio et al., 1997). Since the dominance of cones in the foveola can be observed from the earliest stages, it is unlikely that rods are first produced and then deleted in this particular region. However, going from the cone pure center of the retina towards the periphery there is a gradual increase in incidents of rods and a concomitant increase in their size. In addition, there is a nasal to temporal asymmetry as well as a peak rod density located in the restricted region dorsal to fovea, referred to as dorsal rod peak. Finally, three cone subtypes that express opsins preferentially sensitive to long (LWS or red), middle (MWS or green) and short (SWS or blue) wavelengths are distributed in primates in a species-specific mosaic pattern (Wikler and Rakic, 1990; Curcio et al., 1991). For example, in macaque monkey each SWS cone is surrounded by approximately 10 LWS and MWS cones, depending on retinal eccentricity. This reiterative organization of photoreceptor subtypes is shared by other primates, but the members of the Old and New World orders have a different distribution and ratio of various cell subtypes (Wikler and Rakic, 1990). The development of the mosaic-like distribution of photoreceptors has been reviewed recently (Wikler and Rakic, 1996) and only selected aspects of this complex process will be highlighted here.

As described in the previous section, the order of production of cones and rods results in a transient developmental pattern of unique cell appositions. Thus, because the genesis of cones at any sector of the retina precedes the genesis of rods, cones come transiently in direct apposition with each other, the contact interrupted only at later stages when rods become interposed between them. We have suggested that this transient apposition allows interactions *via* specialized contacts, such as puncta adherentia and/or some putative distance-restricted factors that may influence the determination of the specific phenotypes or position of neighboring cells (Wikler and Rakic, 1991).

Use of the antibodies to cone subtype-specific opsins revealed a subset of about 10% cones that express their opsin two to three weeks before they appear in surrounding cones (Wikler and Rakic, 1991). The precocious differentiation of a small number of cones strategically placed across the immature retina suggested to us that they may be involved in the formation of the photoreceptor mosaic in primate retina (Wikler and Rakic, 1994). As a next step, we examined the degree of early cone differentiation using the monoclonal antibodies, XAP-1(specific to the photoreceptor membranes), and SV2 (specific to synaptic vesicle protein).The analyses revealed a subset of precociously-immunoreactive cones that are distributed in the similar pattern as the precocious red/green-sensitive cones in immature regions of the fetal monkey retina (Wikler and Rakic, 1994). This indicates that the early maturation of subsets of cones is not restricted to opsin expression, but may include other molecules engaged in cone differentiation and their connectivity.

The expression of an antigen recognized by a novel monoclonal antibody, 7G6, which stains all cones in the adult primate retina (Wikler et al., 1997) is restricted to cones even in the youngest specimen examined (E65). However, in contrast to the adult, not all cones are labeled with 7G6, and the ratio of labeled to unlabeled cones changes as a function of retinal eccentricity. The onset of 7G6 immunoreactivity in a sub population of peripheral cones occurs in regions of the retina with active cone genesis, and precedes the expression of the cone opsins and the formation of synaptic contacts in the outer plexiform layer by at least one week (Nishimura and Rakic, 1987a). The early-differentiating 7G6-positive cones are organized into arrays throughout the embryonic retina, indicating that the spatial arrangement of cones emerges during or immediately after cone division, and prior to overt differentiation or formation of connectivity between the retina and the brain (Wikler et al., 1997). Thus, photoreceptor mosaics in the embryonic primate retinae seem to possess a spatially-organized pattern that correlates with cell patterning observed later in development. The early expression of cell class-specific-markers suggests that the retinal mosaic pattern is initiated independently of synaptic contacts with either horizontal or bipolar cells which are established later (see below and Nishimura and Rakic, 1985, 1987). Thus, the species-specific number and ratio of photoreceptors in various primate species (Wikler and Rakic, 1990) is formed independently of the more centrally located structures. It further suggest that local interactions between neighboring cells rather than interaction via synaptic contacts with distant neural centers initiate the species specific pattern. The next challenge is to determine how this mosaic is connected *via* bipolar cells and ganglion cells across the geniculate nucleus to the cerebral cortex. The fact that a subset of cones forms synapses in the outer plexiform layer earlier than the surrounding cones suggests their involvement in the coordination of the photoreceptor mosaic with the other retinal layers (Wikler and Rakic, 1996).

To begin to address the issue of the time of differentiation of wavelength- sensitive cone subtypes, we have used RT-PCR to examine the expressions L/MWS or SWS opsins. This approach so far has not revealed the expression of opsin in monkey retina at E50. However, by E65 L/MWS and SWS opsins can be detected in the fovea, but only the SWS was detected in parafoveal and peripheral regions (Wikler et al., 1995). By E72, both opsin mRNAs were evident at foveal and parafoveal eccentricities, however, only the SWS opsin mRNA was detected in the peripheral retina. By E90 both opsins become detectable throughout the retina. These data indicate that SWS opsin expression precedes the expression of the L/MWS opsin by approximately three weeks and that SWS cones express their photopigment within one week of their final mitotic division. In addition, these results suggest that periodically positioned, early-differentiating cones in the fetal monkey retina may correspond to the SWS phenotype. It remains to be determined whether the early spacing of SWS cones influences the patterning of undifferentiated nascent cones to establish the adult cone mosaic.

GENESIS OF RETINAL GANGLION CELL SUBTYPES

Although in histological preparation all ganglion cells in the primate retina may look rather similar, they nevertheless can be subdivided into several distinct classes on the basis of various characteristics, such as soma size, dendritic field size, pattern of axonal termination and projection targets, as well as conduction velocity and responses to different visual stimuli (Rowe and Stone, 1980; DeMonesterio, 1978; Shapely and Perry, 1986). In both human and Old World primates, visual information is conveyed by two parallel pathways: the magnocellular (M) and parvocellular (P) streams which start with the ganglion cells in the retina and continue via separate neuronal layers of the lateral geniculate nucleus to separate compartments of the visual cortex that are involved primarily in motion and color/form discrimination (Livingston and Hubel, 1988; Maunsell, 1992). In general it is difficult to distinguish between M and P ganglion cell types in histological preparation. However, a simple measure of cell body size, in spite of some overlap, provides a reliable distinction between these two major classes in the macaque retina (e.g., Perry and Cowey, 1984). The systematic analysis of postnatal animals exposed to ^{3}H-TdR at early embryonic ages revealed radiolabeling of cells with diameters in the range of the Pγ and Pβ (Rapaport et al., 1992). These two cell subtypes project to the superior colliculus and P layers of the lateral geniculate nucleus, respectively (Perri and Cowey, 1984; Perri et al., 1984) . In contrast, the Pα or M cells, which project to the M layers of the lateral geniculate nucleus become labeled at successively later ages. Thus, in general, genesis of P ganglion cells is followed by genesis of M cells.

The availability of markers that can distinguish between the M and P ganglion cell populations in developing primates provided an opportunity to analyze the time and mechanisms of ganglion cell diversification. Although perfect cell-class-specific labels have so far not been found, the POU domain family of homeodomain transcription factors, Brn-3a and Brn-3b can serve as useful markers. For example, in the adult macaque monkey the Brn-3a antibody labels heavily about 10% of P-type ganglion cells and lightly approximately 90% of M-cells. In contrast, the Brn-3b antibody stains all P-type cells heavily and M-type cells lightly (Xiang et al., 1996). Thus, although these two antibodies cannot be used to determine unequivocally whether an individual ganglion cell belongs to the M or P class, they can be used to determine whether these two separate populations exist in the retina as a whole.

Our immunocytochemical analysis indicates that Brn-3a and Brn-3b-are expressed in migrating neurons as well as in the ganglion cell layers of the central retina as early as E47 (Meisserel et al., 1997). Labeling in the central regions in older cases becomes confined to the ganglion cell layer, while in the less mature peripheral regions of these retinae Brn antisera continued to stain spindle-shaped cells migrating from the proliferative neuroepithelium toward the ganglion cell layer (e.g., Fig. 3). The presence of adult-like patterns of light and heavy labeled cells with either Brn-3a or Brn-3b antibodies in the embryonic retina suggests that POU domain proteins are expressed differentially in separate ganglion cell classes soon after last cell division. A possibility that differences in labeling intensity signifies a cell's transition from an immature to mature stage is unlikely since we did not observe an increasing percentage of darkly labeled cells at later embryonic ages and the labeling patterns were from the start consistent with those observed in the mature retina (Meisserel et al., 1997). Thus, in harmony with the studies in the avian and rodent retina showing the onset of differentiation of ganglion cells to a generic ganglion cell phenotype immediately after their final mitotic division (McLoon and Barnes, 1989; Trisler et al., 1996), primate ganglion cells also differentiate into M and P neuronal subtypes shortly after being generated.

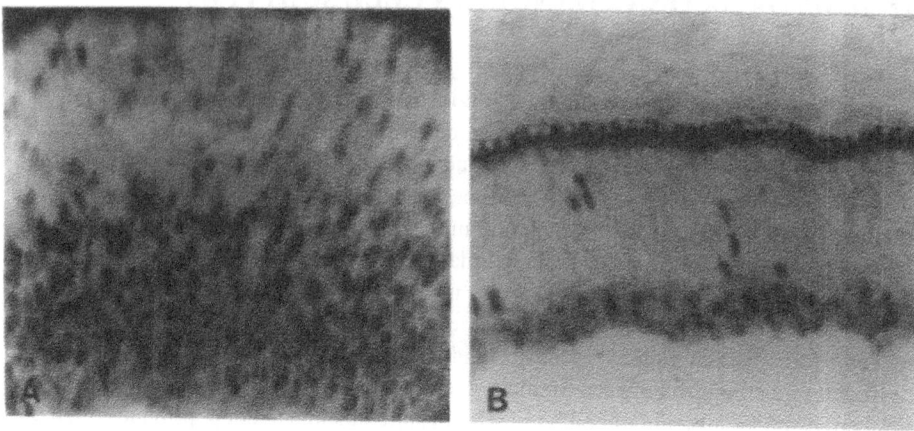

Figure 3. Cross-sections of the central retinae immunostained with an antibody generated against Brn-3b at E47 (A) and the peripheral retina stained with the same antibody at E68 (B). Note that most labeled cells in both specimens are situated in the ganglion cell layer (bottom), but some appear to be in the process of migrating from their sites of origin near the retinal pigment epithelium (top) towards the ganglion cell layer. At E68, neurogenesis for the central region of the retina has been completed and all labeled cells are situated within the ganglion cell layer (not shown) (modified from Meisserel et al., 1997).

In the macaque monkey optic tract, axons from early generated cells are situated deep, those generated later are closer to the pia (Williams et al., 1991; Williams and Rakic, 1985; Reese, 1996). The chronotopic arrangement makes sense if one takes into account that, in general, P cells are generated before M. The early diversification of ganglion cells into M and P subtypes is supported by the finding that their axons in embryonic monkey from the start project directly and selectively to either the M or P moieties of the developing lateral geniculate nucleus (Meisserel et al., 1997). The initial contingent of retinal fibers either bypasses the thalamus and innervates the midbrain or becomes distributed preferentially in the P moiety of the lateral geniculate nucleus (prospective layers 3–6) situated in the medial segment of the nuclear anlagen (Fig. 4). In contrast, the lateral segment which eventually forms M moiety of the geniculate (respective layers 1 and 2) is innervated nearly a month after fibers colonize the medial segment. Based on the outside-to-inside temporal generation sequence of geniculate neurons (Rakic, 1977a), it can be inferred with confidence that the early-innervated medial segment corresponds to the region that will form the P layers, while the later-innervated lateral segment differentiates into the M laminae (Meisserel et al., 1997). Furthermore, axons innervating the M and P moieties of the geniculate nucleus display clear morphological differences from the time that their terminal arbors enter appropriate segments of the lateral geniculate anlage. Taken together, the early expression of cell class-specific molecules, selective innervation of the M and P moieties of the lateral geniculate by retinal axons with characteristic terminal arbors (Meisserel et al., 1997), as well as the target-independent development of distinctive morphological features of their dendrites (Campbell et al., 1997) indicates an early specification of these functional visual subsystems in the primate embryo.

The initial ratio of M and P ganglion cells in a given sector of the retina may be altered in favor of P cells by selective elimination of M ganglion cells (Meisserel et al., 1997). This elimination may occur in response to a change in the number of M-dedicated target neurons in the LGN by E80 (Williams and Rakic, 1988). Activity-mediated segregation of

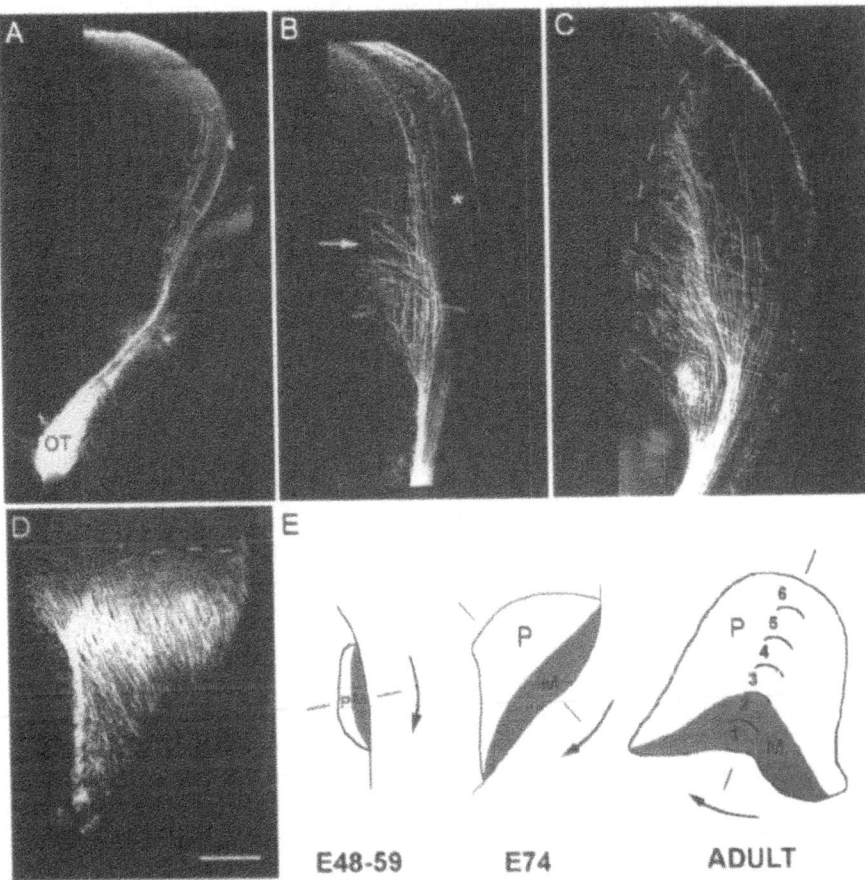

Figure 4. Photomontages of confocal images through representative coronal sections of the diencephalon showing the distribution of DiI-labeled retinal axons in the embryonic thalamus. Dashed lines indicate the border of the dorsal lateral geniculate (dLGN) with a dorsal orientation to the top and a lateral orientation to the right (A–D). A, B and C show the contralateral side to the DiI optic nerve implants. **A:** At E48, retinal axons navigate the contralateral optic tract (OT) before being deflected away from the pial surface as they approach the geniculate. A few axons course dorsally past the dLGN toward the midbrain. **B:** At E53, there is greater ingrowth of retinal fibers, some of them elaborating medially-directed branches. Note that the lateral aspect of the dLGN remains totally devoid of retinal axons (asterisk). Branches derived from the axon trunks are concentrated at the bottom third of the dLGN and in some cases these extend past its medial border in the external medullary lamina (arrow). **C:** At E64, an increasing number of axonal branches invade the medial region of the nucleus. Note that, at this age, the lateral segment of the dLGN is still virtually free of retinal afferents. The section shown is from the rostral part of the dLGN, but essentially the same pattern is observed throughout the rostro-caudal extent of the nucleus. **D:** At E74, DiI crystals were implanted into the optic tract to reduce the distance of diffusion. Virtually the entire extent of the nucleus now receives a retinal innervation. The coronal section is from the caudal aspect of the dLGN. Scale bar: A–C, E, 500 μm, D 400 μm. **E:** Schematic representation of the dLGN rotation (arrows) from E48 to adulthood. The presumptive M layers (shaded area) rotate from a lateral to a ventral position whereas the presumptive P layers rotate from a medial to a dorsal position (Rakic, 1977a; Meisserel et al., 1997).

retinogeniculate terminals axons into ipsi- and contralateral layers of the geniculate nucleus occurs at later stages of development (Rakic, 1981; Rakic and Riley, 1983b; Chalupa and Williams, 1984; Shatz, 1996) and in primates involves elimination of axons innervating inappropriate monocular layers (Rakic, 1981, 1986; L. Chalupa, personal communications), rather than rearrangement of terminal fields, as described in cat (Sretavan and Shatz, 1984). Thus, in primates, with six geniculate layers, M retinogeniculate terminals originating from the left and right eyes eventually compete for only M layers 1 and 2, while P ganglion cell axons compete only for P layers 3 to 6. (Meisserel et al., 1997).

The distinct M and P systems are separated from the start, presumably by a precise genetic program (Meisserel et al., 1997). In contrast, the initially intermixed inputs from two eyes become allocated to the two hemispheres by competition for available synaptic targets (Rakic, 1976, 1977b). This may explain why manipulation of retinal input in primate embryo has a different effect on the development of these two visual subsystems: The monocular enucleation in monkey embryos results in a dramatically enlarged input and territory of the brain subserving the remaining eye (Rakic, 1981; Rakic and Rile, 1983b). In contrast, neither monocular nor binocular enucleation before birth prevents the emergence of basic cyto- and chemo-architectonic features of the M and P systems in monkeys (Rakic, 1981; Dehay et al., 1989; Kennedy and Dehay, 1993, 1997; Kuljis and Rakic, 1991; Rakic and Lidow, 1995). In this respect, factors underlying the formation of the M and P channels in primates may be fundamentally similar to those proposed for guiding the selective innervation of cortical layer IV by thalamic afferents (Boltz et al., 1992; Antonini and Stryker, 1993) and may follow similarly precise molecular markers as observed in the olfactory system (Mombaerts et al., 1996).

The projections of M and P retinogeniculate axons into separate territories in the geniculate nucleus is unique to primates, although parallel visual channels exist in other mammalian species (Sur et al., 1982; Garraghty et al., 1993). In particular, studies in the cat have called attention to the role of competitive interactions among different classes of retinogeniculate terminals in refining immature X and Y pathways (Sur et al., 1982; Garraghty et al., 1993). Such a mechanism may be operative in this species because different cell types are largely intermingled within the A laminae of the geniculate. Even the formation of ocular dominance related projection, which in both species proceed from overlapping into segregated (Williams and Chalupa, 1982; Shatz, 1983, 1996) seems to be achieved by different mechanisms. In the monkey, the segregation occurs by selective elimination (Rakic and Riley, 1983b; L. Chalupa, personal communications), while in the cat it is achieved mainly by rearrangement of axons (Sretavan and Shatz, 1987). The early differentiation and segregation of M and P pathways demonstrated experimentally in an Old World primate is likely to occur during formation of the human visual system.

SYNAPTOGENESIS IN THE PRIMATE RETINA

After retinal neurons in a given radial sector of the retina are generated and begin to express their phenotypes, they start to form synaptic connections. A light microscopic examination of the series of embryonic retinas reveals that the inner plexiform layer (IPL) appears earlier than the outer plexiform layer (OPL) (Nishimura and Rakic, 1985, 1987a,b). The IPL in adult primate contains two principal types of synaptic contacts: conventional and ribbon synapse which form a variety of contacts among four classes of neurons: amacrine (A), bipolar (B), ganglion (G) and interplexiform (I) cells. The conventional synapses are formed between the processes of amacrine cells and the dendrites of ganglion cells (A–G), other amacrine cells (A–A), interplexiform cells (A–I), axons of bipolar cells (A–B), and interplexiform cells contacting bipolar processes (I–B). Ribbon synapses, on the other hand, are established between axons of bipolar cells and dendrites of ganglion and/or amacrine cells, in the form of monads (B–G and B–A) or dyads (B–GG; B–GA; B–AA). In addition, a triad synaptic complex is formed by components of bipolar, ganglion and amacrine cells, but the composition and proportion of the participating elements has not been analyzed in detail. In general, the ribbon synapse represents the "straight signal" pathway between receptors and retinal ganglion cells, whereas conventional synapses form local synaptic circuits between neuronal elements within the IPL.

Analysis of serial sections prepared from the embryonic monkey retina demonstrates that the first sign of synapses in the IPL can be defined by filamentous thickening which appears before any other ultrastructural sign of membrane specialization on the apposing elements (Nishimura and Rakic, 1987a). Our observations indicate ganglion cells in the embryonic retina establish contacts with amacrine cells by means of conventional synapses significantly earlier than the first appearance of ribbon synapses connecting bipolar cells with either ganglion or amacrine cells (Fig. 5). The early formation of lateral connectivity within the ganglion cell layer may provide a cellular substrate for the generation of coordinated action potentials observed in the developing retina before formation of trough-line connections with visual centers in the brain (Skaliora et al., 1993; Wong et al., 1995; Feller et al., 1997). Therefore, paradoxically in this case, local neural circuits as defined by the Society for Neuroscience Research Program (Rakic, 1975) are established before the formation of a central projections, i.e., "straight signal" connections (Nishimura and Rakic, 1985, 1987a). Furthermore, since some ganglion cell axons already form synapses in the developing lateral geniculate nucleus and superior colliculus between E78 and E99, both the basic synaptic complexes within the IPL, as well as neuronal contacts between the retina and brain, are initiated prior to the formation of any contacts with the photoreceptors *via* bipolar cells. The central-to-peripheral sequence of the onset of synapse emergence is also opposite to the sequence of neurogenesis in the primate visual system, which starts with ganglion cells (LaVail et al., 1991) and is followed by the lateral geniculate in the thalamus (Williams and Rakic, 1988) and finally by the neurons of the striate cortex (Rakic, 1974).

The sequence of synaptogenesis within the microcircuitry of the IPL, as well as the finding of the central-to-peripheral sequence of synaptogenesis in the straight signal pathway (Nishimura and Rakic, 1987a,b), stands in contrast to the hypothesis postulating that the sensory periphery determines the pattern of neuronal and synaptic organization in more centrally located structures initially suggested by Bok in 1915. Rather, the results in primates indicate that the onset of synaptogenesis proceeds in a direction opposite to the flow of visual information, making it unlikely that signals from the retinal photoreceptors can initiate the basic pattern of visual connections in the central nervous system which are

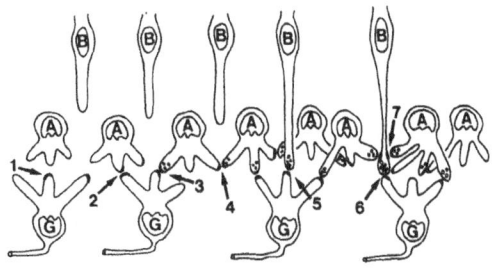

Figure 5. A schematic illustration of the possible sequence of synaptogenesis in the inner plexiform layer of the macaque retina. Abbreviations: A, amacrine cells; B, bipolar cells; G, ganglion cells. (For further information see Nishimura and Rakic, 1985, 1987a,b).

present (Hendrickson and Rakic, 1977; Cooper and Rakic, 1983) and display a correct to-pography (Shatz and Rakic, 1981) before the appearance of the first contacts with the pho-toreceptors in the OPL of the retina. The central-to-peripheral sequence of development is also observed in the timing of cell death in the lateral geniculate nucleus, which starts around E50 and precedes by more than a month the phase of ganglion cell death in the ret-ina (Williams and Rakic, 1988). Finally, independent formation of topographical thalamo-cortical and corticocortical connections has also been observed in the visual system of monkeys with early binocular enucleation (Rakic, 1988; Kuljis and Rakic, 1991; Rakic et al., 1991; Rakic and Lidow, 1995; Kennedy and Dehay, 1993, 1997) and in congenitally anophthalmic mice (Kaiserman-Abramof and Graybiel, 1980; Olivaria and Van Sluyters, 1984). It should, however, be recognized that at later stages of synaptogenesis, particularly at postnatal ages, the periphery begins to play a more significant role in the process of elimination, and maintenance of already formed synapses (Bourgeois and Rakic, 1993; Rakic et al., 1994). Thus, the final pattern of synaptic architecture in the centrally located structures must take into account reciprocal interactions between the photoreceptors, on the one side, and the cerebral cortex, on the other side, of the primary visual pathway.

REFERENCES

Algan, O., and Rakic, P. (1997) Radiation-induced area- and lamina-specific deletion of neurons in the primate visual cortex. *J.Comp. Neurol.* 381: 335–352.

Antonini, A., and Stryker, M.P. (1993) Development of individual geniculocortical arbors in cat striate cortex and effects of binocular impulse blockade. *J. Neurosci.* 13: 3549–5373.

Bok, S.T. (1915) Stimulogenous fibrillation as the cause of the structure of the nervous system. Psychiatrishe en Neurologishe Bladen 19: 393–408.

Bolz, J., Novak, N., and Staiger, V. (1992) Formation of specific afferent connections in organotypic slice cultures from rat visual cortex cocultured with lateral geniculate nucleus. *J Neurosci* 12: 3054–3070.

Bourgeois, J.-P., and Rakic, P. (1993) Changing of synaptic density in the primary visual cortex of the rhesus mon-key from fetal to adult stage. *J. Neurosci.* 13: 2801–2820.

Campbell, G., Ramoa, A.S., Striker, M.P., and Shatz, C.J. (1997) Dendritic development of retinal ganglion cells after prenatal intracranial infusion of tetradotoxin. Vis. Neurosci. 14: 779–788.

Chalupa, L.M., and Williams, R.W., and Henderson, T. (1984) Binocular interaction in the fetal cat regulates the size of the ganglion cell population. *Neurosci., 12*: 1139–1146.

Chalupa, L.M., and Williams, R.W. (1984) Organization of the cat's lateral geniculate nucleus following interrup-tion of prenatal binocular competition. *Human Neurobiology 3*: 103–107.

Cooper, M.L., and Rakic, P. (1983) Gradients of cellular maturation and synaptogenesis in the superior colliculus of the fetal rhesus monkey. *J. Comp. Neurol.* 215: 165–186.

Curcio, C.A., Allen, K.A., Sloan, D.R., Lerea, C.L., Hurley, J.B., Klock, I.B., and Milam, A.H. (1991) Distribution and morphology of human cone photoreceptors stained with antiblue opsin. *J. Comp.* 312: 610–624.

Dehay, C., Horsburgh, O., Berland, M., Killackey, H., and Kennedy, H. (1989) Maturation and connectivity of the visual cortex in monkey is altered by prenatal removal of retinal input. *Nature* 337:265–267.

DeMonasterio, F.M. (1978) Properties of concentrically organized X and Y ganglion cells of macaque retina. *J. Neurophysiol.* 41: 1394–1417.

Easter, S.S., Jr., Purves, D., Rakic, P., and Spitzer, N.C. (1985) The changing view of neural specificity. *Science* 230: 507–511.

Feller, M.B., Butts, D.A., Aaron, H.L., Rokhsat, D.S., and Shatz, C.J. (1997) Dynamic processes shape spatiotem-poral properties of retinal waves. *Neuron, 19*: 293–306.

Garraghty, P.E., and Sur, M. (1993) Competitive interactions influencing the development of retinal axonal arbors in cat lateral geniculate nucleus. *Physiol Rev.* 73 :529–545.

Goodman, C.S., and Shatz, C.J. (1993) Developmental mechanisms that generate precise patterns of neuronal con-nectivity. *Cell* 72: 77–98.

Hendrickson, A., and Rakic, P. (1977) Histogenesis and synaptogenesis in the dorsal lateral geniculate nucleus (LGd) of the fetal monkey brain. *Anat. Rec.* 187: 602.

Holt, C.E., Bertsch, T.W., Ellis, H.M., and Harris, W.A. (1988) Cellular determination in the Xenopus retina is independent of lineage and birth date. *Neuron* 1: 15–26.

Kaiserman-Abramof, I.R., Graybiel, A.M., and Nauta, W.J.H. (1989) The thalamic projection to cortical area 17 in congenitally anophthalmic mouse strain. Neurosci. 5: 41–52.

Kennedy H., and Dehay, C. (1997) The nature and nurture of cortical development. (In: Normal and Abnormal Development of the Cortex. (Galaburda A and Christen Y eds) Springer, Berlin pp. 25–56.

Kennedy, H., and Dehay, C. (1993) Cortical specification of mice and men. Cerebral Cortex 3: 171–186.

Kuljis, R.O., and Rakic, P. (1991) Hypercolumns in primate visual cortex develop in the absence of cues from photoreceptors. *Proc. Nat. Acad. Sci. USA* 87: 5303–5306.

LaVail, M.M., Fletcher, J., Rapaport, D.H., and Rakic, P. (1991) Cytogenesis in the monkey retina. *J. Comp. Neurol.* 309: 86–114.

Linden, D.C., Guillery, R.W., and Cucchiaro, J. (1981) The dorsal lateral geniculate nucleus of the normal ferret and its postnatal development. *J. Comp Neurol.* 203: 189–211.

Livingstone, M.S., and Hubel, D.H. (1988) Segregation of form, color, movement, and depth: Anatomy, physiology and perception. *Science* 240:740–749.

Maunsell, J.H.R. (1992) Functional visual streams. *Current Opinion in Neurobiology* 2: 502–510.

McLoon, S.C., and Barnes, R.B. (1989) Early differentiation of retinal ganglion cells: an axonal protein expressed by premigratory and migrating retinal ganglion cells. *J. Neurosci.* 9: 1424–1432.

Meissirel, C., Wikler, K.C., Chalupa, L.M., and Rakic, P. (1997) Early divergence of M and P visual subsystems in the embryonic primate brain. Proc. *Nat. Acad. Sci.(USA)*, 94: 5900–5905.

Mombaerts, P., Wang, F., Dulac, C., Chao, S.K., Nemes, A., Mendelsohn, M., Edmondson, J., and Axel, R. (1996) Complete record visualizing an olfactory sensory map. Cell. 87: 675–686.

Mrzljak, L., Levey, A.I., and Rakic, P. (1996) Selective expression of m2 muscarinic receptor in parvocellular channel of the primate visual cortex *Proc. Nat. Acad. Sci.(USA)*, 93: 7337–7340.

Nishimura, Y., and Rakic, P. (1985) Development of the rhesus monkey retina: I. Emergence of the inner plexiform layer and its synapses. *J. Comp. Neurol.* 241: 420–434.

Nishimura, Y. and Rakic, P. (1987a) Development of the rhesus monkey retina: II. A three-dimensional analysis of the sequences of synaptic combinations in the inner plexiform layer. *J. Comp. Neurol.* 262: 290–313.

Nishimura, Y. and Rakic, P. (1987b) Synaptogenesis in the primate retina proceeds from the ganglion cells toward the photoreceptors. *Neurosci. Res. Suppl.* 6: 253–268.

Olavarria, J., and Van Sluyters, R.C. (1984) Callosal connections of the posterior neocortex in normal-eyed, congenitally anophthalmic and neonatally enucleated mice. *J. Comp. Neurol.* 230, 249–268.

Perry, V.H., Oehler, R., and Cowey, A. (1984) Retinal ganglion cells that project to the dorsal lateral geniculate nucleus in the macaque monkey. *Neuroscience* 12: 1101–1123.

Perry, V.H., and Cowey, A. (1984) Retinal ganglion cells that project to the superior colliculus and pretectum in the macaque monkey. *Neuroscience* 12: 1125–1137.

Polyak, S.L. (1957) The Vertebrate Visual System. Chicago: University of Chicago Press.

Rakic, P. (1985) Limits of neurogenesis in primates. *Science* 227: 154–156.

Rakic, P. (1972) Mode of cell migration to the superficial layers of fetal monkey neocortex. *J. Comp. Neurol.* 145: 61–84.

Rakic, P. (1973) Kinetics of proliferation and latency between final cell division and onset of differentiation of cerebellar stellate and basket neurons. *J. Comp. Neurol.* 147: 523–546.

Rakic, P. (1975) Timing of major ontogenetic events in the visual cortex of the rhesus monkey. In: *Brain Mechanisms in Mental Retardation*. (N.A. Buchwald and M. Brazier, eds.) Academic Press, New York, pp. 3–40.

Rakic, P. (1976) Prenatal genesis of connections subserving ocular dominance in the rhesus monkey. *Nature* 261: 467–471.

Rakic, P. (1977a) Genesis of the dorsal lateral geniculate nucleus in the rhesus monkey: site and time of origin, kinetics of proliferation, routes of migration and pattern of distribution of neurons. *J. Comp. Neurol.* 176: 23–52.

Rakic, P. (1977b) Prenatal development of the visual system in the rhesus monkey. *Phil. Trans. Roy. Soc. Lond. B.* 278: 245–260.

Rakic, P. (1983) Geniculo-cortical connections in primates: Normal and experimentally altered development. *Progress in Brain Res.* 58: 393–404.

Rakic, P. (1981) Development of visual centers in the primate brain depends on binocular competition before birth. *Science* 214: 928–931.

Rakic, P. (1988) Specification of cerebral cortical areas. *Science* 241: 170–176.

Rakic, P. (1992) Development of primate visual system: From photoreceptors to cortical modules. In: Visual System from Genesis to Maturity. R. Lent, ed. Birkhauser, Boston, pp. 1–17.

Rakic, P., and Lidow, M.S. (1995) Distribution and density of neurotransmitter receptors in the visual cortex devoid of retinal input from early embryonic stages. *J. Neurosci.*,15: 2561–2574

Rakic, P., and Riley, K.P. (1983) Regulation of axon numbers in the primate optic nerve by prenatal binocular competition. *Nature* 305: 135–137.

Rakic, P., and Riley, K.P. (1983) Overproduction and elimination of retinal axons in the fetal rhesus monkey. *Science* 209: 1441–1444.

Rakic, P. (1974) Neurons in the monkey visual cortex: Systematic relation between time of origin and eventual disposition. *Science* 183: 425–427.

Rakic, P., Suner, I., and Williams, R.W. (1991) A novel cytoarchitectonic area induced experimentally within the primate visual cortex. *Proc. Nat. Acad. Sci. (USA)*, 88: 2083–2087.

Rapaport, D.H., LaVail, M.M., and Rakic, P. (1992) Genesis of subclasses of neurons in the retinal ganglion cell layer of the monkey. *J. Comp. Neurol.* 322:577–588.

Rapaport, D.H., Rakic, P., and LaVail M. (1996) Spatiotemporal gradients of cell genesis in the primate retina. *Perspective on Dev. Neurobiol.* 3: 142–159.

Rapaport, D.H., Rakic., P., Yasamura, D., and LaVail, M.M. (1995) Genesis of the retinal pigment epithelium in the macaque monkey. *J. Comp. Neurol.* 363: 359–376.

Reese, F.R. (1996) Chromotopic order of optic axons. *Presp. Dev. Neurobiol.* 3: 233–242.

Rodieck, R.W. (1988) The Vertebrate Visual System. University of Chicago Press. Chicago.

Rowe, M.H., and Stone, J. (1980) The interpretation of variation in the classification of nerve cells. *Brain Behav. Evol.* 17: 123–151.

Shapley, R., and Perry, V.H. (1986) Cat and monkey retinal ganglion cells and their visual functional roles. *Trends Neurosci.* 9: 229–235.

Shatz, C.J. (1983) Prenatal development of cat's retinogeniculate pathway. *J. Neurosci.* 3: 482–499.

Shatz, C., and Rakic, P. (1981) The genesis of efferent connections from the visual cortex of the fetal rhesus monkey. *J. Comp. Neurol.* 196: 287–307.

Shatz, C.J. (1996) Emergence of order in visual system development. *Proc. Nat. Acad. Sci., USA* 93: 602–608.

Sidman, L.R. (1970) Autoradiographic methods and principles for study of the nervous system with thymidine H³. In: Contemporary Research Methods in Neuroanatomy, W.J.H. Nauta and J.O.E. Ebbeson, eds., pp 225–274. Springer, Berlin.

Skaliora, I., Scobey, R.P., and Chalupa, L.M. (1993) Prenatal development of excitability in cat retinal ganglion cells: action potentials and sodium. *J. Neurosci.* 13: 313–323.

Sretavan, D.W., and C.J. Shatz (1984) Prenatal development of individual retinogeniculate axons during the period of segregation. *Nature* 308: 845–848.

Stone, J. (1983) Parallel processing in the visual system. New York: Plenum.

Stryker, M.P., and Harris, W.A. (1986) Binocular impulse blockade prevents the formation of ocular dominance columns in cat visual cortex. *J. Neurosci.* 6: 2117–2133.

Suner, I., and Rakic, P. (1996) Numerical Relationship between Neurons in the Lateral Geniculate Nucleus and Primary Visual Cortex in Adult Macaque Monkeys. *Visual Neurosci.*, 13: 585–590.

Sur, M., Humphrey, A.L., and Sherman S.M. (1982) Monocular deprivation affects X- and Y-cell retinogeniculate terminations in cats. *Nature.* 300: 183–185.

Trisler, D., Rutin, J., and Pessac, B. (1996). Retinal engineering: Engrafted neural cell lines locate in appropriate layers. *Proc Natl Acad Sci USA* 93: 6269–6274.

Turner, D.L., and Cepko, C.L. (1987) A common progenitor for neurons and glia persists in rat retina late in development. *Nature* 328: 131–136.

Wetts, R., and Fraser, S.E. (1988) Multipotent precursors can give rise to all major cell types of the frog retina. *Science* 239: 1142–1145.

Wikler, K.C., and Rakic, P. (1990) Distribution of photoreceptor subtypes in the retina of diurnal and nocturnal primates. *J. Neurosci.* 10: 3390–3400.

Wikler, K.C., and Rakic, P. (1991) Emergence of the photoreceptor mosaic from a protomap of early-differentiating cones in the primate retina. *Nature* 351: 397–400.

Wikler, K.C., and Rakic, P. (1996) Development of photoreceptor mosaic in the primate retina *Perceptive on Developmental Neurobiology,* 3: 161–175.

Wikler, K.C., Rakic, P., and Barnstable, C. (1996) Differential onset of cone opsin expression in the fetal monkey retina. *Invest. Ophthal. Vis. Sci. Abstr.*, 35, S693.

Wikler, K.C., Rakic, P., Bhattacharyya, N., and MacLeish, P.R. (1997) A novel cone-specific monoclonal antibody, 7G6, identifies sub-populations of cones in the fetal monkey retina *J. Comp. Neurol.*, 377: 500–508.

Wikler, K.C., Williams, R.W., and Rakic, P. (1990) The photoreceptor mosaic: Number, distribution, and patterns of rods and cones in the rhesus monkey retina. *J. Comp. Neurol.* 297: 499–508.

Wikler, K.S., and Rakic, P. (1994) An array of early-differentiating cones precedes the emergence of the photoreceptor mosaic in the fetal monkey retina. *Proc. Nat. Acad. Sci.USA* 91: 6534–6538.

Williams, R.W., and Chalupa, L.M. (1982) Prenatal development of retinocollicular projections in the cat: An anterograde tracer transport study. *J. Neurosci.* 2: 604–622.

Williams, R.W., and Rakic, P. (1985) Dispersion of growing axons within the optic nerve of the embryonic monkey. *Proc. Natl. Acad. Sci. USA* 82: 3906–3910.

Williams, R.W., and Rakic, P. (1988) Elimination of neurons in the rhesus monkey's lateral geniculate nucleus during development. *J. Comp. Neurol.* 272: 424–436.

Williams, R.W., Borodkin, M., and Rakic, P. (1991) Growth cone distribution patterns in the optic nerve of fetal monkeys: Implications for mechanisms of axonal guidance. *J. Neurosci.* 11: 1081–1094.

Williams, R.W., and Goldowitz, D. (1992) Structure of clonal and polyclonal arrays in chimeric mouse retina. *Proc. Natl. Acad. Sci. USA* 89: 1164–1188.

Wong, R.O., Chernjavsky, A., Smith, S.J., and Shatz, C.J. (1995) Early functional neural networks in the developing retina. *Nature.* 374: 716–718.

Xiang, M., Zhou, L., Macke, J., Yoshioka, T., Hendry, S.H.C., Eddy, R., Shows, T.B., and Nathans, J. (1995) The Brn-3 family of POU-domains factors: Primary structure, binding specificity, and expression in subsets of retinal ganglion cells and somatosensory neurons. *J. Neuroci.* 15: 4762–4785.

6

DEVELOPMENT OF ON AND OFF RETINAL GANGLION CELL MOSAICS

Leo M. Chalupa,[1] Gayathri Jeyarasasingam,[2] Cara J. Snider,[1] and Stefan R. Bodnarenko[3]

[1]Center for Neuroscience
Section of Neurobiology, Physiology, and Behavior, and Department of Psychology
University of California
Davis, California 95616
[2]The Parkinson's Institute
1170 Morse Ave.
Sunnyvale, California 94089
[3]Clark Science Center
Department of Psychology
Smith College
Northampton, Massachusetts 01063

1. INTRODUCTION

The regularity of cells in the vertebrate retina was first recognized in the mid-nineteenth century by Hannover (1843) who noted that "in many animals, double and single cones of the retina form a definite pattern." Numerous investigators have since described the mosaics formed by cell populations in all three retinal nuclear layers (e.g., Wässle and Riemann, 1978; Young and Vaney, 1991; Cook and Becker, 1991; Hutsler and Chalupa, 1994). Such regular retinal arrays are thought to be necessary for the efficient functioning of the visual system. In particular, computer simulations and mathematical modeling have shown that the orderly distribution of photoreceptors is necessary for the adequate detection of spatial information (French et al., 1977). To preserve the integrity of such information, one would also expect complementary distributions in other retinal layers. Furthermore, as all visual information received by photoreceptors is conveyed to the brain via retinal ganglion cells (RGCs), these cells would be expected to possess a highly organized distribution pattern to ensure topographic input to visual target regions (Wässle and Riemann, 1978; Hirsch and Hylton, 1984). Moreover, as discussed by Jeremy Cook in this volume, it has been suggested that the presence of a regular distribution of cells is a determining factor for neuronal classification in the retina (Peichl, 1991; Wässle and Boycott, 1991).

Development and Organization of the Retina, edited by Chalupa and Finlay.
Plenum Press, New York, 1998.

In view of the prevalence of cell mosaics in the mature retina and the presumed importance of such distributional patterns for visual information processing, it is quite astonishing that, until recently, so little attention has been directed at the development of this fundamental feature of retinal organization. Several years ago, we began to consider this issue (Chalupa, 1995) by addressing two related questions. First, we wondered when mosaic patterns can first be discerned in the developing retina; and second, we sought to gain an understanding of the developmental factors that could underlie the formation of regular cell distributions.

For the most part, our studies have focused on alpha and beta ganglion cells since these two cell classes in the mature cat retina have been extensively studied with respect to their distributional patterns (Wässle et al., 1981a,b; Kirby and Chalupa, 1986). Specifically, it has been well-established that the ON and OFF subclasses of alpha and beta cells form independent arrays, providing complete coverage of the retinal surface by their dendritic arbors. Each subclass responds selectively to light increments (ON) or decrements (OFF), as was initially shown by Kuffler (1953) in his pioneering work dealing with the receptive field properties of cat retinal ganglion cells.

In the mature retina, ON and OFF cells can be differentiated morphologically on the basis of their dendritic stratification patterns within the inner plexiform layer (IPL): ON cell dendrites branch proximal and OFF cell dendrites branch distal to their respective somas situated in the ganglion cell layer (Nelson et al., 1978). This organizational feature is illustrated in Figure 1 which depicts DiI labeled ganglion cells with two clearly defined strata of dendrites within the IPL.

By contrast, early in development ON and OFF cells cannot be differentiated from each other because the dendrites of immature ganglion cells ramify throughout the IPL

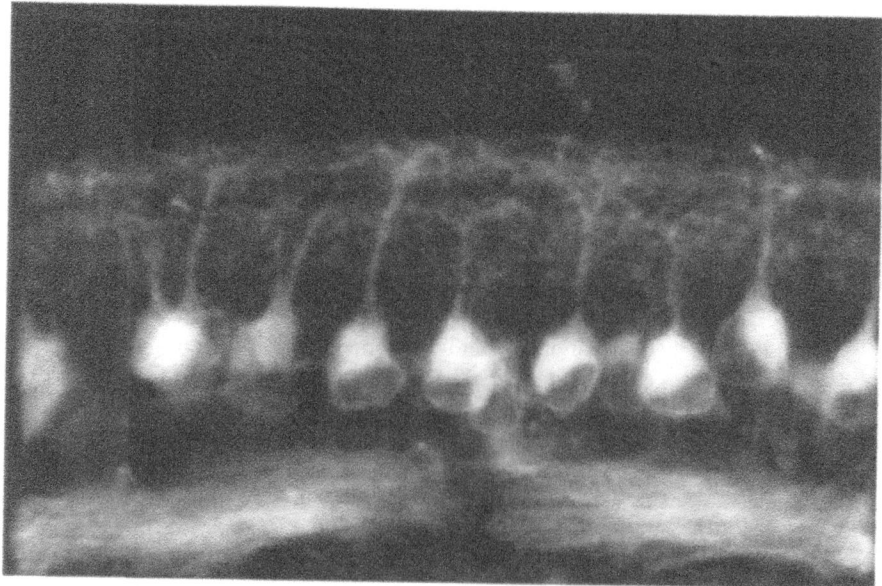

Figure 1. Photomicrograph showing DiI labeled RGCs in transverse section from a P13 cat retina. Note the distinct appearance of two tiers of dendrites within the IPL and the stratification pattern of ON (proximal to the soma) and OFF (distal to the soma) RGC dendrites.

(Maslim and Stone, 1988; Ramoa et al., 1988; Bodnarenko et al., 1995). Thus, a necessary prerequisite for the formation of ON and OFF ganglion cell mosaics is for the dendrites of these neurons to change from a multistratified to a unistratified state.

It has also been well-established that there is a massive loss of ganglion cells during the normal development of the cat (Chalupa, 1988). Some degree of ganglion cell loss is thought to represent the correction of misprojections in developing retinofugal pathways. This could be the case in fetal cats since retinal ganglion cells projecting either to the wrong hemisphere or inappropriate loci within retinorecipient target nuclei have been documented (Williams and Chalupa, 1982). However, by birth, the available evidence indicates that the pattern of retinal projections in carnivores is virtually identical to that found in the mature animal (Chalupa et al., 1996; Chalupa and Snider, 1998). Consequently, the significance of the postnatal loss of ganglion cells presents a conundrum. The results of recent studies we will describe here provide an answer to this puzzle: ganglion cell loss in the postnatal cat retina serves to refine the early distributions of ON and OFF cells to form the regular mosaic patterns essential for the normal processing of visual information. Before considering this issue, we will provide an account of our work dealing with the stratification of dendrites in the developing cat retina.

2. STRATIFICATION OF RETINAL GANGLION CELL DENDRITES

As indicated above, whereas in the mature cat retina, alpha and beta cells can be subdivided into ON and OFF subclasses, developing ganglion cells cannot be differentiated in this manner because of their initially multistratified dendritic branching patterns (Maslim and Stone, 1988; Ramoa et al., 1988; Bodnarenko et al., 1995). Examination of DiI labeled retinal cross-sections at different stages of development has revealed that by embryonic day (E) 50 virtually all beta cells are multistratified. This is two weeks before birth and the youngest age at which the three major ganglion cell classes can be distinguished in the cat retina (Ramoa et al., 1988). The stratification process was found to proceed rapidly so that by the end of the second postnatal week relatively few beta cells were found to be multistratified in the central region of the retina (Bodnarenko et al., 1995). The timing of this event appears to coincide with synaptogenesis in the IPL (Maslim and Stone, 1986), suggesting a role for afferent cells in regulating dendritic stratification. The unique actions of the glutamate analog, 2-amino-4-phosphonobutyrate (APB), which hyperpolarizes rod bipolar and ON-cone bipolar cells selectively (Slaughter and Miller, 1981), enabled us to investigate such a role for these afferent cells. APB blocks the release of glutamate by these interneurons thereby abolishing all visual responses in ganglion cells of dark-adapted animals (Wässle et al., 1991). Intraocular injections of APB, performed during the time period of normal dendritic stratification, resulted in a virtually total arrest of this developmental process.

Figure 2A shows a cross-section of a normal P13 retina in which the dendrites of each ganglion cell can be seen to terminate in either the ON or OFF sublamina of the IPL. In marked contrast, Figure 2B illustrates a similar cross-section of a P13 retina treated with APB since P2. Here, no such sublamination of the IPL is visible; the dendrites of each cell ramify throughout the IPL. As shown in Figure 3, the incidence of multistratified cells was approximately 40% at P2, the age at which APB treatments were initiated. This incidence of multistratified cells was not decreased appreciably when daily APB treatment was continued as late as P13. By contrast, at P13 in the normal retina, only about 12% of the ganglion cells are still multistratified.

These findings provide the first indication that activity plays a role in the dendritic remodeling of developing ganglion cells. More specifically, these results suggest that glutamate-mediated afferent activity regulates the dendritic stratification process. Interestingly, manipulations which are known to alter the formation of eye-specific domains in

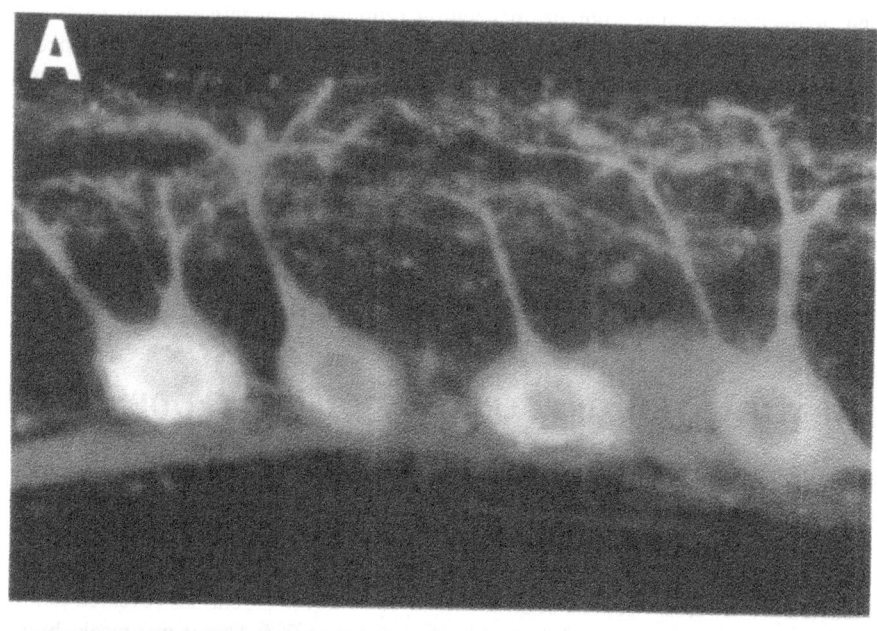

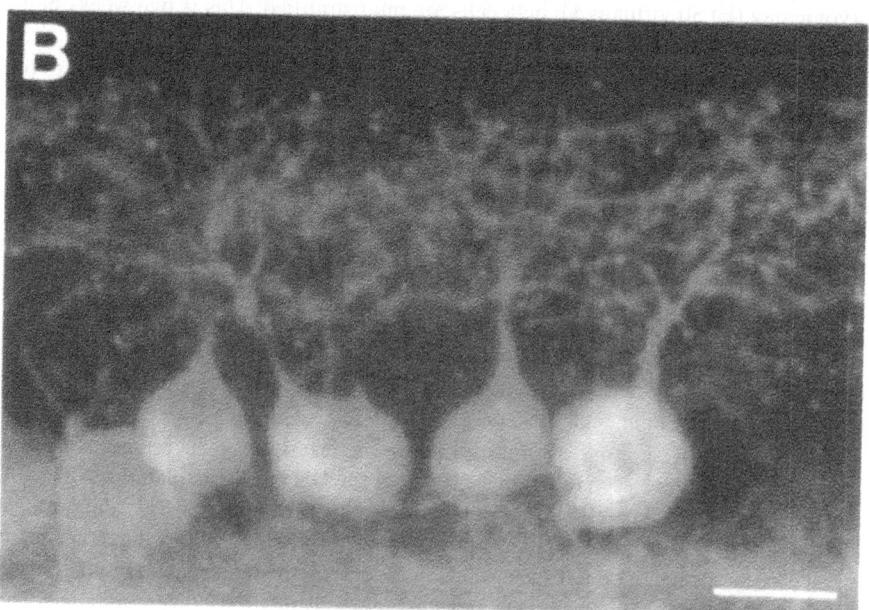

Figure 2. Examples of DiI labeled beta ganglion cells in transverse section from a normal P13 retina (A) and an APB-treated retina (B). Scale bar = 20 μm. Reprinted from *J. Neuroscience* (Bodnarenko et al., 1995).

visual target regions, such as intraocular injection of tetrodotoxin (TTX) and monocular deprivation, do not affect the normal restriction of RGC dendritic processes in the developing cat retina (Leventhal and Hirsch, 1983; Dubin et al., 1986; Lau et al., 1990; Wong et al., 1991). Thus, it is not ganglion cell activity *per se* but rather pre-synaptic afferent activity which appears to regulate the maintenance or elimination of RGC dendritic processes. Moreover, RGC density as well as somal and dendritic field sizes were unaffected following APB treatment, demonstrating that such afferent input has a highly selective impact on RGC dendritic development (Bodnarenko et al., 1995).

When short-term APB treatments (P2 to P13) were terminated, the dendritic stratification process was found to resume. At three months of age there were very few multistratified cells in the treated eyes as is the case in the normal adult cat retina (Bodnarenko et al., 1995). Recently, in a collaborative study with the laboratory of Silvia Bisti in Pisa we have found that APB treatment throughout the first postnatal month results in what appears to be the permanent arrest of dendritic stratification. This provided an opportunity to examine the visual response properties of the APB-treated eye. The obvious question we were interested in addressing was whether or not the presence of ganglion cells with multistratified dendrites resulted in receptive fields with ON–OFF discharge patterns. Extracellular recordings from the A or A1 laminae of the dorsal lateral geniculate nucleus innervated by the APB-treated eye, as well as recordings from the optic tract, revealed that this was indeed the case. Whereas virtually all of the cells driven by the normal eye responded as expected with either ON or OFF discharges, in the case of the treated eye about 40% of the units manifested ON–OFF discharge patterns (Bisti et al., forthcoming). These observations demonstrate a clear-cut functional correlate for the morphological changes observed in ganglion cells following APB treatment of the developing retina. Moreover, they imply that at maturity the dendrites of these multistratified cells are innervated by axon terminals of ON as well as OFF bipolar cells.

3. A ROLE FOR CELL DEATH IN MOSAIC FORMATION

Ganglion cell death has been documented in the developing cat retina by assessment of optic nerve fiber number (Williams et al., 1986) and the presence of pyknotic profiles

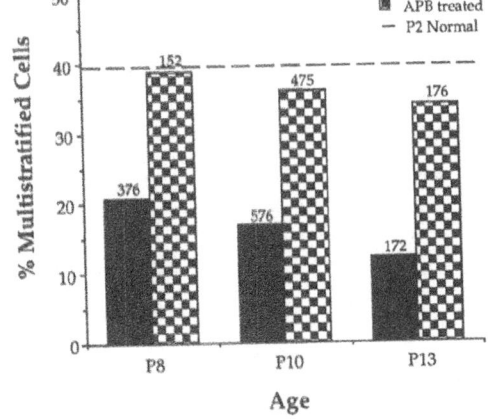

Figure 3. A comparison of the incidence of multistratified RGCs in normal and APB-treated retinas at three postnatal ages. The dotted line denotes the incidence of multistratified cells at P2, the age when treatment was initiated. Reprinted from *J. Neuroscience* (Bodnarenko et al., 1995).

(Wong and Hughes, 1987; Pearson et al., 1993). In particular, optic nerve counts have revealed that most ganglion cells die during embryonic life, and the period of ganglion cell loss continues during the first postnatal month. However, neither of these measures can provide an indication of the degree to which the different classes of cells contribute to the overall magnitude of ganglion cell loss.

By counting all alpha cells within the central region of the developing cat retina, we have recently shown that approximately 20% of these neurons are eliminated during the first postnatal month (Jeyarasasingam et al., 1998). Because the central region of the retina does not expand during this developmental period (Mastronarde et al., 1984; Jeyarasasingam et al., 1998), it can be inferred that this loss of cells must reflect the normal death of these neurons. Moreover, the postnatal period of ganglion cell loss continues after most ganglion cells have completed their stratification process. Thus, ON and OFF subclasses can be differentiated before the final number of ganglion cells is established. This leads to an intriguing question: are regular mosaics present in the retina at a time when there is an "excess" of ganglion cells. In considering this matter, one of two possible scenarios can be envisaged. As depicted in Figure 4: (i) regular distributions of cells might be present even though the number of cells is higher than normal; or (ii) cell regularity could be "masked" by the excess cells. In the latter case, the loss of neurons would contribute to the formation of cell mosaics.

The resolution of this matter seemed rather straightforward: Label all ganglion cells in a large region of the retina so that ON and OFF cells could be distinguished, and then compare the mosaics in the developing retina with those present at maturity. For technical reasons, however, it has been problematic to label the dendrites of a large number of ganglion cells sufficiently well so as to allow classification of these neurons into ON or OFF subtypes. Consequently, it has not been feasible to directly assess mosaic patterns in the developing retina.

To overcome this problem, we relied on the common observation that ON and OFF RGCs of a given class are often situated in close proximity to one another (Wässle et al., 1981a,b). By means of computer simulations we first showed that the superimposition of two regular distributions consistently resulted in around 90% opposite sign pairing (Figure 5). By contrast, the superimposition of two random distributions repeatedly resulted in only 50% of such pairs. This relationship between the incidence of opposite sign cell pairs and the degree of regularity exhibited by two superimposed distributions was remarkably robust over a relatively broad range of cell densities, approximating those found from the central to the peripheral retina.

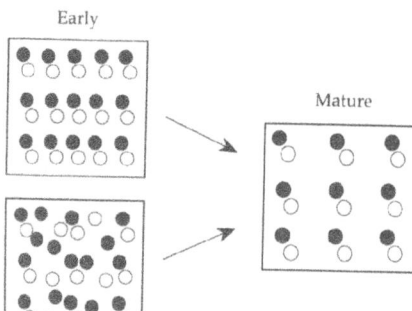

Figure 4. Schematic diagrams showing two alternative scenarios for ON and OFF cell distributions before the period of cell death is completed (left panels) and at maturity (right panel). At the top, the distribution of these neurons is as regular as at maturity, even though cell density is substantially greater. At the bottom, the "excess" cells obscure the regular patterns of ON and OFF cells so that cell mosaics only become evident following a spatially selective pattern of cell loss.

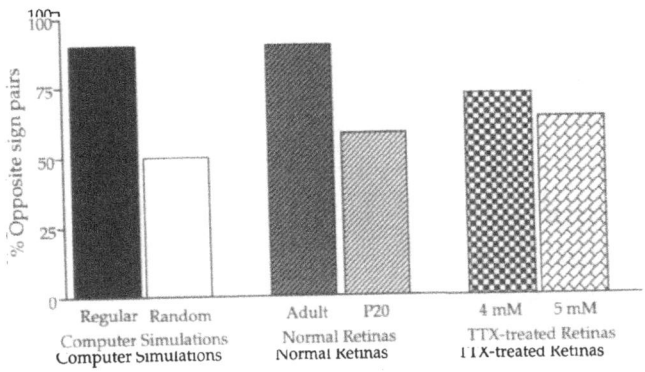

Figure 5. Bar graph depicting the incidence of opposite sign cell pairs resulting from the superimposition of two regularly distributed computer simulated regions (A), two randomly distributed regions (B), the number of such pairs in the normal adult retina (C), in the developing retina (D), and in TTX-treated retinas (E). Note the close correspondence between the incidence of ON-OFF pairs in the regularly distributed simulations and the mature retina as well as that between the random simulations, the developing retina and following TTX-treatment.

Using an *in vitro* wholemount preparation we were able to obtain Golgi-like labeling of a relatively small number of ganglion cells by making focal deposits of horseradish peroxidase (HRP) into the fiber layer. Although, not suitable for assessing mosaic patterns using conventional measure of regularity (see Cook, this volume) this material permitted us to quantify the incidence of opposite sign pairs in the developing cat retina. This approach revealed that only 58% of alpha cell pairs are of opposite sign before the developmental period of cell death has ended, suggesting that at this stage the distribution of these neurons is not appreciably different from random (Figure 5). By contrast, in the mature retina the incidence of such opposite sign pairs was found to be around 90%, as predicted by our computer simulations.

Having demonstrated that ON and OFF alpha cell distributions become more regular during postnatal development, we next considered the possibility that this process could be regulated by sodium voltage-gated retinal activity. This would be the case if the spatial pattern of ganglion cell loss in the developing retina was dependent on activity-mediated mechanisms involving the firing of action potentials (O'Leary et al., 1986a,b; Thompson and Holt, 1989). Accordingly, we treated postnatal cat retinas with TTX beginning at P9, when the density of alpha cells is greater than at maturity and before the adult complement of opposite sign alpha cell pairs is established. (Details of injection protocol provided in Jeyarasasingam et al. in press.) When these animals reached maturity, we examined both the incidence of opposite sign cell pairs and the regularity indices of the resulting distributions. Unlike in the developing retina, it is feasible to label large regions of the mature retina so as to differentiate between ON and OFF ganglion cells, permitting the calculation of regularity indices.

Figure 6A illustrates a region from a control retina in which the adult complement of opposite sign pairs (~90%) is evident (see also Figure 4). In contrast, the TTX treated retinal region in Figure 6B demonstrates only 60% opposite sign pairs, a value comparable to that seen in the developing retina (see also Figure 5). Similarly, the regularity indices for the ON and OFF cell distributions from Figure 6A were 3.64 and 3.59, respectively, whereas the TTX treated retinas display regularity indices of 2.6 for each cell population illustrating a more disorderly pattern. At the same time, the density of alpha cells in the TTX-treated retinas were within normal values suggesting that sodium channel blockade

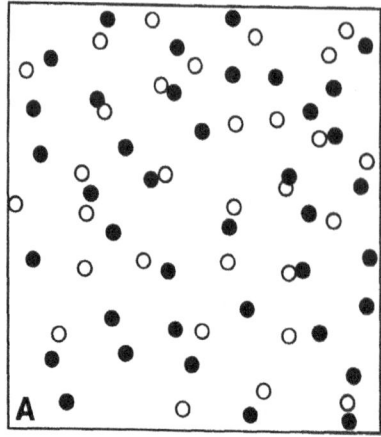

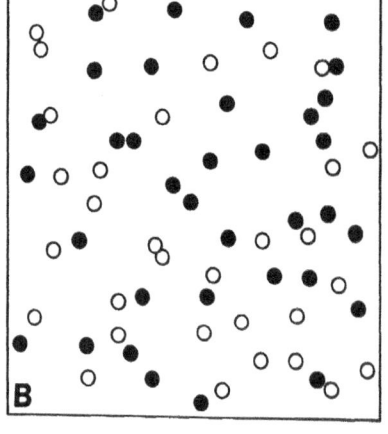

Figure 6. Examples of alpha cell distribution patterns from normal (A) and TTX-treated (B) adult retinas. Note the decreased incidence of opposite sign pairs and the lower regularity indices in (B) as compared to (A). Scale bar = 250 μm.

altered the pattern but not the magnitude of cell loss in the developing retina. In a recent study, we have further shown that the normal magnitude of cell loss observed during RGC development is sufficient to produce a regular distribution pattern from a random one (Jeyarasasingam et al., 1998). Collectively, these findings indicate that spatially selective cell death plays a key role in the formation of RGC mosaics and that this process is regulated by sodium-voltage gated activity.

4. DISCUSSION

The results of the studies summarized above indicate that the formation of retinal ganglion cell mosaics involves two developmental events: (i) the restriction of initially multistratified dendrites to form morphologically distinct ON and OFF cells and (ii) the selective elimination of ganglion cells to change the ON and OFF distribution patterns from random to regular. Figure 7 illustrates these two mechanisms in schematic form.

In the fetal cat, virtually all ganglion cells possess multistratified dendrites (Bodnarenko et al., 1995) and, therefore, ON and OFF cells cannot be morphologically identified (Figure 7A). Beginning in late embryonic life and continuing postnatally, RGCs begin

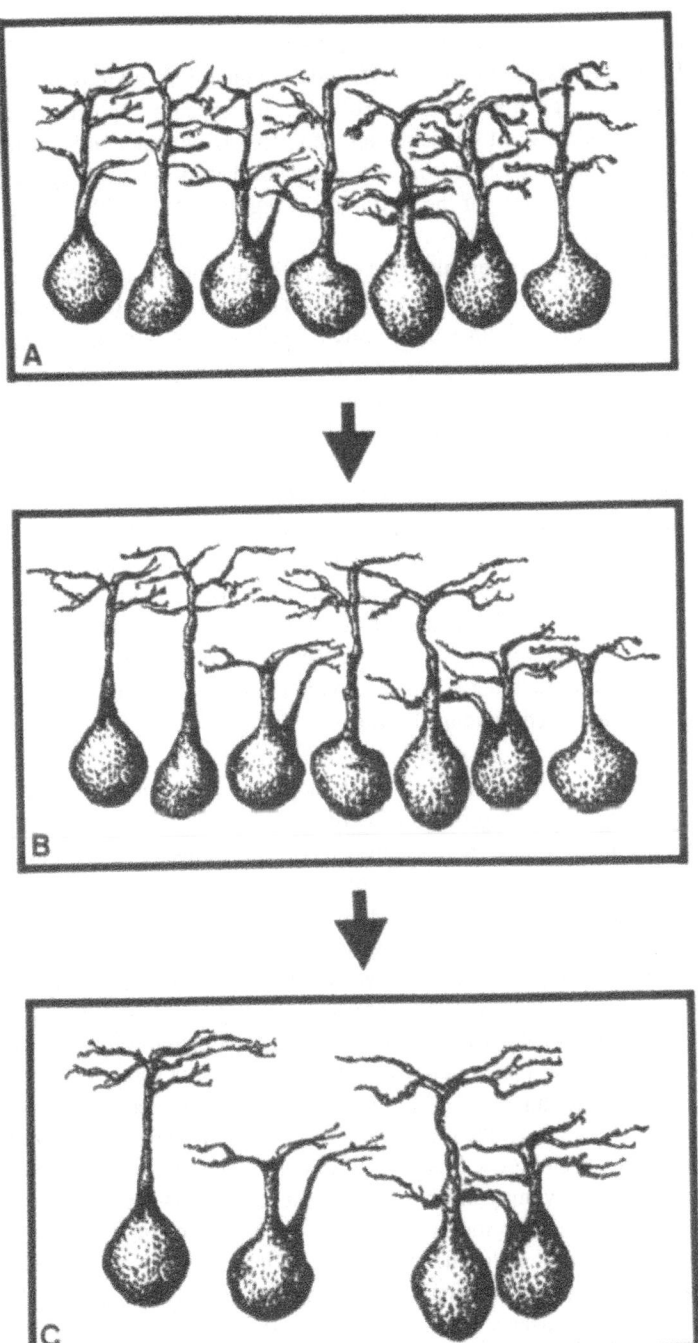

Figure 7. Schematic diagram illustrating the developmental events contributing to the formation of ON and OFF RGC mosaic patterns. In (A), note the multistratified state of RGC dendrites and the larger number of cells present as compared to (C). (B) In the postnatal retina, ganglion cells have stratified to form the ON and OFF subtypes, yet there is still an "excess" of cells present and the incidence of opposite sign pairs is reduced as compared to (C) where the mature complement of cells is organized into ON–OFF pairs signifying regularly distributed ON and OFF RGC populations.

the stratification process establishing the structural signature for ON and OFF cells (Figure 7B). Until this process is completed, ON and OFF ganglion cell mosaics cannot be discerned. During early postnatal life, "excess" cells obscure regular mosaic patterns resulting in a low incidence of opposite sign pairs following dendritic stratification (Figure 7B). Normal retinal activity during this postnatal period induces a spatially selective pattern of cell death, allowing for the formation of regular ON and OFF ganglion cell mosaic patterns (Figure 7C).

These studies have also revealed that two different types of activity-based events are involved in forming ON and OFF ganglion cell mosaic patterns: (i) glutamate-mediated afferent activity and (ii) sodium-voltage gated discharges. It remains to be established, however, how these diverse activity-regulated mechanisms regulate their respective developmental changes. For example, it is clear that blockade of bipolar cell activity with APB prevents ganglion cell dendritic stratification, but how does normal afferent activity direct the retraction of diffusely branching dendrites to allow for the formation of both ON and OFF cells? Perhaps these afferent cells provide selective input to either proximal or distal dendrites of multistratified ganglion cells early in development. If this were the case, the release of glutamate by these afferents could instruct ganglion cells to maintain those processes receiving the necessary input. Alternatively, these ganglion cells may be specified intrinsically as ON or OFF, despite the multistratified dendritic state. In this case, glutamate release from afferents may simply activate a genetic program within a ganglion cell to retract the appropriate dendrites to provide the morphological signature that corresponds to its pre-determined functional state.

These possibilities can be explored by investigating the state of synaptic contacts onto multistratified ganglion cells during development. Localized unistratified synaptic input to ganglion cells would provide evidence for the instructional hypothesis whereas "diffuse" afferent input would more likely support an intrinsic specification hypothesis. Recall that the results of our recent recordings have revealed that multistratified ganglion cells in the APB-treated retina respond to light with ON-OFF discharges. As noted above, this implies that these neurons are innervated by the axonal processes of both ON and OFF bipolar cells. However, it is not known whether this reflects the maintenance of immature bipolar inputs or *de novo* axonal ingrowth in response to the APB treatment. For these reasons, it would be of great interest to establish the pattern of connections between bipolar cells and multistratified ganglion cell dendrites in the developing retina.

Similarly, the mechanisms underlying activity-mediated selective cell death have yet to be explored. In this context, the recent findings of Rachel Wong summarized in this volume, showing that developing ON and OFF ganglion cells in the ferret retina generate separate waves of activity, may be of relevance. Such subclass-specific waves of activity were proposed to underlie the formation of ON and OFF sublaminae in the ferret dorsal lateral geniculate nucleus. Independent waves of ON and OFF cell activity could also serve to regulate the cell loss required to form regular ON and OFF ganglion cell distributions across the retinal surface. Blocking this activity during development alters the pattern of cell death thereby disrupting the formation of ganglion cell mosaics.

The question of how sodium voltage-gated activity regulates the pattern of cell loss, however, remains to be addressed. Three possible activity-mediated mechanisms can be proposed: interactions at the level of ganglion cell afferents, terminals, and/or directly among RGCs. At the level of ganglion cell afferents, it has been shown that neuropeptide Y containing amacrine cells are arranged in mosaics in the inner nuclear layer during the embryonic development of the cat retina, much earlier than we have shown here to be the case for ganglion cells (Hutsler and Chalupa, 1995). The regular distribution pattern of these and other afferent populations could therefore act as a template directing the selective loss of cells resulting in the formation of ganglion cell mosaics. Alternatively, activity-mediated interactions at the level of ganglion cell terminals may direct the appropriate loss of neurons to form RGC mosaics. Retinal activity has been implicated in the removal of inappropriately projecting neurons in the refinement of retinocollicular topography in the rodent visual system (O'Leary et al., 1986a,b; Thompson and Holt, 1989). Though there is a higher degree of topographic precision in the developing cat visual system (Chalupa et al., 1996), a fine tuning of the topographic pattern may occur via activity-mediated interactions. If the inappropriately positioned ganglion cells (which obscure mosaic patterns during development) contribute to topographic imprecision, their removal by such activity-mediated events could refine irregular distribution patterns as well.

Electrical interactions between ganglion cells themselves may also act to regulate the pattern of cell death. For example, alpha ganglion cells are electrically coupled to one another during development (Penn et al., 1994). Perhaps this communication serves to maintain cells within the coupled network at the expense of non-coupled cells. If these non-coupled cells are randomly distributed among a regular array of coupled cells, the removal of these non-coupled cells would result in the formation of mosaics during development.

In order to distinguish between these possibilities, it would be necessary to selectively block activity at each level independently. For example, if activity blockade within retinorecipient nuclei disrupted mosaic formation, this would support a target-mediated mechanism. In contrast, mosaic disruption by blockade of communication between ganglion cells with gap junction inhibitors would suggest that electrical coupling directed the selective loss of cells necessary for mosaic formation.

5. CONCLUDING REMARKS

Mature retinal mosaics are essential for spatial information processing. For this reason alone, it is valuable to understand the mechanisms underlying their formation. The present research has investigated this fundamental feature of retinal organization at a systems level by invoking such ubiquitous developmental phenomena as dendritic restructuring, cell death, and activity-mediated events. It remains for future studies to unravel the cellular and molecular mechanisms behind these events to further our understanding of the formation of retinal mosaics.

ACKNOWLEDGMENTS

We are pleased to thank our colleagues for their valuable contributions to various aspects of this work: Drs. Gimmi Ratto and Silvia Bisti. Supported by National Institutes of Health, National Science Foundation, Fogarty Institute for International Studies and NATO.

REFERENCES

Bodnarenko SR, Jeyarasasingam G, Chalupa LM (1995) Development and regulation of dendritic stratification in retinal ganglion cells by glutamate-mediated afferent activity. Journal of Neuroscience 15(11): 7037–7045.

Chalupa, L.M. Factors underlying the loss of retinal ganglion cells. In: Cell Interactions in Visual Development. Eds. S.R. Hilfer and J.B. Sheffield, Springer Verlag, 1988, 69–86.

Chalupa, L.M. The nature/nuture of retinal ganglion cell development. In: The Cognitive Neurosciences, a Handbook for the Field. Ed. M.S. Gazzaniga, MIT Press, 1995, 37–50.

Chalupa LM, Snider CJ (1998) Topographic specificity in the retinocollicular projection of the developing ferret: An anterograde tracing study. Journal of Comparative Neurology (in press).

Chalupa LM, Snider CJ, Kirby MA (1996) Topographic organization in the retinocollicular pathway of the fetal cat demonstrated by retrograde labeling of ganglion cells. Journal of Comparative Neurology 368: 295–303.

Cook JE, Becker DL (1991) Regular mosaics of large displaced and non-displaced ganglion cells in the retina of the cichlid fish. Journal of Comparative Neurology 306(4): 668–684.

Dubin M, Stark L, Archer S (1986) A role for action potential activity in the development of neuronal connections in the kitten retinogeniculate pathway. Journal of Neuroscience 6: 1021–1036.

French AS, Snyder AW, Stavenga DG (1977) Image degradation by an irregular retinal mosaic. Biological Cybernetics 27: 229–233.

Gargini C, Chalupa LM, and Bisti S (1996) Reorganization of receptive field properties after treatment of the developing retina with APB. Soc. Neuroscience. 22: 1725.

Hannover A (1843) Mikroskopiske undersogelser af nervesystemet. Vid. Sel. Naturvid. Og Mathem. Afh. 10: 9–112.

Hirsch J, Hylton R (1984) Quality of the primate photoreceptor lattice and limits of spatial vision. Vision Research 24: 347–355.

Hutsler JJ, Chalupa LM (1994) Neuropeptide Y immunoreactivity identifies a regularly arrayed group of amacrine cells within the cat retina. Journal of Comparative Neurology 346: 481–489.

Hutsler JJ, Chalupa LM (1995) Development of neuropeptide Y immunoreactive amacrine and ganglion cells in the pre- and postnatal cat retina. Journal of Comparative Neurology 361: 152–164.

Jeyarasasingam G, Snider CJ, Ratto G, Chalupa LM (1998) Activity-regulated cell death contributes to the formation of ON and OFF alpha ganglion cell mosaics. Journal of Comparative Neurology (in press).

Kirby MA, Chalupa LM (1986) Retinal crowding alters the morphology of alpha ganglion cells. Journal of Comparative Neurology 251: 532–541.

Kuffler SW (1953) Discharge patterns and functional organization of mammalian retina. Journal of Neurophysiology 16: 37–68.

Lau K, So K, Tay D (1990) Effects of visual or light deprivation on the morphology and the elimination of the transient features during development of type I retinal ganglion cells in hamsters. Journal of Comparative Neurology 300: 583–592.

Leventhal A, Hirsch H (1983) Effects of visual deprivation upon the morphology of retinal ganglion cells projecting to the dorsal lateral geniculate nucleus of the cat. Journal of Neuroscience 3: 332–344.

Maslim J, Stone J (1986) Synaptogenesis in the retina of the cat. Brain Research 373: 35–48.

Maslim J, Stone J (1988) Time course of stratification of the dendritic fields of ganglion cells in the retina of the cat. Developmental Brain Research 44: 87–93.

Mastronarde DN, Thiebeault MA, Dubin MW (1984) Non-uniform postnatal growth of the cat retina. Journal of Comparative Neurology 228:598–608.

Nelson R, Famiglietti EV, Kolb H (1978) Intracellular staining reveals different levels of stratification for on- and off-center ganglion cells in cat retina. Journal of Neurophysiology 41: 472–483.

O'Leary DDM, Crespo D, Fawcett JW, Cowan WM (1986a) The effect of intraocular tetrodotoxin on the postnatal reduction in the numbers of optic nerve axons in the rat. Developmental Brain Research 30: 96–103.

O'Leary DDM, Fawcett JW, Cowan WM (1986b) Topographic targeting errors in the retinocollicular projection and their elimination by selective ganglion cell death. Journal of Neuroscience 6: 3692–3705.

Pearson HE, Payne BR, Cunningham TJ (1993) Microglial invasion and activation in response to naturally occurring neuronal degeneration in the ganglion cell layer of the postnatal cat retina. Developmental Brain Research 76: 249–255.

Peichl L (1991) Alpha ganglion cells in mammalian retinae: common properties, species differences, and some comments on other ganglion cells. Visual Neuroscience 7: 55–169.

Penn AA, Wong ROL, Shatz CJ (1994) Neuronal coupling in the developing mammalian retina. Journal of Neuroscience 14(6): 3605–3615.

Ramoa AS, Campbell G, Shatz CJ (1988) Dendritic growth and remodeling of cat retinal ganglion cells during fetal and postnatal development. Journal of Neuroscience 8: 4239–4261.

Slaughter MM, Miller RF (1981) 2-amino-4-phosphonobutyric acid: A new pharmacological tool for retina research. Science 211: 182–184.

Thompson I, Holt C (1989) Effects of intraocular tetrodotoxin on the development of the retinocollicular pathway in the syrian hamster. Journal of Comparative Neurology 282: 371–388.

Wässle H, Boycott BB (1991) Functional architecture of the mammalian retina. Physiological Reviews 71(2): 447–480.

Wässle H, Boycott BB, Illing R-B (1981a) Morphology and mosaic of on- and off-beta cells in the cat retina and some functional considerations. Proceedings of the Royal Society of London B 212: 177–195.

Wässle H, Peichl L, Boycott BB (1981b) Morphology and topography of on- and off-alpha cells in the cat retina. Proceedings of the Royal Society of London B 212: 157–175.

Wässle H, Riemann HJ (1978) The mosaic of nerve cells in the mammalian retina. Proceedings of the Royal Society of London B 200: 441–461.

Wässle H, Yamashita M, Greferath U, Grünert U, Müller F (1991) The rod bipolar cell of the mammalian retina. Visual Neuroscience 7: 99–112.

Williams RW, Chalupa LM (1982) Prenatal development of retinocollicular projections in the cat: an anterograde tracer transport study. Journal of Neuroscience 2: 604–622.

Wong ROL, Herrmann K, Shatz CJ (1991) Remodeling of retinal ganglion cell dendrites in the absence of action potential activity. Journal of Neurobiology 22: 685–697.

Wong ROL, Hughes A (1987) Role of cell death in the topogenesis of neuronal distributions in the developing cat retinal ganglion cell layer. Journal of Comparative Neurology 262: 496–511.

Young HM, Vaney DI (1991) Rod-signal interneurons in the rabbit retina: I. Rod bipolar cells. Journal of Comparative Neurology 310: 139–153.

7

GETTING TO GRIPS WITH NEURONAL DIVERSITY

What *Is* a Neuronal Type?

Jeremy E. Cook

Department of Anatomy and Developmental Biology
University College London
Gower Street, London WC1E 6BT, United Kingdom

1. LUMPERS *VS* SPLITTERS

The concept of a neuronal type can be quite slippery. Just when you think you've got it in hand, it can jump out of your grasp like the soap in the shower. This probably explains why so many articles in recent years have claimed to address the molecular mechanisms that generate retinal diversity and yet have ended up focusing on just a few of its many neuronal types. The aim of this article is to set out the problems inherent in the concept of a neuronal type and discuss some of the ways in which a particular kind of spatial organization, the neuronal mosaic, can provide a tool to get to grips with it.

A good way to begin is to consider the question: 'How many neuronal types are there in the vertebrate retina?' At first, it seems that the answer might depend on your own personality—whether you are a 'lumper' or a 'splitter'.

Even if you are a lumper, you should be happy to accept that there are five basic kinds of neurons in the retina: photoreceptors, horizontal cells, bipolar cells, amacrine cells and ganglion cells. Throughout this article, I refer to each of these five divisions as a 'class'. (The term is often used for a finer subdivision than this, as a synonym for what I here call a 'type', but I use 'class' here *only* for the major divisions—see Fig. 1.) A life-long lumper can be happy with the idea of classes if the differences between them are broad and easy to define.

But a sworn splitter—the kind of person who painstakingly documents every little difference in phenotype and gives it a name or a number—sees these classes as bursting at the seams with different types, subtypes and variants. Even those of a less extreme cast of mind accept that many types must exist. Vaney and Hughes (1990) have suggested that there must be at least 70 neuronal types in any given mammalian retina, and "a three-figure total does

Development and Organization of the Retina, edited by Chalupa and Finlay.
Plenum Press, New York, 1998.

91

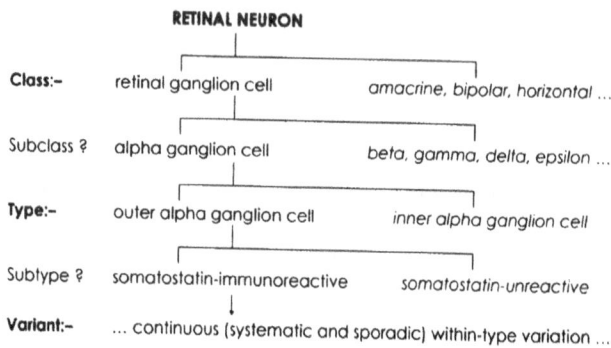

Figure 1. A typical hierarchical classification scheme for retinal neurons showing, with examples, how various terms that lack generally accepted definitions are used in specific ways throughout this article. Neurons are divided into a few broad 'classes', which are in turn subdivided into many individual 'types'. For mammals, an intermediate level is often assumed to exist, containing similar types with complementary stratification patterns grouped together as a 'subclass' (a 'genus' in the sense of Rodieck and Brening, 1983). Such a level remains hypothetical and there is no good evidence for it outside the mammals (Section 5.4) but recent developmental evidence does tend to support its existence in the cat (Section 4.6). Occasionally, variation among the members of a single type appears to divide them into two or more discrete subsets, which individually lack full retinal coverage. Each of these subsets can be termed a 'subtype' but the objective status of such a level remains questionable (Section 5.3).

not seem implausible". Achieving a proper classification must always be near the top of a neuroscientist's agenda, because it is hard to discover anything interesting about neurons by studying heterogeneous populations. Yet a surprising number of retinal neurons, especially in the inner layers, remain unclassified and poorly understood.

So just what *is* a neuronal type, and how can we find out how many there are in this elaborate visual outpost of the central nervous system? It turns out to be a large problem that needs to be taken in stages; but it is also very illuminating. The first step is to slim down the broad, general question to a narrower, more specific form that allows us to get to grips with the basic rules for identifying a type.

2. HOW MANY NEURONAL TYPES IN THE RETINA OF A SINGLE SPECIES?

This is really the question that Vaney and Hughes were addressing, and I guess that most students of the retina would opt, like them, for an answer between 70 and the low hundreds, totalling up the types already known and making intelligent guesses about the number left to be found. Although my aim is to consider the *problem* of counting types rather than the *result*, some readers may find a brief catalogue useful.

In the outer retina, non-mammals have the greatest diversity. There are at least five distinct types of cone, because many non-mammals have UV-sensitive cones as well as the usual blue-, green- and red-sensitive ones, and double or twin cones as well as single ones (Raymond et al., 1993). Among the frogs, even the rods are divided into two spectrally-distinct types (Liebman and Entine, 1968). Then there are up to four different types of horizontal cell, with highly specific rod and cone synaptic patterns (review: Van Haesendonck and Missotten, 1979). The outer retina of placental mammals is less diverse because all of them lack double cones and UV-sensitive cones, and very few have more than two cone pigments. As their outer retinal colour processing is more restricted, mammals also get by with only two types of horizontal cell, one axon-bearing and one (usually) axonless (Peichl, Sandmann and Boycott, this volume).

Linking the outer and inner retina, in mammals and non-mammals alike, are separate bipolar cell subgroups driven by rods and cones. The cone-driven subgroup can be further divided into nine apparently distinct types in the rat (Euler and Wässle, 1995), ten in the primate (Boycott and Wässle, 1991) and eleven in the turtle (Ammermüller and Kolb, 1995) differing in their shapes, the sublaminae where they terminate, and the polarity of their responses to light.

Although the numbers are rising fast, it is the amacrine and ganglion cell types of the inner retina that dominate the counting. There seem to be dozens, and the total in any analysis will depend not only on the species, and on the amount of effort that has gone into looking for these types in that species, but also (and critically) on the *criteria* that have been used in identifying them. In the turtle, Ammermüller and Kolb (1995) described 37 phenotypes for amacrine cells and 24 for ganglion cells. In the cat, Kolb et al. (1981) described 22 for amacrine cells and 23 for ganglion cells, but Vaney and Hughes (1990) argued that others may have been missed.

Given all this complexity, how can we reach a confident answer, even for just one species? The first step must be to distinguish between two different kinds of variation that neurons and all other cells display. On the one hand, there is the natural, continuous variation that we see in all living things, which creates a wide diversity of phenotypes. On the other, there is a discrete, categorical set of developmental and evolutionary dichotomies, which for convenience we can think of as switches, partitioning cell fates in discontinuous ways to create the entities that we call types.

2.1. Within-Type Variation and Between-Type Variation

Rowe and Stone (1977; 1980) advanced our understanding of neuronal types by stressing the need for clear distinctions between these two kinds of phenotypic variation. The discontinuous, role-specific kind that separates a red cone from a green cone, or a wide-field on-centre ganglion cell from a narrow-field off-centre one, they called 'between-type' variation. The continuous kind, that exists even within each of these categories, they called 'within-type' variation. If we can succeed in distinguishing between these two kinds of variation in respect of any particular set of neurons, we have effectively delimited their types.

Without this information, trying to distinguish between cell types is exactly like trying to assess two sets of numerical data by comparing their means without knowing anything about their standard deviations or ranges. The analogy is made all the more potent by the fact that we all know people who frequently and cheerfully do both of these things, without realizing that the conclusions they reach are logically unsustainable!

Understanding neuronal variation can be hard, but our efforts are helped by the fact that within-type variation can itself be split into categories. 'Systematic' within-type variation shows some known trend that allows it to be predicted on the basis of an external factor such as retinal position. 'Residual' (or 'sporadic') variation is whatever remains after all known trends have been accounted for, and so appears arbitrary. As new correlations are discovered, some examples of variation may move between these categories.

A classic example of systematic variation is the gradual change in many neuronal properties that corresponds to a change of location across the retina, especially in species with a fovea (fishes, reptiles and birds as well as primates) or a well-defined *area centralis*

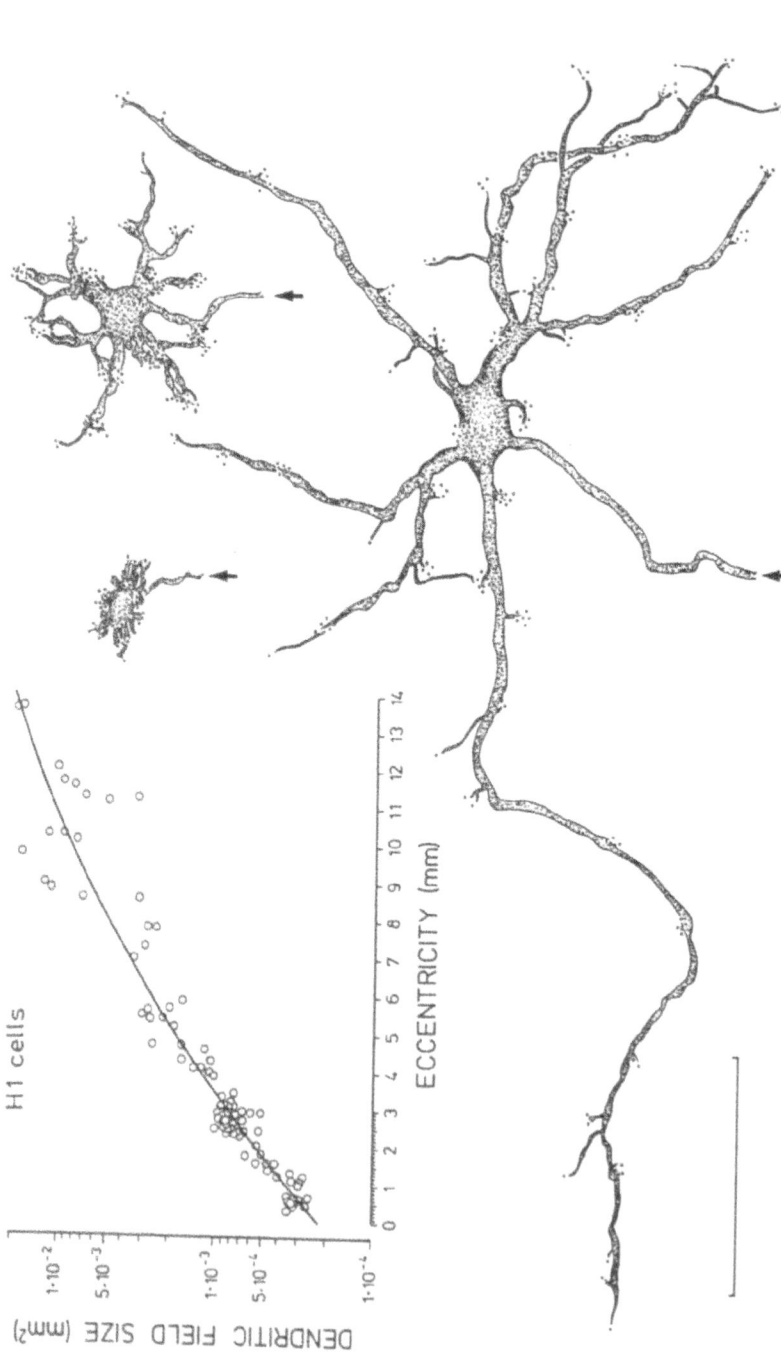

Figure 2. An extreme case of systematic within-type variation (see Section 2.1), affecting the size, general appearance and connectivity of H1 horizontal cells in the macaque monkey. The graph reveals a fifty-fold variation of dendritic field area from the centre to the periphery of the retina. The drawings show a small cell at an eccentricity of 0.36 mm, a medium-sized cell from the nasal horizontal meridian at an eccentricity of 7.8 mm, and a large cell from the upper temporal retina at an eccentricity of 9 mm. The clusters of dots represent contacts with cones. Scale bar = 50 μm. Adapted from Wässle et al. (1989), by permission of Oxford University Press.

or visual streak. In such retinae, two cells of the same type from different locations can look ridiculously different (shown for horizontal cells by Fig. 2, and for ganglion cells by Kier et al., 1995) and yet lie on a continuous morphological spectrum that extends systematically across the retina. To a large extent, our present understanding of the main ganglion cell types of the cat began with the discovery by Boycott and Wässle (1974) that much of their wide variation in size and form correlates directly with location and so can be 'factored out'. As Peichl (1991) has noted: "All of a sudden, the agglomerate of ganglion cell morphologies recognized previously in the cat retina clearly separated into distinct morphological classes . . . the cells changed sizes and detailed branching pattern in a regular way with distance from the central area, but at a given retinal location one could always tell the classes apart."

A clear, if extreme, example of the second kind of within-type variation, the residual or sporadic kind, is the occurrence of individual neurons with their somata in a different retinal layer from others of the same type. Other evidence for sporadic within-type variation abounds but is sometimes misconstrued as evidence for the existence of additional types, so I consider it more fully in Section 5.2, after I have shown how mosaics can provide a framework for identifying types and exploring the variation within them.

2.2. Natural Types Must Be Discovered, Not Defined

The kind of cell type that these ideas point towards is elastic but resilient. We can't mould it to our own will, defining it in arbitrary terms, because it springs back when we release it from the mould. A cell type of this kind is objective: it exists 'out there' in the real world as a functional, developmental and evolutionary entity, and the task of the neuroscientist is not to define it but to *discover* it. This was implicit in the view of Rowe and Stone (1977) that attributions of cell type are modifiable hypotheses rather than definitions; and explicit in a paper on the classification of retinal ganglion cells by Rodieck and Brening (1983), who used the phrases 'natural cell type' and 'natural type' to stress this attribute of objectivity.

2.3. Natural Types Form Discrete Clusters

It is axiomatic that 'natural types' should be discrete. To be considered objective units of classification, they must be so. But when we actually measure properties that vary widely even within a type, like cell size, we usually find extensive overlap between one type and another: the cat's alpha and beta cells were a lucky exception. To isolate each type from every other, Rodieck and Brening (1983) argued that we need to consider not just *one* variable, like 'size', but as many variables as possible: variables of shape, form, and stratification; variables of a functional kind to do with receptive fields and conduction velocities; categorical variables like transmitter content and projection target. They showed how all of these could be viewed as coexisting in a multidimensional space, on an imaginary graph with many separate axes, all at right-angles and thus independent of each other. In such a space, they argued that a natural type can be represented as a cluster of individual cell measurements that is itself a discrete entity.

This multiparametric approach is admirable in theory, and certainly helped me to understand the nature of the problem of discovering cell types. However, it becomes taxing to deal with in practice because it presupposes an interdisciplinary approach and involves making many different measurements on the same set of cells. I know of only a few papers that have taken this approach literally, using statistical methods to identify multidimensional clusters—and only one of those was primarily concerned with the identification of cell types. Kock et al. (1989) made extensive morphometric measurements on the ganglion cells of the frog *Rana temporaria* and applied a statistical technique known as discriminant analysis. Their main findings were summarized in a pair of diagrams (their Fig. 8) in which individual types really did show up as clusters, with each point within a cluster representing an individual cell. Each diagram mapped the data onto a different pair of axes of canonical variation. On the first and second axes, the clusters for ganglion cell types G2 and G3 overlapped extensively, but those for G3 and G4 did not. Exchanging the second axis for the third led to the total separation of G2 and G3, at the cost of some overlap between G3 and G4. I return briefly to this result in Section 6.3.

A few mammals (cats, rabbits and macaque monkeys) have been studied so intensively by neuroscientists of all kinds that the kind of formal analysis outlined by Rodieck and Brening (1983) could have been done for them, too. However, I argue in Section 3.3 for the importance of spreading the net much more widely than this; for the importance of doing *extensive*, as well as *intensive* studies; for the importance of increasing our understanding of the variation in neuronal characteristics *between species*. This can only be done if we have much more cost-effective, time-saving ways of identifying hypothetical natural types, a point to which I return in Section 4.

3. HOW MANY NEURONAL TYPES IN THE RETINA IN ONE VERTEBRATE ORDER?

In Section 2, I restricted the scope of the discussion to a single species. In effect, my answer has been: 'We don't yet know how many types there are in one species, but we have defined our terms and devised some strategies for finding out.' Before even this small success can go to our heads, let us quickly advance to the harder problem presented by a broader question: 'How many neuronal types in the retina in one vertebrate order?'

3.1. Linnaeus vs Darwin in Neuronal Systematics

The cluster-based approach advocated by Rodieck and Brening (1983) is entirely appropriate to the delineation of natural types within a single species. It should receive much more attention that it has received so far—*and yet, it should not satisfy a modern neurobiologist.*

Why? Because classifying neurons by their morphological and functional similarities, however rigorously, is directly analogous to the outmoded Linnaean system of classifying animals and plants by *their* morphological and functional similarities. The Swedish naturalist who laid the foundations of modern taxonomy in 1735 through his book *Systema Naturae* had no conception of the divergent evolution of all life from a common ancestry and was in no position to evaluate the deeper relationships between the organisms that he assigned to genera, families, orders and so forth. As a result, when Darwin's (r)evolutionary view of the world began to prevail more than 120 years later, the Linnaean system needed a radical and comprehensive overhaul. Indeed, the process of replacing similarity-based or 'phenetic' classifications by 'cladistic' classifications (drawing on basically the same empirical data but using them to infer evolutionary relationships) is still going on, as new information steadily emerges from anatomical and molecular analyses.

We neuroscientists cannot hide behind Linnaeus's excuse: we should be as familiar as other biologists with Dobzhansky's dictum "Nothing in biology makes sense except in the light of evolution", and we should not forget that the evolution of organisms goes hand in hand with the evolution of the cell types from which they are built. Perhaps the closing years of the 20th century are a good time to set aside our 18th century perspective on neuronal systematics and taxonomy! To keep up with mainstream biology, we must learn to see neuronal classifications *not only* as hypotheses about the boundaries of structural and functional variation separating natural classes and types, as Rowe and Stone (1977) and later authors taught us to do, *but also* as hypotheses at a broader and yet deeper level, concerning the evolutionary relationships between these natural categories and the developmental programmes that produce them. Evolutionary relationships are the key to any rational hierarchy of subclasses, types or subtypes within a *single* species, as well as to any attempt to find correlations between related types in *different* species.

To argue that it is high time for such a change, however, is not to belittle its difficulty. To achieve it—even to *begin to think* about achieving it—we must extend our investigations much wider, outside the single species or genus and across families, orders, classes and phyla! This last would be an enormous step for which we are clearly not ready. That is why I have opened up the question only as far as the taxonomic order—plenty far enough to let us explore another set of relevant rules.

3.2. The Importance of Finding Homologies between Types

Suppose, for the sake of argument, we finally decide that the cat retina holds 33 distinct types of amacrine cell. Now we turn to the ferret, another carnivore with a fairly similar eye, and work on that in the same way. Maybe (just maybe) it will also have 33 such types. Assuming for now that it does, the problem is this: should the archetypal carnivore retina be considered to hold 33 amacrine types, or 66, or some number in between? Put another way, can we establish *homologies* between the cell types of the cat and those of the ferret that will allow us to say that particular cells from these two species (or, indeed, any others) belong to the same type, in the same confident way that we now say such things for two conspecific animals? Or does the cat have a few types that the ferret lacks, and *vice versa*?

Now suppose that our hypothetical ferret doesn't have 33 amacrine types after all, but as few as 28 or as many as 41? How would we know which ones to match with which? This kind of problem (presented as a worked fictitious example in Fig. 3) matters greatly to those who care about the function of the retina in vision, as well as to those who care how it evolved. Such differences as exist may be clues to different functional demands.

We can *guess* at some answers, using the familiar phenetic (Linnaean rather than Darwinian) mode of thought, by looking carefully at similarities of form and function and grouping similar cells together. From this viewpoint, for example, recognizable alpha ganglion cells can be found in all mammals, and those of any two mammalian species appear to match quite well at equivalent spatial densities, with only a few odd exceptions (see Section 6.2). In contrast, cells with a beta morphology have not been found in the rabbit or rat (review: Peichl, 1991). However, this approach cannot make a formal distinction between superficial similarities due to convergent evolution (analogies) and deep similarities due to shared ancestry (homologies). From an evolutionary (cladistic or phylogenetic) viewpoint, only homologies (see Fig. 3) provide acceptable groupings, mainly because only homologies have any predictive power with respect to those characters that have *not* yet been studied—which, in neuroscience, are usually in a majority! And, as the next section will make clear, information about comparable cells in related lineages (in this case, monotremes and reptiles) is needed before homologies can be established.

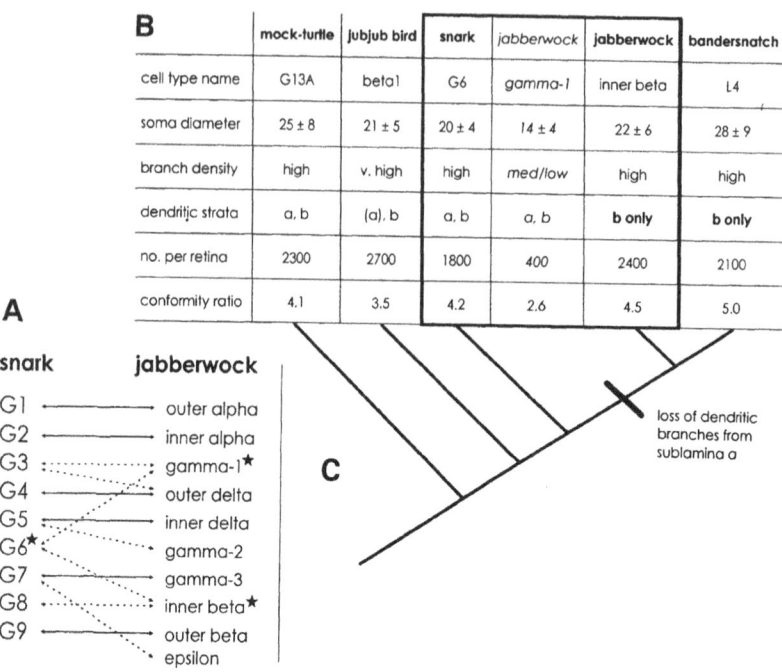

Figure 3. The problem of matching neuronal types across (fictional) species and a generic approach to its solution (see Section 3.4). (A) Attempts to match 9 known neuronal types in the snark to 10 in the jabberwock according to similarities of form and function leave many uncertain pairings, including that of G6 cells with either gamma-1 or inner beta cells (asterisks). The canonical literature describing these species implies that the snark is an elusive marine herbivore, while the jabberwock is fierce, terrestrial and carnivorous. Thus, differences between homologous neuronal types are to be expected, and may be of adaptive significance. (B) Accumulated data for G6-like cells in five species of the subphylum Carrollata. The inner beta type can be seen to resemble the G6 and its known homologues in all respects *except* the functionally important one of bistratification, while the gamma-1 type is bistratified but otherwise less similar. (C) In the context of an independently-derived phylogenetic tree, all the characters that are common to snark G6 and jabberwock inner beta cells are revealed as 'shared-ancestral' (see Section 3.3), whereas the 'b only' monostratification of the inner beta cells is 'shared-derived' within a group (clade) that also includes its close relative, the bandersnatch. The homology of G6 and inner beta cells is confirmed. The smaller, less numerous and much less regular gamma-1 cells of the jabberwock may not constitute a 'natural type' (Section 2.3) but, if they do, they should be considered for possible homology with other snark types.

3.3. The Importance of Being Extensive

The only way to get reliable answers to questions about the matching of neuronal types across species, then, is to establish patterns of common evolutionary ancestry (homology) for the types at issue. Ideally, for the cat and ferret, we should identify the last carnivore that was a common ancestor of both and study *its* retina. But this ancestor is long extinct and its retinae will not have been preserved as fossils, so the ideal approach is doomed from the start.

Instead, we must turn to an indirect analytical approach that relies wholly on studies of the neurons of *living* animals but still permits strong inferences to be made about their evolutionary relationships (review: Northcutt, 1984). To do this, we must compare the various living species of immediate interest (the 'in-group') not only with each other but also with one or more 'out-group' samples, drawn from species that are known to be more distantly related. We are *obliged* to use this approach because, even in the nervous system, superficial similarities do commonly arise by convergent evolution (Northcutt, 1984). We are *able* to use it because the evolutionary relationships between most living vertebrates have already been worked out through cladistic analyses of less evanescent structures that do leave fossils.

The related species that make up the in-group must, clearly, have shared a common ancestor at a relatively recent point in their history. Every out-group species, too, must have shared a common ancestor with the in-group at some point, but that point was further back in time. The period separating these early and late ancestors corresponds to the period in which any evolutionary change would affect the lineage of the in-group but not the out-group: it would occur *after* the out-group and in-group diverged, but *before* the in-group itself diversified.

When the neurons of two or more in-group species are seen to share particular characters that are absent from the out-group, the most economical (parsimonious) explanation is that these characteristics are 'shared-derived' (synapomorphic). They were absent in the first common ancestor, where the out-group split off, but had appeared by the time the in-group began to diversify. A second possibility is that these shared characters arose independently in all the different lineages of the in-group; but this kind of coincidence becomes less and less plausible as the number of different in-group lineages showing the critical characters is increased. A third possibility is that the critical characters were already present before the out-group and in-group diverged, which would make them 'shared-ancestral' (symplesiomorphic) rather than 'shared-derived'. However, we know that the out-group lacks these characters so, if this is the correct explanation, they must have been lost later on within all the different out-group lineages. This kind of coincidence becomes less and less plausible as the number of different *out-group* lineages *lacking* the critical characters is increased.

A key point here is that, in both cases, *the more species that are studied the better*. It doesn't pay, if we want to understand cell types, to concentrate all the research effort on a few experimental models.

The characters that can be used to establish homologies are many and varied, and will vary from type to type. The larger the number of *independent* characters that can be tested, the greater the reliability of the result. (As an example: somatic spacing, tree size and dendritic coverage could not all be considered independent, but any two of them could.) The only stipulation of formal cladistic methodology is that characters should be categorical, rather than numerical. If, for example, 'average number of primary dendrites' is to be included as a character, it might be made into a two-category character by setting a threshold, as in 'average of more than N primary dendrites', or by devising a suitable comparison, as in 'more primary dendrites than any other large cell type', either being a 'true or false' decision. Some dichotomies may arise naturally, such as 'on-centre'/'off-centre' or (if no diffusely stratified cells are present) 'inner-stratified'/'outer-stratified'. Others have implicit thresholds that must be justified by the experimenter: 'large'/'small', 'dense'/'sparse', 'brisk'/'sluggish'. Just as it helps if we study more species, so also it helps if we study more characters.

When the expression patterns of different individual characters suggest contradictory lineages, formal analytical methods beyond the scope of this article can be used to identify the most probable and parsimonious lineages, but it is not yet clear that such methods will be either necessary or appropriate for studies of neuronal types.

3.4. The Importance of Studying Minor Types

Another key question is whether to focus on the large differences that separate the major cell classes, or the smaller differences that separate the types *within* these classes.

It seems intuitively obvious that the large, clear differences between neurons of different classes (cones and ganglion cells, say) will be easier to study than the smaller, more subtle differences between different types of cone, or between different types of ganglion cell. For some kinds of cell biological analysis, this may well be true. Unfortunately, it is probably the reverse of the truth when the issue is one of homologies across different taxa. Why? Because *all* the major neuronal classes of the retina are present in *all* the surviving branches that have come down to us from the very first vertebrates, 450–500 million years ago. Even jawless fishes such as lampreys have recognizable receptor, horizontal, bipolar, amacrine and ganglion cells, although their retinae differ from those of jawed vertebrates in other interesting ways (Fritzsch, 1991). In fact, there are no out-groups to vertebrates that are of any use at all in comparative studies of the eye, to the sustained distress of those who want to study its evolution in a more general way. And, without out-groups, there can be no objective tests of homology: Everything is speculation. Thus, there is no obvious way of studying the emergence of the major retinal classes: their origins, like those of the eye itself, are lost along with those of our earliest chordate ancestors.

So, in fact, our *only* option, if we agree with Dobzhansky that "nothing in biology makes sense except in the light of evolution", is to look at the evolution of minor types over the last half-billion years. If we can "make sense" of that, we may hope that it will throw light on earlier steps.

4. THE NATURE, ASSESSMENT, AND IMPLICATIONS OF MOSAICS

Up to here, I have set out a case for thinking of neuronal types in a more Darwinian way and taking a broad view across vertebrate phylogeny to discover their homologies. For this purpose, I repeat, the more species that are studied, the better. But, of course, spreading the effort more widely in this way makes it harder than ever to apply a multi-disciplinary, cluster-based approach to the discovery of cell types in each individual species, since that approach demands an intensive focus. It seems that we can have *either* the rigour needed to discover natural types in individual species through phenetic cluster analyses, *or* the phylogenetic breadth needed to understand their homologies across related species through cladistic character analyses, but not both.

Is there any way out of this dilemma? It seems so, because for a steadily increasing number of retinal neurons the problem of discovering how they fall into natural types has already been cracked open by the discovery that they exist as regular mosaics.

We already know that the key to discovering types is understanding within-type variation. In practice, many of the complications caused by this form of variation shrivel away to nothing if we can establish that an independent regular mosaic exists, to which some cells belong while others do not. We can legitimately assume, as a potent working hypothesis, that all the variation of form and function that we see in an independent regular mosaic of wide-field neurons should be considered as within-type variation, rather than between-type variation. I substantiate these points below.

This escape route has been recognized by many earlier authors, but only up to a point. For example, Vaney and Hughes (1990) took dendritic coverage into account in their discussion of the number of types in the retina, but stopped short of considering somatic regularity in the same light.

4.1. What Is a Mosaic?

In the retina, the patterning of some neurons is regular enough to allow the position of an individual cell to be predicted fairly accurately from the positions of its neighbours. Usually, these neurons give rise to a tile-like pattern of dendritic trees recalling a Roman ceramic floor (hence the terms 'mosaic' and 'tessellate') but 'mosaic' is now used by extension for any regular neuronal pattern, with or without tessellated trees.

This article focuses on wide-field neurons, where the spacing of neighbours of the same type imposes very little constraint on the positioning of their somata. This category includes examples of ganglion cells, amacrine cells and horizontal cells from all vertebrate classes: fishes, amphibians, reptiles and birds as well as mammals. With these types, when their somata *do* form regular patterns, we can be confident that something non-trivial exists to be explained. Highly regular mosaic patterns are also seen among narrow-field cells, including cone and bipolar cells, but in some or all of these cases their tight packing contributes to their regularity. Even some horizontal cells (for example, goldfish H1 cells: Marc, 1982) are too tightly packed to be anything but uniformly spaced. Rodieck and Marshak (1992) have also pointed out that even types that are no longer tightly packed in the adult may have acquired some of their order from tight packing in the embryo, although this is less likely when many types occupy the same retinal layer.

4.2. Assessment by Nearest-Neighbour Methods

Our own visual systems are good at detecting patterns. Unfortunately, they also report patterns where they don't exist and fail to differentiate low degrees of order from randomness, so objective assessment methods are essential. One familiar group of analytical methods is based on the distance from each cell to its nearest neighbour of the same type, known as the nearest-neighbour distance (NND). Wässle and Riemann (1978) introduced NND analysis of retinal neurons as a simple way of demonstrating spatial order. Even without electronic aids, it is straightforward to measure the distance from each cell in a mosaic to its nearest neighbour on a tracing or photograph with a pair of dividers, and to calculate the mean and standard deviation of those distances. However, some important limitations of NND methods with respect to tests of statistical significance have only recently been addressed, and other limitations remain.

In particular, although the ratio of the mean NND to the standard deviation of the NND has been very widely used for all of those 20 years as an informal indicator of regularity, a derivation for the expected value of this ratio in an infinite random spatial distribution was published only recently (Cook, 1996), and its numerical value (1.913) was found to be almost as high as some values that have been cited in the literature as evidence for limited spatial order!

The need for a simple test of significance based on NND measurements led me, in that same study, to determine the probability distribution of this ratio (the conformity ratio) directly and empirically by computer simulation methods. A major advantage of the empirical approach was that it made it possible to study the robustness of the NND method when presented with small, realistic samples with elongated or irregular borders and a high proportion of missing neighbours, rather than the large, borderless samples universally assumed in theoretical work. At the same time, I was able to compare the statistical properties of the conformity ratio with those of another traditional measure of spatial order, the dispersion index.

In these simulations, the NND distributions of very large random-point samples behaved as expected. However, the NND distributions of small-to-medium samples, where many cells lie close to the sample borders, turned out to be significantly different from the Rayleigh distribution, the 'gold standard' against which NND distributions have always been tested (Wässle and Riemann, 1978). The dispersion index, too, is flawed as a basis for significance tests because it overestimates the regularity of these small-to-medium samples with irregular boundaries that are encountered in real life.

Fortunately, these limitations do not apply to the conformity ratio, which reacts conservatively to extremes of sample geometry and to missing cells and so can form the basis of a useful and safe test. The significance of any apparent regularity in a mosaic sample can now be determined very easily from a 'ready-reckoner' chart (Fig. 2 of Cook, 1996) by finding the point that corresponds to any observed number of cells (between 25 and 6400) and their observed conformity ratio (mean NND/SD).

4.3. Assessment by Spatial Correlogram Methods

Another valuable technique for the assessment of spatial order is the spatial correlogram, introduced by Rodieck (1991). This is analogous to the familiar temporal correlogram of electrical firing in two neurons, but reveals coincidences of points in two-dimensional space rather than events in one-dimensional time. There are two versions: the *auto*correlogram explores the spatial relationships inside a *single* spatial array or mosaic, while the *cross*correlogram looks for spatial interactions *between* two mosaics, to see if they are truly independent. The availability of this independence test is an important advance, because mosaics can only help us to understand neuronal types if we can be sure that any two mosaics we compare with each other are indeed spatially (and thus presumably developmentally) independent.

To construct a spatial autocorrelogram for an array of cells, each cell in turn is placed at the focus of a bull's-eye of concentric rings, at a point called the reference point, and a suitably scaled plot is made of all its neighbours out to the edges of the chart. The rings mark equal increments of distance, and their spacing is normally of the same order as the soma diameter. When all the cells in the array have been treated as reference cells in this way, the pattern of dots representing all the overlaid neighbours can be analysed. If the original array was spatially random then, whichever cell was at the reference point, its neighbours would always fall randomly in relation to it, so the final distribution of dots across the chart would be statistically uniform. If the original array contained any kind of consistent local order, the cumulative dot distribution should reflect this order.

In assessing the uniformity of the dot distribution, each of the annular spaces between the rings is treated as a histogram bin and the number of dots inside it is divided by its area to give a series of estimates of spatial density at increasing distances from the reference point. Density is then plotted against radius on a histogram, giving a cross-section or *density profile* through the chart. A truly random array yields a density profile that is completely flat except for random fluctuations. Neuronal mosaics never yield flat autocorrelogram profiles, because each cell is surrounded by a limited territory from which other cells of the same mosaic are at least partially excluded. This leads to a 'hole' in the dot distribution around the reference point, and a 'well' in the corresponding part of the density profile. The physical width of the well obviously depends on the scale factor used in plotting the correlogram chart, but its scaled width can be used to derive an objective measure of the minimum cell spacing, the *exclusion radius*.

The computation of the exclusion radius is not entirely straightforward, because Rodieck's (1991) procedure makes the demanding assumption that every reference cell is surrounded by neighbours on all its sides, so that the sample boundary never cuts across the chart. In practical work with small samples this assumption is often impossible to fulfil, so Cook and Sharma (1995) introduced approximations that extend the utility of the correlogram approach. Although these changes limit its formal power, empirical tests (Cook, 1996) show that they do not undermine its practical benefits.

The first benefit is simple, but can hardly be overstated. The fact that a random distribution yields a flat density profile makes it easy to see at a glance, without further computation, what spatial relationship each cell in a mosaic has with its neighbours, and how far this relationship extends.

The second benefit, just as important, concerns the effects of a failure to record every cell in a mosaic. This happens for several unavoidable reasons: individual cells may fail to take up a tracer, may be obscured by other cells or by damage during dissection, or may appear atypical in some way that leads the observer to judge it best to omit them from the record. This shortfall, known as 'undersampling', is a fact of life in most practical work with mosaics.

Unfortunately, analytical methods based on the NND respond rather poorly to undersampling, and it can be difficult to get reliable estimates either of spacing or of regularity. Starting with ten real mosaics of ganglion cells, all derived from non-mammalian retinae, and deleting individual cells progressively and at random, I investigated how undersampling affected the NND, the conformity ratio and the exclusion radius. As the undersampling increased, the NND rose, steadily at first but then more rapidly, while the conformity ratio fell steeply, quickly losing significance. The exclusion radius, however, was astonishingly little changed by the deletion of 50% (and sometimes even 80%) of the cells in the original sample (Cook, 1996).

To construct a *cross*correlogram to test the spatial independence of two overlapping cell arrays, the same procedures are followed except that the reference cells are all taken from one array (it makes no difference which) and the neighbours from the other. If the arrays are truly independent, there will be no consistent spatial relationship between the cells in one array and those in the other, so the dots representing neighbours will fall uniformly across the chart to yield a flat density profile. In practice, because two somata cannot share the same space, the profile may show a very narrow, steep-sided well, often only one bin (about one soma width) in radius, representing a physical interaction between the cells at close range.

4.4. Assessment Methods Summarized

Having a choice of analytical tools allows us to match them to the task in hand. If the task is to compare the *spacing* of different mosaics (formed by different cell types, or by the same type in animals of different sizes, ages or species), then sensitivity to undersampling becomes a critical factor. It is always hard to ensure that the same proportion of cells has been plotted in each of the cases being compared, so the exclusion radius is the measurement of choice, even though it gives values that are not directly comparable with mean NNDs. In practice, with fairly complete mosaics, the exclusion radius is usually about two-thirds of the NND (Cook, 1996).

If the task is to discover whether a distribution of cells has enough *regularity* to justify its being considered a mosaic, then the most straightforward approach is to measure NNDs, calculate the conformity ratio and read its significance directly from the 'ready-reckoner' chart (Fig. 2 of Cook, 1996). It is worth noting, though, that a conformity ratio derived from a large sample may test as significant even when that sample contains cells from more than one mosaic, because two overlaid mosaics do not make a fully random distribution. Conversely, the failure of a small sample to reach significance may be due to general undersampling, or to the omission of many nearest neighbours along a long or convoluted boundary. Unless there is independent evidence of undersampling, an array with a conformity ratio in the range 2.3–3.0 should certainly be treated as potentially heterogeneous, even if this ratio passes a test of significance. In such cases, the quality of dendritic tessellation may allow a firm distinction to be made between an irregular single mosaic and a pair of more regular but overlaid ones.

If the task is to demonstrate that two mosaics show spatial *independence*, then this can be done through a comparison of spatial autocorrelograms and crosscorrelograms and their respective density profiles.

4.5. Functional Aspects of Mosaics

The most salient function of a retinal mosaic is inherent in the premise: "If the retina is to sample the visual world faithfully . . . cells dealing with different aspects of the visual scene should be distributed so as to leave no holes in our perceptual world" (Perry, 1989). At an adaptational and functional level, a mosaic can be seen as a crucial mechanism for ensuring that our most favoured prey cannot slip out through such a hole—and neither can our most feared predator slip in! For this purpose, it is the tessellation of trees that matters, rather than the pattern of somata; but a regular somatic mosaic is also likely to improve the reliability of dendritic tessellation and increase the functional and metabolic efficiency of the cells by allowing the tessellated trees to be more compact. Another consideration is the spatial density of a type at a particular location, which is evidently a hard-won compromise between visual resolution and cellular economy: fewer cells are needed when the spacing is regular. A third, less obvious, function is to reduce spatial noise in the transmission of visual information to the brain (Snyder et al., 1990).

All this applies to mosaics whose individual members interact during development but are functionally independent in adult life. In addition, some retinal mosaics show continued functional and metabolic integration into adulthood, through highly selective patterns of gap-junctional coupling.

The best understood of the coupled mosaics are composed of horizontal cells. Each cell type forms its own coupled mosaic which can function as an independent electrical syncytium. The pattern of coupling can be revealed by the spread of an injected tracer molecule of low molecular weight, such as Lucifer Yellow or a biocytin derivative. When this is done for the axon-bearing horizontal cells of mammals, the array of axon terminals can even be shown to form its own, independent, coupled mosaic (review: Vaney, 1994a). The gap junctions that perform the coupling are under tight regulatory control and their conductance and dye permeability are reduced by the transmitter dopamine, which is released within the retina in amounts that vary on a daily basis, partly in response to light and partly to circadian rhythms. The degree of coupling among horizontal cells of the same type controls the magnitude and lateral spread of their graded light responses (Bloomfield et al., 1995) and this, in turn, adjusts the sensitivity and acuity of the retina to high and low light intensities (reviews: Witkovsky and Dearry, 1991; Cook and Becker, 1995).

There is also a clear functional significance to the mosaic coupling of one particular type of dense, narrow-field amacrine cell found in mammals, the AII amacrine cell. These cells pass responses from rod bipolar cells through to the cone bipolar system using a combination of conventional synapses for inhibition and gap junctions for excitation. But, in addition, AII cells are also coupled to each other, much like the members of a horizontal cell mosaic (review: Vaney, 1994a). Also like horizontal cells, their receptive fields expand in dim light and contract in bright light (and also in extreme darkness) as their coupling waxes and wanes (Bloomfield et al., 1997), although little is so far known about the detailed mechanism of regulation. A computer-modelling study (Smith and Vardi, 1995) suggests that the main function of this coupling may be to suppress uncorrelated quantal noise while amplifying weak but spatially correlated signals.

Tracer-coupling has also been found in mosaics where a functional role for it is not yet apparent. For example, the rabbit retina contains bistratified ganglion cells (BiS1 cells) with on-centre, direction-selective, receptive fields. When all the BiS1 cells in a small patch of retina were plotted together, their distribution appeared to be random, even though individually they had compact, sharply defined dendritic trees that would be expected to tessellate precisely. The paradox was resolved when these cells were injected with the tracer Neurobiotin: some were found to be isolated, while others were coupled into a highly regular, beautifully tessellated mosaic that included only just over a third of the BiS1 population (Vaney, 1994b). It seems likely that the cells comprising the coupled, tessellated mosaic all shared the same directional preference, and that the remaining cells also formed tessellated mosaics according to their own directional preferences, but it is still not clear why only one mosaic showed tracer-coupling.

Similarly, the role of gap-junctional coupling is unclear for several other wide-field ganglion and amacrine cell types that are known to be tracer-coupled, including the on-centre and (separately) off-centre alpha ganglion cells of the cat (Vaney, 1991) and the on- and off-centre parasol cells of the macaque monkey (Dacey and Brace, 1992). Mastronarde (1983) found evidence in the adult cat for small but very rapid electrical interactions between neighbouring Y cells (alpha cells) of the same centre sign, and these may well be mediated by the same gap junctions as the tracer coupling, but their functional significance remains obscure. Bloomfield and Xin (1994) have claimed that the receptive fields of all the major amacrine cell types of the rabbit, other than the AII amacrines, are essentially limited to the spread of their dendritic trees even in conditions when injected tracer spreads widely through neighbouring cells. Thus, it is conceivable that some examples of mosaic-specific gap-junctional coupling have no residual function in the adult and must be seen as vestiges of development.

4.6. Developmental Aspects of Mosaics

The formation and gap-junctional coupling of wide-field neuronal mosaics are topics with important developmental implications. Chalupa (this volume) focuses on the emergence of the independent inner- and outer-stratified alpha and beta cell mosaics of the cat, so I consider mammals here only to compare them with non-mammals, where the nature of the developmental problem is (or at least appears) different in three ways.

First, the wide-field neurons of the mammalian retina are all born long before eye-opening (Stone, 1988), whereas mosaics of fishes and frogs go on acquiring new members by neurogenesis at the retinal margin for months or even years after they become fully functional (Straznicky and Gaze, 1971; Johns, 1977).

Secondly, in mammals (and probably also in birds) we need to understand how regular neuronal mosaics can develop, and survive to adulthood, in a retina that undergoes the programmed death of up to 80% of its neurons (cat: Williams et al., 1986). If, as widely believed, fish and frog retinae do not undergo massive neuronal death (Williams and Herrup, 1988; but see Gaze and Grant, 1992), there may be major differences between them and mammals in the way their mosaics are set up. A mosaic from which 80% of the neurons are deleted at random no longer looks like a regular mosaic, or behaves like one when analysed by NND methods (Cook, 1996). Adult mammalian mosaics *do* look and behave like regular mosaics, so they must either be sculpted *by* cell death out of some less orderly but much denser population or (if they arise before the wave of death occurs) be protected in some way from its randomizing effects. An interesting possibility is raised by a study (Williams et al., 1993) comparing the retina of the domestic cat with that of a species of Spanish wildcat, thought to retain many ancestral characteristics. The adult wildcat has many more ganglion cells in total than its domestic relative, but the foetal optic nerves contain similar numbers of axons, suggesting that the adult difference could result from enhanced ganglion cell death in the domestic lineage rather than from any change in neurogenesis. Both species contain very similar numbers of alpha ganglion cells, which do, therefore, appear to be protected, if not from all cell death then at least from this phylogenetic increase in it.

Thirdly, in recent studies of the formation of the inner- and outer-stratified beta and alpha ganglion cell mosaics of the cat (Chalupa, this volume) these *two* mosaics of monostratified neurons were shown to emerge gradually, through an interesting activity-dependent mechanism of dendritic segregation, out of a *single* population of diffusely stratified precursors. It may be intrinsic to this mechanism of mosaic formation that the two mosaics are complementary, sharing similar spatial densities and morphological and functional properties and differing only in their stratification pattern and its functional correlate, the sign of their light response. Again, non-mammals appear to be different. Their ganglion cells do not form complementary on-centre/off-centre mosaics like the alpha and beta mosaics of the cat: indeed, I have yet to see a single clear case of complementarity after studying almost forty different ganglion cell mosaics in fishes, frogs, reptiles and a bird. As it is clear that non-mammals had neuronal mosaics long before the first mammal ever gave suck to its young, the development of alpha and beta cell mosaics, though important in itself, cannot be assumed to serve as a general model for the development of neuronal mosaics.

Another major developmental issue to which mosaics may be highly relevant is the control of neurogenesis, at least in anamniote vertebrates. The greater part of any mosaic in an adult fish or frog is generated post-embryonically by neurogenesis at the retinal margin, as mentioned above, so clear opportunities exist for the generation of new neurons to be regulated by the proximity of older neurons of the same type. Type-specific negative feedback is the most plausible explanation for the type-specific increases in neurogenesis seen at the margins of frog and fish retinae after type-specific neuronal ablations (review: Reh, 1989). If wide-field neurons do indeed inhibit their own production in the adjacent ciliary margin in a distance-dependent way, they could contribute strongly to the regularity of their own mosaic. In a cichlid fish, Cook and Becker (1991) could find no consistent alignment of the regular mosaic of inner alpha (alpha-b) ganglion cells with the ciliary margin, but this does not exclude the possibility of self-regulation because there is no obvious requirement for the generation of such cells to be synchronous.

The last major issue that falls under this general heading is the role of gap-junctional coupling in development. In non-neuronal developing systems involving cells with epithelial properties, gap-junctional coupling seems to be critically important in allowing individual cells to become integrated into a coherent population. Cells that are excluded from a coupled system may also be excluded from the differentiation process (Becker and Davies, 1995). If the same principle holds good in the retina, there may be an intimate connection between the development of mosaics and their observed tracer-coupling. Hitchcock (1993; personal communication) has shown by tracer-injection that the members of one well-characterized ganglion cell mosaic in the goldfish are coupled together (Fig. 4A), just like mammalian alpha cells except that their mosaic-specific coupling extends to young, immature members still differentiating at the retinal margin (Fig. 4B). This is circumstantial evidence, but it is at least consistent with the proposal that type-specific coupling plays a direct role in mosaic development. Hitchcock (1997) has also studied newly generated fusiform amacrine cells in locally lesioned adult goldfish retinae and shown that they, too, become tracer-coupled to pre-existing cells of the same type.

Finally, gap-junctional coupling may also contribute to the generation of the synchronized waves of electrical activity that sweep over the immature retina before the onset of vision and play a critical role in the refinement of retinotopic, functionally segregated, central projections (Wong, this volume).

5. MOSAICS AS AN AID TO FINDING HOMOLOGIES AMONG WIDE-FIELD NEURONS

5.1. The Intimate Relationship between Mosaics and Types

At the opening of Section 4, I wrote that many problems of neuronal classification arise from within-type variation and shrivel away once an independent, regular mosaic has been discovered. I also claimed that it was legitimate, in order to speed up the process of discovering natural types, to adopt the working hypothesis (proposed by Cook and Sharma, 1995) that all the variation of form and function present within a mosaic of wide-field neurons is equivalent to within-type variation.

Now, I know that this link between 'type' and 'mosaic' is viewed by some as a blind leap of faith, a conjecture of such long range that it must be calibrated individually for each cell type before it can strike home. I disagree, preferring to think of it as an operational *redefinition* of the term 'natural type' for the special context of the retina, arising directly from the premise I quoted above: "Cells dealing with different aspects of the visual scene should be distributed so as to leave no holes in our perceptual world." While the premise itself involves an assumption (that holes reduce evolutionary fitness and will have been minimized by natural selection), it is an assumption that convinced Darwinians can live with quite happily until it can be tested, especially if it greatly speeds up their work!

Since all classifications must in any case be regarded as modifiable hypotheses (Rowe and Stone, 1977; 1980), any particular case in which the equation of 'mosaic' with 'type' happens to be misleading will simply result in some wasted effort: no robust natural type (one with predictive power for untested characters) will emerge. The waste in such a case must be set against the effort saved in other cases. Besides, we must consider what it would mean for 'one mosaic, one type' to be untrue. It would be impossible for cells of a single type to participate in two spatially independent mosaics without causing these mosaics to appear statistically dependent to an observer plotting them; and there is no evidence either that this occurs, or that two or more wide-field types ever share in the formation of a mosaic.

For some narrow-field neurons, three special factors make mosaic-sharing an inevitable compromise, preventing the approach being used. First, all cones, whatever their spectral sensitivities, must lie side-by-side in the shallow plane where the optical image is in sharp focus, and all bipolar cells must lie side-by-side as they traverse the inner nuclear layer. Secondly, it is important for the photoreceptor mosaic to be packed solid, to maxi-

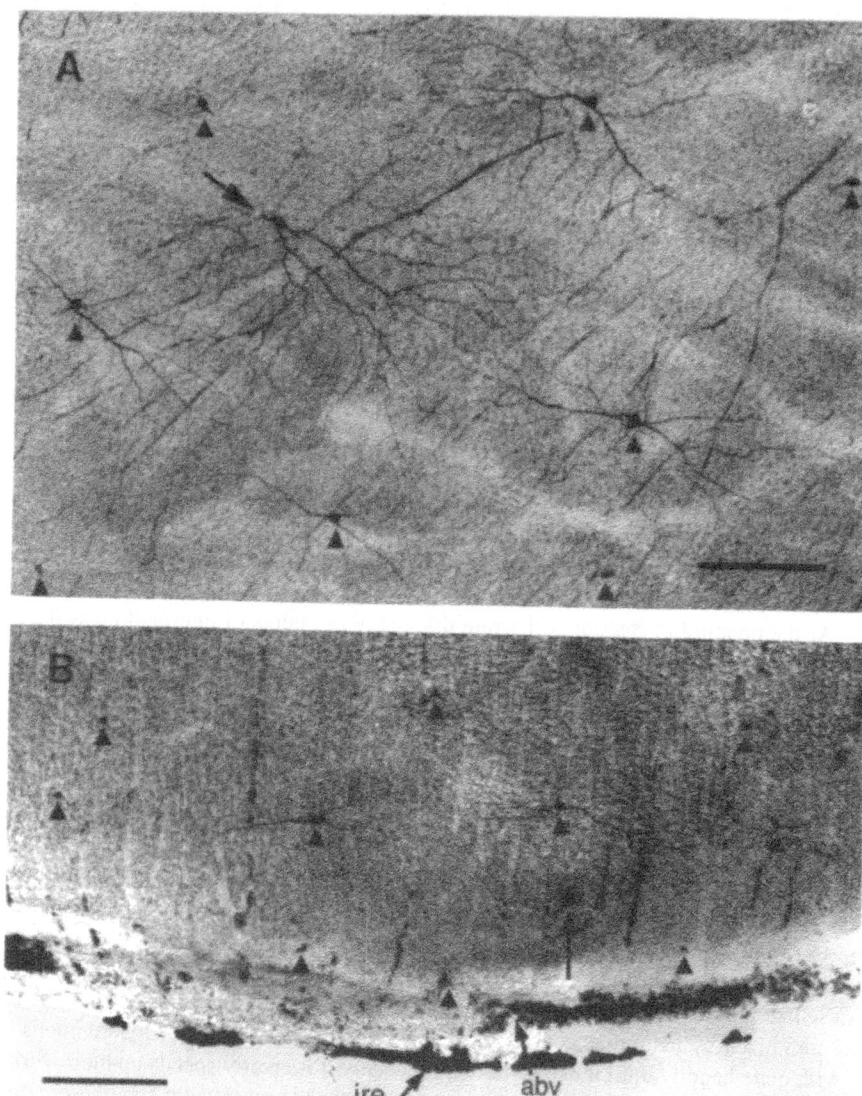

Figure 4. Tracer-coupled ganglion cell mosaics in flatmounted goldfish retinae (see Section 4.6). These previously unpublished images were kindly provided by Dr P. F. Hitchcock. A specially adapted upright microscope was used to make visually guided injections of Neurobiotin™ into the Type 1.2 (alpha-a) ganglion cells indicated by the large arrows (one each in A and B). These cells, and the others of the same type that were tracer-coupled to them (arrowheads), were visualized with peroxidase-conjugated streptavidin and diaminobenzidine and reconstructed digitally at low magnification. (A) Mature, fully differentiated retina. The optic disc is beyond the upper right corner. The retinal margin is beyond the lower left corner. Branched, linear structures running diagonally across the field are capillaries containing erythrocytes with endogenous peroxidase activity. (B) The retinal annulus adjoining the marginal germinal zone. Note three smaller tracer-coupled ganglion cells at the extreme periphery, close to the inner border of the annular blood vessel (abv). Their spacings and thick primary dendrites (parallel to the margin) identify them as immature members of the Type 1.2 (alpha-a) mosaic. Pigmented iris epithelium (ire) lies immediately beyond the annular vessel. Scale bars = 200 μm.

mize photon capture, and the arrangement of bipolar cells (in non-mammals, at least) is highly dependent on that of the cones (Podugolnikova, 1985). Thirdly, the intimate relationship between cone spacing and visual acuity ensures that holes caused by the intermingling of colour-specific cones always fall at the limit of perception. Thus, the principles of organization that create independent mosaics of wide-field, planar neurons may not apply to narrow-field cells, or at least not to all of them.

What about wide-field neurons that are *not known* to form mosaics? Two points can be made about these. First, as yet there are no fully documented natural types that are *known not* to form mosaics (although there are some candidate 'types' characterized by a single feature: for an example, see Hutsler and Chalupa, 1995). Secondly, there are likely to be many more mosaic-forming types than have so far been recognized. The directionally-selective BiS1 ganglion cells of the rabbit were not known to form mosaics until they were injected with Neurobiotin (Vaney, 1994b), and similarly it remains possible that some wide-field amacrine cells with high retinal coverage factors are not single types but families of close relatives, forming independent mosaics. Hidaka et al. (1993) were able to separate two subtypes of sustained depolarizing amacrines in a fish, the dace, only after finding that they formed independent tracer-coupled mosaics. Finally, even if there are indeed wide-field neurons that do not themselves form mosaics, subtracting all mosaic-forming types from the general picture should make their classification easier.

The relationship between mosaics and types is intimate, certainly, but it is not incestuous. A few readers have damned as self-contradictory my claim that mosaics can be used to reveal natural types, telling me that one must already have defined the types to be able to plot their mosaics. This argument has a fine rhetoric about it, but it is flawed. Any initial assignment of provisional selection criteria for a cell type can be treated *right from the start* as a hypothesis to be tested empirically by *attempting* to plot a mosaic, and to be validated (or invalidated) by the spatial regularity, morphological consistency and specimen-to-specimen repeatability of the distribution that can be seen to emerge. When I or my colleagues survey a new retinal specimen and see clusters of wide-field ganglion cell somata piled up close to each other with their trees overlapping extensively, we know before we ever pick up our pencils that we shall need to find criteria that will assign these cells (and others like them) to different mosaics if the results are to satisfy our joint requirements of regularity, consistency and repeatability. As a documented example, grouping large ganglion cells in ranid frogs by predetermined criteria including soma shape and dendritic tree symmetry (Frank and Hollyfield, 1987) produces mosaics of a sort, but they are irregular and contain many "perceptual holes". In contrast, ignoring these prominent characters and grouping equivalent cells by size and dendritic stratification produces regular mosaics with much less internal variation (Shamim et al., 1997b). In other words, feeding back the kind of spatial distribution that would result from any particular set of selection criteria and using it to inform the final choice and weighting of these criteria makes the process of plotting a mosaic inherently iterative, rather than circular.

This closed feedback loop is the key feature distinguishing the revised approach to mosaic studies introduced by Cook and Sharma (1995) from the original 'open loop' approach, now firmly established, where selection criteria for the neuronal type under study are determined in advance by other means and the observation of a mosaic is able to validate them only in retrospect. Using feedback to close the loop relies on the assumption that types and mosaics are related, but it teaches the observer a great deal, not only about the general morphology and within-mosaic variation of the cells in question but also about key features that are well controlled in the development of that mosaic and thus may underlie its key functions.

5.2. What Mosaics Tell Us about Within-Type Variation

Cook and Noden (1997) studied regular ganglion cell mosaics in a reptile, the common house gecko *Hemidactylus frenatus*. In one sample of 187 displaced ganglion cells forming a regular, tessellated dendritic array, a single soma lay in the ganglion cell layer with the main population of ganglion cells rather than at the vitread border of the inner nuclear layer. Wässle et al. (1987) reported a similar case in one of the amacrine cell mosaics of the cat and described the non-displaced deviant as a 'misplaced' cell. As they explained, the crucial point about a 'misplaced' cell is that its integration into the regular mosaic pattern shows it to be a full member of the same cell type, subject to the same developmental control of dendritic form and presumably serving the same function, even though it differs from the other members in this one very conspicuous attribute.

Such cells have another lesson to teach us about variation. Danger lurks in terms like 'displaced' and its converse, 'orthotopic', because we are liable to forget that they are modes on a continuum. Cook and Becker (1991) studied a regular mosaic of 263 outer alpha (alpha-a) ganglion cells in which displaced and orthotopic cells were patchily intermixed (Fig. 5). A systematic analysis of their somatic depth distributions revealed interstitial forms at all possible depths. In other words, although this particular form of within-mosaic variation is often bimodal, it is still in principle continuous.

What about other forms of variation within mosaics, which I argue can also be interpreted as within-type variation? Much has been written about variation in mammalian retinal neurons so I summarize here the main kinds of within-mosaic variation seen among the large ganglion cells of non-mammals (cichlid: Cook and Becker, 1991; goldfish: Cook et al., 1992; catfish: Cook and Sharma, 1995; marine teleosts: Cook et al., 1996; pipid and ranid frogs: Shamim et al., 1997a, b; gecko: Cook and Noden, 1997).

In non-mammals, just as in mammals, much of the variation is systematic and related to retinal location. *Relative soma and tree size* at any particular location are useful selection criteria for mosaic membership, but absolute size is not. *Tree size and symmetry* are regulated by competitive developmental mechanisms in fishes (Hitchcock, 1989), just as in mammals (Finlay, this volume). Such mechanisms help to eliminate "perceptual

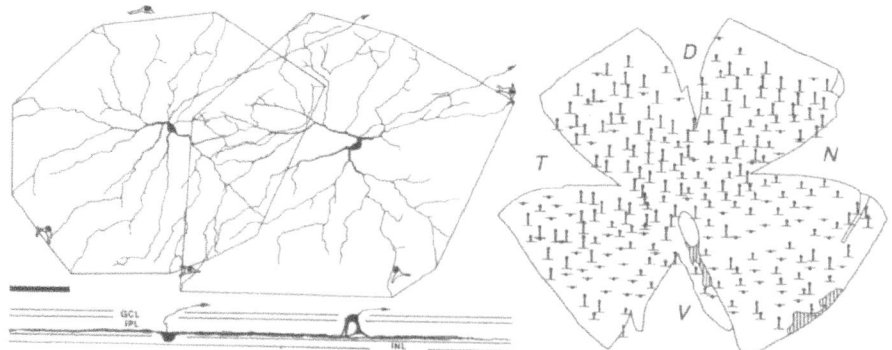

Figure 5. An example of sporadic within-type variation within a regular mosaic (see Sections 2.1 and 5.2). The plan view drawing (left) shows a pair of alpha-a ganglion cells in a flatmounted retina from the cichlid fish *Oreochromis spilurus*, surrounded by outlines of neighbouring members of the same mosaic. Profile views (below), reconstructed from depth measurements and stereoscopic views, show that one of the pair is orthotopic and the other displaced. A plot of the complete, regular, tessellated alpha-a mosaic in a similar flatmount (right) shows that the distribution of soma depth is locally highly variable, and includes intermediate forms. Scale bar = 100 μm. Adapted from Cook and Becker (1991) copyright © 1995 Wiley-Liss, Inc.

holes" but also create sporadic local variation, especially in symmetry. Tree size itself is often impossible to estimate when a whole mosaic of overlapping trees is labelled, but primary *dendrite thickness* is empirically useful as a selection criterion, and is known to be directly related to tree size in spinal cord motoneurons (Ramirez and Ulfhake, 1991). Functionally, the size of the dendritic tree has an important role in setting the size of the receptive field centre but may not otherwise be important: Bloomfield and Hitchcock (1991) have shown that the mosaic-forming Type 1.2 retinal ganglion cells of the goldfish (the wide-field off-centre cells shown in Fig. 4) can grow considerably, as the eye itself grows, without changing their electrotonic cable properties.

We have found that *soma shape*, like *soma displacement*, is highly variable on a sporadic basis and rarely (if ever) effective as a selection criterion for ganglion cells: it seems to be a trivial consequence of the orientations and sizes of the primary dendrites, which dominate the arrangement of the cytoskeleton and in turn depend on the competitive tessellation of the dendritic tree. Regular, radially symmetric trees that arise from several primary dendrites of similar size are associated with spherical or polygonal somata; bilaterally symmetrical trees that arise mainly from two large opposed primary dendrites are associated with spindle-shaped somata, and strongly asymmetrical trees that arise chiefly from one large stem dendrite are associated with pyriform somata.

Selection criteria that emphasize the pattern and level(s) of *dendritic stratification* tend to produce the most regular and consistent mosaics. Even so, regional and sporadic variations have been noted within one of three basically monostratified ganglion cell mosaics in the catfish (Dunn-Meynell and Sharma, 1986; Cook and Sharma, 1995) and more recently within highly regular mosaics of large, bistratified retinal ganglion cells in scorpaeniform fishes (J.E. Cook, T.A. Podugolnikova and S.L. Kondrashev, unpublished data). In these mosaics, although the dendrites are always bistratified in the same two planes, the proportion of terminal dendrites in each plane changes with retinal location. Thus, even this rather stable character has its limits.

One character of retinal neurons that I have not yet discussed is their expression of transmitters, neuromodulators and other functional molecules. The neurochemical approach to the understanding of retinal cell types has been used rather freely, on the assumption (not always made explicit) that neurochemistry should correlate strongly with function. It is clear that there must be correlations, or particular types could not have consistent synaptic effects, and indeed there have been many instances where neurochemistry has led directly to the robust categorization of mosaic-forming cell types.

However, as a student of mosaics, I can't help noticing that neurochemistry has also produced some very odd results. A well-documented example concerns the alpha ganglion cells of the cat, where there is undisputed evidence of several kinds that two independent natural types form regular mosaics covering the entire retina. Chalupa and his colleagues (White et al., 1990; White and Chalupa, 1991) found that these alpha mosaics contained a subset of cells with somatostatin-like immunoreactivity, *not* spread uniformly across the retina but focused mainly on the inferior half (with a precise horizontal cut-off), and avoiding the *area centralis,* the region where the overall alpha cell density is at its peak. Most of these somatostatin-positive cells were members of the outer alpha mosaic, but some were from its inner counterpart. For the moment, what somatostatin might be doing in these cells is not important: the point is that some cells were picked out by this method as showing intrinsic differences from the rest. Are such differences able to divide the labelled and unlabelled cells into different types? No, no more than differences of soma displacement or tree symmetry are. In this case, the encompassing cell types were already very well understood so the observation did not engender a false classification; but the danger is clear and the neurochemical literature overflows with papers in which 'phenotype' is implicitly equated with 'natural type'. Clearly, all phenotypes, including neurochemical ones, can exist as variants *within* a natural type.

5.3. Does Within-Type Variation Have Any Function?

It remains possible that some *within-type* variation—perhaps including the somatostatin variation described above—reflects a *secondary* level of functional differentiation within the enclosing boundaries of the natural type, creating discrete subdivisions that could fairly be called subtypes. In discussing this problem, I like to use an academic analogy.

Let all the institutes in which visual neuroscience is taught throughout the world represent, together, the retinal span of some particular cell type. Let each appoint a Professor of Visual Neuroscience who fulfils the basic function of coordinating teaching. We can imagine these professors as elements in a mosaic with a uniform function, spanning the educational globe. However, they all also have independent research interests that don't need to be carried out in parallel in every institute of learning. They are all members of one major class (teachers in higher education) and one natural type (coordinators of visual neuroscience teaching) but they are not interchangeable clones of each other. As long as there are no "perceptual holes" in the mosaic of *teachers*, the *research* can be done on a more fragmented, local basis, giving rise to a high degree of within-type variation. In keeping with this theoretical analysis, empirical morphological observations on a sample of 57 visual neuroscientists (including both mature and partially differentiated forms: see Preface to this volume) show that they are indeed a pretty mixed bunch.

It is at least conceivable that some visual tasks (reporting general levels of illumination to the circadian system, say, or releasing diffusible neuromodulators) could similarly be performed by the retina on a fragmented, local basis, by specific subtypes operating within a broader framework of natural types.

The other major possibility is that local variation, even of something as potentially significant as neurotransmitter expression, may sometimes have no functional consequences at all; or, if consequences follow, they may be unimportant, neutral ones that have not influenced the evolution of the cell type in question.

For some years there has been discussion in fields outside neuroscience of the idea that gene expression can sometimes be nonfunctional and superfluous. Bowers (1994) has summarized for neuroscientists the idea that all eukaryotic cells may have to tolerate a certain amount of uncontrolled gene expression, in contexts where it is not directly maladaptive, because of constraints imposed by the combinatorial regulation of genes. The principle is analogous to the one that makes it hard for us move our fourth and fifth fingers completely independently because they share common flexor muscles. If the consequences of superfluous expression of a particular gene, in a particular species, at a particular stage of development, *would* actually be maladaptive, then it is thought that the regulatory mechanism has to evolve yet another level of complication to tighten the control, possibly putting another gene at risk of superfluous expression in its turn and almost certainly enlarging the genome. All authors submitting papers in which neurochemistry is implied to reveal cell types (and perhaps all reviewers, too!) should first be obliged to pass a short test on Bowers' timely article.

5.4. Mosaics and Classification Hierarchies

The idea of a classification hierarchy (Fig. 1), with classes divided into subclasses and these in turn divided into types, was implicit in most early classification schemes, whether structural or functional. Even when the approach became multiparametric, the assumption of a hierarchy remained. For example, Rodieck and Brening (1983) considered mammalian alpha cells as forming a broad 'genus' or subclass that was further divided into two types by the congruent evidence of stratification level and centre response sign. However, the reality of such a hierarchy is still hypothetical, and the case for imposing any third taxonomic level between class (e.g. ganglion cell) and type (e.g. outer alpha cell) rests entirely on circumstantial evidence, chiefly the frequent observation of 'complementary', 'paramorphic' or 'matching' relationships of form and function, and sometimes (but not always) spatial density, between inner- and outer-stratified populations, almost exclusively in mammals (see Section 4.6).

Mosaics do not provide any confirmatory evidence for subclasses. Wherever they have been tested, the inner and outer members of such 'complementary' populations have been completely spatially independent (see, for example, Wässle et al., 1981). Neither does there seem to be any evidence to tell us whether genetic or developmental interference with one population of a 'complementary' pair would affect the other. However, some support for subclasses as developmental entities, at least in mammals, comes from recent evidence that the inner- and outer-stratified alpha and beta mosaics of the cat emerge from diffusely stratified precursor populations in which the separate inner- and outer-stratified types of the adult are initially indistinguishable (Chalupa, this volume).

Exceptions to the broad rule of matching populations are also quite common. Again taking ganglion cells as examples, Peichl (1991) noted that inner and outer alpha cells in the periphery of the rat retina can differ so much in size, spacing and form that "it is hard to regard both morphologies as belonging to the same class unless one sees a complete eccentricity series and the gradual changes across the retina". Williams et al. (1993) found a marked dichotomy of dendritic structure between inner and outer beta cells in the Spanish wildcat that turned out to exist in the domestic cat as well, and allowed these types to be distinguished without reference to arbor depth.

It seems likely that mammals differ from non-mammals in respect of complementarity and its consequences, as in so many other ways (see Section 4.6). Perhaps mammals were forced, by the destructive effects of retinal cell death on any spatial order created beforehand, to evolve new ways of creating or refining spatial order after the close of neurogenesis; and perhaps these new ways favoured complementary pairings (see Section 4.6), adding an entirely new level to the classification hierarchy. Perhaps, also, complementary populations with opposite centre signs are able to serve specific functions in the highly derived retino-thalamo-cortical pathways of mammals that would be irrelevant in non-mammals, where more feature extraction occurs at a retinal level: this idea is discussed by Cook and Noden (1997).

Finally, as discussed in Section 3.1, phenetic classifications *can* only ever show superficial and circumstantial similarities. Thus, an understanding of the evolutionary relationships between natural types and the developmental programmes that produce them is a precondition for any rational hierarchy of classes, subclasses, types and subtypes, and the weakness of any argument based on the current status of such hierarchies should be borne in mind.

6. HOW MANY NEURONAL TYPES IN THE ARCHETYPAL VERTEBRATE RETINA?

6.1. The Need for a Step-by-Step Approach

Let us at last peer down gingerly into the deep, dark phyletic abyss that was posed by my first question in its most general form: 'How many neuronal types are there in the vertebrate retina?' It should by now be clear that the only route to an answer involves establishing patterns of evolutionary homology for individual neuronal types, right across the vertebrates. This sounds daunting, I admit. Can it be done, even a step at a time? Is it practical to establish homologies, for example, between retinal neurons in birds and reptiles (I mean 'other reptiles', because to cladistic systematists birds *are* reptiles), or between fishes and frogs? Very probably, and I summarize some initial observations on fishes and frogs in the next section.

How about a broader view across, for example, fishes and humans: do *they* have any neuronal types in common? Yes, they do. Fishes have red and green cones, as do humans, and molecular sequence studies imply that the red cone pigments of fishes, humans and all other vertebrates are probably homologous, along with the cones that express them. However, the green cone pigments of humans and other apes are *not* homologous with those of non-primates: the opsin of the ancestral green cones was lost early in mammalian evolution. When our primate ancestors could again derive an adaptive benefit from having separate red and green cone pigments, they duplicated the red pigment gene and changed just 15 of its 364 amino acid codes to make a new green pigment (Asenjo et al., 1994).

The point of telling this story here is to point out why it can be told at all: the specialized function of a red cone requires it to express large amounts of a highly individual protein, with a structure tightly constrained by its function. If other retinal neurons possess such handy molecular markers, we are only just beginning to find them (Xiang et al., 1996). In fact, the use of gene sequences to study cell type homologies is bedevilled by the fact that many *single* genes are used by *multiple* cell types—including non-neurons—and *don't* often diverge in a cell-type-specific way as the opsins do.

So, until we have made more progress in defining homologies on a local scale (fish with frogs, frogs with reptiles, reptiles with birds, reptiles with monotremes, monotremes with marsupials, marsupials with placentals, and so on) we shall not be able to see the big picture clearly—but it seems likely that a big picture must exist, because evolution rarely invents anything new, preferring to tinker with what has gone before (Jacob, 1977).

6.2. First Steps with Ganglion Cell Mosaics

The first broad, comparative study of neuronal types from the inner retina was that of Peichl et al. (1987), who used a neurofibrillar stain to document large ganglion cells similar to the alpha cells of the cat in 20 mammalian species. Other mammals have been added to this list, which now stands at 32 species across ten orders, including three marsupials (review: Peichl, 1991). Although there is variation, many familiar morphological characters known from the cat are well conserved, especially relative size, proportion of the ganglion cell population, and stratification pattern.

I pointed out in Section 3.3 that, without out-group comparisons, such an approach could not separate similarities due to shared ancestry (homologies) from those due to convergent evolution (analogies). Nevertheless, the lack of an equivalent complementary pair of types in any anamniote, and especially in the only reptile whose mosaics have yet been studied (Cook and Noden, 1997), does suggest that the inner and outer alpha cell types of all mammals may indeed be homologous. Mosaic data from other reptilian out-groups are needed, and also from monotremes if possible, because monotreme eyes have several 'reptilian' features (for references, see Cook and Noden, 1997).

Similar comparative studies from my own laboratory, based on mosaic analyses in 14 species from 12 well-separated anamniote families, have revealed what seems to be an 'anamniote consensus pattern' of three large, spatially independent ganglion cell mosaics, termed alpha-a, alpha-b (or alpha-ab) and alpha-c on grounds of large size and level(s) of stratification. The fishes we studied were all teleosts but ranged from relatively primitive silurids and cyprinids (Cook et al., 1992; Cook and Sharma, 1995) to advanced, highly visual perciforms (Cook and Becker, 1991; Cook et al., 1996; S.L. Kondrashev, T.A. Podugolnikova and J.E. Cook, unpublished data). The frogs were a pipid, two ranids and two bufonoids (Shamim et al., 1997a, b; K.M. Shamim, P. Tóth, D.L. Becker and J.E. Cook, unpublished data).

In all these species, each retina contained an alpha-a mosaic containing 150–1100 cells, representing 0.1–1.3% of the total ganglion cell count. These cells had large somata (often displaced) and thick primary dendrites supporting large, relatively sparse, outer stratified trees. An alpha-b or alpha-ab mosaic (see below) contained 300–1900 cells per retina, representing 0.2–1.2% of the total count. These cells, too, had large somata, but they were never displaced. Their primary dendrites were typically a little thinner than those of alpha-a cells, and their trees were more compact and more elaborately branched. Depending on species and retinal location, the cells of this mosaic were either stratified predominantly in the middle sublamina of the inner plexiform layer (alpha-b) but with minor branches in the outer sublamina, or evenly bistratified across the middle and outer sublaminae (alpha-ab). An alpha-c mosaic contained cells that were more variable in their retrograde labelling but were estimated to number 180–800 per retina and to represent 0.1–1.0% of the total count. They had flattish, often polygonal somata in the ganglion cell layer and long, sparsely branched, dendrites that ran very close to this layer. In some fishes, independent mosaics of biplexiform displaced ganglion cells with additional fine dendrites in the outer plexiform layer, close to the horizontal cells, could also be demonstrated (Cook et al., 1996).

In ranid frogs, the alpha-ab cells were very strictly bistratified and planar. In *Rana esculenta* and *Rana pipiens*, our mosaic analyses (Shamim et al., 1997b) gave results in full agreement with the cluster-based analysis of Kock et al. (1989), which was performed on *Rana temporaria*. Those authors concluded that their type G4 (our alpha-a) included both the class 1 and class 2 cells of Frank and Hollyfield (1987), and that symmetric and asymmetric cells were variants, not distinct types. We reached the same conclusion in a different way by showing that cells with class 1 and class 2 morphology both belonged to the alpha-a mosaic. Types G3 and G5 of Kock et al. (1989) also closely matched the members of our alpha-ab and alpha-c mosaics, respectively.

It should be clear from the preceding descriptions that each of the three non-mammalian 'alpha' (that is, large) ganglion cell mosaics has many characters that stay constant across species, and others that vary. Just as two cells from different places in the same mosaic can look very different and yet lie on a positional spectrum (Fig. 2 and Section 2.1), so also two cells from different species can look very different and yet lie on a phylogenetic spectrum (Fig. 6). However, between-species variation seems to be related as much to behaviour, ecology and eye size as to lineage. At present, the evidence that large ganglion cell mosaics are subject to broad phylogenetic trends across the anamniotes is limited. Terrestrial frogs generally have more ganglion cells of all types than do fishes, and may exert more precise control over their dendritic stratification, but otherwise their mosaics are quite similar. At first, it seemed likely that bistratification of the alpha-ab mosaic might be a shared-derived character in frogs, but our recent observations of bistratified alpha-ab mosaics in marine teleosts call this interpretation into question. The scope for wider out-group comparisons is currently limited by a lack of data, but work on one out-group species, the dogfish, is in progress.

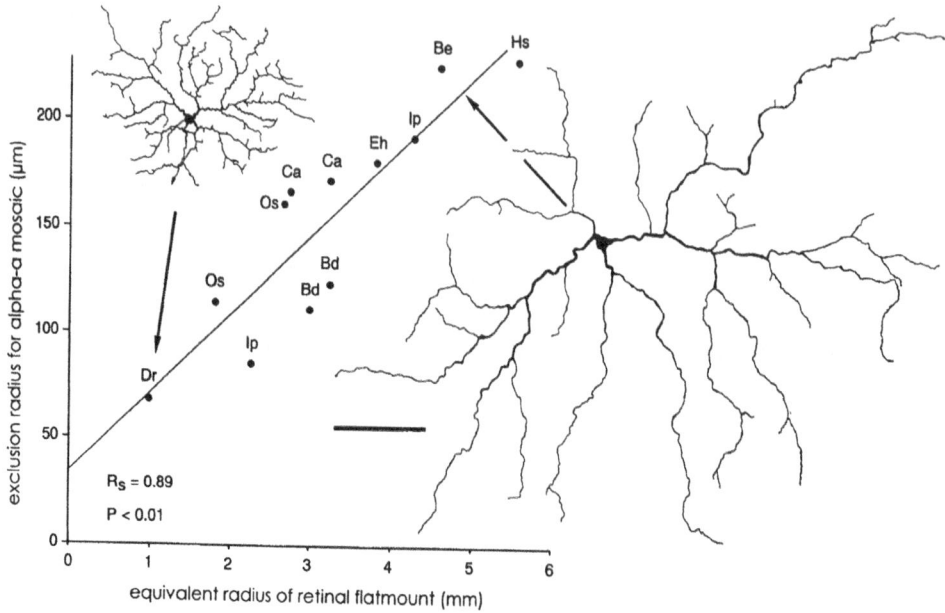

Figure 6. A correlation between retinal size and the spacing of the alpha-a ganglion cell mosaic across eight species of teleost fish, implying broad conservation of alpha-a patterning and function. The eight species (identified by their initials) are the zebrafish *Danio rerio*, the tilapid cichlid *Oreochromis spilurus*, the channel catfish *Ictalurus punctatus*, the goldfish *Carassius auratus*, the ronquil *Bathymaster derjugini*, the six-lined prickleback *Ernogrammus hexagrammus*, the sculpin *Bero elegans*, and the white-spotted greenling *Hexagrammos stelleri*. The zebrafish, catfish and goldfish (Dr, Ip, Ca) are relatively primitive ostariophysians; the others (Os, Bd, Eh, Be, Hs) are advanced neoteleosts. The spacing of the alpha-a mosaic is represented by its exclusion radius (see Section 4.3). In those species where the retina showed clear density gradients, the sample was taken from mid-peripheral retina well away from the region of peak density. Retinal size, expressed as equivalent radius to give both axes the same dimensionality, was determined from the total flatmount area. For four species, retinae of two different sizes are included, giving twelve data points in all. The correlation coefficient and associated P-value are from Spearman's rank correlation test. The small alpha-a ganglion cell (upper left) is from an adult zebrafish, *Danio rerio*; the large cell (lower right) is from an adult sculpin, *Bero elegans*, to the same scale. Scale bar = 100 μm.

The relationship between the large ganglion cells of mammals and non-mammals remains obscure. In particular, for reasons discussed by Cook and Sharma (1995), the outer and inner alpha cells of mammals are *not* strong candidates for one-on-one homology with non-mammalian alpha-a and alpha-b/ab cells. Mammalian inner alpha cells might perhaps be related to alpha-c cells; but their late and activity-dependent emergence, along with outer alpha cells, from a single bistratified precursor population (Chalupa, this volume) suggests another intriguing possibility: that *both* of the mammalian alpha cell mosaics might be descended from the ancestral bistratified alpha-ab lineage. If this is so, then any mammalian descendants of the more sparsely branched alpha-a and alpha-c types might lie within the heterogenous 'gamma' grouping.

7. MOSAICS BEYOND THE RETINA: A SPECULATION

My final question is this: 'How did these beautiful mosaics of wide-field neurons first come into existence?' Clearly, they now serve important roles in the adult, ensuring that coverage of the visual world is both massively parallel and minimally redundant, and at the same time reducing spatial noise in the mapping of visual scenes onto the brain (Snyder et al., 1990). Yet, in evolutionary terms, some of these functional roles may have been latecomers, and the regularity a mere by-product. Mosaics may first have arisen because their establishment efficiently determines other factors just as important as regularity: the total number of neurons of each type in the retina, their global distribution pattern, and the amount of territory allocated to each one. These factors must be important in many parts of the nervous system, not only in the retina, so mosaics may reflect ancestral mechanisms of developmental control that pre-date the vertebrate eye.

Studies of the lamprey retina, which shows many primitive features, point to an evolutionary origin of the retina from a section of diencephalic wall that was little different in laminar structure or cell constitution from the present-day midbrain (Fritzsch, 1991). The logical inference is that wide-field neuronal mosaics may exist elsewhere in the brain, whether or not they have a similar functional role in partitioning sensory-receptive fields among neurons. We would expect to be largely ignorant about these mosaics, of course, because the retina is more accessible and far more often studied in flatmounts than any other part of the brain, and probably no-one has looked for them elsewhere. Indeed, even in the retina, wide-field mosaics are hard to detect in histological sections.

If this speculation is correct, then the spatial patterning of neuronal mosaics is not just of concern to students of visual mechanisms. It could also help to solve conundrums about between-type and within-type variation in other parts of the central nervous system.

8. SUMMARY AND CONCLUSIONS

Any survey of the retinal literature shows 'cell type' taking on meanings as diverse and potentially confusing as those that 'recent' might take on in debates between a palaeontologist and a political historian. This laxity, though understandable, does a disservice to all disciplines and especially to multidisciplinary and evolutionary studies.

Rigorous approaches to classifying neurons and demonstrating their natural types have been available (though underexploited) for many years, but questions of neuronal ancestry and homology have been ignored or rejected as intractable. This is partly because the major classes of the vertebrate retina are themselves of unknown origin, and partly because experimental efforts have been focused on a few model species. There have been very few serious attempts to establish patterns of cross-species homology for individual neuronal types, even though such studies could help us to understand their functional roles and ecological adaptations.

However, for many wide-field neurons, Nature has provided clear demonstrations of natural types in the form of regular mosaics. Advances in analytical technique have made their evaluation easier, and the spatial crosscorrelogram in particular now provides a sound basis for establishing mosaic independence, which is crucial to the demonstration of natural types. By making the plausible Darwinian assumption that mosaics of wide-field neurons have evolved to eliminate 'perceptual holes' in the coverage of individual types, we can view the variation of form and function within each mosaic as a guide to the variation within the associated type, and equate between-mosaic variation with between-type variation. Data obtained on this basis raise interesting questions about possible effects of within-mosaic variation on vision, and challenge the assumption that retinal neurons occupy a multilevel classification hierarchy.

More importantly, the use of mosaics to reveal types allows many new, testable hypotheses of neuronal type to be generated at low cost across a much wider range of species than ever before, opening up new opportunities for testing hypotheses of homology. Empirically, some of the emergent properties of mosaics (relative numbers, spacings and associated levels of regularity) seem promising as additional characters for establishing homologies.

Some first steps have already been taken in this direction for the largest ganglion cells of mammals, and for comparable cells in fishes, frogs and a reptile. It seems likely that all mammals share a consensus pattern involving two mosaics of large monostratified (alpha) ganglion cells in a complementary laminar arrangement. In contrast, many anamniotes and at least one reptile share a different consensus pattern involving three large ganglion cell mosaics and including both bistratified (alpha-ab) and monostratified (alpha-a, alpha-c) cells. The evolutionary relationship between these two consensus patterns is still obscure. However, their differences might be explained either by the destructive effects of neuronal death on any pre-existing spatial order, which may have led mammals to evolve secondary mechanisms of spatial ordering, or by the increasing dominance of the retino-geniculo-cortical pathway, which may have favoured functional complementarity rather than the extraction of specialized visual features.

ACKNOWLEDGMENTS

I thank Steve Easter and Leo Peichl for their comments on a draft of this article, Peter Hitchcock for providing the previously unpublished images shown here as Fig. 4, and Karl Howells, Sergei Kondrashev and Tanja Podugolnikova for preparing some of the previously unpublished material included in Fig. 6. The work on marine teleost mosaics was partially supported by a Joint Project Grant to J.E.C., T.A.P. and S.L.K. from the Royal Society of London (638072.P746), and a grant from the Russian Foundation for Basic Research to T.A.P. and S.L.K. (95-04-11076a). The work on frog mosaics was partially supported by a Scholarship from the Association of Commonwealth Universities to K.M. Shamim, and grants from the Hungarian Scientific Research Council to P. Tóth (ETT T-04, T-568/1990).

REFERENCES

Ammermüller, J. and Kolb, H., 1995, The organization of the turtle inner retina. I. ON- and OFF-center pathways. J. Comp. Neurol. 358: 1–34.

Asenjo, A.B., Rim, J. and Oprian, D.D., 1994, Molecular determinants of human red/green color discrimination. Neuron 12: 1131–1138.

Becker, D.L. and Davies, C.S., 1995, Gap junctions in the mouse embryo. Microsc. Res. Tech. 31: 364–374.

Bloomfield, S.A. and Hitchcock, P.F., 1991, Dendritic arbors of large-field ganglion cells show scaled growth during expansion of the goldfish retina: a study of morphometric and electrotonic properties. J. Neurosci. 11: 910–917.

Bloomfield, S.A. and Xin, D., 1994, Relationship between receptive field and tracer-coupling size of amacrine and ganglion cells in the rabbit retina. Invest. Ophthalmol. Vis. Sci. 35: 1822.

Bloomfield, S.A., Xin, D. and Persky, S.E., 1995, A comparison of receptive field and tracer coupling size of horizontal cells in the rabbit retina. Visual Neurosci. 12: 985–999.

Bloomfield, S.A., Xin, D. and Osborne, T., 1997, Light-induced modulation of coupling between AII amacrine cells in the rabbit retina. Visual Neurosci. 14: 565–576.

Bowers, C.W., 1994, Superfluous neurotransmitters? Trends Neurosci. 17: 315–320.

Boycott, B.B. and Wässle, H., 1974, The morphological types of ganglion cells of the domestic cat's retina. J. Physiol. (Lond.) 240: 397–419.

Boycott, B.B. and Wässle, H., 1991, Morphological classification of bipolar cells of the primate retina. Eur. J. Neurosci. 3: 1069–1088.

Cook, J.E., 1996, Spatial properties of retinal mosaics: an empirical evaluation of some existing measures. Visual Neurosci. 13: 15–30.

Cook, J.E. and Becker, D.L., 1991, Regular mosaics of large displaced and non-displaced ganglion cells in the retina of a cichlid fish. J. Comp. Neurol. 306: 668–684.

Cook, J.E. and Becker, D.L., 1995, Gap junctions in the vertebrate retina. Microsc. Res. Tech. 31: 408–419.

Cook, J.E. and Noden, A.J., 1997, Somatic and dendritic mosaics formed by large ganglion cells in the retina of the common house gecko (*Hemidactylus frenatus*). Brain, Behav. Evol., in press.

Cook, J.E. and Sharma, S.C., 1995, Large retinal ganglion cells in the channel catfish, *Ictalurus punctatus*: three types with distinct dendritic stratification patterns form similar but independent mosaics. J. Comp. Neurol. 362: 331–349.

Cook, J.E., Becker, D.L. and Kapila, R., 1992, Independent mosaics of large inner- and outer-stratified ganglion cells in the goldfish retina. J. Comp. Neurol. 318: 355–366.

Cook, J.E., Kondrashev, S.L. and Podugolnikova, T.A., 1996, Biplexiform ganglion cells, characterized by dendrites in both outer and inner plexiform layers, are regular mosaic-forming elements of the teleost fish retina. Visual Neurosci. 13: 517–528.

Dacey, D.M. and Brace, S., 1992, A coupled network for parasol but not midget ganglion cells in the primate retina. Visual Neurosci. 9: 279–290.

Dunn-Meynell, A.A. and Sharma, S.C., 1986, The visual system of the channel catfish (*Ictalurus punctatus*). I. Retinal ganglion cell morphology. J. Comp. Neurol. 247: 32–55.

Euler, T. and Wässle, H., 1995, Immunocytochemical identification of cone bipolar cells in the rat retina. J. Comp. Neurol. 361: 461–478.

Frank, B.D. and Hollyfield, J.G., 1987, Retinal ganglion cell morphology in the frog, *Rana pipiens*. J. Comp. Neurol. 266: 413–434.

Fritzsch, B., 1991, Ontogenetic clues to the phylogeny of the visual system. In 'The Changing Visual System', Eds. P. Bagnoli and W. Hodos. Plenum Press, London, pp. 33–49.

Gaze, R.M. and Grant, P., 1992, Spatio-temporal patterns of retinal ganglion cell death during *Xenopus* development. J. Comp. Neurol. 315: 264–274.

Hidaka, S., Maehara, M., Umino, O., Lu, Y. and Hashimoto, Y., 1993, Lateral gap junction connections between retinal amacrine cells summating sustained responses. NeuroReport 5: 29–32.

Hitchcock, P.F., 1989, Exclusionary dendritic interactions in the retina of the goldfish. Development 106: 589–598.

Hitchcock, P.F., 1993, Neurobiotin coupling between developing ganglion cells in the retina of the goldfish. Invest. Ophthalmol. Visual Sci. 34: 878.

Hitchcock, P.F., 1997, Tracer coupling among regenerated amacrine cells in the retina of the goldfish. Visual Neurosci. 14: 463–472.

Hutsler, J.J. and Chalupa, L.M., 1995, Development of neuropeptide Y immunoreactive amacrine and ganglion cells in the pre- and postnatal cat retina. J. Comp. Neurol. 361: 152–164.

Jacob, F., 1977, Evolution and tinkering. Science 196: 1161–1166.

Johns, P.R., 1977, Growth of the adult goldfish eye. III. Source of the new retinal cells. J. Comp. Neurol. 176: 343–358.

Kier, C.K., Buchsbaum, G. and Sterling, P., 1995, How retinal microcircuits scale for ganglion cells of different size. J. Neurosci. 15: 7673–7683.

Kock, J., Mecke, E., Orlov, O.Y., Reuter, T., Väisänen, R.A. and Wallgren, J.E., 1989, Ganglion cells in the frog retina: discriminant analysis of histological classes. Vision Res. 29: 1–18.

Kolb, H., Nelson, R. and Mariani, A., 1981, Amacrine cells, bipolar cells and ganglion cells of the cat retina: a Golgi study. Vision Res. 21: 1081–1114.

Liebman, P.A. and Entine, G., 1968, Visual pigments of frog and tadpole (*Rana pipiens*). Vision Res. 8: 761–775.

Marc, R.E., 1982, Spatial organization of neurochemically classified interneurons of the goldfish retina. I. Local patterns. Vision Res. 22: 589–608.

Mastronarde, D.N., 1983, Interactions between ganglion cells in the cat retina. J. Neurophysiol. 49: 350–365.

Northcutt, R.G., 1984, Evolution of the vertebrate central nervous system: Patterns and Processes. Amer. Zool. 24: 701–716.

Peichl, L., 1991, Alpha ganglion cells in mammalian retinae: common properties, species differences, and some comments on other ganglion cells. Visual Neurosci. 7: 155–169.

Peichl, L., Ott, H. and Boycott, B.B., 1987, Alpha ganglion cells in mammalian retinae. Proc. Roy. Soc. (Lond.) B 231: 169–197.

Perry, V.H., 1989, Dendritic interactions between cell populations in the developing retina. In 'Development of the Vertebrate Retina' Eds. B.L. Finlay and D.R. Sengelaub, Plenum Press, New York, pp. 149–172.

Podugolnikova, T.A., 1985, Morphology of bipolar cells and their participation in spatial organization of the inner plexiform layer of jack mackerel retina. Vision Res. 25: 1843–1851.

Ramírez, V. and Ulfhake, B., 1991, Postnatal development of cat hind limb motoneurons supplying the intrinsic muscles of the foot sole. Dev. Brain Res. 62: 189–202.

Raymond, P.A., Barthel, L.K., Rounsifer, M.E., Sullivan, S.A. and Knight, J.K., 1993, Expression of rod and cone visual pigments in goldfish and zebrafish: A rhodopsin-like gene is expressed in cones. Neuron 10: 1161–1174.

Reh, T.A., 1989, The regulation of neuronal production during retinal neurogenesis. In 'Development of the Vertebrate Retina' Eds. B.L. Finlay and D.R. Sengelaub, Plenum Press, New York, pp. 43–67.

Rodieck, R.W., 1991, The density recovery profile: A method for the analysis of points in the plane applicable to retinal studies. Visual Neurosci. 6: 95–111.

Rodieck, R.W. and Brening, R.K., 1983, Retinal ganglion cells: Properties, types, genera, pathways and trans-species comparisons. Brain Behav. Evol. 23: 121–164.

Rodieck, R.W. and Marshak, D.W., 1992, Spatial density and distribution of choline acetyltransferase immunoreactive cells in human, macaque, and baboon retinas. J. Comp. Neurol. 321: 46–64.

Rowe, M.H. and Stone, J., 1977, Naming of neurons: Classification and naming of cat retinal ganglion cells. Brain Behav. Evol. 14: 185–216.

Rowe, M.H. and Stone, J., 1980, The interpretation of variation in the classification of nerve cells. Brain, Behav. Evol. 17: 123–151.

Shamim, K.M., Tóth, P. and Cook, J.E., 1997a, Large retinal ganglion cells in the pipid frog *Xenopus laevis* form independent, regular mosaics resembling those of teleost fish. Visual Neurosci., in press.

Shamim, K.M., Scalia, F., Tóth, P. and Cook, J.E., 1997b, Large retinal ganglion cells that form independent, regular mosaics in the ranid frogs *Rana esculenta* and *Rana pipiens*. Visual Neurosci., in press.

Smith, R.G. and Vardi, N., 1995, Simulation of the AII amacrine cell of mammalian retina: functional consequences of electrical coupling and regenerative membrane properties. Visual Neurosci. 12: 851–860.

Snyder, A.W., Bossomaier, T.J. and Hughes, A.A., 1990, The theory of comparative eye design. In 'Vision: Coding and Efficiency', Ed. C. Blakemore, Cambridge University Press, Cambridge, pp. 45–52.

Stone, J., 1988, The origins of the cells of vertebrate retina. Prog. Ret. Res. 7: 1–19.

Straznicky, K. and Gaze, R.M., 1971, The growth of the retina in *Xenopus laevis*: an autoradiographic study. J. Embryol. Exp. Morphol. 26: 67–79.

Van Haesendonck, E. and Missotten, L., 1979, Synaptic contacts of the horizontal cells in the retina of the marine teleost, *Callionymus lyra* L. J. Comp. Neurol. 184: 167–192.

Vaney, D.I., 1991, Many diverse types of retinal neurons show tracer coupling when injected with biocytin or Neurobiotin. Neurosci. Lett. 125: 187–190.

Vaney, D.I., 1994a, Patterns of neuronal coupling in the retina. Prog. Ret. Eye Res. 13: 301–355.

Vaney, D.I., 1994b, Territorial organization of direction-selective ganglion cells in rabbit retina. J. Neurosci. 14: 6301–6316.

Vaney, D.I. and Hughes, A.A., 1990, Is there more than meets the eye? In 'Vision: Coding and Efficiency', Ed. C. Blakemore, Cambridge University Press, Cambridge, pp. 74–83.

Wässle, H. and Riemann, H.J., 1978, The mosaic of nerve cells in the mammalian retina. Proc. Roy. Soc. (Lond.) B 200: 441–461.

Wässle, H., Boycott, B.B. and Röhrenbeck, J., 1989, Horizontal cells in the monkey retina: cone connections and dendritic network. Eur. J. Neurosci. 1: 421–435.

Wässle, H., Chun, M.H. and Müller, F., 1987, Amacrine cells in the ganglion cell layer of the cat retina. J. Comp. Neurol. 265: 391–408.

Wässle, H., Peichl, L. and Boycott, B.B., 1981, Morphology and topography of on- and off-alpha cells in the cat retina. Proc. Roy. Soc. (Lond.) B 212: 157–175.

White, C.A., Chalupa, L.M., Johnson, D. and Brecha, N.C., 1990, Somatostatin-immunoreactive cells in the adult cat retina. J. Comp. Neurol. 293: 134–150.

White, C.A. and Chalupa, L.M., 1991, Subgroup of alpha ganglion cells in the adult cat retina is immunoreactive for somatostatin. J. Comp. Neurol. 304: 1–13.

Williams, R.W., Bastiani, M.J., Lia, B. and Chalupa, L.M., 1986, Growth cones, dying axons, and developmental fluctuations in the fiber population of the cat's optic nerve. J. Comp. Neurol. 246: 32–69.

Williams, R.W. and Herrup, K., 1988, The control of neuron number. Ann. Rev. Neurosci. 11: 423–453.

Williams, R.W., Cavada, C. and Reinoso-Suárez, F., 1993, Rapid evolution of the visual system: A cellular assay of the retina and dorsal lateral geniculate nucleus of the Spanish wildcat and the domestic cat. J. Neurosci. 13: 208–228.

Witkovsky, P. and Dearry, A., 1991, Functional roles of dopamine in the vertebrate retina. Prog. Ret. Res. 11: 247–292.

Xiang, M., Zhou, H. and Nathans, J., 1996, Molecular biology of retinal ganglion cells. Proc. Natl. Acad. Sci. USA 93: 596–601.

8

GLIO–NEURONAL INTERACTIONS IN RETINAL DEVELOPMENT

Andreas Reichenbach, Angela Germer, Andreas Bringmann,
Bernd Biedermann, Thomas Pannicke, Mike Francke, Heidrun Kuhrt,
Winfried Reichelt, and Andreas Mack

Paul Flechsig Institute for Brain Research
Department of Neurophysiology
Leipzig University
Jahnallee 59, D-04109 Leipzig, Germany

1. INTRODUCTION: MÜLLER CELLS IN THE MATURE RETINA

Müller cells have been found in the retinae of all vertebrates where they constitute the dominant type of macroglia. They have a bipolar morphology ("radial glia") with their vitread (inner) trunk terminating in a conical endfoot adjacent to the vitreous body, and their opposite end extending apical microvilli into the subretinal space which is a main source of nutrients and oxygen delivered by the choriocapillary circulation. In the adult retina, their side branches form elaborate sheaths around neuronal somata, dendrites and synapses, and fascicles of optic axons (cf. Reichenbach and Robinson, 1995a).

This intimate morphological contact with retinal neurons is supplemented by the physiological features of mature Müller cells. Their membranes display a variety of voltage-gated ion channels including K^+, Na^+, and Ca^{2+} channels. In particular, the inwardly rectifying K^+ channels ($I_{K(IR)}$) of Müller cells seem to play a crucial role for (i) the maintenance of the characteristic high resting membrane potential, and (ii) the extracellular ionic homeostasis of the retina, by mediating spatial buffering currents (left part of Fig. 1). Further, ligand-gated channels have been observed for glutamate, γ-aminobutyric acid (GABA), and other signal molecules. Moreover, the repertoire of membrane transport proteins involves a glutamate uptake system and a GABA transporter, a cysteine/glutamate exchanger as well as a Na^+/HCO_3^- cotransporter and other acid/base transporters. Most of these transporters are dependent on the high membrane potential maintained by the presence of a large $I_{K(IR)}$, and by the action of the Na^+, K^+ pump which also constitutes a second tool of extracellular K^+ homeostasis. Müller cells also possess enzymes (e.g., glutamine synthetase) for degradation of neuroactive substances. These uptake carriers and enzymes are involved in the control of extracellular neuroactive substances, as well as in a series of mechanisms summarized as "transmitter recycling" (middle part of Fig. 1).

Development and Organization of the Retina, edited by Chalupa and Finlay.
Plenum Press, New York, 1998.

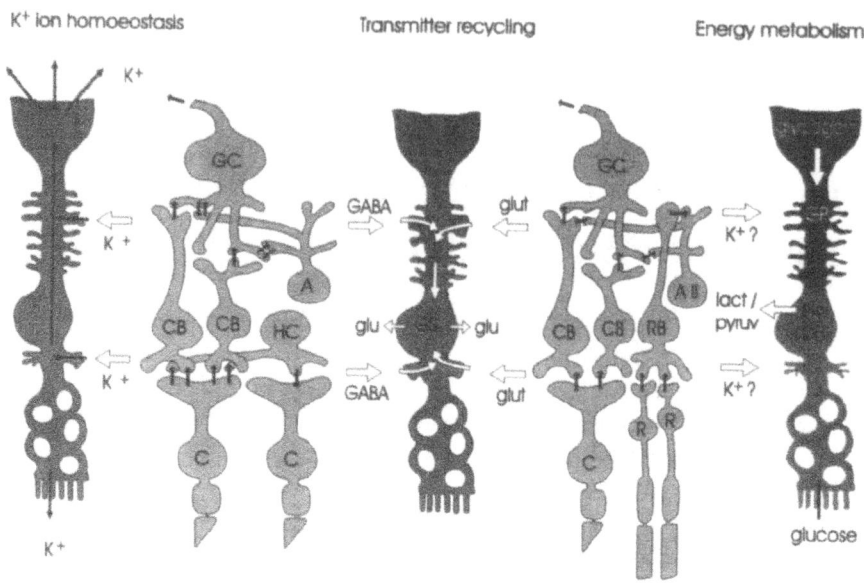

Figure 1. Summary of glio-neuronal interactions in the retina. Three Müller cells are drawn, each representing one of the main glial functions. Among them, two simplified neuronal assemblies are shown, containing retinal ganglion cells (GC), unspecified (A) or type II (A II) amacrine cells, cone (CB) and rod (RB) bipolar cells, a horizontal cell (HC), and cone (C) and rod (R) photoreceptor cells. The neurons release signals such as K^+ ions, and neurotransmitters such as glutamate (glut) and γ-aminobutyric acid (GABA). The arrows through the left Müller cell represent the fast redistribution of excess K^+ ions into the vitreous (top) and/or the subretinal space (bottom), by K^+ channel-mediated "K^+ siphoning" (Newman et al., 1984). The middle Müller cell is shown to possess active uptake carriers for glutamate and GABA, and the enzyme glutamine synthetase (GS) which is involved in converting the neuroactive substrates into the inert molecule glutamine (glu). Glutamine is then released by Müller cells, taken up by neurons, and used for re-synthesis of neurotransmitter molecules (Pow and Robinson, 1994). The right Müller cell represents the glial glycogen stores. Release of signals (K^+?) by active neurons activates the glial glycogen phosphorylase (GP), causing enhanced glycogenolysis (and glycolysis). By means of the glia-specific enzymes pyruvate kinase (PK) and lactate dehydrogenase (LDH), Müller cells release pyruvate and/or lactate to fuel the Krebs cycle of the neurons (Poitry-Yamate et al., 1995).

Finally, the glycogenolytic and glycolytic metabolism of Müller cells is thought to provide lactate/pyruvate fueling the Krebs cycle of retinal neurons (right part of Figure 1; for recent reviews of retinal glio-neuronal interactions, see Reichenbach and Robinson 1995b; Newman and Reichenbach, 1996; Reichenbach et al., 1997).

Considerations about these elaborate functional interactions between retinal neurons and Müller cells have led to the suggestion that the retinal tissue is composed of radial "units". Each unit consists of one Müller cell and a defined number of neighbouring retinal neurons which are ensheathed, and functionally supported, by this Müller cell (Reichenbach et al., 1993, 1994). The size of these units seems to be rather uniform within the retina of a given species (Reichenbach et al., 1994) but may vary greatly among different vertebrates. A range between about 6 and more than 80 neurons per Müller cell has been reported (cf. Reichenbach and Robinson, 1995c). Small units are characteristic for poorly developed or primitive retinae, but also for cone-dominant mammalian retinae, whereas large units may occur in teleosts (Mack et al., 1997) and amphibia, and in rod-dominant mammalian retinae. Specifically in mammalian retinae, a strict correlation was observed between the size of units (i.e., neuron-to-Müller cell ratios), and the rod-to-cone

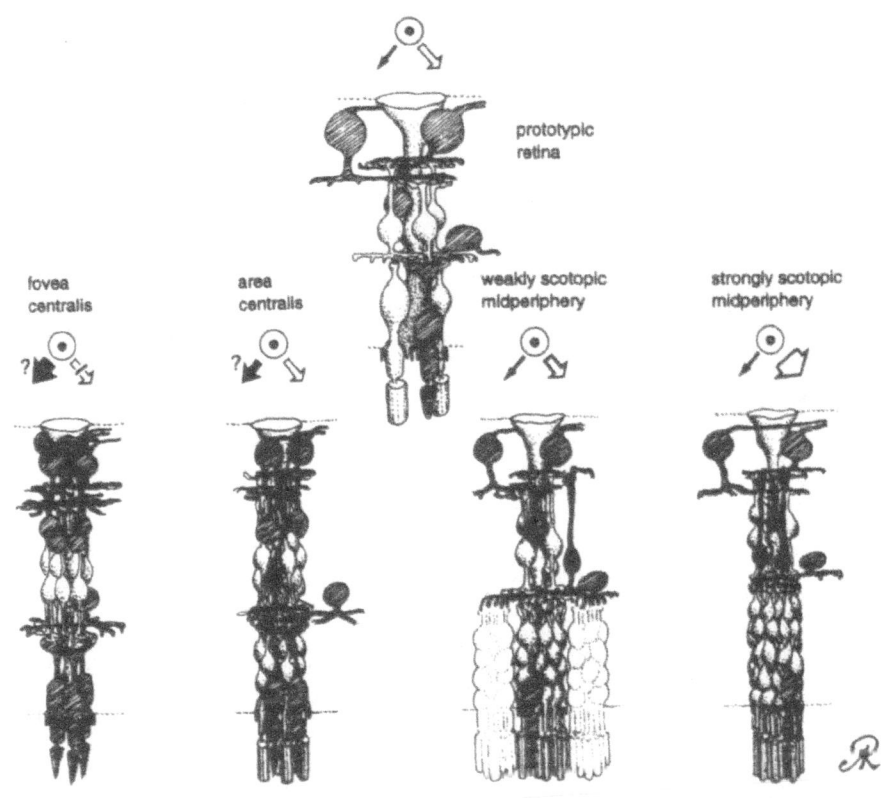

Figure 2. Schematic view of the cellular compositions of glio-neuronal units in different types of mammalian retinae (or retinal regions). Originating from a hypothetical prototypic retina (top), increasing photopic specialisation is demonstrated to the left (area and fovea centralis), and increasing scotopic (nocturnal) specialization to the right. Hatched cells are the progeny of early progenitors (see Fig. 4), and comprise the key elements of photopic vision (ganglion cells, most amacrine cells, horizontal cells and cones). White cells are generated by late progenitors (see Fig. 4); among them are Müller cells, bipolar cells, and some amacrine cells, as well as the key elements of scotopic vision, the rods. Rod bipolar cells are shown in black. The trend towards nocturnality involves an enhanced number of rods per unit, that is, a greater input from the late progenitors (increased thickness of the white arrows). The cellular composition of the area centralis and fovea seems to require more "photopic circuit neurons" (e.g., ganglion cells and cones) per unit (see Fig. 3), due to additional replication of the early progenitors (indicated by the black arrows). The crossed white arrow represents inhibition of rod genesis at the fovea. The numerical size of the units varies between about 8 (prototypic retina) and more than 30 (strongly scotopic midperiphery) neurons per Müller cell. With permission, from Reichenbach and Robinson (1995c).

ratios (Reichenbach and Robinson, 1995c; Chao et al., 1997). This means that in "radial units" of mammals, a Müller cell is confronted to the metabolic needs of a fairly uniform number (5–8) of "non-rod neurons" plus a widely varying number (<1–>30) of rods (Fig. 2). As rods are known to be metabolically very active cells (cf. Ames et al., 1992), the metabolic burden of mature Müller cells should greatly differ among mammalian species. A particular case is the fovea centralis of primates, which is devoid of rods but contains an enhanced number of ganglion cells and cones per Müller cell, in contrast to the rod-dominated retinal periphery (Fig. 3). Further metabolic differences may be due to the fact that the retina is partially or entirely vascularized in many mammals, but not in others (cf.

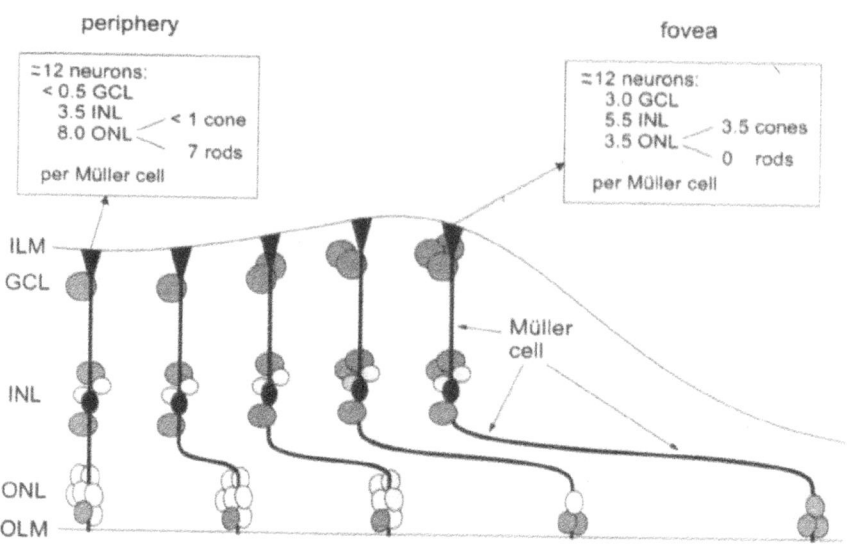

periphery

≈12 neurons:
< 0.5 GCL
3.5 INL   < 1 cone
8.0 ONL
               7 rods
per Müller cell

fovea

≈12 neurons:
3.0 GCL
5.5 INL   3.5 cones
3.5 ONL
               0 rods
per Müller cell

ILM
GCL

Müller
cell

INL

ONL
OLM

Figure 3. Neuron-to-Müller cell ratios in the foveal and peripheral regions of an adult rhesus monkey retina. Müller cells were labeled by vimentin-immunocytochemistry, and cell nuclei were counter-stained by hematoxylin. Local densities of Müller cells and neuronal cell nuclei were counted along the course of Müller cell processes, assuming that these indicate the arrangement of units which changes in the course of foveation (Kuhrt et al., unpublished results). The total numbers of neurons per Müller cell (about 12) are strikingly similar across the retina. However, peripheral units are dominated by rods (≈60% of neurons) whereas the foveal units are devoid of rods but contain strongly enhanced numbers of cones and ganglion cells. The darkly hatched cells represent early-born neurons whereas lightly hatched cells are the progeny of the late progenitors. ILM = inner limiting membrane, GCL = (cell nuclei in the) ganglion cell layer, INL = (neuronal cell nuclei in the) inner nuclear layer, ONL = (cell nuclei in the) outer nuclear layer, OLM = outer "limiting membrane".

Michaelson, 1954), or to species- or region-specific differences in the abundance of certain types of synapses and/or neurotransmitters (cf. Dowling, 1987).

In addition to the wealth of (specific) functional demands of neuronal information processing within normal adult retinae, various retinal injuries or diseases constitute a challenge to Müller cells. In contrast to retinal neurons which are permanently postmitotic cells, Müller cells may respond to pathophysiological stimuli by cell proliferation (Erickson et al., 1983; Geller et al., 1995) and, thus, contribute to the formation of specific glial scars ("preretinal membranes") that may cause retinal detachment and subsequent blindness (e.g., Kono et al., 1995).

Even this short introduction raises several intriguing problems related to the development of Müller cells. This article is an attempt to summarize currently available, albeit preliminary, answers to the following questions:

1. what is the developmental origin of Müller cells?
2. how are the specific types of radial units (e.g., neuron-to-Müller cell ratios) generated?
3. how do Müller cells adjust their functional capacities to the demands of the neurons within the developing units?
4. what do Müller cells contribute to the development of the neighbouring neurons?
5. what can we learn from normal development to understand the behavior of Müller cells in retinal pathology?

2. ORIGIN OF MÜLLER CELLS AND EARLY DEVELOPMENTAL INTERACTIONS

The origin of Müller cells is still poorly understood. One of the challenging problems is that presumptive Müller cells have been recognized very early in ontogenetic development by immunocytochemistry and in Golgi preparations, but ^{3}H-thymidine autoradiographic cell "birthday" dating and retroviral cell lineage studies have only succeeded in labeling permanent Müller cells toward the end of cytogenesis (for review, see Robinson, 1991). Müller cells are generated by the same type of "late" progenitor cells that also generate rods, bipolar cells, and (subtypes of) amacrine cells (Turner and Cepko, 1987; Germer et al., 1997a). There are many intriguing similarities between these late progenitor cells and Müller cells. Both are long radially oriented cells that express the intermediate filament-protein vimentin and the enzyme carbonic anhydrase, as well as other cellular antigens; furthermore, both are capable of (re-) entering the cell cycle (for reviews, see Linser and Perkins, 1987; Polley et al., 1989; Reichenbach and Robinson, 1995a). Studies involving the neurogenic gene *Xotch* have shown that the final *Xotch*-positive precursor cells mostly differentiate as Müller glia, suggesting that this is the last available fate of cells in the frog retina (Dorsky et al., 1995). Summarizing these data, a picture evolves that may solve some of the inconsistencies. It seems to be safe to say that (at least newborn immature) Müller cells are closely related to (late) progenitor cells, from which they arise by a transition that is barely detectable at the phenotypic level (and might even be reversible in some lower vertebrates: Braisted et al., 1994). The result of this transition, termed a retinal "radial glial cell", will usually differentiate as a mature Müller cell which is phenotypically distinct from both the late progenitor and early radial glial cells, by (i) expressing glial cell-specific enzymes and membrane properties, and (ii) having a high threshold to mitogenic stimuli (see following sections).

These peculiarities of the origin of Müller cells have been considered in constructing a schematical diagram of cytogenesis and cell fate decision in the (mammalian) retina (Fig. 4). The diagram is aimed to account both for the way different sizes and types of Müller cell-supported radial units may be generated, and the particular roles played by Müller cells in later stages of retinal development and proliferative retinal diseases.

The details of the proposed mode(s) of progenitor cell proliferation have been described and discussed earlier (Reichenbach and Robinson, 1995c). Briefly, in a first phase of cell proliferation (I), identical replication generates a species-specific number of omnipotent stem cells that (at least, in mammals) more or less determines the surface area of the future retina, and, thus, the eye size. The results of early regeneration and transplantation experiments (Spemann, 1912; Mangold, 1931) suggest that the extent of proliferation in this phase is species-specific, and under genetic control ("genetic clock").

In a next phase (II), a few rounds of asymmetrical cell division give rise to early progenitor cells that each divide just once to produce early postmitotic neurons which then differentiate into the "key cell types" of the photopic pathway, *viz.* ganglion cells, cones, horizontal cells, and (subtypes of) amacrine cells. There are reasons to believe that the extent of this phase of proliferation is fairly similar throughout the wide variety of mammals or even vertebrates (Reichenbach and Robinson, 1995c). However, the proliferation of these early progenitor cells can be stimulated by certain growth factors *in vitro* (Fig. 4) and probably is, indeed, stimulated *in vivo* in cases when the "photopic power" of the retina is to be enhanced. Particularly, radial units of the primate fovea contain enhanced numbers of early-born ("photopic") neurons per Müller cell (Figs. 2, 3). Thus, the generation of a fovea (or even an area) centralis may require a local stimulation of early progenitor cell proliferation.

The fate decision of postmitotic neurons generated by this phase (II) is not well understood. There is evidence that such cells may be determined to differentiate as ganglion cells by, e.g., laminin, and to differentiate as cones by, e.g., retinoic acid (Fig. 4) but less is known about factors that would drive such cells to differentiate as horizontal or amacrine cells (Alexiades and Cepko, 1997). In any case, there seems to be a negative feed-back from determined neurons of a given type onto the neighbouring undetermined postmitotic cells (Fig. 4) which may be essential for the generation of the regular distribution mosaics of the various retinal cell types (e.g., Reh, 1987).

In a further phase of proliferation (III), which probably proceeds by symmetrical divisions, a "new" distinct ("late") type of progenitor cell generates late postmitotic neurons and Müller cells (for reviews, see Robinson, 1991; Reichenbach and Robinson, 1995c). This phase of proliferation is apparently controlled by elaborate regulatory mechanisms. *In-vitro* experiments have shown that several factors, different from those which are mitogenic for the early progenitor cells, can stimulate the proliferative activity of these late progenitors (Fig. 4). Considering the fate of the offspring neurons, such a regulation is very likely to occur also *in vivo*. The majority of these neurons differentiate as rods, the remaining as either bipolar or (subtypes of) amacrine cells. It has been shown (Chao et al., 1997; Reichenbach and Robinson, 1995c) that the number of rods per radial unit is high (up to 35) in mammals with nocturnal life style (scotopic dominance), but low (<1) in diurnally active mammals (photopic dominance). This difference could easily be achieved by (species-specific) different degrees of stimulation of the late progenitor proliferation (Reichenbach, 1993). In the foveolar region of primate retinae, this proliferation seems to be actively suppressed since (i) foveolar units contain no rods, and much less progeny of late progenitors, than peripheral units of the same retina (Figs. 2, 3), and (ii) the relative duration of cytogenesis is about half that of other mammals, or that of the periphery (Robinson, 1991). In albinotic humans, this suppression fails to occur and the (morphologically poorly developed) foveolar region contains many rods (e.g., Elschnig, 1913; Naumann et al., 1976; Fulton et al., 1978). This has led to the suggestion that a product of the tyrosinase (the gene of which is defect in albinos) reaction, probably DOPA, is respon-

---

Figure 4. Scheme describing the proposed hypothesis accounting for cell proliferation (left vertical column) and cell fate decision (right branch-lines) in the mammalian retina. The bold frames indicate the various (transient or permanent) cell types generated, the faint frames indicate factors known (or assumed) to stimulate (+) or inhibit (−) certain steps of proliferation and/or transition. Data are shown of experiments on various classes of vertebrates although the scheme is principally adjusted for mammalian retinae (Robinson and Reichenbach, 1995c). For details, see text. The abbreviations used are, aFGF = acidic fibroblast growth factor, bFGF = basic fibroblast growth factor, CNTF = ciliary neurotrophic factor, EGF = epidermal growth factor, glut = glutamate, IGF-I = insulin-like growth factor I, NGF = nerve growth factor, PDGF = platelet-derived growth factor, RA = retinoic acid, RPE = retinal pigment epithelium, TGFα = transforming growth factor α, TGFβ-2/-3 = transforming growth factor β-2/-3. The following references are given in brackets, [1] Adler and Hatlee (1989); [2] Altshuler et al. (1993); [3] Anchan and Reh (1995); [4] Braisted et al., (1994), Rentsch (1973), Anderson et al. (1986), Erickson et al. (1983); [5] Browman and Hawryshin (1994); [6] Cepko (1993); [7] DeCurtis and Reichardt (1993); [8] Dorsky et al. (1997); [9] Calvaruso et al. (1992); [10] Guillemot and Cepko (1992); [11] Harris and Holt (1990); [12] Hitchcock and Vanderyt (1994); [13] Hunter et al. (1992); [14] Hyatt et al. (1996); [15] Ikeda and Puro (1995); [16] Jeffery (1997), Jeffery et al. (1997); [17] Jensen and Wallace (1997); [18] Kelley et al. (1994); [19] Kirsch et al. (1997), Ezzeddine et al. (1997); [20] Layer and Willbold (1993); [21] Layer et al. (1997); [22] Lewis et al. (1992); [23] Lillien and Cepko (1992); [24] Mack and Fernald (1993); [25] Puro (1995); [26] Reh (1991); [27] Reh (1992); [28] Repka and Adler (1992); [29] Reichenbach and Robinson (1995c); [30] Spemann (1912), Mangold (1931).

sible for the normal foveolar suppression of late progenitor cell proliferation (Jeffery 1997; Jeffery et al., 1997; Fig. 4). Another candidate regulator, an as yet uncharacterized peptide, has been identified *in vitro* (Calvaruso et al., 1992). Whatever factor may act *in vivo*, the regulation of the late progenitor proliferation seems to be a very important issue

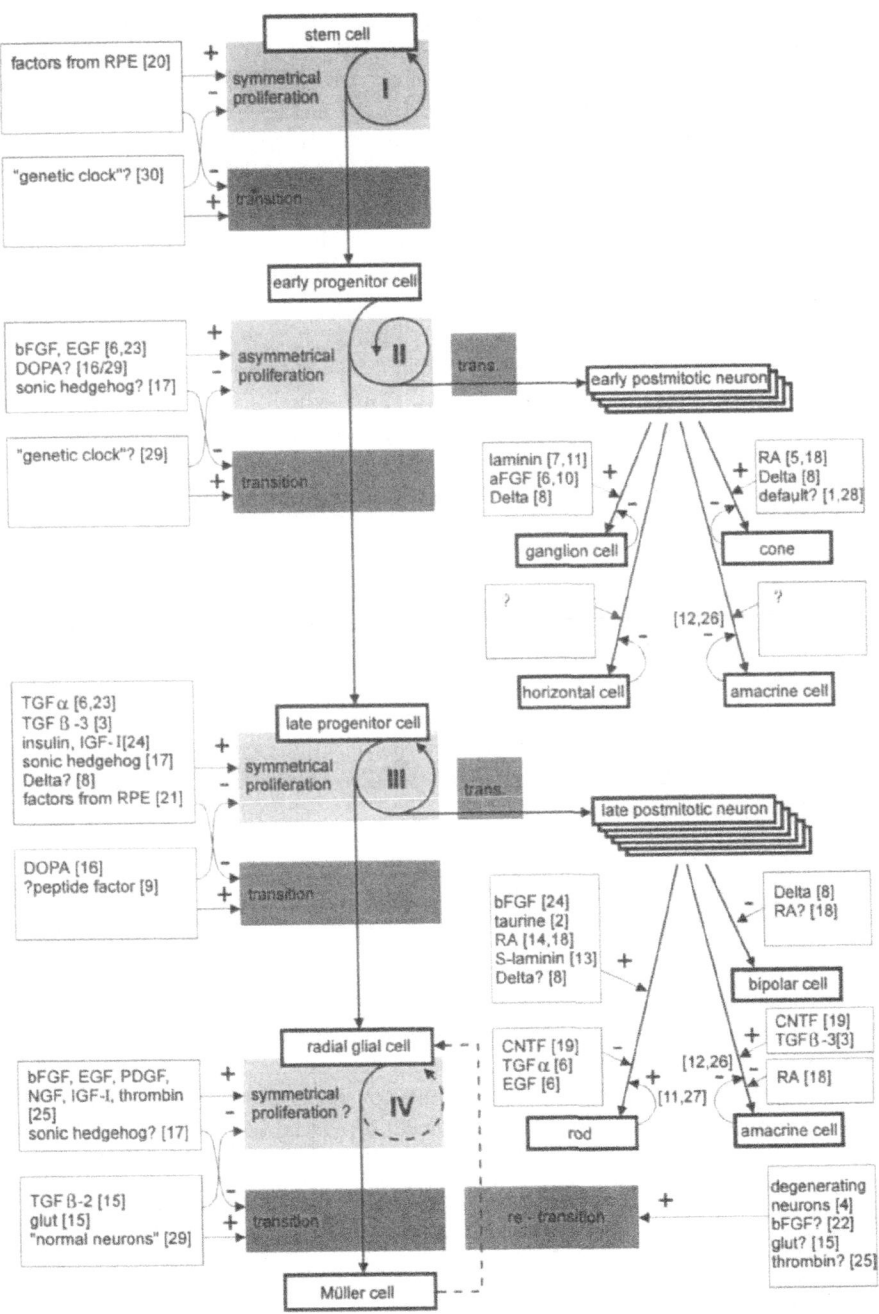

in species- and region-specific retinal organization. The end of this third phase of proliferation is accompanied, or even induced, by the transition of one of the final (Dorsky et al., 1995) daughter cells into a radial glial cell (Fig. 4).

In recent years, many data have been accumulated concerning cell fate decision of the late postmitotic neurons; a variety of factors have been found that modulate these processes, at least *in vitro* (Fig. 4). This is also the stage when Müller cells (which are born almost simultaneously) may exert the first demonstrable influence on developmental decisions of their "sisterly" newborn neurons. Ciliary neurotrophic factor (CNTF) has been shown to reduce the number of cells developing as rods (Kirsch et al., 1997; Ezzeddine et al., 1997), and to enhance the number of cells that become bipolar (Ezzeddine et al., 1997) or perhaps also amacrine (Kirsch et al., 1997) cells, within an unchanged total pool of late postmitotic neurons. It has also been shown that (at least mature) Müller cells are a source of endogenous CNTF (Kirsch et al., 1997). Thus, a scenario becomes feasible in which the young Müller cells stop the emergence of more rods but stimulate the last undetermined postmitotic neurons to differentiate as bipolar or amacrine cells. Such a mechanism could be particularly important for the foveolar region where less late-type neurons are generated (and no rods are required) but the formation of "private pathways" (Rodieck, 1988) for a large number of cones requires an enhanced density of bipolar (and amacrine) cells.

A final phase of proliferation (IV) involving radial glial cells does not occur during normal development, at least not in mammals (Reichenbach and Robinson, 1995c). This is apparent from retroviral cell lineage studies, from double-labeling experiments (for glial-cell specific proteins and proliferations markers), and from numerical relations between adult retinal cell types. Probably, there are signals released from maturing neurons which suppress this proliferation, and enforce transition and differentiation as Müller cells (Fig. 4). There is however ample evidence that in certain cases of retinal (neuronal) degeneration, mature Müller cells may de-differentiate, and re-enter the mitotic cell cycle (Erickson et al., 1983; Fisher et al., 1991; Lewis et al., 1992; Geller et al., 1995; Kono et al., 1995). Although several factors have been found that may initiate this process (Fig. 4), intensive research remains to be done to identify the signal(s) acting under non-experimental conditions. The clinical implications of this problem have already been mentioned, and will be further illustrated in the last section.

3. NEURONAL DIFFERENTIATION: THE ROLE(S) OF MÜLLER CELLS

After the postmitotic neurons have been generated and determined to a distinct cell fate, further important developmental processes are necessary to create a mature retina. These involve, among others: (i) the migration of newborn neurons from the site of their birth (the ventricular—or sclerad—retinal margin) to the site of their final destiny (e.g., the inner—or vitread—retinal layers), (ii) the growth of neurites contacting the prospective synaptic targets, (iii) synaptic competition involving "programmed" cell death of the "loser" cells, and (iv) establishment of the cell type-specific biochemical pathways of energy and transmitter metabolism (for reviews, see Young, 1983; Robinson, 1991). There are several crucial contributions of Müller cells to these processes.

Early-born neurons, such as ganglion cells, seem to emerge from a bipolar stage that allows their soma to arrive at the layer of their destination just by translocation of the nucleus within the radial cytoplasmic "tube" (Hinds and Hinds, 1974, 1979; Morest, 1979).

By contrast, late-born neurons, such as (subtypes of) amacrine and bipolar cells and even rods destined for the inner (vitread) zone of the outer nuclear layer, are born within a tissue which became rather thick. Thus, these cells need guidance for their migration away from the ventricular surface. As is the case for radial glial cells of the brain (Rakic, 1971), Müller cells are well-suited to provide such a guidance for migrating young neurons (Meller and Tetzlaff, 1976; Craft et al., 1983; Raymond and Rivlin, 1987; Reichenbach et al., 1994). Application of a glia cell-specific toxin (α-aminoadipic acid) to retinal explant cultures destroys the regular pattern of Müller cells, and causes severe disturbances in retinal layer formation (Willbold et al., 1997; Germer et al., 1997a) and in the columnar arrangement of retinal cell clones (Willbold et al., 1995). Neuronal ectopy (i.e., disturbed migration of young neurons) has been observed in a tiger retina with Müller cell anomalies (Reichenbach et al., 1992).

Several studies have unequivocally demonstrated a guidance function of Müller cells for growing neurites of retinal neurons. Specifically, the (membranes of) radial glial (Müller) cell endfeet provide an excellent growth substrate for ganglion cell axons (Goldberg, 1977; Halfter and Deiss, 1986; Stier and Schlosshauer, 1995) whereas the non-endfoot (soma) membranes of radial glial cells inhibit the growth of ganglion cell axons (Kljavin and Reh, 1991; Stier and Schlosshauer, 1997) but support neurite extension of rods (Kljavin and Reh, 1991; Hicks et al., 1994). Growth cones of ganglion cell axons collapse when exposed to glial somatic membranes; this collapsing activity was found to be specific for these axons, and was destroyed by heat treatment of glial membranes (Stier and Schlosshauer, 1997). Furthermore, the growth of ganglion cell dendrites is supported by somatic, but inhibited by endfoot-derived, membranes of Müller cells or their immediate precursors (Schlosshauer et al., 1996). Thus, polarized radial glia are likely to affect the development of the neuronal cytoarchitecture, including the polarity of retinal neurons. The expression of cell surface adhesion molecules such as N-CAM (Drazba and Lemmon, 1990; Stier and Schlosshauer, 1995) and L1 or N-cadherin (Drazba and Lemmon, 1990) by Müller cell membranes may be involved in these functions, but additional mechanisms may be required as well (Stier and Schlosshauer, 1995). Another candidate (extracellular matrix) glycoprotein is tenascin-R; it was shown to react with CALEB, a novel member of the EGF family of differentiation factors which is associated with the surfaces of retinal neurites and Müller cells (Schumacher et al., 1997).

In the mammalian retina, over 50% of ganglion cells generated normally die, and substantial cell death has also been observed in the inner and outer nuclear layer (for review, see Robinson, 1991). The majority of the (debris of) dying cells seems to be ingested by microglial cells (Hume et al., 1983; Thanos, 1991; Egensperger et al., 1996), but there is clear evidence that Müller cells are involved in phagocytosis of dying retinal neurons in the period of "programmed" cell death (Penfold and Provis, 1986; Provis and Penfold, 1988; Egensperger et al., 1996). Both in situ and in vitro, Müller cells have a high capacity of rapid phagocytosis, even of rather large particles, and a fast intracytoplasmic transport system for the ingested material (Stolzenburg et al., 1992).

Less is known about the (presumptive) role(s) of Müller cells in the metabolic maintenance and maturation of neurons. In-vitro experiments have shown that (factors from) Müller cells support the survival and differentiation of various types of retinal neurons (Raju and Bennett, 1986; Armson et al., 1987; Hicks et al., 1994). However, (at least certain types of) retinal neurons may achieve a substantial degree of differentiation in dissociated glia-free cultures (e.g., Adler and Hatlee, 1989; Watanabe and Raff, 1990). In adult retinae, the expression of several enzymes such as aldose reductase and carboanhydrase, as well as the presence of glycogen stores, are shared by (some types of) early-born neurons and Müller cells whereas these features are lacking in late-born neurons (for review, see Reichenbach et al., 1993). It has been speculated that this difference may be due to different glio-neuronal relationships during initial stages of neuronal development. The late-born neurons develop in association with (and may "rely upon") their "sisterly" Müller cells which provide necessary metabolites, whereas the early-born neurons emerge within an environment devoid of mature Müller cells, and therefore depend on their own metabolic capacities. Unfortunately, it has not been tested whether rods or bipolar cells express carboanhydrase or aldose reductase, or accumulate glycogen stores, in dissociated glia-free cultures. Further work is required to learn more about the developmental control of metabolic features of retinal neurons.

4. MÜLLER CELL DIFFERENTIATION: THE ROLE(S) OF NEURONS

4.1. Transmitter Recycling

This section will return to the question how Müller cells manage to meet the demands of the neurons within the (greatly variable) units of their "responsibility". A very suitable model to study this question was found in the fish retina. In teleosts, the retina grows virtually life-long by both peripheral addition of new cells and mechanical stretching of the mature major retinal area. The latter process causes a decrease of the cell densities per surface area of the retina, with the exception that the density of rods is kept constant by continuous insertion of newborn rods from specific rod progenitor cells (Johns, 1982; Mack and Fernald, 1993, 1995). This increasing number of rods should constitute a metabolic challenge to the Müller cells. For instance, rods release the neurotransmitter glutamate which must be transformed into glutamine by the glutamine synthetase of Müller cells (Fig. 1). There are two principal ways how Müller cells might meet enhanced demands: (i) by cell proliferation which would keep the neuron-to-Müller cell ratio (i.e., the size of the supported units) at a constant level, or (ii) by elevation of the metabolic capacities of the existing Müller cells (within growing units). This problem has been studied in growing fish of various body lengths, by counting vimentin-immunolabeled Müller cells, and performing quantitative Western blots of glutamine synthetase (GS) protein expression (Mack et al., 1997). The results are shown in Figure 5. There is no generation of new Müller cells in the differentiated major part of the retina; as a result, the number of rods per Müller cell increases about twofold during the period studied (whereas the number of "non-rod" neurons per Müller cell remains constant). However, the GS content per Müller cell roughly doubles within the same period. This means that an unchanged amount of retinal GS is available to every glutamate-releasing rod (Fig. 5).

These data suggest that Müller cells are able to adjust their GS content to the actual demands, depending on their neighbouring neurons. It fits well with the idea that in cases of retinal (rod) degenerations, the GS expression by Müller cells is greatly reduced (LaVail and Reif-Lehrer, 1971; Lewis et al., 1989, 1994; Härtig et al., 1995; Grosche et al., 1995). This raises the question how this regulation is mediated. This has been intensively studied in the chicken retina (Piddington and Moscona, 1967; Linser and Moscona, 1979; among others). It has been shown that the embryonic steep rise in GS expression closely follows the increase in cortisol plasma levels (Fig. 6A), and that GS expression can be precociously induced by external application of cortisol. Whereas much effort has been invested to understand the molecular mechanisms of this hormone-mediated regula-

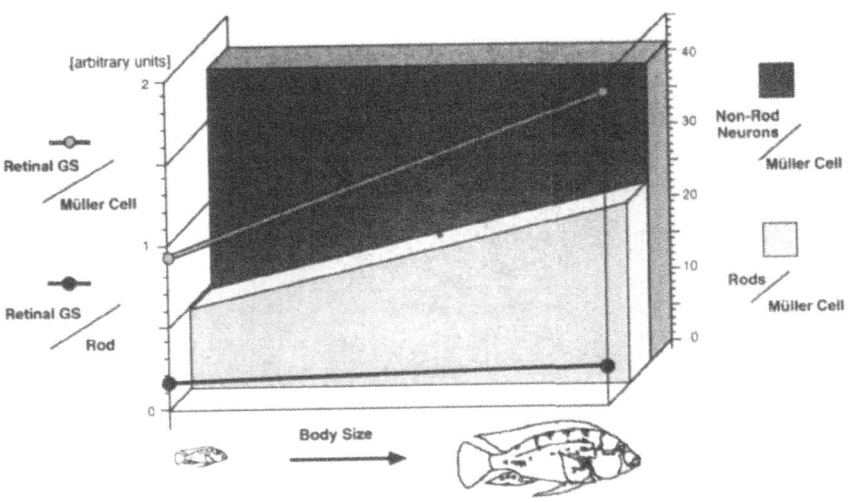

Figure 5. Growth-related changes of glio-neuronal interactions in a teleost (*Haplochromis burtoni*) retina. While the fish grow from 10 to 50 mm body length, new rod photoreceptor cells (but no "non-rod neurons" or Müller cells) are generated in the central retina, which causes a duplication of the rod-to-Müller cell ratio from 10 to 22 (block diagrams, right scale). Within the same period, the glutamine synthetase (GS) content per Müller cell is roughly duplicated (line diagram, left scale). As a result, a constant amount of GS remains available per rod (line diagram, left scale). Data from Mack et al. (1997).

tion (e.g., Vardimon et al., 1993), recent data suggest that it might be specific for chicken (or avian) development but not applicable to mammals (Fig. 6B). In the rabbit retina, a steep rise in GS expression by Müller cells occurs after the second week of postnatal life (Germer et al., 1997b) whereas the plasma concentrations of cortisol remain essentially unchanged from midgestation to adulthood (references in Fig. 6). There was, however, a good correlation between glial GS expression and maturation of neuronal (ribbon) synapses, constituting the sources of glutamate release (Fig. 6B). In both chicken and rabbit retina, the increase of glial GS expression follows the synapse formation with a delay of one or two days (Fig. 6). A similar correlation exists also in another mammalian species (the rat), where the postnatal development of GS expression (Riepe and Norenberg, 1978) and glutamate- and GABA-uptake (Fletcher and Kalloniatis, 1997) have been studied.

The problem was further studied on rabbit retinal explant cultures. In such cultured retinae, the GS expression of Müller cells was greatly reduced (about an order of magnitude), compared to corresponding stages grown *in vivo* (Germer et al., 1997b). The accumulation of GS protein in Müller cells of the cultures was stimulated by application of cortisol (1 ng/ml). This effect was similar (but less pronounced) to what had been earlier described in chicken retinae. Interestingly, a similar degree of stimulation (roughly, a duplication) was also achieved by exposure to enhanced levels of glutamate (0.1 mM) or ammonia (0.1 mM). These observations (Fig. 7) strongly suggest that (mammalian) Müller cells display some kind of "substrate regulation" of GS expression. This means, glial GS is up regulated when the cells are challenged by increased amounts of its substrates glutamate and ammonia (Fig. 7), as in normal development when neuronal glutamate-releasing synapses become established (Fig. 6), or in cases of hepatic failure when the blood ammonia increases (Reichenbach et al., 1995a, b). Conversely, in cases of retinal degeneration (i.e., loss of glutamate-releasing neurons) Müller cells have been shown to down-regulate

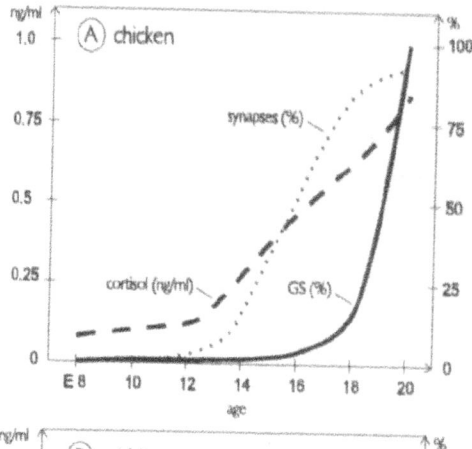

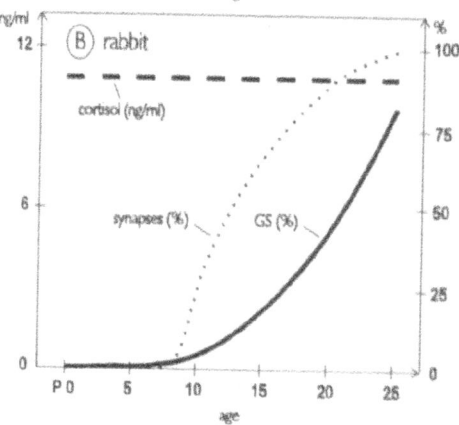

Figure 6. Comparison of glutamine synthetase (GS) development in chicken (A) and rabbit (B) retina. Retinal GS concentration (continuous lines) is given in percent of the levels achieved at hatching (chicken: Piddington and Moscona, 1967) or adulthood (rabbit: Germer et al., 1997b), respectively. It is compared with the plasma levels of cortisol (in ng/ml; chicken: Tanabe et al., 1986; rabbit: means of data from Mulay et al., 1973; Barr et al., 1980; Hümmelink and Ballard, 1986; and O'Loughlin et al., 1990), and with the density of ribbon synapses in the inner plexiform layer (in % of adult levels; chicken: Daniels and Vogel, 1980; rabbit: McArdle et al., 1977). Whereas a correlation between synaptogenesis and GS can be seen in both species, a correlation between plasma cortisol and GS seems to exist in chicken but certainly not in rabbit.

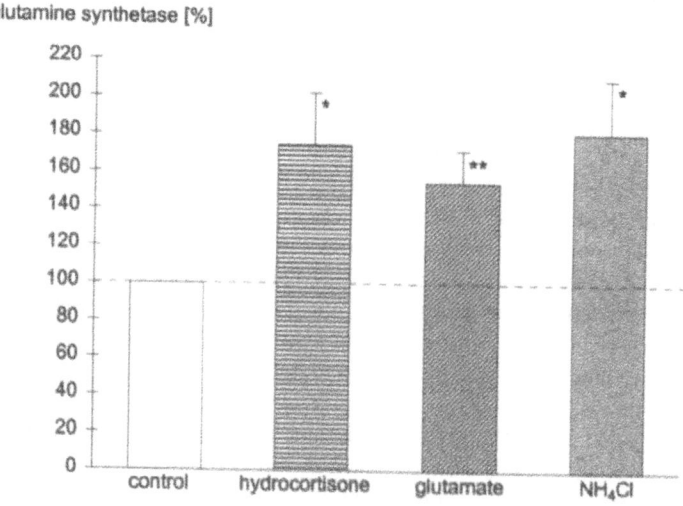

Figure 7. Glutamine synthetase levels (pixel counts on Western blots from tissue homogenates) in rabbit retinal organ cultures. Both hydrocortisone and the two substrates glutamate and ammonia cause an increase of retinal glutamine synthetase (data from Germer et al., 1997b); * = significant with $p \le 0.05$; ** = significant with $p \le 0.01$.

their GS content (LaVail and Reif-Lehrer, 1971; Lewis et al., 1989; Härtig et al., 1995; Grosche et al., 1995). In astrocytes, GS stimulation by glutamate exposure has been related to the activation of non-NMDA glutamate receptors (Fleischer-Lambropoulos et al., 1996). It remains to be elucidated how Müller cells adjust their GS to varying levels of glutamate and ammonia.

4.2. Energy Metabolism and Retinal Vascularization

Glio-neuronal metabolic interactions are thought to occur also in carbohydrate and energy metabolism of mammalian retinae. Specifically, Müller cells are the sites of retinal glycogen stores (Kuwabara and Cogan, 1961; Eichner and Themann, 1962), and they contain the glycogenolytic enzyme glycogen phosphorylase (Pfeiffer et al., 1994) which is activated by conditions that mimic enhanced neuronal activity, such as increased extracellular K^+ ion concentrations (Reichenbach et al., 1993). Elegant experiments have shown that the glycolytic (i.e., anaerobic) metabolism of Müller cells results in the release of lactate/pyruvate which is then taken up by (rod photoreceptor) neurons, and used as a substrate for their (aerobic) Krebs cycle (Poitry-Yamate et al., 1995). Furthermore, Müller cells contain the enzyme carbonic anhydrase (Sarthy and Lam, 1978; Moscona, 1983) and specific Na^+/HCO_3^- exchange carriers (Newman, 1991, 1994) necessary to buffer the CO_2 produced by the oxidative metabolism of neurons ("CO_2 siphoning": Newman, 1994). This is a very convincing case of retinal glio-neuronal interaction, and might thus provide another suitable case of developmental regulation.

However, there is evidence that (at least some of) the above-mentioned Müller cell features constitute a primitive rather than a derived condition. Both accumulation of glycogen (Uga and Smelser, 1973) and expression of carbonic anhydrase (Linser and Moscona, 1981) occur very early in retinal development, and are probably shared by immature radial glial (or even progenitor) cells and differentiated Müller cells (see also section "origin of Müller cells") although some quantitative changes may occur (Linser and Moscona, 1981). Furthermore, virtually all evidence for Müller cell-neuron interaction in glucose (anaerobic) vs. lactate/pyruvate (aerobic) metabolism comes from studies on avascular mammalian retinae, mostly guinea-pig and rabbit. For instance, glycogen stores and glycogen phosphorylase immunoreactivity are easily demonstrable in guinea-pig Müller cells but hardly in Müller cells from rat retina which is well vascularized (Pfeiffer et al., 1994; own unpublished observations).

These considerations have led to the following hypothesis on energy metabolism in mammalian retinae. During the functional and (oxidative) metabolic maturation of postmitotic retinal neurons, young Müller cells retain (and probably even intensify) the metabolic (e.g., glycolytic) features of their precursor cells. This allows the establishment of the above-mentioned glio-neuronal "symbiosis" in carbohydrate metabolism which may be sufficient to support the energy demands of the cells in avascular retinae during the entire life of animals such as the guinea-pig or the horse. By contrast, in species where Müller cells must support large units (many neurons per Müller cell), or units with highly elaborate and active neuronal circuits (e.g., with a thick inner plexiform layer), the basic glio-neuronal metabolic cooperation may be insufficient to maintain the functional activity, or even the survival, of retinal neurons. In these cases (i.e., in most mammals), intraretinal vascularization is thought to support the neuronal metabolism by facilitating the supply of nutrients and/or the removal of waste products. In this context, intraretinal blood vessels can be imagined as "assistants" of Müller cells, by assuming some of their responsibilities in neuronal care.

Indeed, it has recently been demonstrated that vascular endothelial growth factor (VEGF; Stone *et al.*, 1995, Hata *et al.*, 1995) as well as SPARC, another mediator of angionesis (Kaplan *et al.*, 1995) are produced and released by Müller cells. Likewise, Müller cells were shown to express renin (Berka *et al.*, 1995) and angiotensin (Datum & Zrenner, 1991) which have been implicated in the proliferation of retinal blood vessels under certain conditions. Thus, enhanced metabolic needs of maturing retinal neurons apparently stimulate the ingrowth of retinal blood vessels indirectly, via the young retinal glia. There is evidence that the occurrence of an absolute or relative hypoxia induces neurons to release signals which then stimulate the production of substances such as VEGF or SPARC by retinal glial cells (Chan-Ling, 1994; Chan-Ling *et al.*, 1995; Stone *et al.*, 1995; Amin et al., 1997).

The nature of the neuronal "SOS" signal(s) is not yet known. The involvement of specific signals, such as adenosine, has been proposed (Aiello et al., 1994; Takagi et al., 1996). On the other hand, there might be a rather unspecific release of neuronal "waste products", simply causing a metabolic overload of Müller cells. As the metabolic situation of cells cannot be easily measured, a morphometric approach has been proposed (Chao et al., 1997). Assuming that the energy (i.e., adenosine triphosphate = ATP) demands of a cell are predominantly caused by the active transport proteins within the cell membrane, and that the (glycolytic) ATP production is confined to the cytoplasm, the surface-to-volume ratio (SVR) of a Müller cell may serve as a rough indicator of its metabolic burdening. A comparative study on Müller cells from various mammalian species (Chao et al., 1997) has shown that cells from avascular retinae (or retinal regions) have low SVRs (less than about 7 μm^{-1}) whereas cells from vascularized retinae have higher SVRs of more than about 10 μm^{-1} (Fig. 8). Although these data were estimated from adult retinae, and no comparative data are available from neonatal stages when blood vessels occupy the retina, they support the view that Müller cells stimulate vascularization when the demands of glio-neuronal interaction exceed their metabolic capacity.

4.3. Membrane Properties and K$^+$ Clearance

It has already been pointed out that the membrane of mature Müller cells faces the narrow extracellular space surrounding all neuronal compartments which are almost completely enveloped by glial sheaths. Furthermore, the membrane of differentiated Müller cells is a site of transport proteins mediating many glio-neuronal interactions (Fig. 1). This adult condition results from radical developmental processes.

Young postmitotic Müller cells resemble bipolar radial glia, and basically consist of two long, slender, cylindrical stem processes and an ovoid soma (left photograph in Fig. 9). Their surface is smooth, and structural relationships to neurons are poorly developed; no side branches or perineuronal sheaths are present (Uga and Smelser, 1973; Meller and Tetzlaff, 1978; Reichenbach and Reichelt, 1986; Reichenbach et al., 1991b). Thus, their surface membrane area is small. In young postnatal rabbits, a surface area of about 2000 μm^2 has been estimated for dissociated Müller cells (Fig. 9; Biedermann et al., unpublished results). Within the next two weeks of life, the cells grow many side branches bearing perineuronal sheaths (right photograph in Fig. 9; Uga and Smelser, 1973; Reichenbach and Reichelt, 1986), and their surface area increases more than threefold (Fig. 9).

Within the same period, the very low K$^+$ conductance of the young cells (about 2 nS) becomes enhanced by a factor of 10 (Fig. 9; Biedermann et al., unpublished results). This is probably due to the insertion of many new (inwardly rectifying) K$^+$ channel molecules into the cell membrane. It is interesting to note that there is some developmental delay

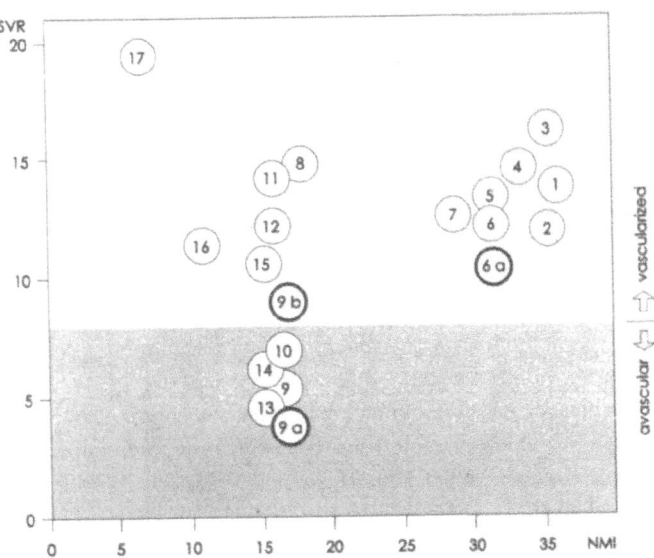

Figure 8. Diagram showing the correlation between the neuron-to-Müller cell indices (MNI) and the surface-to-volume ratios (SVR) of Müller cells from mature retinae of various mammalian species. The data were evaluated morphometrically (cell volume) and electrophysiologically (membrane capacity – surface area) on isolated Müller cells (Chao et al., 1997) from cat (1), mouse (2), fox (3), dog (4), polecat (5), rat (6), mink (7), *Monodelphis* (8), rabbit (9), zebra (10), barbary sheep (11), pig (12), horse (13), guinea pig (14), man (15), gerbil (16), and tree shrew (17). Some data (in bold circles) have been obtained by electron microscopical stereology [rat (6a)]: Rasmussen, 1973, 1975; rabbit avascular periphery (9a): Reichenbach *et al.*, 1988; rabbit vascularized medullary rays (9b): Reichenbach *et al.*, unpublished observations). The grey area includes all species (or retinal areas) where the tissue is avascular (data on vascularization: Johnson, 1901, 1968; Chase, 1982; Müller and Peichl, 1989; own unpublished observations). There is a certain level of SVRs (about 8) that seems to delimit avascular from vascularized retinae. With permission, from Chao et al. (1997).

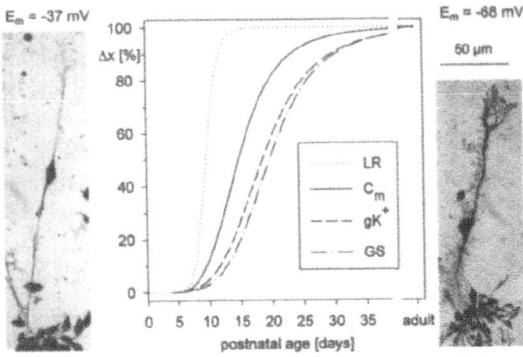

Figure 9. Postnatal development of rabbit Müller cells. The microphotographs show enzymatically isolated Müller cells from 6th (left, p6) and 24th (right, p24) postnatal day, together with their representative membrane potentials; the calibration bar is valid for both cells. The diagram shows the relative increase (between p6 and adult) of the membrane capacity (C_m; data from Biedermann et al., in preparation) and the K^+ conductance of the membrane (gK^+; data from Biedermann et al., in preparation). These data are compared to the percentage of light-responsible ganglion cells (LR: Masland, 1977), as a measure of synaptic maturation, and to the glutamine synthetase levels (GS: Germer et al., 1997b), as another indicator of Müller cell differentiation. Between p6 and p24, C_m increases threefold from 20 to 60 pF, reflecting an increase of the membrane surface area from about 2000 to 6000 μm^2. During the same period but with some delay, the membrane conductance increases more than tenfold, from about 2 to 20 nS; thus, the specific conductance of the membrane (as a rough indicator of the channel density) increases more than threefold.

between the surface area and the K^+ conductance (gK) of the membrane (Fig. 9). This suggests that the cell surface is increased first by the addition of a "naked" lipid bilayer membrane whereas the channel proteins are inserted by a later secondary process. In any case, the increasing K^+ conductance of the membrane is accompanied by elevated "resting" membrane potentials (E_m) which are close to −40 mV in immature cells but increase to almost adult levels of more than −70 mV in cells from the 24th postnatal day (Fig. 9; Biedermann et al., unpublished results).

This increase in gK is likely to be a key event in glial development. First, it is a crucial precondition for the flux of spatial buffering K^+ currents through Müller cells (left part of Fig. 1) which are thought to provide an important contribution to the clearance of excess extracellular K^+ ions (Newman, 1984; Newman et al., 1984). As any neuronal activity is accompanied by K^+ release, this glial buffer mechanism should soon become a limiting factor when the neuronal circuits are established. Second, a high gK is a precondition for a high E_m. High E_m values are not just a characteristic feature of differentiated Müller cells (e.g., Chao et al., 1997) but are also required for the proper function of most transport carriers in their membrane. The activity of the uptake carriers for glutamate (Barbour et al., 1991) and GABA (Biedermann et al., 1994) is voltage-dependent. This means that low E_m values reduce or even prevent the important action(s) of Müller cells in the transmitter recycling pathways (middle part of Fig. 1). For instance, the establishment of high GS levels during the third week of life (Fig. 9) would make no sense if a high E_m had not been stabilized as a precondition for an effective uptake of glutamate.

Again, the question arises what kind(s) of signals drive the differentiation of glial membrane properties. On the one hand, because glio-neuronal interactions are essential for retinal functioning it can be assumed that there are several parallel signalling pathways ensuring that the necessary glial properties become established at the right time. On the other hand, (at least some of) these signals are likely to arise from the differentiating neurons, since glial differentiation enables optimal interactions with these partner cells. This latter assumption is supported by various experimental observations, such as the findings that glial differentiation closely follows neuronal maturation (e.g., Fig. 9), and that Müller cells in purified neuron-free cultures fail to differentiate properly (e.g., Wolburg et al., 1990).

More specifically, such a signalling function may be "piggybacked" by molecules that are released in the normal process(es) of neuronal activity (compare also Fig. 7). Likely candidate molecules are K^+ ions, since enhanced neuronal activity causes an increase in the extracellular K^+ concentration ($[K^+]_e$), which can influence the glial cells e.g. by depolarizing their E_m. When purified Müller cell cultures were exposed to elevated $[K^+]_e$ levels (10 mM) over 48 hours, their Na^+,K^+-ATPase activity (measured in normal $[K^+]_e$) was almost tripled (Fig. 10; Reichelt et al., 1989). This was probably due to synthesis of new ATPase molecules. Indeed, the protein synthesis of such cells was accelerated under the same conditions (Fig. 10; Reichelt et al., 1989). Similar mechanisms might be involved in the postnatal up-regulation of (certain types of) glial K^+ channels, but this matter is still far from being understood.

5. MÜLLER CELL DE-DIFFERENTIATION: THE ROLE(S) OF NEURONS, MEMBRANE PROPERTIES, AND REACTIVE GLIOSIS

In contrast to their "sisterly" retinal neurons, Müller cells seem to retain their capability to undergo mitotic proliferation. Although there is ample evidence supporting this

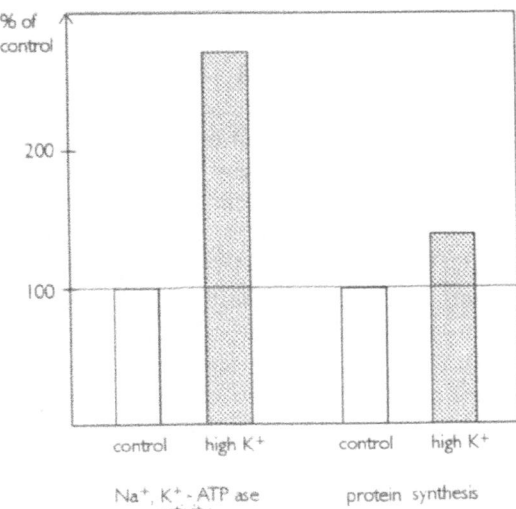

Figure 10. Effects of elevated $[K^+]_e$ in the medium of purified Müller cell cultures. After 48 hrs in 10 mM K^+ ("high K^+"), both the standard Na^+, K^+-ATPase activity (measured at control $[K^+]_e$) and the protein synthesis (measured as L-[^{3}H]-lysine incorporation) are significantly elevated above control levels (data from Reichelt et al., 1989).

view (Erickson et al., 1983; Fisher et al., 1991; Geller et al., 1995), Müller cell proliferation is certainly not a common or frequent response to retinal injuries. There are various inherited or acquired "slow" retinal degenerations that may cause a complete loss of all photoreceptor cells (i.e., the vast majority of retinal neurons) or all ganglion cells (i.e., the vast majority of action potential-generating neurons of the retina). In such cases, Müller cells respond by the (enhanced) expression of "injury-marker proteins" such as GFAP (Bignami and Dahl, 1979) or other proteins such as β-amyloid precursor protein, Bcl-2 protooncogene protein, β-amyloid precursor protein, cathepsin D, and CD-44, as well as by hypertrophic growth of irregular cell processes and increased cytoplasmic volume, but not by mitotic activity (e.g., Härtig et al., 1995; own unpublished results). Müller cell proliferation is observed only in cases of rapid and massive neuronal injury, such as after retinal detachment (Erickson et al., 1983; Fisher et al., 1991; Geller et al., 1995) or intravitreal bleedings (Kono et al., 1995). It seems as if differentiated Müller cells have a high threshold against mitogenic stimuli, and need to undergo a re-transition into an immature phenotype in order to (re-)acquire proliferative capacities. Whereas a series of mitogenic substances are known to act on such immature (cultured) retinal radial glia (Puro, 1995), the nature of retransition-promoting factors remains enigmatic.

There are three possible ways how Müller cell de-differentiation might be achieved: (i) by a loss of factors permanently released from healthy neurons, necessary to maintain the differentiated state of Müller cells; (ii) by a (neuron-, RPE-, or blood vessel-derived) release of pathogenic signals, inducing the de-differentiation of Müller cells; or (iii) by enhanced neuronal release of (threshold-dependent) ambivalent factors, stabilizing differentiation at physiological (low) concentrations but enforcing de-differentiation at pathological (high) concentrations. Factors such as TGF-β may be involved in maintenance of differentiation (case [i]; Ikeda and Puro, 1995). As a breakdown of the blood-retina barrier often occurs in severe cases of retinal injury, the mitogenic effects of blood-derived factors such as thrombin (Puro et al., 1990) may be responsible for de-differentiation (case [ii]). Finally (case [iii]), the neurogenic transmitter glutamate may, at physiological concentrations, contribute to Müller cell differentiation (Ikeda and Puro, 1995; Germer et al., 1997b) but also, at elevated concentrations, promote de-differentiation and proliferation (Uchihori and Puro,

1993). Intravitreally injected bFGF has been shown to bind to receptors of Müller cells (Lewis et al., 1996) and to induce Müller cell proliferation *in situ* (Lewis et al., 1992) but the possible role and source of bFGF in Müller cell reaction (e.g., to retinal detachment) remains to be elucidated.

There are some recent data suggesting that K^+ channels are involved in the signal cascade between the (unknown) trigger and final glial proliferation. Typical potassium currents of a normal adult Müller cell are compared with those of a young and a pathologically altered cell in Figure 11. The normal mature current pattern comprises outward (upward deflections in Fig. 11B) as well as inward (downward deflections in Fig. 11B) currents. The latter show inactivation at stronger hyperpolarizing voltages, and have been shown to flow through inwardly rectifying K^+ channels (Newman, 1993; Chao et al., 1994). When Müller cells were studied that had been isolated from diseased retinae, the current pattern was dramatically changed (Fig. 11C), and resembled that of immature young cells (Fig. 11A). In particular, inward K^+ currents were strongly reduced or even completely missing. The missing presence and/or activity of inwardly rectifying K^+ channels was accompanied by a drastic reduction of E_m, sometimes up to close to -20 mV. Similar changes have been observed in Müller cells from pathologically altered human retinae (Francke et al., 1997; Reichelt et al., 1997) and in frog Müller cells after optic nerve cut (Skatchkov et al., 1996). An inhibition of inwardly rectifying K^+ channels has also been observed in response to thrombin (Puro and Stuenkel, 1995) or glutamate (activation of NMDA receptors: Puro et al., 1996), agents which stimulate Müller cell proliferation *in vitro* (Puro et al., 1990; Uchihori and Puro, 1993). Furthermore, currents

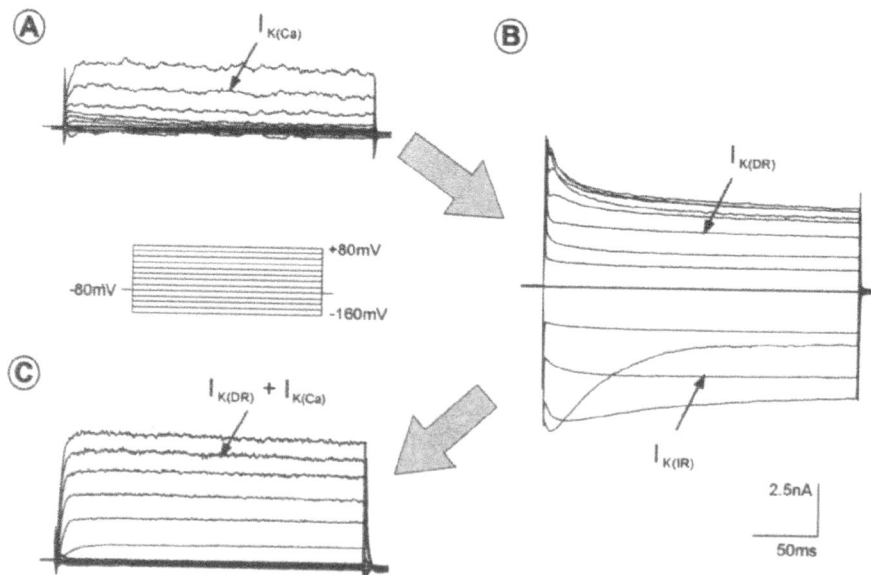

Figure 11. Whole-cell current patterns evoked by voltage steps applied to enzymatically isolated rabbit Müller cells (voltage-clamp experiments). Inward currents are represented by downward deflections ($I_{K(IR)}$), outward currents by upward deflections. The large outward currents mediated by the big Ca^{2+}-dependent K^+ channels are labeled as "$I_{K(Ca)}$". When the cells were isolated from normal adult retinae (B), large inward currents ($I_{K(IR)}$) were recorded but Ca^{2+}-dependent K^+ outward currents were hardly observable. By contrast, both young Müller cells from 6th postnatal day (A) and cells from severely injured retinae (C: experimental proliferative vitreoretinopathy) displayed dominant $I_{K(Ca)}$ but almost no $I_{K(IR)}$. Unpublished data from Biedermann, Francke, Pannicke, Bringmann et al.

through large conductance Ca^{2+}-dependent K^+ channels (**Big K^+** = BK channels) contributed much to the sum current pattern in young (Fig. 11A) and diseased rabbit (Fig. 11C) and human (Reichelt et al., 1997) Müller cells but were hardly recordable in normal adult cells (Fig. 11B).

The BK channels of Müller cells have peculiar properties (Puro et al., 1989; Bringmann et al., 1997a,b). The activity of these channels is dependent on the presence (and concentration) of intracellular Ca^{2+}, requires strong depolarization of E_m, and is blocked by low concentrations (1 mM) TEA from the outside of the membrane (upper trace in Fig. 12). This is in contrast to inwardly rectifying K^+ channels which are open at high E_m levels (that are maintained by their activity), and are largely insensitive to TEA in concentrations of up to 50 mM (Chao et al., 1994). In cultured Müller cells, cell proliferation (measured as ^{3}H-thymidine incorporation) can be stimulated by the depolarizing action of high $[K^+]_e$ (lower diagram in Fig. 12; Reichelt et al., 1989). This stimulation was completely blocked by simultaneous application of 1 mM TEA (Fig. 12). Thus, there are several findings which strongly support the view that BK channels are involved in Müller cell proliferation (Puro et al., 1989; Puro, 1995) even *in vivo*: (i) membrane depolarization activates both the BK channels and cell proliferation, (ii) 1 mM TEA inhibits both the BK channels and cell proliferation, and (iii) both dominant BK channel-mediated currents and proliferative activity are observed in neonatal (Fig. 11A; Reichenbach et al., 1991a) and pathologically altered retinae (Fig. 11C; Francke et al., 1997; Reichelt et al., 1997; Erickson et al. 1983; Geller et al., 1995, Kono et al., 1995).

Summarizing these data, one of the first Müller cell responses to retinal injury seems to be a down-regulation of their inwardly rectifying K^+ channels, causing a depolarization of E_m (Francke et al., 1997; Reichelt et al., 1997). This depolarization facilitates the activation of the BK channels, which may then be opened by a variety of stimuli including arachidonic acid (Bringmann et al., 1997b), altered protein kinase activation (Bringmann et al., 1997a), changes in intracellular Ca^{2+} and pH (Puro et al., 1989; Bringmann et al., 1997a) and others. Such stimuli are known to occur under developmental and pathophysiological conditions, and may then, via BK channel activation, induce cellular proliferation. It is obvious that this line of research has the potential for substantial clinical impact.

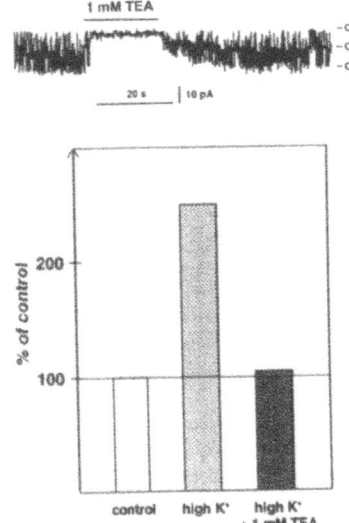

Figure 12. Both the activity of the big Ca^{2+}-dependent K^+ channels of dissociated Müller cells (A: single-channel recordings from a human cell: data from Bringmann et al., 1997a) and the ^{3}H-thymidine incorporation of cultured Müller cells (B: quantitative autoradiography on rabbit monolayer cultures: Reichenbach et al., unpublished results) are inhibited by 1 mM tetraethylammonium (TEA).

ACKNOWLEDGMENTS

The presented work was supported by grants of the Deutsche Forschungsgemeinschaft (Re 849/3-2) and of the Bundesministerium für Bildung, Forschung und Technologie (WR: 0316916A; HK: Interdisciplinary Center for Clinical Research at the University of Leipzig, IZKF Leipzig, Project C5). The authors wish to thank Sabine Kutan, Markus Reuter, and Sven Thomas for their active work on the computer graphics.

REFERENCES

Adler, R. and Hatlee, M., 1989, Plasticity and differentiation of embryonic retinal cells after terminal mitosis. Science 243: 391–393.

Aiello, L. P., Avery, R. L., Arrigg, P. G., Keyt, B.A., Jampel, H. D., Shah, S. T., Thieme, H., Iwamoto, M. A., Park, J. E., Nguyen, H. V., Aiello, L. M., Ferrara, N.a nd King, G. L., 1994, Vascular endothelial growth factor in ocular fluid of patients with diabetic retinopathy and other retinal disorders. N. Engl. J. Med. 331: 1480–1487.

Alexiades, M. R. and Cepko, C. L., 1997, Subsets of retinal progenitors display temporally regulated and distinct biases in the fates of their progeny. Development 124: 1119–1131.

Altshuler, D., Loturco, J. J., Rush, J. and Cepko, C., 1993, Taurine promotes the differentiation of a vertebrate retinal cell type in vitro. Development 119: 1317–1328.

Ames, A. III, Li, Y. Y., Heher, E. C. and Kimble, C. R., 1992, Energy metabolism of rabbit retina as related to function: high cost of Na$^+$ transport. J. Neurosci. 12: 840–853.

Amin, R. H., Frank, R. N., Kennedy, A., Eliott, D., Puklin, J. E. and Abrams, G. W., 1997, Vascular endothelial growth factor is present in glial cells of the retina and optic nerve of human subjects with nonproliferative diabetic retinopathy. Invest. Ophthalmol. Vis. Sci. 38: 36–47.

Anchan, R. M. and Reh, T. A., 1995, Transforming growth factor-beta-3 is mitogenic for rat retinal progenitor cells in vitro. J. Neurobiol. 28: 133–145.

Armson, P. F., Bennet, M. R. and Raju, T. R., 1987, Retinal ganglion cell survival and neurite regeneration requirements: the change from Müller cell dependence to superior colliculi dependence during development. Dev. Brain Res. 32: 207–216.

Barbour, B., Brew, H., and Attwell, D., 1991, Electrogenic uptake of glutamate and aspartate into glial cells isolated from the salamander (Ambystoma) retina. J. Physiol. (Lond.) 436: 169–193.

Barr, H. A., Lugg, M. A. and Nicholas, T. E., 1980, Cortisone and cortisol in maternal and fetal blood and in amniotic fluid during the final ten days of gestation in the rabbit. Biol. Neonate 38: 214–220.

Berka, J. L., Stubbs, A. J., Wang, D. Z. M., DiNicolantonio, R., Alcorn, D., Campbell, D. J. and Skinner, S. L., 1995, Renin-containing Müller cells of the retina display endocrine features. Invest. Ophthalmol. Vis. Sci. 36: 1450–1458.

Biedermann, B., Eberhardt, W. and Reichelt, W., 1994, GABA uptake into isolated retinal Müller glial cells of the guinea pig detected electrophysiologically. Neuroreport 5: 438–440.

Bignami, A. and Dahl, D., 1979, The radial glia of Müller in the rat retina and their response to injury. An immunofluorescence study with antibodies to glial fibrillary acidic (GFA) protein. Exp. Eye Res. 28: 63–69.

Braisted, J. E., Essman, T. F. and Raymond, P. A., 1994, Selective regeneration of photoreceptors in goldfish retina. Development 120: 2409–2419.

Bringmann, A., Faude F., and Reichenbach, A., 1997a, Mammalian retinal glial (Müller) cells express large-conductance Ca^{2+}-activated K$^+$ channels that are modulated by Mg^{2+} and pH, and activated by protein kinase A. Glia 19: 311–323.

Bringmann, A., Skatchkov, S. N., Biedermann, B., Faude F., and Reichenbach, A., 1997b, Alterations of potassium channel activity in retinal Müller glial cells induced by arachidonic acid. Submitted to Neuroscience.

Browman, H. I. and Hawryshyn, C. W., 1994, Retinoic acid modulates retinal development in the juveniles of a teleost fish. J. exp. Biol. 193: 191–207.

Calvaruso, G., Vento, R., Taibi, G., Giuliano, M. and Tesoriere, G., 1992, A factor derived from chick embryo retina which inhibits DNA synthesis of retina itself. Neurochem. Res. 17: 1041–1048.

Cepko. C. L., 1993, Retinal cell fate determination. Progr. Retinal Res. 12: 1–12.

Chan-Ling, T., 1994, Glial, neuronal and vascular interactions in the mammalian retina. Progr. Retinal Res. 13: 357–389.

Chan-Ling, T., Bock, B. and Stone, J., 1995, The effect of oxygen on vasoformative cell division - evidence that 'physiological hypoxia' is the stimulus for normal retinal vasculogenesis. Invest. Ophthalmol. Vis. Sci. 36: 1201–1214.

Chao, T. I., Grosche, J., Friedrich, K. J., Biedermann, B., Francke, M., Pannicke, T., Reichelt, W., Wulst, M., Mühle, C., Pritz-Hohmeier, S., Kuhrt, H., Faude, F., Drommer, W., Kasper, M., Buse, E. and Reichenbach, A., 1997, Comparative studies on mammalian Müller (retinal glial) cells. J. Neurocytol., in press.

Chao, T. I., Henke, A., Reichelt, W., Eberhardt, W., Reinhardt-Maelicke, S., and Reichenbach, A., 1994, Three distinct types of voltage-dependent K$^+$ channels are expressed by Müller (glial) cells of the rabbit retina. Pflüger's Arch. 426: 51–60.

Chase, J., 1982, The evolution of retinal vascularization in mammals. A comparison of vascular and avascular retinae. Ophthalmology 89: 1518–1525.

Craft, J. L., Fulton, A. B., Silver, J., and Albert, D. M., 1983, Development of the outer plexiform layer in albino rats. Curr. Eye Res. 2: 295–299.

Daniels, M. P., and Vogel, Z., 1980, Localization of alpha-bungarotoxin binding sites in synapses of the developing chick retina. Brain Res. 201: 45–56.

Datum, K.-H. & Zrenner, E., 1991, Angiotensin-like immunoreactive cells in the chicken retina. Exp. Eye Res. 53: 157–165.

DeCurtis, I. and Reichardt, L. F., 1993, Function and spatial distribution in developing chick retina of the laminin receptor alpha-6-beta-1 and its isoforms. Development 118: 377–388.

Dorsky, R. I., Chang, W. S., Rapaport, D. H. and Harris, W. A., 1997, Regulation of neuronal diversity in the Xenopus retina by delta signalling. Nature 385: 67–70.

Dorsky, R. I., Rapaport, D. H. and Harris, W. A., 1995, Xotch inhibits cell differentiation in the Xenopus retina. Neuron 14: 487–496.

Dowling, J. E., 1987, The Retina. An Approachable Part of the Brain. Harvard Univ. Press, Cambridge, Massachusetts & London.

Drazba, J., and Lemmon, V., 1990, The role of cell adhesion molecules in neurite outgrowth on Müller cells. Dev. Biol. 138: 82–93.

Egensperger, R., Maslim, J., Bisti, S., Holländer, H. and Stone, J., 1996, Fate of DNA from retinal cells dying during development: uptake by microglia and macroglia (Müller cells). Dev. Brain Res. 97: 1–8.

Eichner, D. and Themann, H., 1962, Zur Frage des Netzhautglykogens beim Meerschweinchen. Z. Zellforsch. 56: 231–246.

Elschnig, A., 1913, Zur Anatomie des menschlichen Albionoauges. Graefes Arch. Ophthalmol. 84: 401–419.

Erickson, P. A., Fisher, S. K., Anderson, D. H., Stern, W. H., and Borgulla, G. A., 1983, Retinal detachment in the cat: the outer nuclear and outer plexiform layers. Invest. Ophthalmol. Vis. Sci. 24: 927–942.

Ezzeddine, Z. D., Yang, X., DeChiara, T., Yancopoulos, G. and Cepko, C. L., 1997, Postmitotic cells fated to be rod photoreceptors can be respecified by CNTF treatment of the retina. Development 124: 1055–1067.

Fisher, S. K., Erickson, P. A., Lewis, G. P. and Anderson, D. H., 1991, Intraretinal proliferation induced by retinal detachment. Invest. Ophthalmol. Vis. Sci. 32: 1739–1748.

Fleischer-Lambropoulos, E., Kazazoglou, T., Geladopoulos, T., Kentroti, S., Stefanis, C. and Vernadakis, A., 1996, Stimulation of glutamine synthetase activity by excitatory amino acids in astrocyte cultures derived from aged mouse cerebral hemispheres may be associated with non-N-Methyl-D-aspartate receptor activation. Int. J. Dev. Neurosci. 14: 523–530.

Fletcher, E. L. and Kalloniatis, M., 1997, Localisation of amino acid neurotransmitters during postnatal development of the rat retina. J. Comp. Neurol. 380: 449–471.

Francke, M., Pannicke, T., Biedermann, B., Faude, F., Wiedemann, P., Reichenbach, A. and Reichelt, W., 1997, Loss of inwardly rectifying potassium currents by human retinal glial cells in diseases of the eye. Glia, in press.

Fulton, A. B., Albert, D. M. and Craft, J. L., 1978, Human albinism. Light and electron microscopy study. Arch. Ophthalmol. 96: 305–310.

Geller, S. F., Lewis, G. P., Anderson, D. H. and Fisher, S. K., 1995, Use of the MIB-1 antibody for detecting proliferating cells in the retina. Invest. Ophthalmol. Vis. Sci. 36: 737–744.

Germer, A., Kühnel, K., Grosche, J., Friedrich, A., Wolburg, H., Price, J., Reichenbach, A. and Mack, A., 1997a, Development of the neonatal rabbit retina in organ culture. 1. Comparison with histogenesis in vivo, and the effect of a gliotoxin (α-aminoadipic acid). Anat. Embryol., in press.

Germer, A., Mack, A. and Reichenbach, A., 1997b, Mammalian Müller (glial) cell glutamine synthetase activity is low in retinal organ cultures but can be stimulated by several factors. Submitted to NeuroReport.

Goldberg, S., 1977, Unidirectional, bidirectional and random growth of embryonic optic axons. Exp. Eye Res. 25: 399–404.

Grosche, J., Härtig, W. and Reichenbach, A., 1995, Expression of glial fibrillary acidic protein (GFAP), glutamine synthetase (SG), and Bcl-2 protooncogene protein by Müller (glial) cells in retinal light damage of rats. Neurosci. Lett. 185: 119–122.

Guillemot, F. and Cepko, C. L., 1992, Retinal fate and ganglion cell differentiation are potentiated by acidic FGF in an *in vitro* assay of early retinal development. Development 114: 743–754.

Halfter, W. and Deiss, S., 1986, Axonal pathfinding in organ-cultured embryonic avian retinae. Devel. Biol. 114: 269–310.

Harris, W. A. and Holt, C. E., 1990, Early events in the embryogenesis of the vertebrate visual system: Cellular determination and pathfinding. Annu. Rev. Neurosci. 13: 155–169

Härtig, W., Grosche, J., Distler, C., Grimm D., El-Hifnawi, E. and Reichenbach, A., 1995, Alterations of Müller (glial) cells in dystrophic retinae of RCS rats. J. Neurocytol. 24: 507–517.

Hata, Y., Nakagawa, K., Ishibashi, T., Inomata, H., Ueno, H. and Sueishi, K., 1995, Hypoxia-induced expression of vascular endothelial growth factor by retinal glial cells promotes in vitro angiogenesis. Virchows Arch. 426: 479–486.

Hicks, D., Forster, V., Dreyfus, H. and Sahel, J., 1994, Survival and regeneration of adult human photoreceptors in vitro. Brain Res. 643: 302–305.

Hinds, J. W. and Hinds, P. L., 1974, Early ganglion cell differentiation in the mouse retina: an electron microscopic analysis utilizing serial sections. Dev. Biol. 37: 381–416.

Hinds, J. W. and Hinds, P. L., 1979, Differentiation of photoreceptors and horizontal cells in the embryonic mouse retina: an electron microscopic, serial section analysis. J. Comp. Neurol. 187: 495–512.

Hitchcock, P. F. and Vanderyt, J. T., 1994, Regeneration of the dopamine-cell mosaic in the retina of the goldfish. Visual Neurosci. 11: 209–217.

Hume, D. A., Perry, V. H. and Gordon, S., 1983, Immunohistochemical localization of a macrophage-specific antigen in developing mouse retina: Phagocytois of dying neurons and differentiation of microglial cells to form a regular array in the plexiform layers. J. Cell Biol. 97: 253–257.

Hümmelink, R. and Ballard, P. L., 1986, Endogenous corticoids and lung development in the fetal rabbit. Endocrinology 118: 1622–1629.

Hunter, D. D., Murphy, M. D., Olsson, C. V. and Brunken, W. J., 1992, S-Laminin expression in adult and developing retinae: A potential cue for photoreceptor morphogenesis. Neuron 8: 399–414.

Hyatt, G. A., Schmitt, E. A., Fadool, J. M. and Dowling, J. E., 1996, Retinoic acid alters photoreceptor development *in vivo*. Proc. Natl. Acad. Sci. USA 93: 13298–13303.

Ikeda, T. and Puro, D. G., 1995, Regulation of retinal glial cell proliferation by antiproliferation molecules. Exp. Eye Res. 60: 435–444.

Jeffery, G., 1997, The albino retina: an abnormality that provides insight into normal retinal development. TINS 20: 165–169.

Jeffery, G., Brem, G. and Montoliu, L., 1997, Correction of retinal abnormalities found in albinism by introduction of a functional tyrosinase gene in transgenic mice and rabbits. Dev. Brain Res. 99: 95–102.

Jensen, A. M. and Wallace, V. A., 1997, Expression of sonic hedgehog and its putative role as a precursor cell mitogen in the developing mouse retina. Development 124: 363–371.

Johns, P. R., 1982, Formation of photoreceptors in larval and adult goldfish. J. Neurosci. 2: 178–198.

Johnson, G. L., 1901, Contributions to the comparative anatomy of vertebrates, chiefly based on ophthalmoscopic examination. Phil. Trans. Roy. Soc. Lond. B 194: 1–82.

Johnson, G. L., 1968, Ophthalmoscopic studies on the eyes of mammals. Phil. Trans. Roy. Soc. Lond. B 254: 207–220.

Kaplan, H., Jasoni, C., Gariano, R., Hendrickson, A. and Sage, E.H., 1995, SPARC, a mediator of angiogenesis, is found in astrocytes and Müller cells of the primate retina. Invest. Ophthalmol. Vis. Sci. 36: S648.

Kelley, N. W., Turner, J. K. and Reh, T. A., 1994, Retinoic acid promotes differentiation of photoreceptors in vitro. Development 120: 2091–2102.

Kirsch, M., Lee, M.-Y., Meyer, V., Wiese, A. and Hofmann, H.-D., 1997, Evidence for multiple, local functions of ciliary neurotrophic factor (CNTF) in retinal development: expression of CNTF and its receptor and in vitro effects on target cells. J. Neurochem. 68: 979–990.

Kljavin, I. J. and Reh, T. A., 1991, Müller cells are a preferred substrate for in vitro neurite extension by rod photoreceptor cells. J. Neurosci. 11: 2985–2994.

Kono, T., Kohno, T. and Inomata, H., 1995, Epiretinal membrane formation. Light and electron microscopical study in an experimental rabbit model. Arch. Ophthalmol. 113: 359–363.

Kuwabara, T. and Cogan, D. G., 1961, Retinal glycogen. Arch. Ophthalmol. 66: 94–104.

LaVail, M. M. and Reif-Lehrer, L., 1971, Glutamine synthetase in the normal and dystrophic mouse retina. J. Cell Biol. 51: 348–354.

Layer, P. G., Rothermel, A., Hering, H., Wolf, B., deGrip, W. J., Hicks, D. and Willbold, W., 1997, Pigmented epithelium sustains cell proliferation and decreases expression of opsins and acetylcholinesterase in reaggregated chicken retinospheroids. Submitted to Europ. J. Neurosci.

Layer, P. G. and Willbold, E., 1993, Histogenesis of the avian retina in reaggregation culture: From dissociated cells to laminar neuronal networks. Int. Rev. Cytol. 146: 1–47.

Lewis, G. P., Erickson, P. A., Guérin, C. J., Anderson, D. H. and Fisher, S. K., 1989, Changes in the expression of specific Müller cell proteins during long-term retinal detachment. Exp. Eye Res. 49: 93–111.

Lewis, G. P., Erickson, P. A., Guérin, C. J., Anderson, D. H. and Fisher, S. K., 1992, Basic growth factor: a potential regulator of proliferation and intermediate filament expression in the retina. J. Neurosci. 12: 3968–3978.

Lewis, G. P., Guérin, C. J., Anderson, D. H., Matsumoto, B. and Fisher, S. K., 1994, Rapid changes in the expression of glial cell proteins caused by experimental retinal detachment. Am. J. Ophthalmol. 118: 368–376.

Lewis, G. P., Fisher, S. K. and Anderson, D. H., 1996, Fate of biotinylated basic fibroblast growth factor in the retina following intravitreal injection. Exp. Eye Res. 62: 309–324.

Lillien, L. and Cepko, C., 1992, Control of proliferation in the retina: temporal changes in responsiveness to FGF and TGFα. Development 115: 253–266.

Linser, P. and Moscona, A. A., 1979, Induction of glutamine synthetase in embryonic neural retina: Localization in Müller fibers and dependence on cell interactions. Proc. Natl. Acad. Sci. USA 76: 6476–6480.

Linser, P. and Moscona, A. A., 1981, Carbonic anhydrase C in the neural retina: transition from generalized to glia-specific cell localization during embryonic development. Proc. Natl. Acad. Sci. USA 78: 7190–7194.

Linser, P. J. and Perkins, M. S., 1987, Regulatory aspects of the in vitro development of retinal Müller glial cells. Cell Differentiation 20: 189–196.

Mack, A. F. and Fernald, R. D., 1993, Regulation of cell division and rod differentiation in the teleost retina. Dev. Brain Res. 76: 183–187.

Mack, A. F. and Fernald, R. D., 1995, New rods move before differentiating in adult teleost retina. Dev. Biol. 170: 136–141.

Mack, A. F., Germer, A., Janke, C. and Reichenbach, A., 1997, Müller (glial) cells in the teleost retina: consequences of continuous growth. Submitted to Glia.

Mangold, O., 1931, Das Determinationsproblem. Dritter Teil. Das Wirbeltierauge in der Entwicklung und Regeneration. Ergebn. Biol. 7. 193–403.

Masland, R. H., 1977, Maturation of function in the developing rabbit retina. J. Comp. Neurol. 175: 275–286.

McArdle, C. B., Dowling, J. E., and Masland, R. H., 1977, Development of outer segments and synapses in the rabbit retina. J. Comp. Neurol. 175: 253–274.

Meller, K., and Tetzlaff, W., 1976, Scanning electron microscopic studies on the development of the chick retina, Cell Tiss. Res. 170: 145–159.

Michaelson, I. C., 1954, Retinal Circulation in Man and Animals. Thomas, Springfield, USA.

Morest, D. K., 1979, The pattern of neurogenesis in the retina of the rat. Z. Anat.Entwickl.-Gesch. 131: 45–67.

Moscona, A. A., 1983, On glutamine synthetase, carbonic anhydrase and Müller glia in the retina. Progr. Retinal Res. 2: 111–135.

Mulay, S., Giannopoulos, M. and Solomon, S., 1973, Corticosteroid levels in the mother and fetus of the rabbit during gestation. Endocrinology 93: 1342–1348.

Müller, B. and Peichl, L., 1989, Topography of cones and rods in the tree shrew retina. J. Comp. Neurol. 282: 581–594.

Naumann, G. O. H., Lerche, W. and Schroeder, W., 1976, Foveola-Aplasie bei Tyrosinase-positivem oculocutanen Albinismus. Graefes Arch. Clin. Exp. Ophthalmol. 200: 39–50.

Newman, E. A., 1984, Regional specialization of retinal glial cell membrane. Nature 309: 155–157.

Newman, E. A., 1991, Sodium-bicarbonate cotransport in retinal Müller (glial) cells of the salamander. J. Neurosci. 11: 3572–2983.

Newman, E. A., 1993, Inward-rectifying potassium channels in retinal glial (Müller) cells. J. Neurosci. 13: 3333–3345.

Newman, E. A., 1994, A physiological measure of carbonic anhydrase in Müller Cells. Glia 11: 291–299.

Newman, E. A., Frambach, D. A. and Odette, L. L., 1984, Control of extracellular potassium levels by retinal glial cell K$^+$ siphoning. Science 225: 1174–1175.

Newman, E. A. and Reichenbach, A., 1996, The Müller cell: a functional element of the retina. TINS 19: 307–312.

O'Loughlin, E. V., Hunt, D. M. and Kreutzmann, D., 1990, Postnatal development of colonic electrolyte transport in rabbits. Am. J. Physiol. 258: G447-G453.

Penfold, P. L. and Provis, J. M., 1986, Cell death in the development of the human retina: phagocytosis of pycnotic and apoptotic bodies by retinal cells. Graefes Arch. Clin. Exp. Ophthalmol. 224: 549–553.

Pfeiffer, B., Grosche, J., Reichenbach, A. and Hamprecht, B., 1994, Immunocytochemical demonstration of glycogen phosphorylase in Müller (glial) cells of the mammalian retina. Glia 12: 62–67.

Piddington, R., and Moscona, A. A., 1967, Precocious induction of retinal glutamine synthetase by hydrocortisone
in the embryo and in culture. Age-dependent differences in tissue response. Biochim. Biophys. Acta 141:
429–432.

Poitry-Yamate, C. L., Poitry, S. and Tsacopoulos, M., 1995, Lactate released by Müller glial cells is metabolized
by photoreceptors from mammalian retina. J. Neurosci. 15: 5179–5191.

Polley, E. H., Zimmermann, R. P. and Fortney, R. L., 1989, Neurogenesis and maturation of cell morphology in
the development of the mammalian retina. In: Finlay, B. L. and Sengelaub, D. R. (eds.): Development of
the Vertebrate Retina. pp.3–29, Plenum Press, New York London.

Pow, D. V. and Robinson, S. R., 1994, Glutamate in some retinal neurons is derived solely from glia. Neuroscience
60: 355–366.

Provis, J. M. and Penfold, P. L., 1988, Cell death and elimination of retinal axons during development. Progr.
Neurobiol. 31: 331–347.

Puro, D. G., 1995, Growth factors and Müller cells. Progr. Retinal Eye Res. 15: 89–101.

Puro, D. G. and Stuenkel, E., 1995, Thrombin-induced inhibition of potassium currents in human retinal glial
(Müller) cells. J. Physiol. (Lond.) 485: 337–348.

Puro, D. G., Roberge, F., and Chan, C.-C., 1989, Retinal glial cell proliferation and ion channels: a possible link.
Invest. Ophthalmol. Vis. Sci. 30: 521–529.

Puro, D. G., Mano, T., Chan, C.-C., Fukuda, M. and Shimada, H., 1990, Thrombin stimulates the proliferation of
human retinal glial cells. Graefe's Arch. Clin. Exp. Ophthal. 228: 169–173.

Puro, D. G., Yuan, J. P. and Sucher, N. J., 1996, Activation of NMDA receptor-channels in human retinal Müller
glial cells inhibits inward-rectifying potassium channels. Vis. Neurosci. 13: 319–326.

Raju, T. R. and Bennet, M. R., 1986, Retinal ganglion cell survival requirements: a major but transient dependence
on Müller glia during development. Brain Res. 383: 165–176.

Rakic, P., 1971, Guidance of neurons migrating to the fetal monkey neocortex. Brain Res. 33: 471–476.

Rasmussen, K. E., 1973, A morphometric study of the Müller cells, their nuclei and mitochondria, in the rat retina.
J. Ultrastruct. Res. 44: 96–112.

Rasmussen, K. E., 1975, A morphometric study of the Müller cell in rod and cone retinas with and without retinal
vessels. Exp. Eye Res. 20: 151–166.

Raymond, P. A., and Rivlin, P. K., 1987, Germinal cells in the goldfish retina that produce rod photoreceptors.
Dev. Biol. 122: 120–138.

Reh, T. A., 1987, Cell-specific regulation of neuronal production in the larval frog retina. J. Neurosci. 7:
3317–3324.

Reh, T. A., 1991, Determination of cell fate during retinal histogenesis: Intrinsic and extrinsic mechanism. In: Development of the visual system. Lam, D. M. and Shatz, C. J. (eds), MIT Press, Boston, pp.79–94.

Reh, T. A., 1992, Cellular interactions determine neuronal phenotypes in rodent retinal cultures. J. Neurobiol. 23:
1067–1083.

Reichelt, W., Dettmer, E., Brückner, G., Brust, P., Eberhardt, W. and Reichenbach, A., 1989, Potassium as a signal
for both proliferation and differentiation of rabbit retinal (Müller) glial growing in cell culture. Cell. Signalling 1: 187–194.

Reichelt, W., Pannicke, T., Biedermann, B., Francke, M. and Faude, F., 1997, Comparison between functional
characteristics of healthy and pathological human retinal Müller glial cells. Survey Ophthalmol., in press.

Reichenbach, A., 1993, Two types of neuronal precursor cells in the mammalian retina - a short review. J. Hirnforsch. 34: 335–341.

Reichenbach, A. and Reichelt, W., 1986, Postnatal development of radial glial (Müller) cells of the rabbit retina.
Neurosci. Lett. 71: 125–130.

Reichenbach, A. and Robinson, S. R., 1995a, Ependymoglia and ependymoglia-like cells. In Neuroglia Cells, B.
Ransom and H. Kettenmann, eds., pp.58–84. New York - Oxford: Oxford University Press.

Reichenbach, A. and Robinson, S. R., 1995b, The involvement of Müller cells in the outer retina. In: Neurobiology
and Clinical Aspects of the Outer Retina, M. B. A. Djamgoz, S. N. Archer and S. Vallerga, eds., pp.
395–416. London: Chapman and Hall.

Reichenbach, A. and Robinson, S. R., 1995c, Phylogenetic constraints on retinal organisation and development.
Progr. Retinal Eye Res. 15: 139–171.

Reichenbach, A., Hagen, E., Schippel, K. and Eberhardt, W., 1988, Quantitative electron microscopy of rabbit
Müller (glial) cells in dependence of retinal topography. Z. mikroskop.-anat. Forsch. 102: 721–755.

Reichenbach, A., Schnitzer, J., Friedrich, A., Ziegert, M., Brückner, G. and Schober, W., 1991a, Development of
the rabbit retina. I. Size of eye and retina, and postnatal cell proliferation. Anat. Embryol. 183: 287–297.

Reichenbach, A., Schnitzer, J., Friedrich, A., Knothe, A. K. and Henke, A., 1991b, Development of the rabbit retina. II. Müller cells. J. Comp. Neurol. 311: 33–44.

Reichenbach, A., Baar, U., Petter, H., Schaaf, P., Osborne, N. N. and Buse, E., 1992, Neuronal ectopia in tiger retina. J. Hirnforsch. 33: 585–593.

Reichenbach, A., Stolzenburg, J.-U., Eberhardt, W., Chao, T. I., Dettmer, D., and Hertz, L., 1993, What do retinal Müller (glial) cells do for their neuronal 'small siblings'? J. Chem. Neuroanat. 6: 201–213.

Reichenbach, A., Ziegert, M., Schnitzer, J., Pritz-Hohmeier, S., Schaaf, P., Schober, W. and Schneider, H., 1994, Development of the rabbit retina. V. The question of 'columnar units'. Dev. Brain Res. 79: 72–84.

Reichenbach, A., Kasper, M., El-Hifnawi, E., Eckstein, A.-K. and Fuchs, U., 1995a, Hepatic retinopathy: morphological features of retinal glial (Müller) cells accompanying hepatic failure. Acta Neuropathol. 90: 273–281.

Reichenbach, A., Stolzenburg, J.-U., Wolburg, H., Härtig, W., El-Hifnawi, E., and Martin, H., 1995b, Effects of enhanced extracellular ammonia concentration on cultured mammalian retinal glial (Müller) cells. Glia 13: 195–208.

Reichenbach, A., Skatchkov, S. N. and Reichelt, W., 1997, The retina as a model of glial function in the brain. In: Laming, P., E. Sykova, A. Reichenbach, G. Hatton and H. Bauer (eds.): Glial Cells and their Role in Behaviour, Cambridge University Press, New York, in press.

Repka, A. and Adler, R., 1992, Differentiation of retinal precursor cells born in vitro. Devel. Biol. 153: 242–249.

Riepe, R. E. and Norenberg, M. D., 1978, Glutamine synthetase in the developing rat retina: an immunohistochemical study. Exp. Eye Res. 27: 435–444.

Robinson, S. R., 1991, Development of the mammalian retina. In: Dreher, B. and S. R. Robinson (eds.): "Neuroanatomy of the visual pathways and their development"; vol. 3 of Vision and Visual Dysfunction, J. R. Cronly-Dillon (Series Ed.), pp. 69–128. Macmillan, U. K.

Rodieck, R. W., 1988, The primate retina. In: Comparative Primate Biology, vol. 4: Neurosciences, Alan R. Liss, Inc., New York, pp 203–278.

Sarthy, P. V. and Lam, D. M. K., 1978, Biochemical studies of isolated glial (Müller) cells from the turtle retina. J. Cell Biol. 78: 675–684.

Schlosshauer, B., Bauch H., and Stier, H., 1996, Polarized radial glia differentially affects axonal versus dendritic outgrowth. Soc. Neurosci. Abstr. 22, p. 1717.

Schumacher, S., Volkmer, H., Buck, F., Otto, A., Tárnok, A., Roth, S. and Rathjen, F. G., 1997, Chicken acidic leucine-rich EGF-like domain containing brain protein (CALEB), a neural member of the EGF family of differentiation factors, is implicated in neurite formation. J. Cell Biol. 136: 895–906.

Skatchkov, S. N., Vyklicky, L., Clasen, T., and Orkand, R. K., 1996, Effect of cutting the optic nerve on K^+ currents in endfeet of Müller cells isolated from frog retina. Neurosci. Lett. 208: 81–84.

Spemann, H., 1912, Zur Entwicklung des Wirbeltierauges. Zool. Jahrb. 32: 1–98.

Stier, H. and Schlosshauer, B., 1995, Axonal guidance in the chicken retina. Development 121: 1443–1454.

Stier, H., and Schlosshauer, B., 1997, Different cell surface areas of polarized radial glia having opposite effects on axonal outgrowth. Submitted to J. Neurosci.

Stolzenburg, J.-U., Haas, J., Härtig, W., Paulke, B.-R., Wolburg, H., Reichelt, W., Chao, T.-I., Wolff, J. R. and Reichenbach, A., 1992, Phagocytosis of different kinds of latex beads by rabbit retinal Müller (glial) cells in vitro. J. Hirnforsch. 33: 557–564.

Stone, J., Itin, A., Alon, T., Peer, J., Gnessin, H., Chan-Ling, T. and Keshet, E., 1995, Development of retinal vasculature is mediated by hypoxia-induced vascular endothelial growth factor (VEGF) expression by neuroglia. J. Neurosci. 15: 4738–4747.

Takagi, H., King, G. L., Ferrara, N. and Aiello, L. P., 1996, Hypoxia regulates vascular endothelial growth factor receptor KDR/Flk gene expression through adenosine A(2) receptors in retinal capillary endothelial cells. Invest. Ophthalmol. Vis. Sci. 37: 1311–1321.

Tanabe, Y., Saito, N., and Nakamura, T., 1986, Ontogenetic steroidogenesis by testes, ovary, and adrenals of embryonic and postembryonic chickens (Gallus domesticus)[1]. General Comp. Endocrinol. 63: 456–463.

Thanos, S., 1991, The relationship of microglial cells to dying neurons during natural neuronal cell death and axotomy-induced degeneration of the rat retina. Eur. J. Neurosci. 3: 1188–1207.

Turner, D. L. and Cepko, C. L., 1987, A common progenitor for neurons and glia persists in rat retina late in development. Nature 328: 131–136.

Uchihori, Y. and Puro, D.G., 1993, Glutamate as a neuron-to-glial signal for mitogenesis: role of glial N-methyl-D-aspartate receptors. Brain Res. 613: 212–220.

Uga, S. and Smelser, G. K., 1973, Electron microscopic study of the development of retinal Müllerian cells. Invest. Ophthalmol. 12: 295–307.

Vardimon, L., Ben-Dror, I., Havazelet, N. and Fox, L. E., 1993, Molecular control of glutamine synthetase expression in the developing retina tissue. Devel. Dynamics 196: 276–282.

Watanabe, T. and Raff, M. C., 1990, Rod photoreceptor development in vitro: Intrinsic properties of proliferating neuroepithelial cells change as development proceeds in the rat retina. Neuron 2: 461–467.

Willbold, E., Berger, J., Reinicke, M. and Wolburg, H., 1997, On the role of Müller cells in histogenesis: only retinal spheroids, but not tectal, telencephalic and cerebellar spheroids develop histotypical patterns. J. Brain Res., in press.

Willbold, E., Reinicke, M., Lance-Jones, C., Lagenaur, C., Lemmon, V. and Layer, P., 1995, Müller glia stabilizes cell columns during retinal development: lateral cell migration but not neuropil growth is inhibited in mixed chick-quail retinospheroids. Europ. J. Neurosci. 7: 2277–2284.

Wolburg, H., Reichelt, W., Stolzenburg, J.-U., Richter, W. and Reichenbach, A., 1990, Rabbit retinal Müller cells in cell culture show gap and tight junctions which they do not express in situ. Neurosci. Lett. 11: 58–63.

Young, R. W., 1983, The life history of retinal cells. Trans. Am. Ophthalmol. Soc. 81: 193–228.

9

COMPARATIVE ANATOMY AND FUNCTION OF MAMMALIAN HORIZONTAL CELLS

Leo Peichl,[1] Daniele Sandmann,[1] and Brian B. Boycott[2]

[1]Max-Planck-Institut für Hirnforschung
Deutschordenstr. 46
D-60528 Frankfurt a. M., Germany
[2]Department of Visual Science
Institute of Ophthalmology
11-43 Bath Street
London EC1V 9EL, United Kingdom

1. INTRODUCTION

Near the end of his life Ramón y Cajal (1933) summarized aspects of his view of the retina and, under the section heading "*The paradox of vertebrate retinal horizontal cells*," admitted defeat in understanding their role in visual processing. By then, horizontal cells had been identified and studied anatomically for more than six decades; today, six decades later, they are still amongst the most enigmatic neurons (historical reviews, e.g., Wässle et al., 1978a; Gallego, 1986; Piccolino, 1986, 1988). Numerous studies, using an ever-increasing arsenal of methods, have modified some of Cajal's observations and added many new ones. For technical reasons, there has been a concentration on horizontal cell physiology and cellular biology in non-mammalian vertebrate retinae (reviewed in Dowling, 1987; Djamgoz et al., 1995; Kamermans & Spekreijse, 1995), and the results have been generalized to deduce mammalian horizontal cell function. Most recently, however, an increasing number of studies are examining mammalian horizontal cells with physiological and immunocytochemical approaches; and comparative anatomical studies are demonstrating previously unsuspected differences between species. More specific questions can now be asked although, as yet, answers are still scant. This review summarizes some earlier views of mammalian horizontal cell morphology and connectivity, then focuses on how some of the newer findings have modified the issues, and tries to suggest where answers may be sought.

Development and Organization of the Retina, edited by Chalupa and Finlay.
Plenum Press, New York, 1998.

2. GENERAL MORPHOLOGY, CONNECTIVITY AND VIEWS ON FUNCTION

2.1. Basic Horizontal Cell Properties

Horizontal cells are the largest neurons found in the outer retina. Their somata are located in the outer part of the inner nuclear layer and their processes ramify in the outer plexiform layer. In mammals Cajal (1893) recognized two types based on their dendritic morphology and relative differences in the level of their somata ("inner" and "outer" horizontal cells). He claimed that both types had an axon. Cajal (1893, 1933) thought both types were connected to rods and surmised their role to be to join particular groups of rods, through the axons, with more distant rod groupings. Further studies have confirmed the presence of two horizontal cell types in a range of mammals, but also qualified some of Cajal's claims. Now mammals are thought to have one axon-bearing type, commonly called B-type, but the second type is axonless and called the A-type. The relative position of their somata does not differ significantly between the types. The B-type has a smaller dendritic tree with many relatively fine dendrites. The A-type has a larger, more sparsely branched dendritic tree with fewer and stouter primary dendrites (Fig. 1, top). The dendrites of both types carry clusters of terminals (terminal aggregates) that synapse exclusively with cones. The only direct connection with rods is at the axon terminal system of the B-type cell (Fig. 1, middle and bottom). This classification and other terminologies and past reviews have recently been summarized and discussed by Sandmann et al. (1996a, b).

The horizontal cell terminals, together with the dendrites of some of the types of bipolar cells, insert into the base of the photoreceptor terminals (cone pedicle or rod spherule). The synaptic complexes thus formed are called triads; there are more triads per cone pedicle than per rod spherule (Fig. 1 bottom). Postsynaptically each triad has a central bipolar cell terminal (two at rod spherules) flanked by two horizontal cell terminals. The presynaptic site is marked by a synaptic ribbon. It is generally supposed the continuous release of the photoreceptor transmitter glutamate is vesicular at the site of the ribbon. The position and some ultrastructural specializations of the horizontal cell dendritic terminals suggest that, at least in cones, they may modulate transmission between photoreceptor and invaginating bipolar cell processes (Raviola & Gilula, 1975). A concise summary of the synaptic arrangement is given in Sterling (1997).

There is now some evidence that horizontal cells in mammals use the inhibitory transmitter GABA (gamma-amino butyric acid) as do certain horizontal cell types in non-mammals; however, the evidence is by no means definitive (recent reviews: Freed, 1992; Sterling et al., 1995; Yazulla, 1995; Sterling, 1997). The horizontal cell response is a graded hyperpolarization to light stimuli (Fig. 1, bottom), or a depolarization to the photoreceptor transmitter glutamate. It is supposed with the ensueing release of GABA, that horizontal cells provide a negative feedback onto the photoreceptors (review: Wu, 1992). They are thought to (at least in part) create the antagonistic surround of bipolar cells and ganglion cells (Dowling 1987; Mangel, 1991). Thus horizontal cells are now seen as sharpening contrast sensitivity and acuity by acting as lateral inhibitory pathways—a very different view from that of Cajal (1893, 1933) who, apparently not considering synaptic inhibition, was disconcerted with the wide lateral spread of signals through horizontal cells that seemed to spoil the crispness of signal processing by the fine-grain matrices of photoreceptors and bipolar cells.

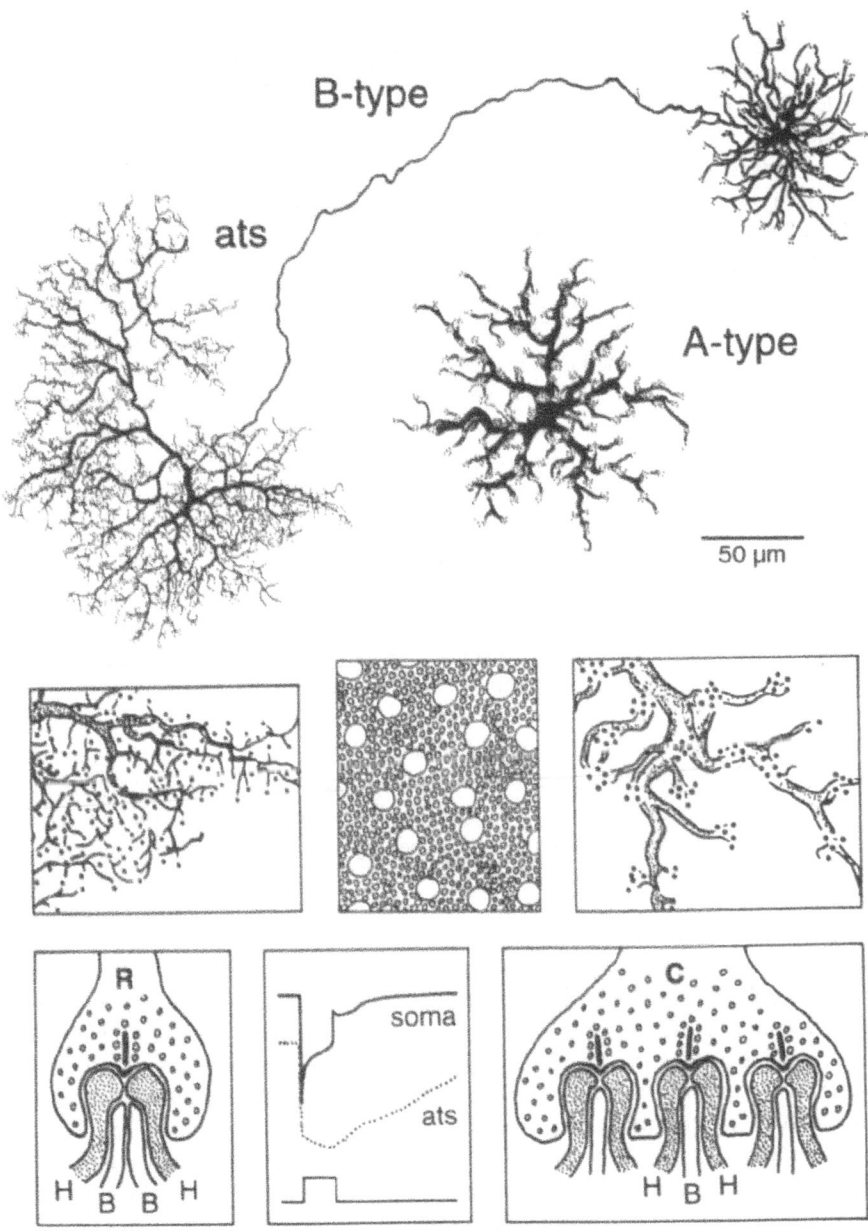

Figure 1. General overview of mammalian horizontal cell properties. **Top:** Flat views of Golgi-stained B-type horizontal cell with axon terminal system (ats) and axonless A-type horizontal cell of cat (adapted from Boycott et al., 1978). **Middle row:** Left, enlarged part of the B-type ats with single terminals that are the contacts with rods; right, part of an A-type dendrite with clustered terminals that are the contacts with cones; middle, schematic photoreceptor pattern with the rather regularly spaced cones surrounded by the more numerous and smaller rods. **Bottom row:** Schematic triad arrangements at a rod spherule (R, left) and a cone pedicle (C, right), H = horizontal cell process, B = bipolar cell process; middle, physiological responses of (A- or B-type) soma and B-type ats to light stimulus (bottom trace). For details see text.

2.2. Contacts with Rods and Cones

In the mammalian retina, as Cajal emphasized, the rod and cone signals are separately carried by cone bipolar cells and rod bipolar cells. The rod and cone pathways only converge at the level of AII amacrine cells and ganglion cells (review: Wässle & Boycott, 1991). Unlike some other vertebrates, mammals have no type of horizontal cell that is exclusively connected to rods; nevertheless, mammalian horizontal cells also separately serve the rod and cone pathways (see above). The B-type axon does not conduct from the soma to the axon terminals as Cajal proposed. Electrophysiological studies of cat and rabbit B-type cells indicate that the axon terminal system is electrically uncoupled from the dendritic part, apparently the axon is too thin and long to conduct the graded potentials that horizontal cells use for signalling (Nelson et al., 1975; Bloomfield & Miller, 1982). The axonal synaptic connections with the rods resemble those of the horizontal cell dendrites with the cones (Fig. 1 bottom). It is now thought that the B-type axon terminal system is the independent rod horizontal cell of mammals. Thus, while being one metabolic entity, the B-type cell may represent two functional units. Actually, a small rod input is found in soma recordings of both B-type and A-type cells (Fig. 1, bottom). But this is attributed to direct rod/cone coupling and not to signal transfer through the B-type axon (Nelson, 1977; Bloomfield & Miller, 1982).

In their cone connections, the two horizontal cell types share common input. Most mammals are cone dichromats with a majority of medium-to-long wavelength sensitive M/L-cones and a minority of short wavelength sensitive S-cones (see Section 4). Anatomically, cat and rabbit A- and B-type horizontal cells contact the vast majority of cones within their dendritic fields; this was taken as evidence that each cone contacted each type of horizontal cell (Wässle et al., 1978b; Raviola & Dacheux, 1990). This matches with the physiological data. Recordings of cat and rabbit horizontal cells have demonstrated that the A-type and the somatic part of the B-type hyperpolarize to light stimuli and show no obvious differences in their response characteristics (e.g., Nelson, 1977; Dacheux & Raviola, 1982; Raviola & Dacheux, 1983); with spectral stimuli, both types hyperpolarize to all wavelengths (De Monasterio, 1978; Nelson, 1985). This starkly contrasts with the findings in goldfish and turtle horizontal cells (review: Djamgoz et al., 1995). Here different types of cone horizontal cell show characteristically different reactions to chromatic stimuli, including wavelength-dependent hyperpolarization and depolarization (see Section 5.3.3). Hence many authors have allocated the chromatic horizontal cell types of non-mammalian vertebrates a major role in colour vision, whereas the common view of mammalian horizontal cells is that they are not involved in colour processing. Since the availability of cone opsin-specific antibodies and the realization that most mammals are cone dichromats it has become important to know whether S-cones connect to both types of horizontal cells. This is discussed in Sections 4 and 5.3.3.

As far as we know, all the rods in all mammals have horizontal cell contacts. These are with the axonal terminals of the B-type cells; conversely, all B-type axonal terminals connect only to rods. The only reported exception is the cone-dominated gray squirrel retina, where the axon of an H1 cell (B-type equivalent) was described to contact cones and perhaps rods (West, 1978). Given the general observation that the horizontal cell contact with rods involves only one type of process (the B-type axonal terminals), one has to ask what detailed functional difference there is between this pathway and that of the cones which commonly involves A- and B-type processes. An answer is currently not available. It is known that the receptive field surround of the ganglion cells becomes broader and shallower with decreasing light levels, and this is interpreted as a useful image processing strategy (summarized in Sterling, 1997). But it is unknown whether and how the rod/horizontal cell connections may be involved.

Details of the morphology of the B-type axon terminal system vary significantly across mammalian species. In some, e.g. cat, rabbit and rat, the axon terminal system is rather densely branched and appears to innervate most of the rods in its field (Figs. 1, 3F). In other species, e.g. horse, the branching is less dense or even sparse (Fig. 3E). Here only a minority of the rods present are innervated by any one terminal system. However, as a population, the overlapping axons of several cells ensure full coverage of the rods. Presumably, in their scotopic working range, the gap junctions between the thicker branches of the neighbouring axon terminal systems are open and the overlapping terminals form a functional syncytium (see Section 2.3).

2.3. Horizontal Cell Populations and Gap Junctional Coupling

The A- and B-type horizontal cells form populations that completely cover the retinal surface and thus provide their signals at all points of the retina. The horizontal cells, like many other neurons, decrease their population density from central to peripheral retina and conversely increase the size of the individual cells such that the dendritic overlap, or coverage factor, remains approximately constant across the retina (Wässle et al., 1978a; Mills & Massey, 1994). Definition of populations of cells is also important for classification purposes (see Sections 3.2 & 4.2, and Cook, 1998). Several staining methods are known to stain entire horizontal cell populations. Neurofibrillar stains and more recently immunocytochemical staining of neurofilament proteins have been widely used (see: Wässle et al., 1978a; Löhrke et al., 1995). In many mammals these methods specifically stain the A-type population (Fig. 2B), but in horse the B-type population is specifically stained (Sandmann et al., 1996a). Antibodies against the calcium binding protein calbindin (CaBP 28kDa) stain both horizontal cell populations in various mammals (Fig. 2A; Röhrenbeck et al., 1987; Peichl & González-Soriano, 1994; Sandmann et al., 1996a).

The dendrites of mammalian horizontal cells are connected by gap junctions, i.e., are presumably electrically coupled. The coupling is homotypical, there are no gap junctions between A- and B-type cells, and the gap junctions are larger between A-type than B-type cells (review: Vaney, 1994). Thus with respect to electrical coupling, A- and B-type cells form separate functional networks although they largely share their chemical input and output partners, the cones. The rod connectivity is similar through the coupling of B-type horizontal cell axon terminals. In some non-mammalian vertebrates it has been shown that the gap junctions are regulated by the ambient light level: In dim light, the horizontal cells are strongly coupled and their signals spread over large distances; with bright light, gap junctions are closed and the horizontal cells act as smaller inhibitory units (review: Piccolino, 1986). In mammals, adaptation-regulated gap junctions exist on AII amacrine cells (Mills & Massey, 1995), but so far such regulation of horizontal cell coupling has not been demonstrated.

None of the basic features listed so far have provided compelling arguments to explain why there should be two horizontal cell types in mammals. Are there really two types in every mammal? Which morphological differences are of functional significance? Are there differences in their synaptic connection with the cones? These questions have recently become more tractable and led to some unexpected observations.

3. DIVERSITY OF THE BASIC PATTERN OF HORIZONTAL CELLS

Cajal always looked for the fundamental patterns of cell types in nervous systems. Thus, in 1893 he stated that "*the retinal structure of all [mammals] is virtually identi-*

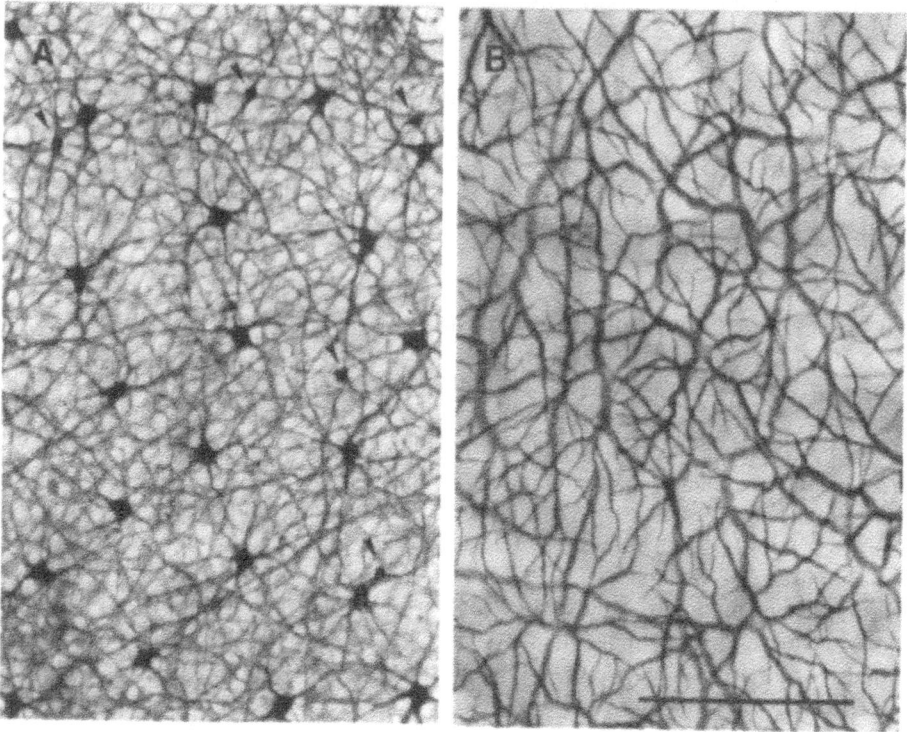

Figure 2. Population stains of horizontal cells, flat views. (A) Calbindin immunostaining of horse A- and B-type cell populations (arrowheads point to A-type somata; from Sandmann et al., 1996a); (B) Bodian neurofibrillar staining of rabbit A-type cell population (from Löhrke et al., 1995). Scale bar 100 μm for A & B.

cal,...[and] the morphology of every type of cell...remains absolutely constant." Until recently this seemed to hold for horizontal cells as it does for the alpha-type ganglion cells (Peichl, 1991). But this was because cat and rabbit horizontal cells were the primary model; primate retina had not been adequately studied and Cajal´s descriptions in ungulates were taken for granted. Now horizontal cells have been studied in some detail in about two dozen species representing seven orders of placental mammals and two species of marsupial. While earlier work mostly relied on Golgi staining, more recent studies also used dye injections (Lucifer yellow, horseradish peroxidase, Neurobiotin or DiI). This approach gives a more reliable yield of well stained cells, and a wider range of species can be readily studied.

The results can be summarized as follows. A- and B-type horizontal cells have been identified in a range of orders including primates, carnivores, lagomorphs, rodents and ungulates; they are both present in rod-dominated and cone-dominated retinae. These data have recently been summarized in Sandmann et al. (1996a, b). A- and B-type horizontal cells are also present in the two marsupials studied (brush-tailed possum: Harman, 1994; quokka: Harman & Ferguson, 1994). On this basic pattern there is superimposed an unexpected diversity in morphology and connectivity which has warranted a reassessment of the criteria for a general definition of mammalian A- and B-type cells.

3.1. Variations of A- and B-Type Shape

Old World primates were early perceived as deviating from the general mammalian pattern (Dogiel, 1891; Boycott & Kolb, 1973). Both the H1 cell (B-type equivalent) and the H2 cell have an axon (Fig. 5); hence Gallego (1986) regarded the H2 cell as unique to primate retina. While the axon of the H1 cell connects to rods, that of the H2 cell connects to cones; the dendrites of both types only synapse with cones (Kolb et al., 1980). Because the H2 axon has no rod contacts this cell can be equated to A-type cells of other mammals and not regarded as unique to primates (review: Boycott et al., 1987). However, H2 cells have finer dendrites than H1 cells, this is the reverse of the once presumed characteristic distinction between the A- and B-type cells (Figs. 1, 5). H1 and H2 cells with the same morphologies exist in the marmoset, a New World monkey (Chan et al., 1997). Moreover, the cone contacts of H1 and H2 cells in both Old World and New World primates differ from those of the B- and A-type cells of other mammals (see Section 4.2).

Primates are the only known example where both horizontal cell types have an axon and thus, ironically because he never reported primate retina, conform to Cajal's general mammalian scheme. However, since Cajal's time artiodactyl retinae, which formed the main data source of his description of two axon-bearing horizontal cell types, have hardly been studied. So we decided to look again at these retinae (Sandmann et al., 1996b). In contrast to rabbit or carnivores, artiodactyl (ox, sheep, pig, deer) B-type cells have a robust dendritic tree while the A-type dendritic tree is delicate (Figs. 3B, 5); in this they resemble primates. The B-type cells have a single axon ending in an axon terminal system. Contrary to Cajal, we never found an axon on an A-type cell. Sometimes there were one or a few dendritic processes that extended beyond the perimeter of the dendritic field (Fig. 3B). We think it is these processes that Cajal interpreted as axonal and used to conclude that both horizontal cell types have an axon. Our evidence indicates they are conventional dendrites connected to cones. Hence we have suggested that artiodactyl A-type cells represent an intermediate shape between the roughly symmetric A-type dendritic fields of many mammals and the singular asymmetry of the primate H2 cell's axon, thus placing primate H2 cells at one end of a spectrum of A-type morphologies (Sandmann et al., 1996b).

In perissodactyls (horse, ass, mule and zebra; Sandmann et al., 1996a) the B-type cells also have a robust dendritic tree, each has one axon which is very long and unusually thick (Fig. 3C, E). The A-type cells' dendrites are very fine and sparsely branched and there is no indication of an axon (Fig. 8C, D). This is very different from Cajal's description that included the horse as being like other mammals. However, he could not have known the most interesting feature of this A-type cell, which is its selective connection to the S-cones (see Section 4.2).

The cone-dominated retina of the tree shrew (order Scandentia) is a further instance where the horizontal cells have unusual morphologies (Fig. 4; Müller & Peichl, 1993). The B-type cells have a conventional dendritic tree, but their axon is very sparsely branched and has only a few terminals. This is to be expected because in this retina less than 10% of the photoreceptors are rods. The second, larger horizontal cell type has stout primary dendrites that radiate from the soma like spokes from a hub. The dendrites rarely branch until the periphery of the dendritic field, where some ramify into unique bushy arborizations. Mariani (1985), who first described these cells, termed them multiaxonal, because the arborizations reminded him of B-type axon terminal systems. However, Müller and Peichl (1993) have shown that all connections of these cells, including those at the peripheral arborizations, are with cones. The cells thus conform to the basic mammalian A-type connectivity and can be interpreted as a further variety of A-type cell shape.

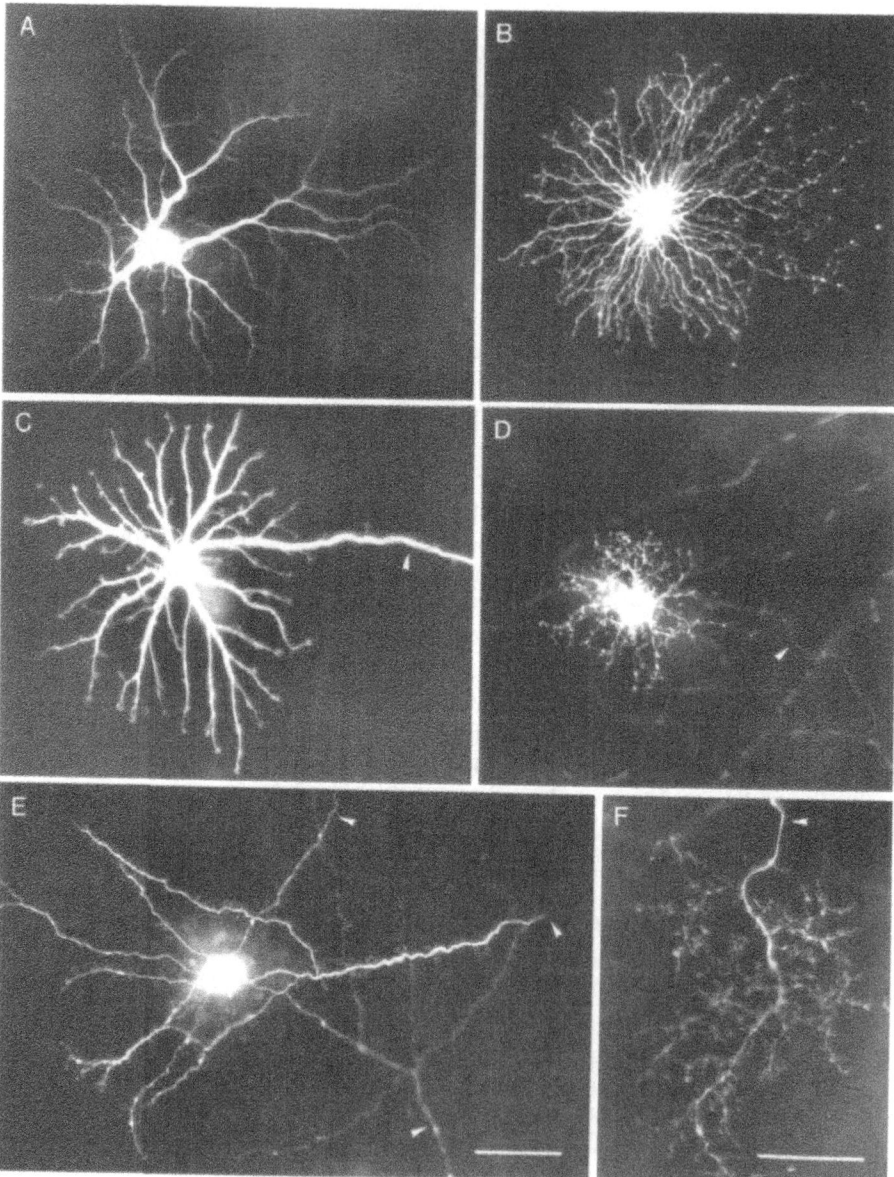

Figure 3. Species variations in horizontal cell morphology. Lucifer Yellow (LY) injected cells, flat view. (A) Guinea pig A-type; (B) pig A-type; (C) horse B-type; (D) gerbil B-type; (E) three horse B-type axon terminal systems; (F) rat B-type axon terminal system. Axons are arrowed; scale bar in E is 50 μm for A-D and 100 μm for E, scale bar in F is 100 μm. A, D, F from Peichl & González-Soriano, 1994; B from Sandmann et al., 1996b; C, E from Sandmann et al., 1996a.

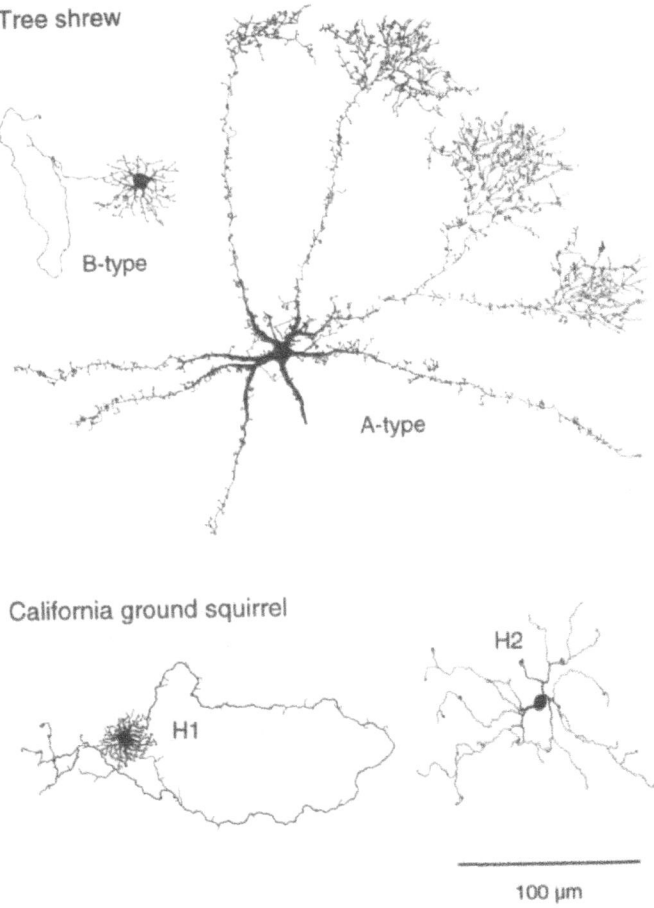

Figure 4. Drawings of A- and B-type cells from the cone-dominated retinae of tree shrew (top; LY injected cells) and California ground squirrel (bottom; Golgi-stained cells, from Linberg et al., 1996). The B-type/H1 cells in the two species have a similar dendritic tree and a sparsely branched axon terminal system, the ground squirrel H1 axon has terminal processes along its length. However, the A-type/H2 cells have a very different branching pattern and a different spacing of dendritic terminals. All cells reproduced at the same magnification. For details see text.

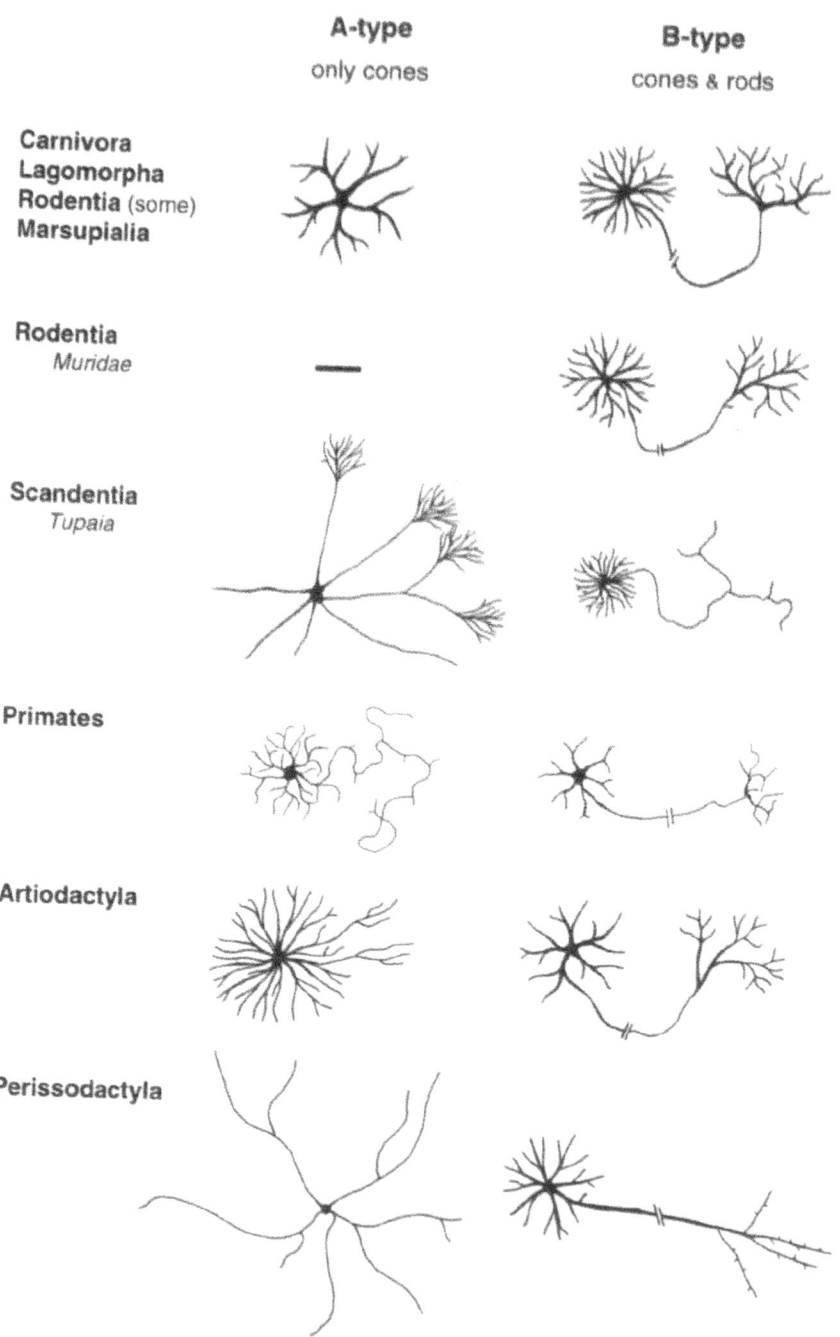

Figure 5. Schematic drawings to show inter-ordinal variations in mammalian A- and B-type horizontal cell morphology. The diagram shows the basic branching patterns but not the synaptic terminals. Interruptions on B-type cell axons indicate that the axons are longer than drawn.

This shape is not in any straightforward way associated with the high cone density in tree shrew retina, because the H2 cell (A-type equivalent) of the equally cone-dominated ground squirrel retina has a rather conventional shape (Fig. 4; Linberg et al., 1996).

The rodents are the most diverse of the mammalian orders, and their horizontal cells have provided the largest surprise. A study of mouse retinal horizontal cells by Suzuki and Pinto (1986) reported a failure to impale and record any A-type cells. This stimulated us to assess the retinae of rodents more closely. In the murid species rat, mouse and gerbil we only found one type of horizontal cell, the axon-bearing B-type (Fig. 3D, F); intracellular injections and population analysis gave no evidence for a further type of horizontal cell (Peichl & González-Soriano, 1994). Recently the Syrian hamster was also shown to have only B-type cells (Glösmann et al., 1997). So far the rodent family of murids is the only known instance where the basic mammalian pattern of two horizontal cell types has not been found; in some other rodent families, e.g. sciurids and several caviomorph families, both types have been shown to be present (Fig. 3A; reviews: Peichl & González-Soriano, 1994; Sandmann et al., 1996b). Perhaps murid ancestors also had A-type horizontal cells but for unknown functional reasons have lost them at some point in evolution. Their absence does not appear to be associated with nocturnality in these species, as suggested by Sterling et al. (1995), since the gerbil has active phases at both day and night and possesses a rather high cone proportion of 10–20%. Rat and mouse have a lower cone proportion (1% and 3% respectively) which is, however, comparable to the cone proportions found in cat (ca. 2%) and rabbit (ca. 4%).These data also show there is no correlation between a low cone/rod ratio and the absence of the A-type cell (for data sources and a discussion see Peichl & González-Soriano, 1994).

From these examples it is clear that the morphology of horizontal cells varies more across orders of mammals than was anticipated by Cajal and many of his successors (Fig. 5). The variation is considerable for the A-type cell while the B-type cell morphology is more conservative. As yet there is little clarity as to the meaning of these variations (see Section 5).

3.2. Are There Further Types of Horizontal Cell?

For some mammals the existence of a third type of horizontal cell has been proposed. In humans an H3 type was described from Golgi-stained material: Its axonal morphology is like that of the H1 cell but its dendritic tree is larger, more sparsely branched and it has been suggested to avoid S-cones; some H3 individuals also have processes descending to the inner retina (Kolb et al., 1994). However, recent recordings and stainings of macaque horizontal cells conclude that the physiological, morphological and connectivity features of H3 cells are encompassed in the normal variation of the H1 population (Dacey et al., 1996). The presence of a descending process is not a sufficent criterion for the definition of a separate type because, in artiodactyls, horizontal cells with descending processes are part of the regular mosaic formed by the B-type population (see Sandmann et al., 1996b)—a conclusion also reached by Cajal (1893).

A third type of horizontal cell has also been claimed for the rabbit retina. Famiglietti (1990) has described a C-type horizontal cell in addition to the conventional A- and B-type cells. On the basis of two Golgi-stained cells with very long and sparse dendritic branches he proposed that some of the processes are axonal, and that the cell is exclusively connected to S-cones. There are no further reports of rabbit horizontal cells with this C-type morphology in the literature. In a recent unpublished study of the cone connections of rabbit horizontal cells, we came across one individual cell that had some morphological resemblance to Famiglietti´s C-type, but it had no axon-like processes and connected to both the S-cones and the L/M-cones (Iris Hack and Leo Peichl, in preparation). So we interpret this cell as an abnormal A-type individual, and the same may be true for Famiglietti´s two cells.

A further example is the South American opossum (*Didelphis marsupialis aurita*). In this species, a brief note described one type of axon-bearing and two types of axonless horizontal cells (Hokoç et al., 1993). The two axonless types appear to differ in dendritic branching pattern and in the density and size of the dendritic terminal aggregates they form. A full description including population data are needed before a definitive decision on the existence of three types can be made. In summary: If individually stained horizontal cells with new or unusual morphological features are to be classified as a new type, it should also be demonstrated that this type exists as a population and adequately covers the retina (see Section 2.3). At present there is no mammalian species where the presence of more than two horizontal cell types has been established beyond doubt.

4. DIVERSITY OF CONE CONNECTIONS OF HORIZONTAL CELLS

Most mammals are cone dichromats (for a comprehensive systematic review see Jacobs, 1993). One cone type has its peak sensitivity in the medium to long wavelength range (M/L-cones, red or green sensitivity depending on species) and represents an approximately 90% majority of the cones (Fig. 6A). The second cone type is most sensitive in the short wavelength part of the spectrum (S-cones, blue or near ultraviolet sensi-

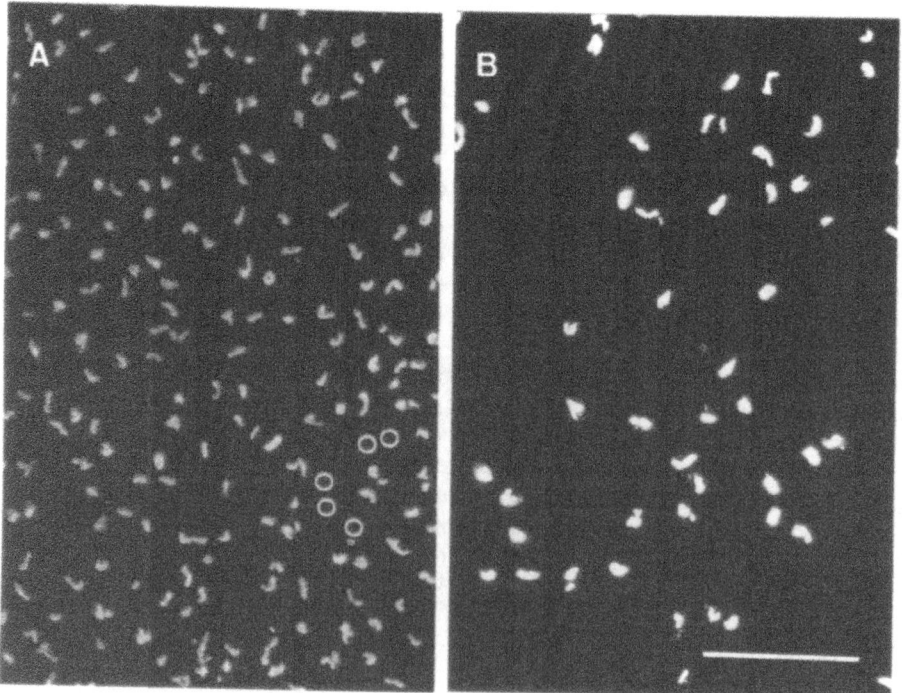

Figure 6. Cone types and cone mosaics in horse retina. Double immunofluorescence labelling of the M/L-cone opsin (A) and the S-cone opsin (B); flat view focused on the cone outer segments. S-cones are an irregularly arrayed minority filling the gaps in the M/L-cone mosaic; the position of a group of five S-cones in B is shown by five open circles in A (adapted from Sandmann et al., 1996a). Scale bar 50 μm.

tivity depending on species) and commonly represents around 10% of the cones (Fig. 6B). The mosaic formed by S-cones is rather irregular in most species. In man, apes and Old World monkeys the mammalian M/L-opsin gene has diverged into separate genes for the M- (green) and L- (red) cone opsins, making these species cone trichromats. Most New World monkeys are cone dichromats by genotype, but marmosets and squirrel monkeys, for example, have an M/L-opsin polymorphism that results in a trichromatic phenotype in many females while the males are dichromats.

Initially quantitative studies of the cone connections of cat horizontal cells showed that individual A-type as well as B-type horizontal cells contact 80% and more of the cones in reach (Wässle et al., 1978b). Equally extensive contacts with the cone population were shown for the A-type cell of the rabbit (Raviola & Dacheux, 1990) and the H1 and H2 cells of the macaque monkey (Boycott et al., 1987; Wässle et al., 1989). Given the considerable overlap of horizontal cell dendritic trees it was concluded for these species that every cone contacts both types of horizontal cell. *Nota bene* in those studies the spectral identity of the cones could not be anatomically determined. Since then several antibodies/antisera have become available against the S-cone opsin and against the M/L-cone opsin of mammals (Fig. 6; see Szél, 1998), and it is now possible to directly identify the spectral types of cone that are in contact with a given horizontal cell. This is particularly important for the S-cone contacts. In most mammals, the S-cones are a minority, so a horizontal cell contacting 90% of the cones in its reach could still be conceived to selectively avoid the S-cones. And sure enough, recent detailed studies have revealed non-selective connections in some species and chromatically specific connections in others.

4.1. Species with Non-Selective Cone/Horizontal Cell Connections

The first mammal where a cone-type specific analysis of the horizontal cell contacts was done was the dichromatic tree shrew (Müller & Peichl, 1993). Here the S-cones happen to specifically label with an antiserum to arrestin ("S-antigen", a molecule involved in the phototransduction process). By injecting A- and B-type horizontal cells with Lucifer Yellow and counterstaining the tissue with the arrestin antiserum, we showed that both types connect to the M/L-cones as well as the S-cones.

The second mammal that is now proven to have no spectral selectivity in its A- and B-type contacts is the dichromatic rabbit. With the technique of Lucifer Yellow injections into individual horizontal cells and immunolabeling for the cone opsins, we have recently demonstrated that both horizontal cell types connect to the M/L-as well as to the S-cones (Hack & Peichl, 1997; Iris Hack and Leo Peichl, in preparation). The proportion of contacted S-cones corresponds to the low (on average 10%) proportion of S-cones present. This is compatible with physiological results, where rabbit horizontal cells showed uniform responses to different spectral stimuli, with a dominant green cone input (Fig. 10; De Monasterio, 1978; Bloomfield & Miller, 1982). Peculiarly, in inferior peripheral rabbit retina the S-cones by far outnumber the M/L-cones (Juliusson et al., 1994). In this region, A- and B-type horizontal cells maintain their morphology, and they contact the many S-cones as well as the few M/L-cones. Nowhere in the rabbit retina did we find evidence for the presence of an S-cone selective C-type horizontal cell as proposed by Famiglietti (1990; see Section 3.2).

4.2. Species with Selective Cone/Horizontal Cell Contacts

Early on, the retinae of trichromatic primates were an interesting place to investigate whether there are chromatically selective horizontal cells, and different positions emerged.

One group of researchers proposed a role for horizontal cells in colour processing and subsequently reported that (ultrastructurally identified) S-cones were specifically avoided or undersampled by the H1/H3 cells (Kolb et al., 1980; Kolb, 1991; Ahnelt & Kolb, 1994a,b). Another group argued on the basis of population data that both horizontal cell types in principle connect to all spectral cone types, i. e. they are not specifically selective; but that individual H1 cells in central retina, where they contact only a small number of cones, could accidentally be in a position without access to an S-cone (Boycott et al., 1987; Wässle et al., 1989). This issue has recently been settled by the physiological and anatomical demonstration of a different weighting in the cone connections of the two horizontal cell types.

Dacey et al. (1996) recorded the responses of H1 and H2 cells to chromatic stimuli in the isolated living macaque retina and demonstrated that H2 cells show the same hyper-

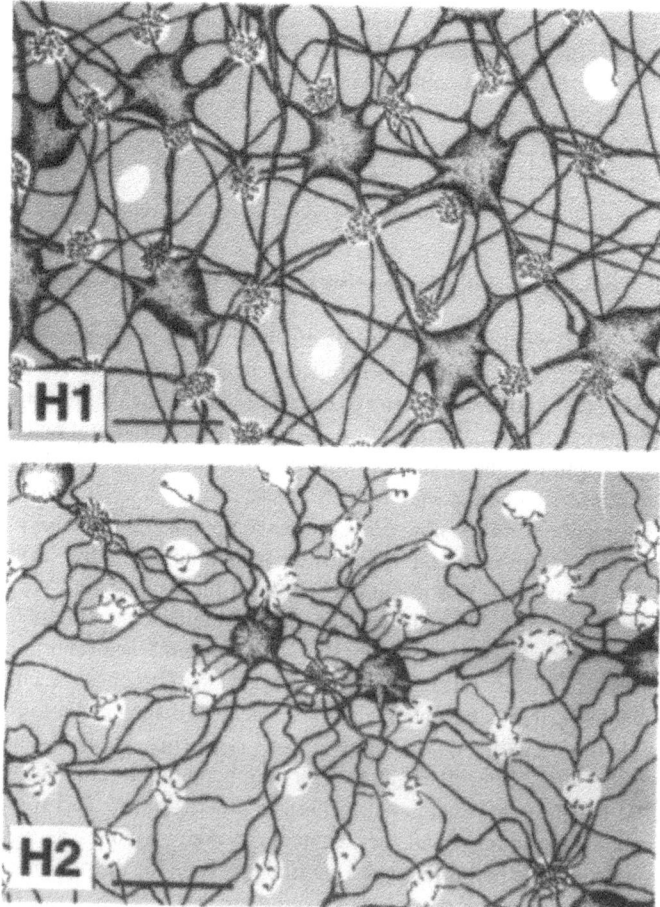

Figure 7. Cone contacts of macaque H1 and H2 horizontal cells. In these two fields, the H1 and H2 populations, respectively, were stained by intracellular injection of neurobiotin; the overlaying cone pedicles are represented by white patches. H1 cell dendrites densely innervate most cone pedicles with their terminals but nearly completely miss the three presumed S-cones. H2 cells strongly innervate the three presumed S-cones but also contact the other (M- and L-) cones. Scale bars 20 μm. Drawings kindly provided by Dennis Dacey.

polarization at all wavelengths, whereas H1 cells hyperpolarize to red or green stimuli but do not respond to blue stimuli (Fig. 10). These authors also injected the recorded cells with the tracer neurobiotin, thus labelling all members of patches of H1 or H2 cells around an injected cell through the homotypic gap junctions. The population of H1 cells inner-vates the majority of the cones, but hardly contacts the fraction of cones, which by their density and spacing were concluded to be the S-cones (Fig. 7 left). The H2 horizontal cell population, on the other hand, connects to all cones, but makes particularly dense contacts with the small fraction of presumed S-cones (Fig. 7 right). This corresponds to the connec-tivity reported for human and monkey H1 and H2 cells in the Golgi study of Ahnelt and Kolb (1994a, b). By dye-labelling horizontal cells in macaque retina and directly identify-ing the S-cones with an antiserum to the S-cone opsin, Goodchild et al. (1996) confirmed this connectivity pattern. They showed that only about 15% of the H1 cells studied con-tacted S-cones and then only sparsely. The H2 cells contacted all cones within reach and on average had more synapses with each S-cone than with each of the M- and L-cones contacted. The axon of the H2 cell definitely contacts S-cones (Ahnelt & Kolb, 1994a, b; Goodchild et al., 1996) but the suggestion that it does so exclusively (Ahnelt & Kolb, 1994a, b) has yet to be confirmed by analysis of completely stained axons.

Two New World monkeys, the marmoset (where there are dichromatic and trichro-matic individuals; see above) and the tamarin, have the same cone connectivity pattern of H1 and H2 cell dendrites as macaque and man (Chan & Grünert, 1998). This suggests that the special horizontal cell connectivity with S-cones is not correlated with the evolution of trichromacy in the Old World primates. Both horizontal cell types of trichromatic primates are non-selective in their M-cone and L-cone contacts. Apparently when the red/green processing pathway evolved from a presumably dichromatic early primate retina this did not involve alteration of the connectivity of the horizontal cells.

The view that a differential connectivity between the two types of horizontal cell and the S-cones is not linked to trichromacy is supported by our findings in the horse ret-ina. When we analysed the horizontal cell types of the dichromatic horse, we were sur-prised by the large dendritic field size, the sparse branching and the paucity of dendritic terminal aggregates of the A-type cells (Sandmann et al., 1996a). For one Lucifer Yellow filled horse A-type cell, counterstaining with an S-cone antiserum revealed that all but one of its 45 contacts were with S-cones (Fig. 8C, D). Further A-type cells could not be obtained for a connectivity analysis, because these cells were rarely impaled and difficult to fill. But immunolabelling of the horizontal cell populations with a calbindin antiserum showed that the A-type, like the B-type, is a consistently occuring cell population that covers the horse retina (Fig. 2A). So the horse possesses an A-type cell that is practically exclusively directly connected to S-cones, while the B-type connects to both types of cone (Fig. 8A, B); the same probably holds for other equids (Sandmann et al., 1996a).

A similarly S-cone selective horizontal cell may be present in the cone-dominated retinae of the dichromatic sciurids. There is an axon-bearing H1 cell (B-type equivalent) and an axonless H2 cell (A-type equivalent) in the red squirrel (Mariani, 1985) and in the ground squirrel (Fig. 4; Linberg et al., 1996). The density of dendritic terminal aggregates on the H1 cell is high enough to contact all cones present. In contrast, the terminal aggre-gates on H2 dendrites are spaced so far apart that they can only contact a small fraction of the cones, which Linberg et al. (1996) speculate to be the S-cones. It would be a very worthwhile and now feasible enterprise to check this experimentally.

In summary, there are now as many known examples of species with a chromatic bias or selectivity in their cone/horizontal cell connections as there are of species con-forming to the proposed non-selectivity. The conclusion is that the cone connections of

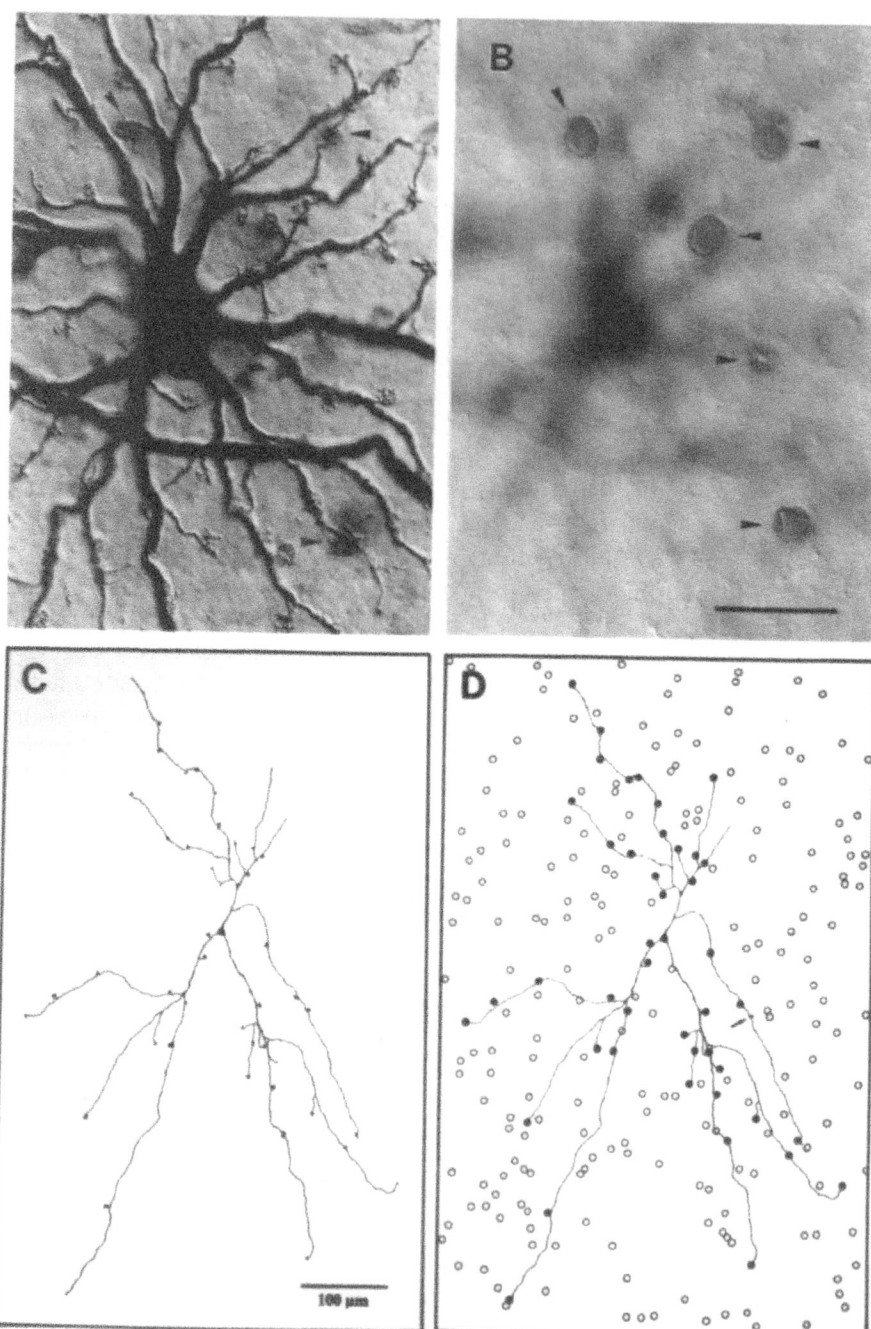

Figure 8. Cone contacts of perissodactyl horizontal cells. (A, B) Mule: Immunolabelled S-cones (B) overlaying a B-type cell (A) which was LY injected and DAB-reacted with a LY antibody. Labelled S-cones can be followed to their cone pedicles and shown to contact horizontal cell terminal aggregates (arrow heads); the other terminal aggregates of this horizontal cell are contacted by M/L-cones. Scale bar in B is 25 μm for A & B. (C, D) Horse: (C) Drawing of a LY injected A-type horizontal cell and its widely spaced terminal aggregates, (D) mosaic of overlaying immunolabelled S-cones. Filled circles indicate S-cones whose pedicles are congruent with terminal aggregates of this A-type cell, open circles are S-cones not in contact with this cell. The arrow marks the only terminal aggregate of this cell with no matching S-cone. From Sandmann et al., 1996a.

Presumed general pattern in dichromats,
confirmed in rabbit & tree shrew

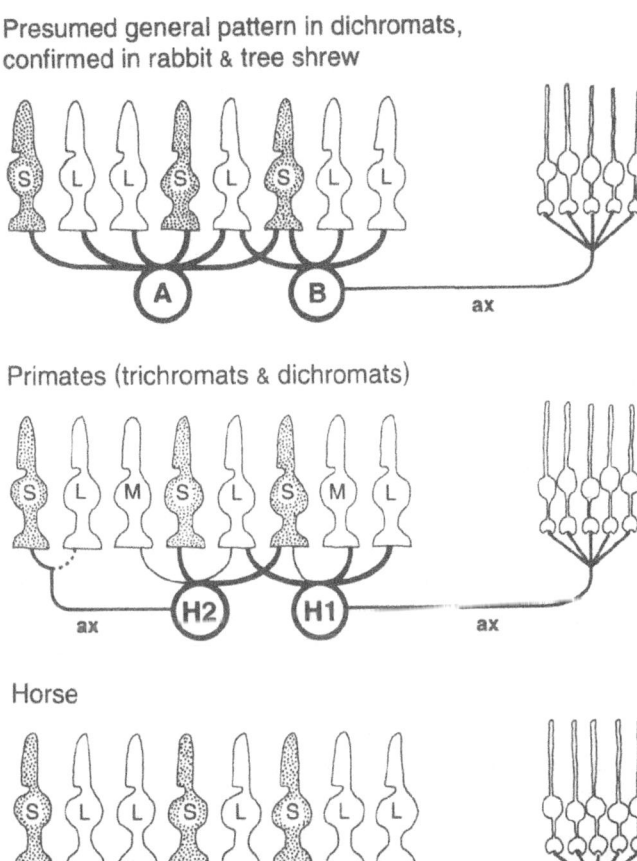

Primates (trichromats & dichromats)

Horse

Figure 9. Scheme of the cone and rod contacts of horizontal cells in different mammals. **Top:** Connectivity pattern in rabbit and tree shrew, presumably also present in many other dichromats. A- and B-type cells indiscriminately contact S-cones (S) and M/L-cones (L), the B-type axon terminal system contacts rods. **Middle:** Connectivity pattern in primates. H1 cells contact M-and L-cones but largely avoid S-cones (thin contact line), their axon contacts rods. H2 cells contact S-cones strongly (thick contact line) and M/L-cones less strongly; their axon contacts S-cones, but may also contact M- and L-cones. In dichromatic primates M- and L-cones are only one spectral M/L-type. **Bottom:** Connectivity pattern in horse (and probably other equids). The A-type contacts S-cones exclusively, whereas the B-type makes indiscriminate contacts with S-cones and M/L-cones.

mammalian horizontal cells vary between species (Fig. 9). Cone/horizontal cell connectivity studies in more species and orders will be necessary to establish whether non-selective contacts are the rule or the exception. Whether the varying connectivity has any impact on colour processing will be addressed in Section 5.3.3.

5. CONCLUSIONS AND PERSPECTIVES

The review makes it plain that mammalian horizontal cells are more diverse across species than previously suspected. This diversity encompasses morphological features as well as details of connectivity. In the final section we shall concentrate on questions such as: Can a common basic plan for mammalian horizontal cells be recognized? What functional consequences do the morphological differences have? At what level may answers be sought?

5.1. A Reassessment of Basic Morphological Properties

The principal dichotomy, of an axon-bearing B-type horizontal cell that serves rods and cones and a commonly axonless A-type that only serves cones, still holds for most mammals as diverse as marsupials and primates. But some qualifications have to be made, as summarized in Figure 5. (1) Dendritic thickness and dendritic branching pattern are no longer a defining characteristic of either type of cell, unless the order of mammals is specified. Within a given species, the two types differ in their dendritic morphology, which suggests some as yet unknown physiological difference. Does it matter whether the finer dendrites are on the B-type (as in cat and rabbit) or on the A-type (as in artiodactyls)? (2) Across species and orders, the A-type is more variable than the B-type. A-type variability ranges from axon-like processes in primates to S-cone selectivity in the horse; it also includes a complete lack of A-type cells in some rodents. B-type variability includes dendritic and axonal branching patterns and the near-avoidance of S-cones by primate H1 cells. (3) To date, no mammalian species has been proven to have more than two horizontal cell types, with the possible exception of the South American opossum.

5.2. Horizontal Cell Morphology, Phylogeny and Lifestyle

One question is whether the observed horizontal cell variations are correlated with the phylogenetic distance of the corresponding mammalian orders (for a recent discussion of the mammalian phylogenetic tree see Novacek, 1992). Disappointingly, at present, we cannot detect any obvious correlation. For example, Carnivora and Lagomorpha are phylogenetically less close than Primates and Scandentia, but the former have very similar horizontal cell morphologies while those of the latter differ significantly. The peculiarity of a B-type/H1 cell with fine dendrites and an A-type/H2 cell with stout dendrites is shared by artiodactyls, perrissodactyls and primates although these are not particularly closely related. Within any one order, horizontal cell features are commonly more conserved across species. But within the admittedly diverse but supposedly monophyletic order Rodentia, there are several families that have two horizontal cell types while members of the murid family have only one.

Are the horizontal cell variations specific adaptations to different visual requirements and linked to other retinal specializations? The evidence so far argues against this. The proposition of Sterling et al. (1995), that A- and B-type cells are present in all diurnal and crepuscular mammals whereas the A-type is missing in strongly nocturnal mammals, has been dismissed in Section 3.1. The proportion of cones is certainly not the decisive factor in determining the number of horizontal cell types present. But what about the types of cones present? Djamgoz et al. (1995) have postulated a trend across vertebrates, that in species with highly developed colour vision specific types of horizontal cells engage in selective contacts with chromatic cone types. In mammals this suggestion does not hold. A-type cells in the dichromatic horse are more cone type selective than the H1 and H2 cells of the trichromatic macaque. And the cone selectivity of H1 and H2 cells is the same in New World and Old World primates, which have different levels of colour vision.

5.3. The Roles of Mammalian Horizontal Cells

Even though we cannot at present interpret the species differences, it appears clear that the A- and B-type horizontal cells represent basic components of the mammalian retinal plan with indispensable functions. If both cones and rods are to be served by horizontal cells, one can see the necessity for the B-type cell in all species that have cones and rods—although it is unclear why mammals do not possess a separate rod horizontal cell to keep rod and cone signals segregated. But why is there an additional A-type horizontal cell in the cone pathway? What functional differences are there between A- and B-type cells? To address this question, let us reiterate the proposed basic roles of mammalian horizontal cells. They are based on the assumption of a negative feedback into the photoreceptors and largely adapted from the situation in non-vertebrates (reviews: Wu, 1992; Sterling, 1997): (1) A "global" contribution to the retinal adaptation to different mean levels of illumination. (2) A local contribution to spatial processing and contrast enhancement by a spectrally broadband but spatially restricted feedback to create the inhibitory receptive field surrounds of photoreceptors, bipolar cells and ganglion cells. (3) A local contribution to chromatic processing by a chromatically selective feedback to create colour-opponent receptive fields of cones, bipolar cells and ganglion cells (proposed for the horizontal cells of trichromatic primates; e.g. Kolb, 1991).

5.3.1. Retinal Adaptation. If there is global retinal adaptation to a given ambient illumination level, e.g. for regulating pupil size, it probably involves the sheet of horizontal cells and their adaptation-dependent electrical coupling across the entire retina. If spatial patterns are not important in this global adaptation, the receptive field size of individual horizontal cells would not matter and the network of one type of horizontal cell would appear sufficient. In fact, this seems to be the situation in the dark-adapted state as the rods have contacts with only B-type axon terminal systems. It has to be noted, however, that the role of the horizontal cells in scotopic (rod) vision is not known.

On the other hand, there is good evidence that adaptation is a highly localized process (demonstrated, e.g., in cat retinal ganglion cells; Cleland & Freeman, 1988) and that it is spectrally specific (demonstrated in primate horizontal cells; Lee et al., 1997). For such adaptation, spatial and spectral processing differences between two horizontal cell types may matter.

5.3.2. Spatial and Temporal Processing. The retinal processing of spatial information has been proposed to require more than one horizontal cell type. Experimentally, horizontal cells have been shown to contribute to the centre/surround balance of concentrically organized ganglion cell receptive fields in rabbit (Mangel, 1991). While the principal characteristics of the light responses are comparable in A-type cells and the somatic parts of

B-type cells, they differ in their area summation properties (for cat and rabbit see, e.g., Nelson, 1977; Dacheux & Raviola, 1982). For cat, Smith (1995) has calculated that the interaction of two horizontal cell types with different spatial summation properties, i. e. different dendritic field sizes and a different degree of electrical coupling, can explain the assumed receptive field characteristics of the cones, and hence the receptive field organization of bipolar cells and ganglion cells (see also Sterling, 1997). But the hypothesis is not conclusive across species. Despite the absence of A-type cells, rat and mouse have ganglion cells with center/surround receptive field organizations very similar to rabbit and cat (discussed in Peichl & González-Soriano, 1994). Certainly more physiological data are necessary to establish the detailed contribution of horizontal cells to receptive field surrounds.

The horizontal cell contributions to spatial processing are frequently discussed, but an equally important task of vision is to render temporal patterns as acutely as spatial patterns. Foerster and colleagues (1977) have analyzed the frequency transfer properties of cat horizontal cells and reported three distinct types of flicker response. The study could not specify whether and how these three types correlated with the two morphological types of cat horizontal cell (plus the B-type axon terminal system), but it is clear that the horizontal cells differ in their temporal processing characteristics. More recently, the influence of flicker stimuli on the temporal, spectral and spatial processing properties of cat horizontal cells has been further scrutinized (Pflug et al., 1990; Nelson et al., 1990). The requirements for temporal processing may be one reason for having more than one horizontal cell type.

5.3.3. Chromatic Processing. The physiological and anatomical data on the cone connections of primate horizontal cells have re-kindled a discussion about the involvement of mammalian horizontal cells in colour processing. Traditionally the reference system in this discussion is the goldfish retina, where three types of cone connectivity and spectral response are found in horizontal cells: Monophasic cells (H1) hyperpolarize irrespective of wavelength, diphasic cells (H2) hyperpolarize at short and medium wavelengths and depolarize at long wavelengths, and triphasic cells (H3) hyperpolarize at short and long wavelengths while depolarizing at medium wavelengths; the monophasic type hence signals luminance whereas the other two types provide chromatically opponent signals (Fig. 10; review: Djamgoz et al., 1995). Since the cascade model of Stell and Lightfoot (1975) the interaction of these three horizontal cell types is commonly regarded as a major factor in creating the colour-opponent receptive fields of bipolar cells and ganglion cells. However, there is now controversy about the detailed mechanisms (see, e.g., Burkhardt, 1993; Kamermans & Spekreijse, 1995), and very recent simulations suggest that in goldfish the horizontal cell feedback to cones as a whole may contribute to colour constancy but not to colour discrimination (Kamermans et al., 1997).

In the retinae of trichromatic primates, two basic colour-opponent systems exist at the level of the ganglion cells: red-green antagonism and blue-yellow antagonism (where yellow is the added input of the M- and L-cones). In dichromatic mammals, where colour-coding ganglion cells are less frequently recorded, only one type of colour opponency has been found, reflecting an antagonistic input of S-cones versus M/L-cones (commonly blue-green antagonism; review: Daw, 1973).

Taking the goldfish model for colour opponency one would conclude that primate horizontal cells cannot contribute to red-green opponency, because both H1 and H2 cells hyperpolarize to red as well as green stimuli; nor can they contribute to blue-yellow opponency, because to blue stimuli the H2 cell hyperpolarizes as well, and the H1 cell does not react at all (Fig. 10; Dacey, 1996; Dacey et al., 1996). For the blue-yellow opponent gan-

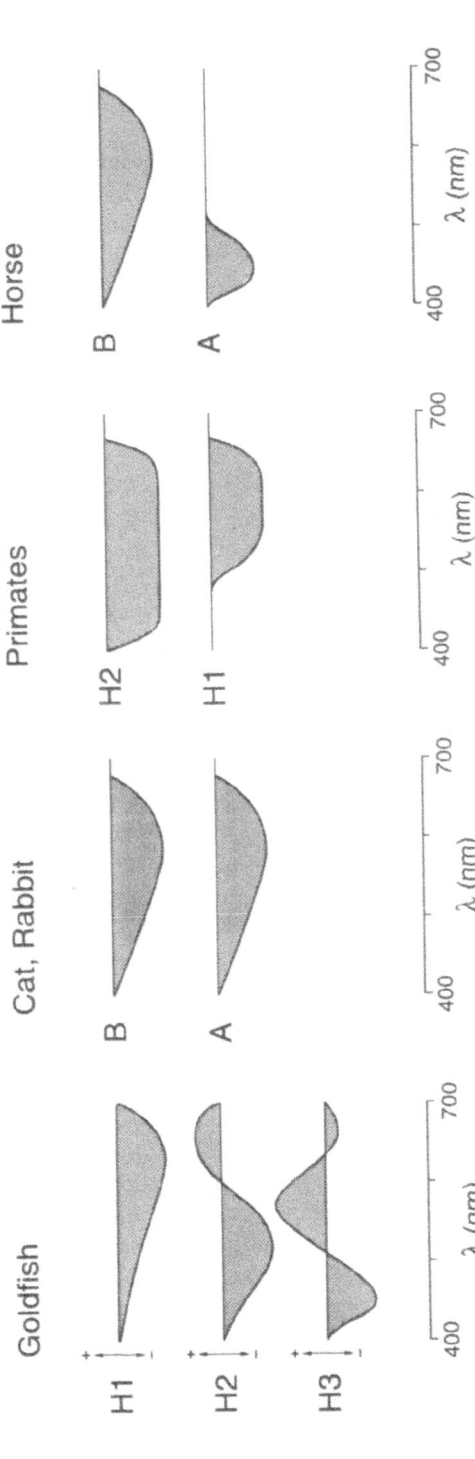

Figure 10. Scheme of horizontal cell responses to chromatic stimuli. The three types of cone horizontal cell in goldfish show monophasic (luminosity type), diphasic and triphasic responses, respectively (see text). Cat and rabbit horizontal cells show a monophasic response which is similar for A- and B-type cells; due to the low proportion of S-cones the response is weaker to blue stimuli. Primate horizontal cells show a monophasic response, H2 cells are depolarized by all spectral stimuli, H1 cells only by longer wavelengths (green and red). Although S-cones are a minority, their H2 contacts are more intense than those of M/L-cones (see Fig. 7), such that the H2 cell response supposedly is as strong to blue stimuli as to green/red stimuli. Horse horizontal cell responses are fictive and predicted from the morphological connectivity pattern. B-type responses are supposedly hyperpolarizing to all wavelengths, but due to the low proportion of S-cones the response should be weaker to blue stimuli; A-type cells supposedly respond to blue stimuli only. Response conventions: + = depolarizing, − = hyperpolarizing; spectral stimuli from blue (400 nm) to red (700 nm). For references and details see text.

glion cells there is further evidence that the antagonism is created in the inner plexiform layer, and for the red-green opponent ganglion cells it is still open whether the surround really is spectrally pure and has to be implemented by chromatically antagonistic lateral connections (reviews: Calkins & Sterling, 1996; Dacey, 1996; Lee, 1996; Masland, 1996). Be that as it may, the consensus now is that primate horizontal cells have no specific part in colour opponency.

In horse, the conclusion is similar. Assuming that both the A- and B-type hyperpolarize to spectral stimuli as do all mammalian horizontal cells recorded so far, the B-type is likely to give a broad band luminosity signal. The A-type, by virtue of its selective contact to S-cones, should hyperpolarize to blue light and not respond to other spectral stimuli (Fig. 10). But it cannot contribute directly to S-cone versus M/L-cone antagonism, because it has no M/L-cone contacts to feed an S-cone signal to them (or vice versa). The only conceivable contribution could come through indirect paths to the M/L-cones, e.g. through the B-type cell with which it shares input from the S-cones.

This leads to a general remark. A horizontal cell contribution to chromatic opponency is mainly considered in those cases where the horizontal cells make type-specific connections with the cones. But in fact, a horizontal cell participating in chromatic opponency has to communicate (directly or indirectly) with all cone types involved, in order to provide the required feedback signal across cone types. Furthermore, simulations of primate ganglion cells have shown that strong and well-balanced chromatic opponency emerges if the excitatory centre of the receptive field is driven by a single cone type and the inhibitory surround is driven indiscriminately by all cone types (Lennie et al., 1991). Hence, there appears to be no need for lateral inhibition by chromatically selective horizontal cells. An unpublished simulation study of primate H1 and H2 cell feedback into cones also concludes that the spectrally broadband negative feedback from the horizontal cells convoluted with the spectrally narrower response profile of a given cone will result in a biphasic response of that cone, e.g., a green cone will hyperpolarize to green light but slightly depolarize to red light (Wolfgang Möckel and Jürgen Röhrenbeck, personal communication). If these conclusions hold, the broadband A- and B-type horizontal cells of other mammals (Fig. 10) could also lead to a chromatically opponent cone response. This makes it all the more puzzling that some species have gone for spectral differences in their cone/horizontal cell connections.

5.4. Perspectives

The morphological and functional properties of A- and B-type horizontal cells discussed so far have not provided any compelling explanation of why two separate networks of horizontal cells are present in most mammals. Neither does the ultrastuctural analysis of the cone triad synapse give any clues for a functional difference between A- and B-type cells, and for a hypothesis explaining why two lateral elements are needed per triad. However, more recently, immunocytochemical studies on the ultrastructural localization of neurotransmitter receptors have increased our knowledge of the receptors present on the different neuronal processes in the triad. There is no space here to review these studies. But we should like to use one study as an illustration of the insights likely to be provided by this type of analysis.

The neurotransmitter of photoreceptors is glutamate. Brandstätter and colleagues (1997) have now shown that in the triad synapses of cones and rods, one of the two horizontal cell processes expresses the high-affinity kainate receptor GluR6/7 while the other does not (Fig. 11). It seems that at the triads of both cones and rods one lateral process

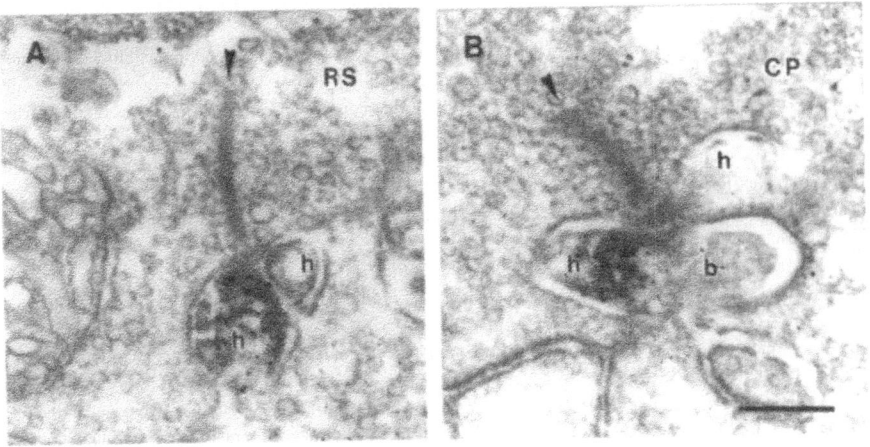

Figure 11. Electron micrographs of the outer plexiform layer of rat retina, showing immunolabelling for the kainate receptor GluR 6/7 (DAB, silver intensification). (A) Triad of a rod spherule (RS), only one of the two ultrastructurally identified horizontal cell processes (h) is immunoreactive (black dots). (B) Triad of a cone pedicle (CP), again only one of the horizontal cell processes (h) is immunoreactive. The presynaptic ribbons are arrowed; the section in B happens to also contain a postsynaptic central bipolar cell process (b) in the triad; scale bar 0.2 μm. For details see text. Micrographs kindly provided by Johann H. Brandstätter.

may respond to glutamate differently from its companion in the same triad. The observations were made in rat retina. Thus both processes are from the same horizontal cell type since the rat retina has only B-type cells. It would seem that individual horizontal cells in this homotypic network can be molecularly different. So it can no longer be assumed that each member of a horizontal cell network will respond in the same way to a given transmitter. Six decades after Cajal's death the role and mode of operation of horizontal cells remain difficult to understand.

ACKNOWLEDGMENTS

We thank our collaborators Juncal González-Soriano, Iris Hack, John Hopkins, Brigitte Müller and Heinz Wässle, who over the years contributed to the data and views presented here. Particular thanks go to Johann Brandstätter, Ulrike Grünert, Iris Hack, Paul Martin, Wolfgang Möckel and Jürgen Röhrenbeck for providing unpublished information and/or helpful comments, and to Heide Ahmed for excellent technical assistance over the years.

REFERENCES

Ahnelt, P. and Kolb, H., 1994a, Horizontal cells and cone photoreceptors in primate retina: A Golgi-light microscopic study of spectral connectivity. Journal of Comparative Neurology 343: 387–405.
Ahnelt P. and Kolb, H., 1994b, Horizontal cells and cone photoreceptors in human retina: A Golgi-electron microscopic study of spectral connectivity. Journal of Comparative Neurology 343: 406–427.
Bloomfield, S. A. and Miller, R. F., 1982, A physiological and morphological study of the horizontal cell types of the rabbit retina. Journal of Comparative Neurology 208: 288–303.

Boycott, B. B. and Kolb, H., 1973, The horizontal cells of the rhesus monkey retina. Journal of Comparative Neurology 148: 115–139.

Boycott, B. B., Peichl, L. and Wässle, H., 1978, Morphological types of horizontal cell in the retina of the domestic cat. Proceedings of the Royal Society (London) B 203: 229–245.

Boycott, B. B., Hopkins, J. M. and Sperling, H. G., 1987, Cone connections of the horizontal cells of the rhesus monkey's retina. Proceedings of the Royal Society (London) B 229: 345–379.

Brandstätter, J. H., Koulen, P. and Wässle, H., 1997, Selective synaptic distribution of kainate receptor subunits in the two plexiform layers of the rat retina. Journal of Neuroscience 17: 9298–9307.

Burkhardt, D. A., 1993, Synaptic feedback, depolarization, and color opponency in cone photoreceptors. Visual Neuroscience 10: 981–989.

Cajal, S. R., 1893, La rétine des vertébrés. La Cellule 9: 119–257.

Cajal, S. R., 1933, Les problèmes histophysiologiques de la rétine. XIV Concilium Ophthalmologicum Hisp. 2: 11–19.

Calkins, D. J. and Sterling, P., 1996, Absence of spectrally specific lateral inputs to midget ganglion cells in primate retina. Nature 381: 613–615.

Chan, T. L. and Grünert, U., 1998, Horizontal cell connections with short wavelength sensitive cones in the retina: A comparison between New World and Old World primates. Journal of Comparative Neurology, in press.

Chan, T. L., Goodchild, A. K. and Martin, P. R., 1997, The morphology and distribution of horizontal cells in the retina of a New World monkey, the marmoset *Callithrix jacchus*: A comparison with macaque monkey. Visual Neuroscience 14: 125–140.

Cleland, B. G. and Freeman, A. W., 1988, Visual adaptation is highly localized in the cat's retina. Journal of Physiology 404: 591–611.

Dacey, D. M., 1996, Circuitry for color coding in the primate retina. Proceedings of the National Academy of Science USA 93: 582–588.

Dacey, D. M., Lee, B. B., Stafford, D. K., Pokorny, J. and Smith, V. C., 1996, Horizontal cells of the primate retina: cone specificity without spectral opponency. Science 271: 656–659.

Dacheux, R. F. and Raviola, E., 1982, Horizontal cells in the retina of the rabbit. Journal of Neuroscience 2: 1486–1493.

Daw, N. W., 1973, Neurophysiology of colour vision. Physiological Reviews 53: 571–611.

De Monasterio, F. M., 1978, Spectral interactions in horizontal and ganglion cells of the isolated and arterially-perfused rabbit retina. Brain Research 150: 239–258.

Djamgoz, M. B. A., Wagner, H.-J. and Witkovsky, P., 1995, Photoreceptor-horizontal cell connectivity, synaptic transmission and neuromodulation. In: Neurobiology and Clinical Aspects of the Outer Retina. Djamgoz, M. B. A., Archer, S. N. and Vallerga, S. (Eds.), Chapman & Hall, London, pp. 155–193.

Dogiel, A. S., 1891, Über die nervösen Elemente in der Retina des Menschen. Archiv für Mikroskopische Anatomie 38: 317–344.

Dowling, J. E., 1987, The Retina: an Approachable Part of the Brain. Harvard University Press, Cambridge, Massachusetts.

Famiglietti, E. V., 1990, A new type of wide-field horizontal cell, presumably linked to blue cones, in rabbit retina. Brain Research 535: 174–179.

Foerster, M. H., van de Grind, W. A. and Grüsser, O.-J., 1977, Frequency transfer properties of three distinct types of cat horizontal cells. Experimental Brain Research 29: 347–366.

Freed, M. A., 1992, GABAergic circuits in the mammalian retina. Progress in Brain Research 90: 107–131.

Gallego, A., 1986, Comparative studies on horizontal cells and a note on microglial cells. Progress in Retinal Research 5: 165–206.

Glösmann, M., Reitsamer, H., Ahnelt, P. K. and Pflug, R., 1997, Horizontal cells of the Syrian hamster retina with a note on proximal axon-like projections. Investigative Ophthalmology & Visual Science 38: S618.

Goodchild, A. K., Chan, T. L. and Grünert, U., 1996, Horizontal cell connections with short-wavelength-sensitive cones in macaque monkey retina. Visual Neuroscience 13: 833–845.

Hack, I. and Peichl, L., 1997, The cone connections of horizontal cells in the rabbit retina. In: Göttingen Neurobiology Report 1997. Elsner, N. and Wässle, H. (Eds.), Georg Thieme Verlag, Stuttgart, p. 506.

Harman, A. M., 1994, Horizontal cells in the retina of the brush-tailed possum. Experimental Brain Research 98: 168–171.

Harman, A. M. and Ferguson, J., 1994, Morphology and birth dates of horizontal cells in the retina of a marsupial. Journal of Comparative Neurology 340: 392–404.

Hokoç, J. N., de Oliveira, M. M. M. and Ahnelt, P., 1993, Three types of horizontal cells in a primitive mammal, the opossum (*Didelphis marsupialis aurita*): A Golgi-LM study. Investigative Ophthalmology & Visual Science 34: 1152.

Jacobs, G. H., 1993, The distribution and nature of colour vision among the mammals. Biological Reviews 68: 413–471.

Juliusson, B., Bergström, A., Röhlich, P., Ehinger, B., van Veen, T. and Szél, Á., 1994, Complementary cone fields of the rabbit retina. Investigative Ophthalmology & Visual Science 35: 811–818.

Kamermans, M. and Spekreijse, H., 1995, Spectral behavior of cone-driven horizontal cells in teleost retina. Progress in Retinal and Eye Research 14: 313–360.

Kamermans, M., Kraaij, D. and Spekreijse, H., 1997, The functional implications of the broad spectral sensitivity of the feedback signals to cones. Investigative Ophthalmology & Visual Science 38: S1163.

Kolb, H., 1991, Anatomical pathways for color vision in the human retina. Visual Neuroscience 7: 61–74.

Kolb, H., Mariani, A. and Gallego, A., 1980, A second type of horizontal cell in the monkey retina. Journal of Comparative Neurology 189: 31–44.

Kolb, H., Fernandez, E., Schouten, J., Ahnelt, P., Linberg, K. A. & Fisher, S. K., 1994, Are there three types of horizontal cell in the human retina? Journal of Comparative Neurology 343: 370–386

Lee, B. B., 1996, Receptive field structure in the primate retina. Vision Research 36: 631–644.

Lee, B. B., Dacey, D. M., Smith, V. C. and Pokorny, J., 1997, Time course and cone specificity of adaptation in primate outer retina. Investigative Ophthalmology & Visual Science 38: S1163.

Lennie, P., Haake, P. W. and Williams, D. R., 1991, The design of chromatically opponent receptive fields. In: Computational Models of Visual Processing. Landy, M. S. and Movshon, J. A. (Eds.), MIT Press, Cambrigde, Mass., pp. 71–82.

Linberg, K. A., Suemune, S. and Fisher, S. K., 1996, The retinal neurons of the California ground squirrel, *Spermophilus beecheyi*: A Golgi study. Journal of Comparative Neurology 365: 173–216.

Löhrke, S., Brandstätter, J. H., Boycott, B. B. and Peichl, L., 1995, Expression of neurofilament proteins by horizontal cells in the rabbit retina varies with retinal location. Journal of Neurocytology 24: 283–300.

Mangel, S. C., 1991, Analysis of the horizontal cell contribution to the receptive field surround of ganglion cells in the rabbit retina. Journal of Physiology 442: 211–234.

Mariani, A. P., 1985, Multiaxonal horizontal cells in the retina of the tree shrew, *Tupaia glis*. Journal of Comparative Neurology 233: 553–563.

Masland, R. H., 1996, Unscrambling color vision. Science 271: 616–617.

Mills, S. L. and Massey, S. C., 1994, Distribution and coverage of A- and B-type horizontal cells stained with neurobiotin in the rabbit retina. Visual Neuroscience 11: 549–560.

Mills, S. L. and Massey, S. C., 1995, Differential properties of two gap junctional pathways made by AII amacrine cells. Nature 377: 734–737.

Müller, B. and Peichl, L., 1993, Horizontal cells in the cone-dominated tree shrew retina: Morphology, photoreceptor contacts, and topographical distribution. Journal of Neuroscience 13: 3628–3646.

Nelson, R., 1977, Cat cones have rod input: A comparison of the response properties of cones and horizontal cell bodies in the retina of the cat. Journal of Comparative Neurology 172: 109–136.

Nelson, R., 1985, Spectral properties of cat horizontal cells. Neuroscience Research Supplement 2: S167–183.

Nelson, R., v. Lützow, A., Kolb, H. and Gouras, P., 1975, Horizontal cells in the cat retina with independent dendritic systems. Science 189: 137–139.

Nelson, R., Pflug, R. and Baer, S. M., 1990, Background-induced flicker enhancement in cat retinal horizontal cells. II. Spatial properties. Journal of Neurophysiology 64: 326–340.

Novacek, M. J., 1992, Mammalian phylogeny: shaking the tree. Nature 356: 121–125.

Peichl, L., 1991, Alpha ganglion cells in mammalian retinae: Common properties, species differences, and some comments on other ganglion cells. Visual Neuroscience 7: 155–169.

Peichl, L. and González-Soriano, J., 1994, Morphological types of horizontal cell in rodent retinae: A comparison of rat, mouse, gerbil and guinea pig. Visual Neuroscience 11: 501–517.

Pflug, R., Nelson, R. and Ahnelt, P. K., 1990, Background-induced flicker enhancement in cat retinal horizontal cells. I. Temporal and spectral properties. Journal of Neurophysiology 64: 313–325.

Piccolino, M., 1986, Horizontal cells: Historical controversies and new interest. Progress in Retinal Research 5: 147–163.

Piccolino, M., 1988, Cajal and the retina: a 100-year retrospective. Trends in Neurosciences 11: 521–525.

Raviola, E. and Gilula, N. B., 1975, Intramembrane organization of specialized contacts in the outer plexiform layer of the retina. Journal of Cell Biology 65: 192–222.

Raviola, E. and Dacheux, R. F., 1983, Variations in structure and response properties of horizontal cells in the retina of the rabbit. Vision Research 23: 1221–1227.

Raviola, E. and Dacheux, R. F., 1990, Axonless horizontal cells of the rabbit retina: synaptic connections and origin of the rod aftereffect. Journal of Neurocytology 19: 731–736.

Röhrenbeck, J., Wässle, H. and Heizmann, C. W., 1987, Immunocytochemical labelling of horizontal cells in mammalian retina using antibodies against calcium-binding proteins. Neuroscience Letters 77: 255–260.

Sandmann, D., Boycott, B. B. and Peichl, L., 1996a, Blue-cone horizontal cells in the retinae of horses and other *Equidae*. Journal of Neuroscience 16: 3381–3396.

Sandmann, D., Boycott, B. B. and Peichl, L., 1996b, The horizontal cells of artiodactyl retinae: a comparison with Cajal's descriptions. Visual Neuroscience 13: 735–746.

Smith, R. G., 1995, Simulation of an anatomically defined local circuit: the cone-horizontal cell network in cat retina. Visual Neuroscience 12: 545–561.

Stell, W. K. and Lightfoot, D. O., 1975, Color-specific interconnections of cones and horizontal cells in the retina of the goldfish. Journal of Comparative Neurology 159: 473–501.

Sterling, P., 1997, Chapter 6: Retina. In: Synaptic Organization of the Brain, 4th Edition. Shepherd, G. (Ed.), Oxford University Press, New York, in press.

Sterling, P., Smith, R. G., Rao, R. and Vardi, N., 1995, Functional architecture of mammalian outer retina and bipolar cells. In: Neurobiology and Clinical Aspects of the Outer Retina. Djamgoz, M. B. A., Archer, S. N. and Vallerga, S. (Eds.), Chapman & Hall, London, pp. 325–348.

Suzuki, H. and Pinto, L. H., 1986, Response properties of horizontal cells in the isolated retina of wild-type and pearl mutant mice. Journal of Neuroscience 6: 1122–1128.

Vaney, D. I., 1994, Patterns of neuronal coupling in the retina. Progress in Retinal and Eye Research 13: 301–355.

Wässle, H. and Boycott, B. B., 1991, Functional architecture of the mammalian retina. Physiological Reviews 71: 447–471.

Wässle, H., Peichl, L. and Boycott, B. B., 1978a, Topography of horizontal cells in the retina of the domestic cat. Proceedings of the Royal Society (London) B 203: 269–291.

Wässle, H., Boycott, B. B. and Peichl, L., 1978b, Receptor contacts of horizontal cells in the retina of the domestic cat. Proceedings of the Royal Society (London) B 203: 247–267.

Wässle, H., Boycott, B. B. and Röhrenbeck, J., 1989, Horizontal cells in the monkey retina: cone connections and dendritic network. European Journal of Neuroscience 1: 421–435.

West, R. W., 1978, Bipolar and horizontal cells of the gray squirrel retina: Golgi morphology and receptor connections. Vision Research 18: 129–136.

Wu, S. M., 1992, Feedback connections and operation of the outer plexiform layer of the retina. Current Opinion in Neurobiology 2: 462–468.

Yazulla, S., 1995, Neurotransmitter release from horizontal cells. In: Neurobiology and Clinical Aspects of the Outer Retina. Djamgoz, M. B. A., Archer, S. N. and Vallerga, S. (Eds.), Chapman & Hall, London, pp. 249–271.

10

PARALLEL PATHWAYS OF PRIMATE VISION

Sampling of Information in the Fourier Space by M and P Cells

Luiz Carlos L. Silveira and Harold D. De Mello, Jr.

Departamento de Fisiologia
Universidade Federal do Pará
66075-900 Belém, Pará, Brazil

1. INTRODUCTION

The visual system is organized in distinct parallel pathways connecting the retina to visual nuclei in the diencephalon and midbrain (Lennie, 1980a, b; Stone, 1983). This seminal principle about the organization of the visual system emerged from anatomical, physiological, and psychophysical experiments carried out in the last three decades, and is reminiscent of classical studies on the organization of the somatic sensory system (Stone, 1983). Notwithstanding, the role of each parallel pathway, and how several pathways share the duties of information processing, remain controversial and still represent a major challenge for those bent on understanding vision and visual dysfunction.

In common with all other vertebrates, the primate retina has several classes of ganglion cells, each one with distinct patterns of inputs from bipolar and amacrine cells and specific outputs to targets in the dorsal thalamus, hypothalamus, pretectum, superior colliculus and the accessory optic system (Rodieck et al., 1993). Two of these circuits, the M and P pathways, are of considerable interest, because they represent major information channels (Kaplan et al., 1990), extending from the retina to the lateral geniculate nucleus (LGN) and primary visual cortex (V1) (Lund, 1988). The functional properties of M and P neurons indicate that they each might play important roles in both spatial and temporal achromatic vision, so it is of considerable theoretical and practical significance to understand how these two pathways cooperate in the generation of the visual percept.

We have reviewed the anatomy and physiology of the M and P pathways from the perspective of the division of labor between these two pathways for the solution of visual tasks in daily life. More comprehensive reviews of the M and P pathways are available elsewhere (Wässle and Boycott, 1991; Kaplan et al., 1990; Merigan and Maunsell, 1993; Lee, 1996).

Development and Organization of the Retina, edited by Chalupa and Finlay.
Plenum Press, New York, 1998.

2. THE ANATOMY OF THE M AND P PATHWAYS

2.1. The M and P Pathways at the Level of the Retinal Ganglion Cell Layer

The M and P pathways can be identified at the level of the retinal ganglion cell layer due to distinct anatomical and physiological properties of the neurons of each pathway. The M and P ganglion cells, also called parasol and midget ganglion cells, respectively, lie side-by-side in the retinal ganglion cell layer, forming intermingled, but independent mosaics that pave the whole retinal extent. They are readily recognizable in retinal sections or flat-mounts stained with a variety of methods, such as the classical methods of Ehrlich (Dogiel, 1891), Golgi (Polyak, 1941; Boycott and Dowling, 1969), and Gros-Schultze (Silveira and Perry, 1991), as well as with neurotracer retrograde transport (Leventhal et al., 1981; Perry and Cowey, 1981) and dye intracellular injection (Watanabe and Rodieck, 1989) (Figs. 1–2).

The M and P ganglion cells have been identified in all primates so far studied, including man (Dogiel, 1891; Rodieck et al., 1985; Dacey and Petersen, 1992; Kolb et al., 1992; Dacey, 1993), other catarrhines (Polyak, 1941; Boycott and Dowling, 1969; Leventhal et al., 1981; Perry and Cowey, 1981; Perry et al., 1984; Watanabe and Rodieck, 1989; Silveira and Perry, 1991; Dacey and Brace, 1992), diurnal and nocturnal platyrrhines (Leventhal et al., 1989; Lima et al., 1993, 1996; Silveira et al., 1994; Ghosh et al., 1996; Goodchild et al., 1996; Yamada et al., 1996a, b, 1998), and prosimians (Yamada et al., 1997). The morphologies of M and P ganglion cells are distinct, allowing them to be recognized at every retinal location, although there are important quantitative differences for cells of the same class as a function of retinal eccentricity (Fig. 1). There are also differences for cells of the same class, located at equivalent eccentricities, but from different primate species (Fig. 2). M ganglion cells have large cell bodies, thick axons, and large dendritic trees with a radial branching pattern. On the other hand, P ganglion cells have small cell bodies, thin axons, and small dendritic trees with a more bushy and dense branching pattern. The M and P ganglion cells comprise the majority of ganglion cells of the primate retina, corresponding, respectively, to 10% and 80% of the ganglion cell population (Perry et al., 1984; Silveira and Perry, 1991; Lima et al., 1993, 1996). The M and P ganglion cells occur in two morphological varieties, one ramifying in the outer half of the inner plexiform layer (IPL), the outer cells, and the other ramifying in the inner half, the inner cells (Perry et al., 1984; Watanabe and Rodieck, 1989). The outer cells correspond to the off-center and the inner cells to the on-center varieties (Dacey and Lee, 1994). Every point of the photoreceptor matrix is connected with at least one inner and one outer ganglion cell of the two cell classes, M and P, in such a way that the entire visual field is sampled by four different mosaics, M-on, M-off, P-on, and P-off.

2.2. The M and P Pathways at the Level of the Inner Nuclear Layer

Are the M and P properties already present prior to the ganglion cell layer? The M and P ganglion cells receive input from distinct bipolar cell classes and thus it is possible that many features of the M and P channels are already present in the early elements of retinal wiring (Boycott and Wässle, 1991). On the other hand it is possible that differential input from amacrine cells can substantially modify the direct bipolar cell input, originating some of the characteristic physiological properties of the M and P ganglion cells. This is a question to be answered by intracellular recording and staining techniques of primate retinal cells (Dacey, 1996).

The bipolar cells most likely to synapse onto M ganglion cells are diffuse bipolar cells that have axon terminals branching in the middle strata of the IPL, at the same level as the M ganglion cells dendrites: DB2 and DB3 bipolar cells for M-off and DB4 and DB5 bipolar cells for M-on ganglion cells, respectively (Boycott and Wässle, 1991; Jacoby and Marshak, 1995, 1996). P ganglion cells receive synapses from midget bipolar cells (Kolb and DeKorver, 1991; Calkins et al., 1995). There are two varieties of midget bipolar cells (Polyak, 1941; Boycott and Dowling, 1969; Kolb et al., 1969; Kolb, 1970; Wässle et al., 1994), one for each variety of P ganglion cells: the flat midget bipolar cells (FMB cells), which make synapses with the P-off ganglion cells, and the invaginant midget bipolar cells (IMB), which make synapses with the P-on ganglion cells.

2.3. The M and P Pathways at the LGN Level

The primary LGN has several layers of neurons, each one corresponding to an independent representation of the visual field. This complex anatomical and functional organization reflects the existence of two superimposed parallel processing systems at this level (Kaas and Huerta, 1986; Casagrande and Norton, 1991). There are three or more sets of layers, each one comprised of neurons with distinct morphological and physiological properties. In addition, alternate LGN layers of every set receive information from the ipsi- or the contralateral eye. The properties of the neurons found in each LGN layer indicate that these compartments are homogeneous relay stations of several retino-geniculo-striate pathways (Kaas and Huerta, 1986; Casagrande and Norton, 1991; Casagrande, 1994). In this regard, the anatomy of the parallel pathways differs in the retina and LGN. In the retina, they form intermingled mosaics, an ideal situation for each mosaic to sample the entire visual field (described by Cook in this volume), while in the LGN they segregate in distinct layers, an arrangement probably important to simplify the neuronal wiring.

The axon terminals of M and P retinal ganglion cells have distinct morphology and make synapses in the magno- and parvocellular layers, respectively (Conley et al., 1987; Lachica and Casagrande, 1988; Michael, 1988; Conley and Fitzpatrick, 1989). M axon terminals are radially symmetric, have stout telodendria and large, ovoid boutons, while P axon terminals are elongated, have slender telodendria and round medium-sized boutons. Axon terminals with different morphologies originate from other classes of retinal ganglion cells and make synapses in a heterogeneous group of layers collectively called koniocellular (K) layers (Casagrande, 1994; Hendry and Yoshioka, 1994; Hendry and Casagrande, 1996).

The neurons of the magno- and parvocellular layers, the M and P LGN neurons, roughly match the number, size and morphology of their afferents. Thus the M LGN neurons are less numerous, corresponding to 11% of the total, and have large, radially symmetric dendritic fields. P LGN neurons are more numerous, corresponding to 85% of the total, and have medium-sized and elongated dendritic fields (Connoly and Van Essen, 1984; Michael, 1988). The receptive field properties of the M and P LGN neurons are also very similar to those of M and P ganglion cells that provide their inputs (Wiesel and Hubel, 1966; Dreher et al., 1976; Sherman et al., 1976; Schiller and Malpeli, 1978; Shapley et al., 1981; Norton and Casagrande, 1982; Hicks et al., 1983; Derrington and Lennie, 1984; Derrington et al., 1984). Taken together these findings strongly support the existence of at least two major anatomical and functional divisions of the visual system at the LGN level, the magnocellular (M) and parvocellular (P) pathways.

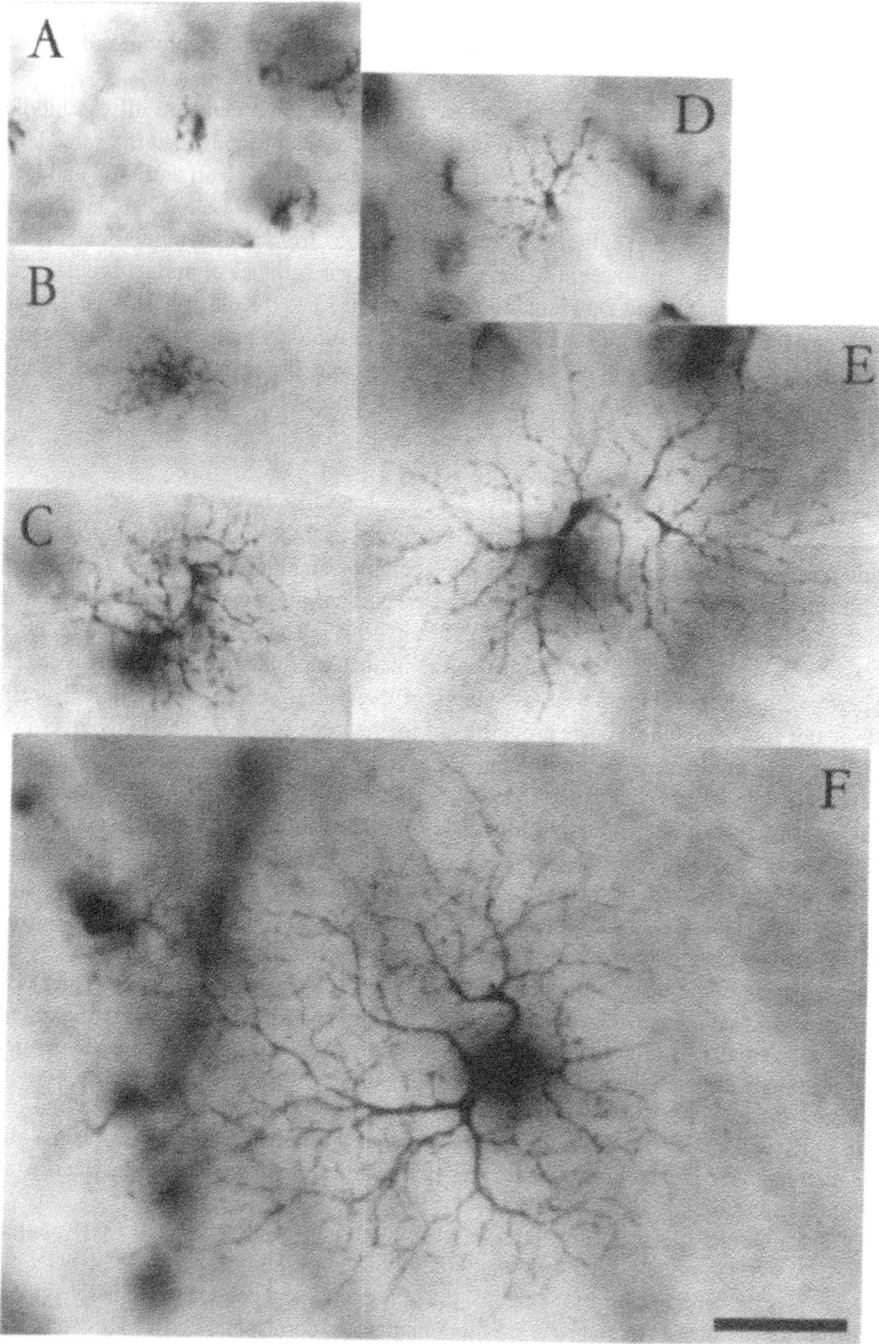

Figure 1. M and P retinal ganglion cells of the common marmoset, *Callithrix jacchus*, from different retinal eccentricities. The ganglion cells were retrogradely labelled from optic nerve deposits of biocytin (original material from Yamada et al., 1996b). The M and P ganglion cells, also called parasol and midget ganglion cells, respectively, lay side-by-side in the retinal ganglion cell layer of all primates so far investigated, forming intermingled, but independent mosaics that pave the whole retinal extent. In all retinal locations M ganglion cells have larger dendritic trees than P ganglion cells. **A.** P-off ganglion cell, 0.3 mm temporal. **B.** P-off ganglion cell, 1.8 mm ventral. **C.** P-off ganglion cell, 4.2 mm temporal. **D.** M-off ganglion cell, 0.3 mm temporal. **E.** M-off ganglion cell, 2.2 mm ventral. **F.** M-off ganglion cell, 3.7 mm ventral to the fovea. Focus on dendritic trees. Scale bar = 25 μm.

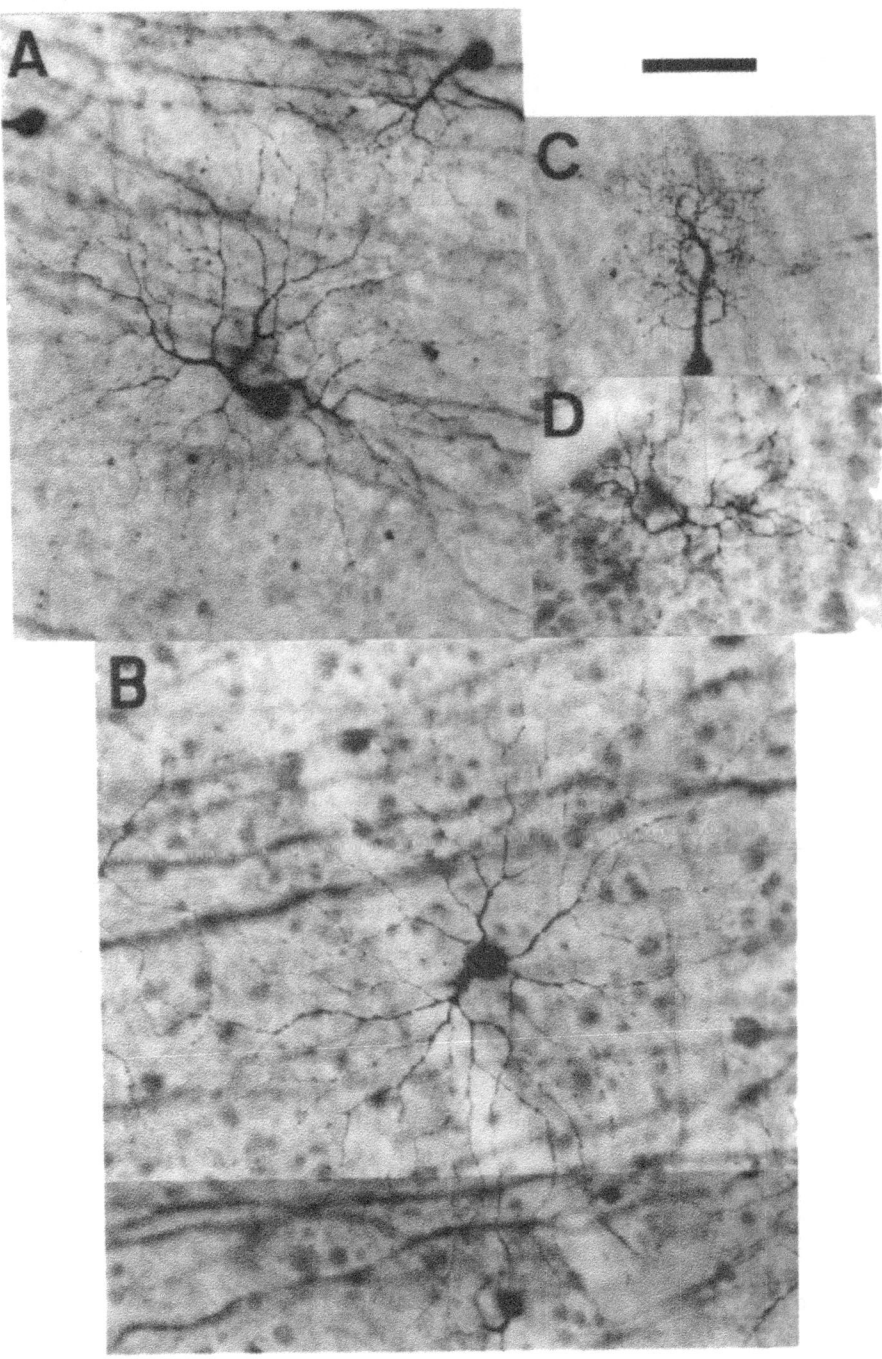

Figure 2. M and P ganglion cells of diurnal and nocturnal anthropoids. Retinal ganglion cells were retrogradely labelled from optic nerve deposits of biocytin (original material from Yamada et al., 1996a, 1998). The morphologies of M and P ganglion cells are distinct, allowing them to be recognized in the retina of all primates so far studied, although there are quantitative differences for cells of the same class from different primate species (Silveira et al., 1994; Yamada et al., 1996a,b, 1998). **A.** M-off ganglion cell from a diurnal dichromatic platyrrhine, a male capuchin monkey, *Cebus apella*; 11 mm dorsal. **B.** M-on ganglion cell from a nocturnal monochromatic platyrrhine, the owl monkey, *Aotus azarae*; 10 mm dorsal. **C.** P-off ganglion cell, same animal as in A, 10.7 mm dorsal. **D.** P-off ganglion cell, same retina as in B; 10.2 mm dorsal to the fovea. Focus on dendritic trees. Scale bar = 50 μm.

2.4. The M and P Pathways at the Cortical Level

The regions of the primary visual cortex (V1) that receive projections from the LGN are those that show the strongest activity of the mitochondrial enzyme cytochrome oxidase. They comprise layer 4A, layer 4C and the "blobs" in layer 3, but there is also additional LGN inputs to layer 1 and layer 6 (Lund, 1988). M and P axon terminals from the LGN have distinct morphologies and make synapses in different layers of V1. Layer 4Cα is the principal target of M LGN axons, which also provide a sparse input to the deeper half of layer 6. The main targets for P LGN axons are layers 4Cβ and 4A, with a sparse input also present at the superficial half of layer 6 (Hubel and Wiesel, 1972; Hendrickson et al., 1978; Blasdel and Lund, 1983; Lachica and Casagrande, 1992). The K LGN axons project to layer 3 "blobs" and layer 1 (Livingstone and Hubel, 1982; Fitzpatrick et al., 1983; Weber et al., 1983; Diamond et al., 1985; Lachica and Casagrande, 1992; Hendry and Yoshioka, 1994; Ding and Casagrande, 1997). These findings indicate that the M and P pathways, as well as the other retino-geniculo-cortical pathways, remain segregated at the level of their entry in V1.

The segregation of the M and P pathways beyond the above targets in the striate cortex is more controversial. There is no agreement on the progress of M and P pathways beyond the second-order V1 compartments and prestriate visual areas. Earlier studies found some degree of segregation of M and P pathways in the output of layers 4Cα and 4Cβ to second-order layers of V1 (Lund and Boothe, 1975), as well as in the pathways connecting V1 to V2 and other prestriate visual areas (Livingstone and Hubel, 1983, 1984a, b, 1987; DeYoe and Van Essen, 1985; Hubel and Livingstone, 1985, 1987; Shipp and Zeki, 1985). However, more recent anatomical (Lachica et al., 1992; Yoshioka et al., 1994) and electrophysiological observations (Nealey and Maunsell, 1994) have shown early convergence of M and P inputs, which seems to begin within the V1 compartments. At later stages, there is electrophysiological evidence both for a fair amount of M and P segregation in some visual areas, such as MT (Maunsell et al., 1990), as well as a high degree of M and P convergence in other visual areas, such as V4 (Ferrera et al., 1992, 1994). Therefore, it is unclear what possible relations the M and P pathways might have with the cortical streams of visual processing, such as the pathways for object recognition and perception of spatial relationships of Ungerleider and Mishkin (1982), or the pathways for perception and action of Milner and Goodale (1995). These pathways, connecting V1 and V2 to ventral or dorsal prestriate visual areas, respectively, appear to have blended inputs from both the M and P pathways. It has been proposed, using computational arguments, that for the performance of many tasks, higher levels of the visual system need access to information from both the P and the M pathways, performing some kind of concurrent processing (DeYoe and Van Essen, 1988; Felleman and Van Essen, 1991; Van Essen et al., 1992).

3. THE PHYSIOLOGY OF THE M AND P PATHWAYS

It is of great theoretical and practical interest to investigate whether the M and P pathways can be distinguished by their physiological properties. If this distinction proves possible, one may be able to hypothesize on the roles played by each pathway by comparison of their physiological properties in the visual performance of normal and abnormal subjects. For this purpose, single unit recordings were made in the retina, optic nerve and tract, LGN, and visual cortex. The receptive fields of neurons of the M and P pathways were investigated by visual stimulation with simple spatiotemporal patterns and measuring neuronal activity.

In the LGN of macaque monkey Wiesel and Hubel (1966) distinguished four types of receptive fields based in the presence of spatial and chromatic antagonisms. Type I receptive fields were those with spatial and chromatic antagonisms between center and surround, and were found only in the parvocellular layers. Type II receptive fields were those with chromatic antagonism without spatial antagonism, i.e., the on and off regions of the receptive fields were co-extensive. Type III had spatial antagonism between center and surround but no chromatic antagonism, while Type IV exhibited spatial antagonism with little chromatic antagonism. Later studies showed that Types III and IV form a single continuum (Derrington et al., 1984). The original observations of Wiesel and Hubel (1966) were extended in subsequent systematic studies of macaque M and P cells at the level of both the LGN (Padmos and van Norren, 1975; Dreher et al., 1976; Schiller and Malpeli, 1978; Creutzfeldt et al., 1979; Hicks et al., 1983; Derrington et al., 1984; Crook et al., 1987; Lee et al., 1987; Hubel and Livingstone, 1990) and the retinal ganglion cell layer (Gouras, 1968; de Monasterio and Gouras, 1975; de Monasterio et al., 1995a, b; Lee et al., 1988, 1989a, b, c; Shapley et al., 1991). These studies have consistently shown that retinal and thalamic cells have similar physiology, except for a higher maintained firing rate of retinal cells. All retinal and thalamic M cells have Type III/IV receptive fields while all P cells have Type I receptive fields.

The M cells receive additive inputs of long- and middle-wavelength sensitive cones (LWS- and MWS-cones) both in the center and surround regions of their receptive fields and have a relative spectral sensitivity close to the human photopic luminosity curve, V_λ (Lee et al., 1988). The P cells have chromatic opponent inputs from LWS- and MWS-cones to the center and surround of their receptive fields, in such way that there are four subclasses of them, two kinds of P-on and two kinds of P-off cells: red-on/green-off, green-on/red-off, red-off/green-on, and green-off/red-on P cells. Thus, P cells are excited by some light wavelengths and inhibited by others. These findings led to the hypothesis that P cells might be primarily involved in chromatic vision, providing the mechanism for visual performance in psychophysical tasks, such as detection of chromatic modulation (Lee et al., 1989a, b, c, 1990, 1993; Kremers et al., 1992, 1993). On the other hand, M cells would form the basis for visual performance in achromatic tasks, such as heterochromatic flicker photometry (Lee et al., 1989a, c; Smith et al., 1992), detection of luminance modulation (Lee et al., 1989a, b, c, 1990, 1993; Kremers et al., 1992, 1993), minimally distinct border (Kaiser et al., 1990; Valberg et al., 1992), and hyperacuity (Lee et al., 1993, 1995).

There are complementary hypotheses for the roles of M and P pathways in visual processing. P cells have shown to be responsible only for red-green color opponency (Dacey and Lee, 1994). Blue-yellow color opponency appear to be the task of a completely different pathway, connecting SWS-cones, SWS-cone bipolar cells (Mariani, 1984; Kouyama & Marshak, 1992; Ghosh et al., 1997), small-field bistratified ganglion cells (Rodieck, 1991; Dacey and Lee, 1994; Ghosh et al., 1997) and LGN relay neurons of K layers (White et al., 1997). The axons of these K LGN neurons establish synapses in V1 compartments, the "blobs" of layers 2/3 (Hendry and Yoshioka, 1994), that are distinct from those where the M and P LGN neurons project. Another qualification for the roles of the M and P pathways, stresses the evidence that the M and P pathways can be identified by anatomical criteria in the visual system of all primates so far studied, including diurnal trichromatic and dichromatic simians, nocturnal monochromatic simians, and also nocturnal monochromatic prosimians (Kaas and Huerta, 1988; Casagrande and Norton, 1991; Casagrande, 1994; Silveira et al., 1994, 1997; Ghosh et al., 1996; Yamada et al., 1996a, b, 1997, 1998). Their M cells respond similarly to achromatic stimuli and the same can be said of their P cells. The only major difference across species appears, then, to be the cell response to chromatic stimuli, which is absent in P cells of dichromatic and monochromatic primates (Jacobs and De Valois, 1965; Jacobs, 1983; Sherman et al., 1976; Norton and Casagrande, 1982; Yeh et al., 1995; Lee et al., 1996). Thus one may advance that the M and P pathways are ancient features of the primate visual system, which evolved before red-green color opponency and thus were primarily involved in achromatic vision (Mollon, 1989, 1991; Mollon et al., 1990; Wässle and Boycott, 1991). According to this hypothesis, primate trichromatic color vision appeared in two steps during the course of evolution. Ancient primates were dichromats, possessing two opsin genes, one in the chromosome 7 for the SWS-opsin and another in the X chromosome for a single LWS/MWS-opsin. These primates had a single color vision pathway, one specialized for blue–yellow opponency, a feature that their visual system shared with most other mammals. After a point mutation in the X-chromosome opsin gene, followed or preceded by gene duplication, central P cells, bearing single-cone dendritic fields, were able to perform red-green color opponency. Then, trichromatism appeared at the photoreceptor level and based on two color opponent mechanisms at the post-receptor level, one already extant for blue–yellow opponency and another recently acquired for red–green color opponency.

If one accepts the preceding logic the next question then becomes why did both M and P cells evolved for achromatic vision? Why do we need both types and how do they share the duties of visual processing? What could be the consequences of impairing one system alone? The M and P cell responses to achromatic stimuli differ more quantitatively than qualitatively, and are related to three main aspects: M cells have larger receptive fields than P cells (de Monasterio and Gouras, 1975; Derrington and Lennie, 1984; Crook et al., 1988; Croner and Kaplan, 1995), M cells respond more transiently than P cells (Gouras, 1968; Hicks et al., 1983; Derrington and Lennie, 1984; Purpura et al., 1990; Lee et al., 1994), and M cells are more sensitive to spatial or temporal contrast than P cells (Shapley et al., 1981; Kaplan and Shapley, 1982, 1986; Derrington and Lennie, 1984; Purpura et al., 1988). Thus, it is possible that P cells evolved for detection of small stimuli and M cells for fast stimuli, or in other words P cells are mainly concerned with spatial aspects of vision, while M cells are devoted to temporal or spatiotemporal vision (Derrington and Lennie, 1984; Derrington et al., 1984). It is also possible that the M pathway is the predominant information channel in scotopic vision due to their superior contrast sensitivity at this illumination level (Purpura et al., 1988). A more general statement could be that the M and P pathways complement each other, occupying different regions of the spatiotemporal domain (Schiller et al., 1990a, b). There is some support for these hypotheses, as well as for the P cell role in color vision, from experiments dealing with selective ablation of the M or P pathway in macaque monkeys (Merigan, 1989; Merigan and Maunsell, 1990; Schiller et al., 1990a, b; Merigan et al., 1991a, b; Lynch et al., 1992).

An alternative way to understanding the division of labor between the M and P pathways will be presented in the next sections.

4. CONFLICTING REQUIREMENTS AND COMPROMISE IN THE INFORMATION PROCESSING BY THE VISUAL SYSTEM

There are conflicting requirements to be met by the visual system in analyzing the environment. The visual system has to extract information carried by light emanating from objects located in the visual field with enough precision in space and time for efficient solution of animal behavioral tasks. Any gain in the capacity of locating more precisely an object in space is accompanied by a loss in the ability to precisely measure the moment of occurrence of a visual change. In addition, natural visual stimuli exhibit simultaneous spatial or temporal singularities and periodicities, so the visual system has to evaluate not only the stimulus spatiotemporal coordinates, but also its spatiotemporal frequency content. The simultaneous precision in time, temporal frequency, space and spatial frequency is a general, basic problem in the design of measuring devices.

Theoretically, infinitely precise analysis of the visual world can be achieved separately in the spatial or temporal domains with an infinitesimal spatial sampling or an infinitesimally short time sampling. In a theoretical situation like this, the discrimination of the spatial and temporal frequency content of the visual scenes would be impossible. Conversely an infinite precision in the spatial or temporal frequency domains can be theoretically achieved with filters each tuned for a single frequency. Such filters would sample through an infinitely extended aperture and time window and would be unable to precisely mark the spatial localization and occurrence in time of visual patterns. These two extremes of spatial or temporal accuracy as well as spatial and temporal spectral accuracy do not exist in the physical world. All devices, either natural or man made, built to store, transmit or analyze visual information represent different degrees of compromise between precision space-time domain and precision in spatiotemporal frequency domain.

The compromise for simultaneous precision between two related domains, such as time and temporal frequency, was formulated by Gabor (1946) in a fundamental paper on theory of communication. He developed a quantitative description of our auditory experiences that favors a simultaneous description of stimuli in terms of both time and temporal frequency. He showed that the amount of information that can be transmitted by a limited frequency band in a limited time-interval can be analyzed in terms of elementary quanta of information belonging to a plane where time and temporal frequency are orthogonal coordinates. In this plane, the diagram of information, also called Fourier space, the shapes and sizes of the elementary quanta of information depend on the characteristics of the device that is performing the time-frequency analysis. In other words they depend on the frequency band-pass and time sampling properties of the device.

The uncertainties in the simultaneous definition of time and temporal frequency of a phenomenon, the so-called joint entropy, are related in the mathematical identity

$$\Delta t \cdot \Delta f \geq 1/2 \tag{1}$$

Gabor (1946) demonstrated that the quanta of information have areas not less than one-half, due to the impossibility of simultaneous increase of time and temporal frequency accuracy. The demonstration of this identity establishes that precision about time and temporal frequency can not be attained simultaneously beyond a certain limit of order unity.

Gabor (1946) also showed that there are some elementary functions that occupy the smallest possible area in the information diagram, and thus have the smallest possible joint entropy. An elementary function, or Gabor function, $\Psi(t)$, is a harmonic oscillation modulated by a probability function:

$$\Psi(t) = e^{[-\alpha^2 (t-t_0)^2]} \cdot e^{i[2\pi f_0 (t-t_0) + \theta]} \tag{2}$$

where α and t_0 represent the probability function sharpness and peak, while f_0 and θ represent oscillation frequency and phase. We represent $\Psi(t)$ as points of the complex plane associated with real values of time as third coordinate. As time elapses $\Psi(t)$ describes a spiral around the time axis in the surface of a solid of revolution that has the shape of a probability function.

The Fourier Transform (Bracewell, 1986) of the elementary function gives its spectra, $\Phi(f)$, which has the same analytical form of the original time function

$$\int \Psi(t) \cdot e^{-2\pi f t} \cdot dt = \Phi(f) \tag{3}$$

$$\Phi(f) = e^{[-(\pi/\alpha)^2 (f-f_0)^2]} \cdot e^{i[-2\pi u_0 (f-f_0)+\theta]} \tag{4}$$

The time and spectral uncertainties, defined as the effective duration and effective frequency bandwidth, are the product of a constant by the root mean square of the deviation of the signal from the mean in each domain

$$(\Delta t)^2 = 2\pi \cdot [[[\int \Psi^* \, t^2 \Psi \, dt] / [\int \Psi^* \, \Psi \, dt]] - [[\int \Psi^* \, t \Psi \, dt] / [\int \Psi^* \, \Psi \, dt]]^2] \tag{5}$$

$$(\Delta f)^2 = 2\pi \cdot [[[\int \Phi^* \, f^2 \Phi \, df] / [\int \Phi^* \, \Phi \, df]] - [[\int \Phi^* \, f \Phi \, df] / [\int \Phi^* \, \Phi \, df]]^2] \tag{6}$$

The time and spectral uncertainties are related to the probability function sharpness by the relations

$$\Delta t = 1/\alpha \cdot \sqrt{(\pi/2)} \tag{7}$$

$$\Delta f = \alpha \cdot 1/\sqrt{(2\pi)} \tag{8}$$

Gabor (1946) also showed that any function can be expanded in terms of elementary functions, $\Psi(t)$, and that the Fourier and Time Analyses are just the two extremes of this process, in which, respectively, $\alpha \to 0$ or $\alpha \to \infty$, the elementary function becoming a sine-wave in the first case and an impulse in the second case.

The original analysis of Gabor was used for the auditory system and time analyzing devices in general. To extend the same approach to the visual system it is necessary to consider that signals arising from one eye are represented in three dimensions, two spatial and one temporal. However, Gabor theory can be used to study joint entropy of any two domains that are related to each other by the Fourier transform, and it can also be generalized to any number of dimensions (MacLennan, 1991). The study of Gabor was initially used in vision to explain how cortical neurons could have receptive fields localized simultaneously in space and spatial frequency (Marcelja, 1980; Daugman, 1980, 1983; Kulikowski and Bishop, 1981; Kulikowski et al., 1982). The first studies analyzed the case of one dimension of space and one dimension of spatial frequency (Marcelja, 1980; Kulikowski and Bishop, 1981) but subsequent work further elaborated this approach to include the two- and three-dimensional cases (Daugman, 1983, 1984, 1985; Wang et al., 1988, 1993; MacLennan, 1991). Three-dimensional Gabor functions minimize six-dimensional joint entropy, being those that have the smallest hypervolume in the appropriate six-dimensional hyperspace formed by two dimensions of space, one dimension of time, two dimensions of spatial frequency, and one dimension of temporal frequency. The theory can satisfactorily explain the empirical measurements of two-dimensional receptive field sensitivity profiles and two-dimensional spatial frequency spectra of cat simple and complex cortical cells (Daugman, 1985, 1989; Palmer et al., 1985, 1991; Webster and De Valois, 1985; Field and Tolhurst, 1986; Jones and Palmer, 1987; McLean and Palmer, 1994;

McLean et al., 1994), and some attempts have been made to generalize a model of visual neurons to include time as a third dimension (Wang et al., 1988, 1993; Jasinschi, 1991; Palmer et al., 1991; McLean and Palmer, 1994; McLean et al., 1994).

5. THE ROLE OF M AND P CELLS IN ACHROMATIC VISION: M AND P CELLS OCCUPY DISTINCT LOCI OF THE FOURIER SPACE

5.1. The Impulse Response of M and P Cells to Spatial and Temporal Luminance Contrast

A complete functional description of M and P cells in space and time can be obtained by measuring their responses to discrete spatial and temporal visual stimuli, such as punctiform luminous objects and instantaneous luminance changes. In formal terms these are Dirac functions, that is temporal $\delta(t)$ or spatial $^2\delta(x,y)$ impulses, or some feasible physical forms of them.

The response time course to discrete temporal stimuli of the M and P cells differ (Gouras, 1968; de Monasterio and Gouras, 1975; Lee et al., 1994). M cells discharge transiently to luminance steps or pulse functions and, for that reason, were originally called phasic cells. P cells exhibit more sustained discharges to the same kind of stimuli and were called tonic. M and P cell impulse functions can also be obtained with Gaussian white noise input and Wiener analysis (Gielen and van Gisbergen, 1980; Gielen et al., 1981, 1982) or by Fourier transform of M and P response to sine-wave stimuli (Purpura et al., 1990; Kremers et al., 1993; Lee et al., 1994). Data comparison for M and P cells from the same retinal location is still limited, but it can be stated, from a survey of the literature, that M cell impulse response is generally more transient, and thus more limited in time, than P cell impulse response. Thus, it can be proposed that M cells represent time domain events more precisely than P cells. M cells appear to have the necessary machinery to signal the moment of occurrence of a phenomenon with a high degree of precision.

The M and P cell responses differ in their spatial properties, and this can be quantified by measuring the structure and size of their receptive fields. For this purpose different approaches have been used, such as the energy determination of small spots of light required to elicit a threshold response across the receptive field (de Monasterio and Gouras, 1975), the measurement of area-threshold curves (Crook et al., 1988), the stimulation of cell receptive field with bipartite fields modulated sinusoidally in counterphase (Kremers and Weiss, 1997; Lee et al., 1997), or the measurement of the cell contrast sensitivity as a function of spatial frequency (Derrington and Lennie, 1984; Crook et al., 1988; Croner and Kaplan, 1995). In the latter case, the receptive field profile has to be obtained by Fourier transform of the frequency response. Some of these studies have shown that M cells have larger receptive fields than P cells, a three-fold difference across all the visual field (de Monasterio and Gouras, 1975; Derrington and Lennie, 1984; Croner and Kaplan, 1995). This is consistent with anatomical measurements made in retinal flat-mounts of the dendritic tree diameter as a function of retinal eccentricity (Perry et al., 1984; Watanabe and Rodieck, 1989) (Fig. 3). However, there are also studies showing that the difference between M and P cell receptive fields is smaller, specially in the foveal region, and that there is quite a lot of variation among cell populations

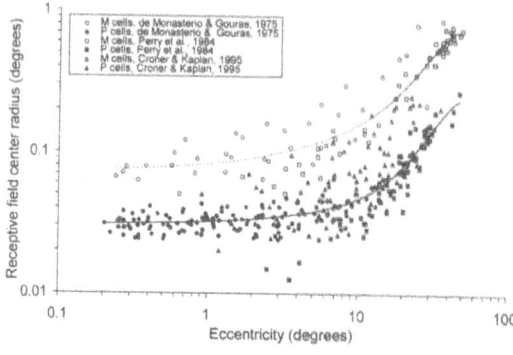

Figure 3. Receptive field center radius of macaque M and P ganglion cells as a function of retinal eccentricity. Data replotted from de Monasterio and Gouras (1975) and Croner and Kaplan (1995). de Monasterio and Gouras measured receptive field using small spots of light, while Croner and Kaplan used extended sine-wave gratings. de Monasterio and Gouras collected data equally from all the four retinal quadrants and thus some cells located in the nasal quadrant might have smaller receptive fields than cells from the same class of other quadrants. Data from Croner and Kaplan were converted by the authors to equivalent temporal retinal eccentricity. For comparison, we plotted data from Perry et al. (1984) for dendritic tree radius of temporal M and P ganglion cells retrogradely labelled from horseradish peroxidase deposits in the optic nerve. We corrected cell eccentricity of Perry et al.'s data to account for cell lateral displacement due to foveal excavation. Correction was applied for the central 3 mm from fovea using the equation published by Goodchild et al. (1996); eccentricity was converted to degrees using the equation of Dacey and Petersen (1992). Fourth-order polynomials were fitted to the pooled data points. Modified with permission from de Monasterio and Gouras (1975), Perry et al. (1984), and Croner and Kaplan (1995). Data from de Monasterio and Gouras (1975) were replotted with authors' permission and kind permission from The Physiological Society (Copyright 1975). Data from Perry et al. (1984) were replotted with authors' permission and kind permission from Elsevier Science Ltd. (Copyright 1984). Data from Croner and Kaplan (1995) were replotted with authors' permission and kind permission from Elsevier Science Ltd.

(Kremers and Weiss, 1997; Lee et al., 1997). Receptive field sensitivity profiles can be considered approximations of the actual two-dimensional spatial impulse responses of M and P cells. Thus, as P cells have smaller receptive fields than M cells, P cell impulse responses are more localized in space than M cell impulse responses, and so P cells transmit more precisely events in the space domain than M cells. In other words, the necessary machinery to signal the place of occurrence of a phenomenon with a high degree of precision appears to be present in P cells.

5.2. The Frequency Response of M and P Cells to Luminance Contrast

Visual stimuli extended in space and time, such as spatial and temporal sinusoidal luminance modulations, have been used to measure the frequency response of M and P cells to luminance contrast. Both retinal (Kaplan and Shapley, 1986; Crook et al., 1988; Lee et al., 1989a, b, 1990; Purpura et al., 1990; Kremers et al., 1992, 1993; Croner and Kaplan, 1995) and thalamic (Kaplan and Shapley, 1982; Hicks et al., 1983; Derrington and Lennie, 1984) M and P cells have been studied in the temporal and spatial frequency domains. P cell spatial frequency responses cover a wider range of spatial frequencies than M cells. This difference is predictable from the spatial impulse response of these two cell classes. P cell responses are more restricted in the spatial domain than M cells, and thus, it is expected that it is more extended in the spatial frequency domain. There are no systematic studies addressing the question of how temporal frequency responses of M and P cells change as a function of retinal eccentricity. However, for the little data available in the literature, it seems that M cell spatial frequency responses cover a wider range of temporal frequencies than P cells. This difference is also predictable from the temporal impulse response of these two cell classes. M cell responses are more restricted in the time domain than P cells, and thus, it is expected that it is more extended in the temporal frequency domain.

5.3. Representing M and P Cell Precision in the Fourier Space

We propose to describe the spatiotemporal precision of M and P cells by the product of three distinct sampling windows by three different spectral bandwidths. The sampling windows are defined in the two dimensions of space and one dimension of time present n the retinal image. The spectral bandwidths are defined in the two dimensions of spatial frequency and one dimension of temporal frequency obtained by Fourier transform of the retinal image. Thus, the spatiotemporal precision of an M or P cell is represented by a six-dimensional hypervolume, the six-dimensional joint entropy locus, occupying a certain place in Fourier space. The smaller this hypervolume, the smaller the uncertainty or error, and the greater the precision of stimulus representation by the cell activity. This hyper-volume is minimum if cell spatiotemporal response function, $r(x, y, t)$, has the shape predicted by generalizing to three dimensions the one-dimensional Gabor function (Wang et al., 1988, 1993; McLennan, 1991). As the M and P receptive fields have circular or quasi-circular spatial symmetry, the $r(x, y, t)$ projection in the spatial domain can have the form of a circularly symmetrical two-dimensional Gabor function, in which the oscillation is a circular sine-wave attenuated by a circular Gaussian function (Fig. 4). As time elapses, cell responses for each receptive field spatial coordinate oscillate sinusoidally and are attenuated by the Gaussian function (Fig. 5). The analytical three-dimensional form representing cell responses is reminiscent of the one-dimensional case:

$$r(x, y, t) = e^{-[\alpha^2(x-x_0)^2 + \beta^2(y-y_0)^2 + \gamma^2(t-t_0)]} \cdot e^{i[2\pi(u_0(x-x_0)+v_0(y-y_0)+f_0(t-t_0))+\theta]} \tag{9}$$

and its spectrum, representing cell frequency response, is given by its three-dimensional Fourier transform (Bracewell, 1986):

$$\iiint r(x, y, t) \cdot e^{-i2\pi(ux+vy+ft)} \, dx \, dy \, df = \rho(u, v, f) \tag{10}$$

$$\rho(u, v, f) = e^{-\pi^2[(u-u_0)^2/\alpha^2 + (v-v_0)^2/\beta^2 + (f-f_0)^2/\gamma^2]} \cdot e^{i[-2\pi(x_0(u-u_0)+y_0(v-v_0)+t_0(f-f_0))+\theta]} \tag{11}$$

For three-dimensional Gabor functions, the joint entropy reaches a minimum:

$$\Delta x \cdot \Delta u \cdot \Delta y \cdot \Delta v \cdot \Delta t \cdot \Delta f = 1/8 \tag{12}$$

This identity represents the three-dimensional condition of joint entropy minimization (Wang et al., 1993). It is analogous to the one-dimensional case theoretically analyzed by Gabor (1946) and similar to the two-dimensional case discussed by Daugman (1980, 1984, 1985) for visual cortical neurons. M and P cells response can be also described by other functions, that are simultaneously localized in the space-time domain and its spectrum, but the joint entropy should be always larger than 1/8.

It is possible to construct a family of functions, including Gabor functions, with each member of the family minimizing joint entropy with distinct compromise for precision in the space-time and spectral domains. It is from this point of view that we propose to understand the visual system parallel processing. Each pathway represents a particular trade-off to reduce joint entropy. Of course, M and P cell responses can be modeled by Gabor functions with only a limited degree of correlation, and other functions can be found that might provide a similar or better fit to the experimental results (Marr, 1982; Gielen et al., 1980, 1981, 1982; Derrington and Lennie, 1984; Purpura et al., 1990; Croner

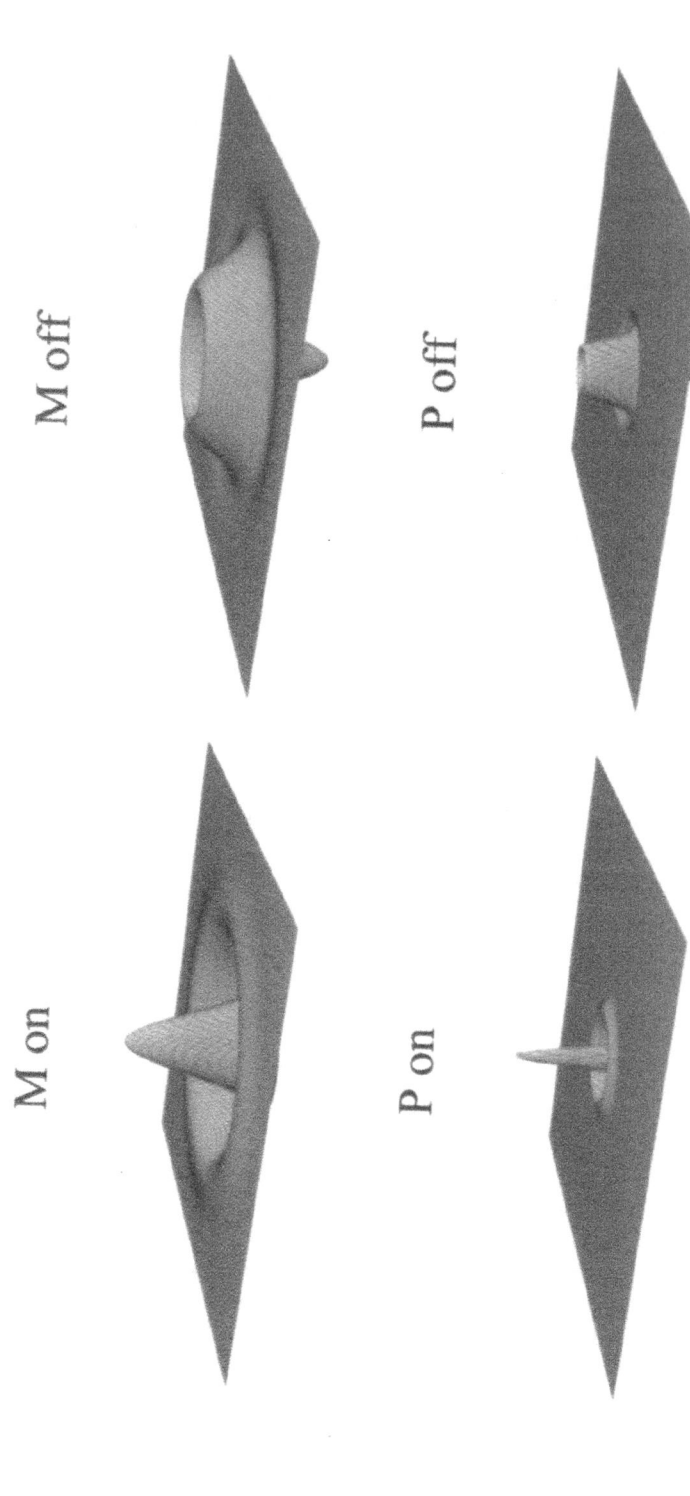

Figure 4. M and P receptive fields as circularly symmetrical Gabor functions. The spatial profile of M and P receptive fields can be described by a number of circularly symmetrical functions, such as DOG functions, Marr-Hildreth functions and Gabor functions. Two-dimensional Gabor functions comprises a circular sine-wave attenuated by a circular Gaussian function (Wang et al., 1988, 1993; McLennan, 1991). There is a 180° difference of phase between on and off pairs of the same ganglion cell class, corresponding to cosine and −cosine Gabor functions. Modified with permission from Silveira (1996). Reprinted with modifications with kind permission of Academia Brasileira de Ciencias (Copyright 1996).

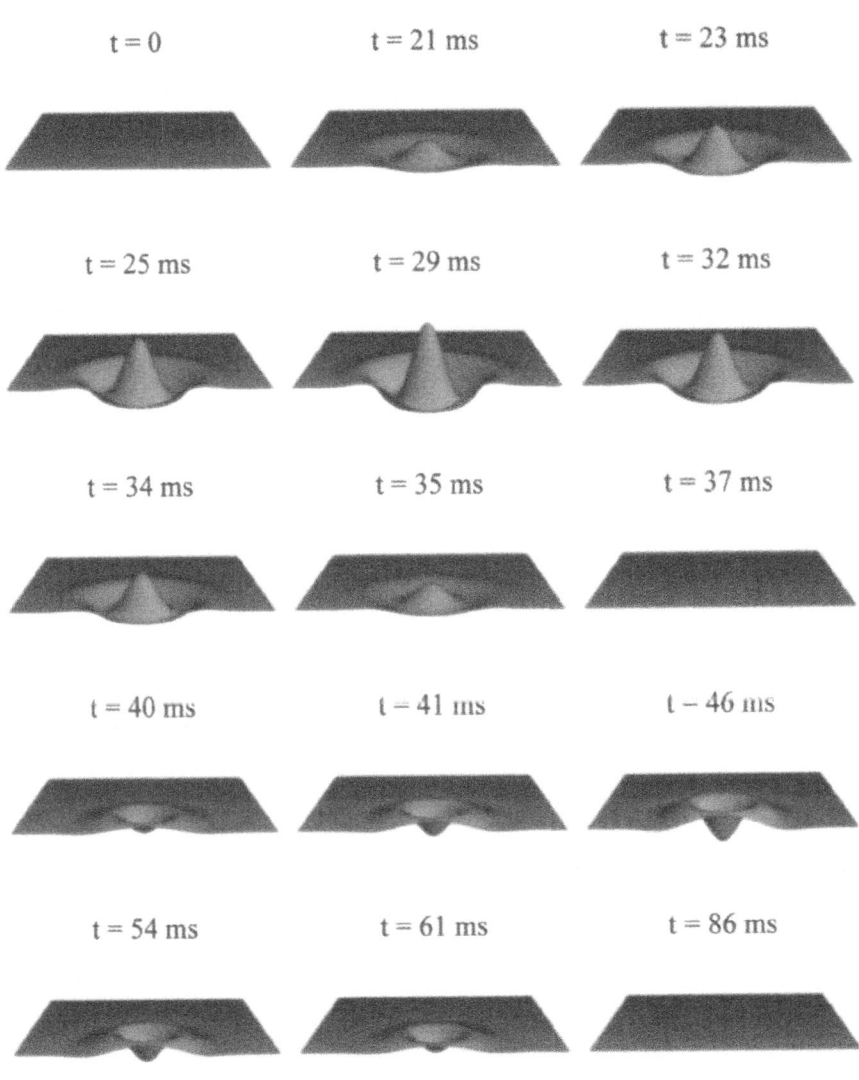

$t = 0$ $t = 21$ ms $t = 23$ ms

$t = 25$ ms $t = 29$ ms $t = 32$ ms

$t = 34$ ms $t = 35$ ms $t = 37$ ms

$t = 40$ ms $t = 41$ ms $t = 46$ ms

$t = 54$ ms $t = 61$ ms $t = 86$ ms

M on

Figure 5. M and P spatiotemporal response. The spatiotemporal response of M and P cells can be described by a generalization to three dimensions of the one-dimensional Gabor function (Wang et al., 1988, 1993; McLennan, 1991). Spatial profiles such as those of Figure 4 can be modified to include time as the third dimension. As time elapses, cell response for each receptive field spatial coordinate oscillates sinusoidally as it is being attenuated according to a Gaussian function. We used the M-on impulse response published by Lee et al. (1994) to estimate the change with time of the Gabor function depicted in the figure. Equation (9) represents the analytical three-dimensional form of cell response.

and Kaplan, 1995; Kremers et al., 1997; Lee et al., 1997). The M and P cell properties in space, time, spatial frequency and temporal frequency are complementary and result in different loci in Fourier space. The only difference in using other functions will be a proportional increase in entropy locus hypervolume for both M and P cells. One way to visualize the division of labor between M and P cells is to represent their entropy loci by elements of a 6 × 6 matrix, that are orthogonal two-dimensional projections of these loci (Silveira, 1996) (Fig. 6). This representation renders explicit the possibility of being able

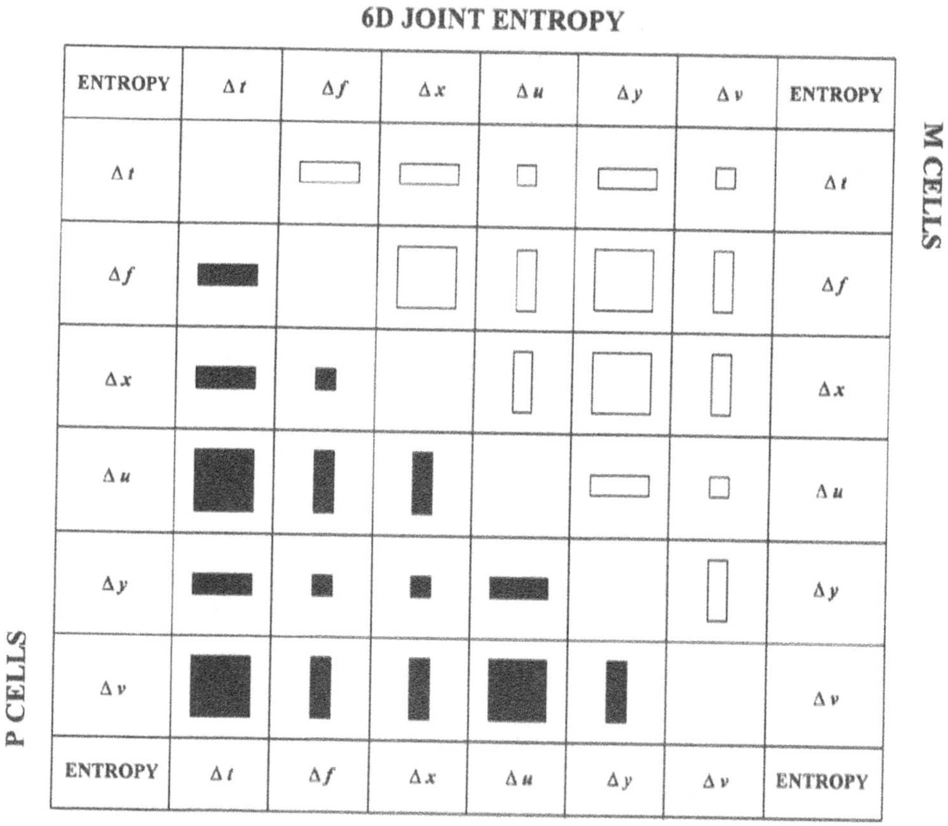

Figure 6. Representation of M and P cell response in Fourier space. M and P cell joint entropy have to be evaluated in six dimensions of the spatiotemporal information hyperspace: two dimensions of space and one dimension of time, Δx, Δy, and Δt, respectively, plus their Fourier transform, corresponding to two dimensions of spatial frequency and one dimension of temporal frequency, Δu, Δv, and Δf, respectively. The actual six-dimensional joint entropy is a hypervolume, but it can be visualized by the means of its orthogonal two-dimensional projections. These are represented schematically in the figure as elements of a 6 × 6 matrix. The projection of the M (empty rectangles) and P cell (filled rectangles) entropy loci were placed on each side of the diagonal. The side of each rectangle is proportional to cell entropy in a given dimension. In this figure we assumed, based on measurements of spatial performance of M and P cells, that M and P cell entropy differ from each other by a three-fold difference in each dimension. Quantitative data for some elements of the matrix are presented in Figures 7 and 8. The matrix representation allows one to predict that the M pathway sends highly precise information to the visual cortex concerning time of occurrence and spatial frequency content of visual patterns, while the P pathway does the same concerning the spatial location and temporal frequency content of visual patterns. Modified with permission from Silveira (1996). Reprinted with modifications with kind permission of Academia Brasileira de Ciencias (Copyright 1996).

to improve precision in a given dimension at the expenses of the others without changing the overall entropy. This may be an important guide to understanding the complementariness of M and P cell responses, and why the visual system needs these two pathways working in parallel. For each cell class, precision trade-off between different dimensions minimizes some areas in the matrix, while other areas increase. For instance, M cells have better precision than P cells in the time dimension due to their more transient impulse response. This could be compensated by poorer precision of M cells, when compared to P cells, in some other dimension, such as temporal frequency and space, leaving room for M cells' better performance in spatial frequency.

It can be seen that M ganglion cells perform with greater accuracy, tasks involving simultaneous measurement of the moment of occurrence and spatial periodicity of visual stimuli, or those involving two-dimensional measurements of spatial periodicity of visual stimuli. On the other hand P ganglion cells have greater precision in tasks involving simultaneous measurement of the place of occurrence and the temporal periodicity of visual stimuli, as well as those involving accurate two-dimensional measurements in space. From the diagram of Figure 6 we can hypothesize that M cells discriminate better than P cells between stimuli consisting of briefly presented, two-dimensional spatial periodicities. On the other hand, P cells perform better than M cells in tasks demanding discrimination between flickering, two-dimensional small spatial targets. Thus, we define M and P cells as devices whose tuning can be better characterized by using a multidimensional approach. Moreover, M and P cells differ from each other when the probing stimuli are optimized for cell sensitivity. This is opposite to the most commonly used approach in which M and P cell sensitivities are compared by probing stimuli located at the extremes of their response range, either high or low frequencies in the spatial or temporal domain.

Figure 6 is a schematic representation of M and P joint entropy loci, assuming that one cell class is more precise than the other by a factor of three in each dimension. This assumption might be valid at least for spatial entropy once there is a three-fold difference in receptive field size between the two cell classes across eccentricity (Fig. 3). As both cell classes are spatially linear, at least in some circumstances, the same proportion is expected for their precision in the spatial frequency domain.

We used data from the literature to estimate joint entropy values for macaque M and P cells. The results are presented in Figures 7–8 as areas in Fourier space, and in Figure 9 as

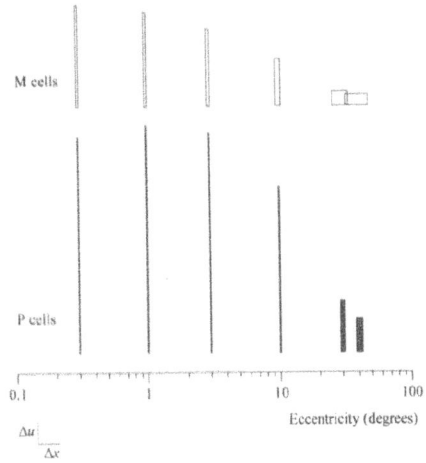

Figure 7. Joint entropy loci of macaque M and P cells, at different retinal eccentricities, for the domains of space and spatial frequency. The area and shape of the empty and filled rectangles correspond to the product $\Delta x \Delta u$ for M and P cells, respectively, as estimated from published data (de Monasterio and Gouras, 1975; Perry et al., 1984; Croner and Kaplan, 1995) as described in the text. Central M and P cells have smaller entropy in the spatial domain than in the spatial frequency domain. At increasing eccentricities both cell classes change their $\Delta x \Delta u$ trade-off improving precision in spatial frequency at expenses of spatial precision, but only for peripheral M cells the trade-off reaches a point where the relationship inverts and cells become more precise in spatial frequency than in space. At all eccentricities, M cells have smaller entropy in the spatial frequency domain and larger entropy in the space domain, when compared with P cells. Scale: $\Delta x = \Delta u = 1$ unity.

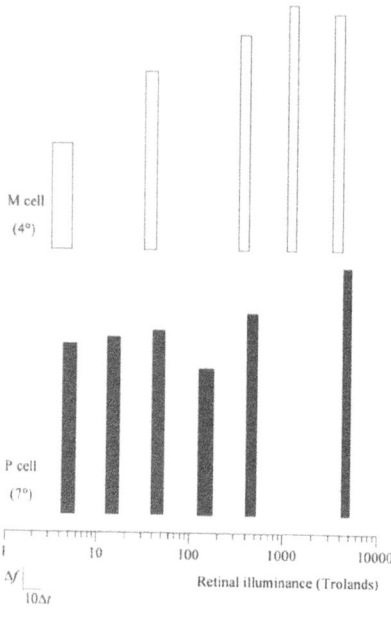

Figure 8. Joint entropy loci of macaque M and P cells, at different retinal luminances, for the domains of time and temporal frequency. The area and shape of the empty and filled rectangles correspond to the product $\Delta t \Delta f$ for M and P cells, respectively, and were estimated, as described in the text, from published data of Purpura et al. (1990) for two cells, M7/8 and P8/25. At all luminance levels, both M and P cells have much smaller entropy in the time domain than in the temporal frequency, but when luminance decreases there is a change in the trade-off of both cell classes improving precision in temporal frequency at expenses of time. Scale: $10\Delta t = \Delta f = 1$ unity.

points in xy coordinates. Figure 7 represents the M and P joint entropy loci for space and spatial frequency at different eccentricities. The rectangles correspond to the product $\Delta x \Delta u$ and were estimated using published data for M and P receptive fields (center radius from Figure 3; surround radius, center and surround sensitivities from Croner and Kaplan, 1995). We used the Levenberg-Marquardt method (Press et al., 1992) to find Gabor functions that best matched Croner and Kaplan (1995) DOG functions for M and P cells, and estimate entropy values the Gabor (1946) equations. Figure 7 shows that central M and P cells have smaller entropy in the spatial domain than in the spatial frequency domain. At increasing

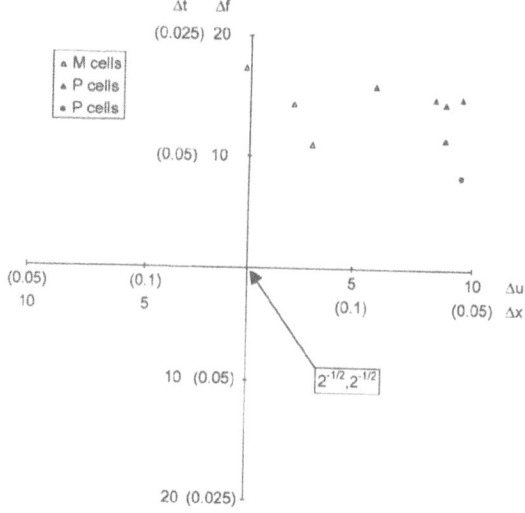

Figure 9. Spatiotemporal trade-off for M and P cells located at different retinal eccentricities. The horizontal axis represents a continuum of different forms of joint entropy minimization for space and spatial frequency, while the vertical axis represents a similar continuum of different forms of joint entropy minimization for time and temporal frequency. All the points of each axis have the same value for two-dimensional joint entropy, and for Gabor functions this value corresponds to 1/2. The point where the two axes cross each other correspond to equal values of single entropy in the four dimensions, $1/\sqrt{2}$. We plotted the values for Δx, Δu, Δt, and Δf for five P cells ($\blacktriangle$) and three M cells ($\triangle$) studied by Purpura et al. (1990) and one P cell ($\bullet$) studied by Gielen et al. (1982). With increasing eccentricity, M and P cells progressively become more precise in the time and spatial frequency domains.

eccentricities both cell classes change their $\Delta x \Delta u$ trade-off, improving precision in spatial frequency at the expense of spatial precision. While for P cells precision remains higher for space than spatial frequency at all eccentricities, for peripheral M cells the trade-off reaches a point where the relation inverts and cells become more precise in spatial frequency than in space. At all eccentricities, M cells have smaller entropy in the spatial frequency domain and larger entropy in the space domain, when compared with P cells.

Figure 8 represents the M and P joint entropy loci for time and temporal frequency at different retinal luminances. The rectangles correspond to the product $\Delta t \Delta f$ and were estimated from published data of Purpura et al. (1990) for the temporal modulation transfer function (TMTF) of two cells, M7/8 and P8/25. We used the inverse Fourier transform to obtain the temporal impulse function from the TMTF, and then used the Levenberg-Marquardt method to find the Gabor function that best matched the descending part of the cell impulse function. Finally we estimated time and temporal entropies using Gabor equations. At all luminance levels, both M and P cells have much smaller entropy in the time domain than in temporal frequency, but when luminance decreases there is a change in the trade-off of both cell classes, improving precision in temporal frequency at the expenses of time.

Figure 9 represents the spatiotemporal trade-off for M and P cells located at different retinal eccentricities. The horizontal axis represents a continuum of different forms of joint entropy minimization for space and spatial frequency, while the vertical axis represents a similar continuum of different forms of joint entropy minimization for time and temporal frequency. The axes extend from a region of maximum spectral precision and minimum spatial and temporal precision, in the lower left quadrant towards a region of opposite properties in the upper right quadrant. For the sake of clarity only one space-spatial frequency pair was considered, assuming circular symmetry for M and P receptive fields. All the points of each axis have the same value for two-dimensional joint entropy and for Gabor functions this value corresponds to 1/2. The point where the two axes cross each other correspond to equal values of single entropy in the four dimensions, $1/\sqrt{2}$. Using similar procedures to those used for Figures 7 and 8, we estimated the values for Δx, Δu, Δt, and Δf for five P cells and three M cells studied by Purpura et al. (1990) and one P cell studied by Gielen et al. (1982). With increasing eccentricity there is a tendency of M and P cells to progressively change their spatiotemporal trade-off, moving towards lower values of Δt and Δu, becoming more precise to convey information about time events and spatial frequency content.

In Figure 9, most M and P cells have their entropy loci in the first quadrant. This is consistent with the common view that both cell classes are much better at localizing events in time and space than in their respective transform domains. Only peripheral M cells cross the limit between the first and second quadrants, and are slightly more precise in the spatial frequency domain than in the space domain.

6. CONCLUSIONS

A Gabor expansion uses the same window, independent of its center frequency, to span the whole information diagram. This does not optimize the analysis of signals with both very high and very low frequencies (Chui, 1992). When translating this property to visual system organization, this rigid scheme would be represented by cells all having the same properties at all visual field locations (Kulikowski and Bishop, 1981; Kulikowski et al., 1982; Field, 1987). Natural scenes have a large range of spatiotemporal frequencies and some visual tasks demand simultaneous analysis of very high and very low spatiotem-

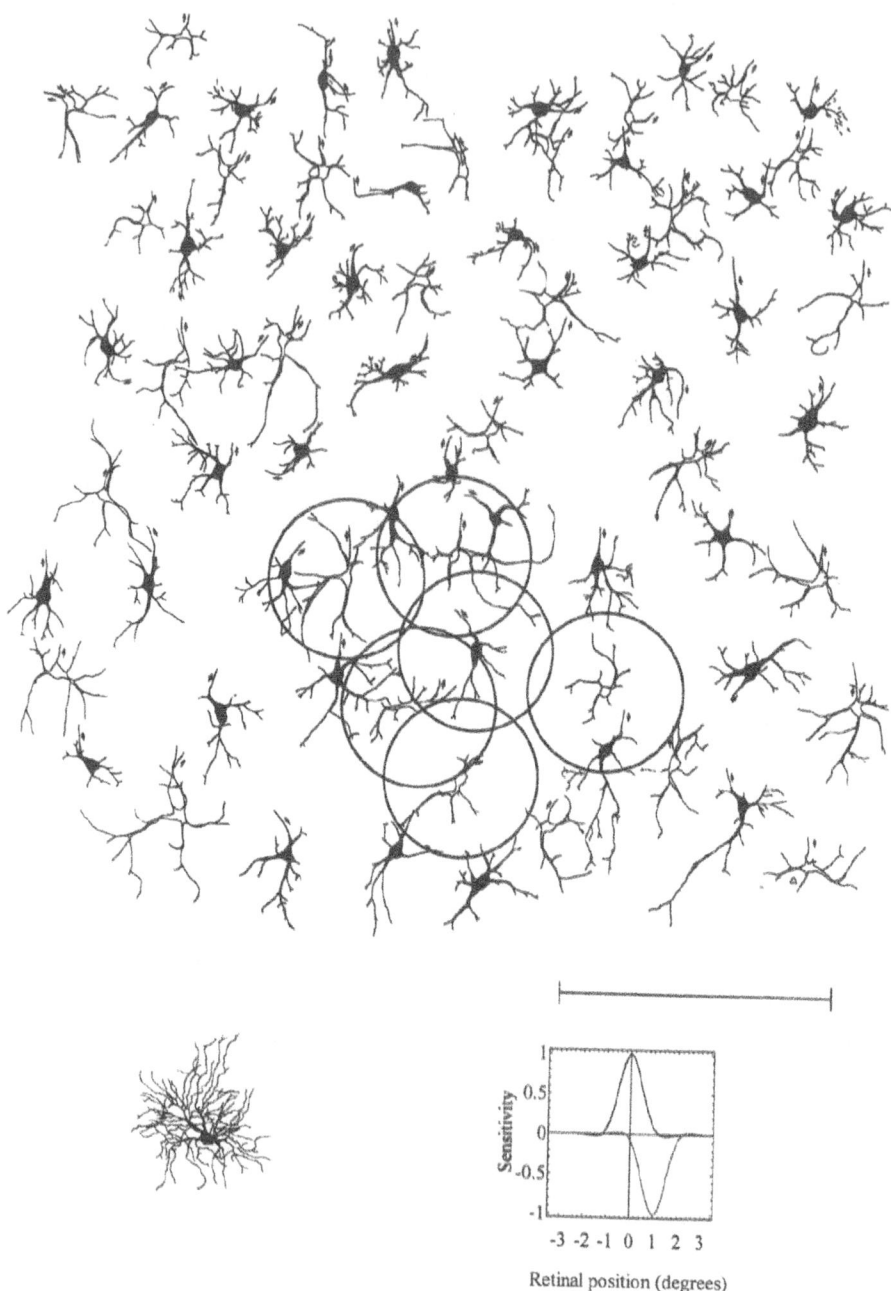

Retinal position (degrees)

Figure 10. Spatial phase difference for M-on and M-off cells. The upper part of the figure illustrates the mosaics formed by macaque monkey M-on and M-off ganglion cells located 10 mm (46°) temporal to the fovea. Cells were stained with the method neurofibrillar of Gros-Schultze (Silveira and Perry, 1991). For six M-on ganglion cells, we estimated the mean distance between each one of them and all the M-off cells intercepted by a circle corresponding to one dendritic tree diameter. It corresponds to a retinal distance of 0.209 mm (s.d. = 0.026 mm) or 0.994° (s.d. = 0.129°) in the visual field. Two Gabor functions of opposite polarities shifted by this amount would have spatial phase difference of 106° (bottom right). Thus higher level interaction between on and off subclasses of both M and P ganglion cells may account for an approximate form of phase quadrature. On the bottom left, we drew in the same scale of the mosaic the dendritic tree of an M-on ganglion cell of the same size of those depicted in the mosaic. The cell was from a capuchin monkey retina retrogradely labelled from the optic nerve with biocytin. Scale bar = 500 μm.

poral frequencies. The visual system can sample more efficiently the spatiotemporal content of natural scenes by using two or more channels having different entropy loci in the information diagram, such as the M and P cells. The distance between M and P cells in the information diagram seems to be necessary for efficient coding. Shorter distances are probably inefficient at encoding the range of spatiotemporal properties of the stimuli. Longer distances probably preclude higher order interactions between the two channels.

In a single visual field region, such as central vision, M or P cells operate within the constraints of a rigid compromise of space-time and spectral precision, each cell class analyzing with its own metrics the whole mixture of spatiotemporal singularities and periodicities present in the visual scenes. The visual system uses these two channels to twice measure the luminance contrast variation present in the visual scenes. Each measurement is made with a certain joint entropy value, which is minimum if cell response follows a Gabor function, but with different precision, according to cell class, in space, time, spatial frequency and temporal frequency.

Neuronal response functions to visual stimuli can only be real functions. Thus one important constraint to Gabor expansion of visual stimuli in the neuronal response concerns the need of paired quadrature-phase receptive fields centered on the same visual field location (Marcelja, 1980; Kulikowski and Bishop, 1981). This kind of paired receptive field has been shown to exist in the primary visual cortex but data available are limited (Pollen and Ronner, 1981). Both M and P cells occur in subclasses with opposite responses to certain stimuli, the on and off subclasses, which represent pairs 180° out of phase. Hypothetically a smaller phase difference would provide the needs for M and P subclasses to operate in phase quadrature, but there is no evidence that such hypothetical quadrature-phase pairs exist in the retina or LGN (Silveira, 1996). However, there is an alternative way to examine this question, at least for the space domain. Figure 10 illustrates the mosaics of macaque monkey M-on and M-off ganglion cells located 10 mm (46°) temporal to the fovea. Cells were stained with the neurofibrillar method of Gros-Schultze (Silveira and Perry, 1991). For six M-on ganglion cells, we estimated the mean distance between each one of them and all the M-off cells intercepted by a circle corresponding to one dendritic tree diameter. We obtained a mean distance of 0.994° (s.d. = 0.129°) in the visual field. Two Gabor functions of opposite polarities shifted by this amount would have a spatial phase difference of 106°. Thus, higher level interactions between on and off subclasses of both M and P ganglion cells may account for an approximate form of phase quadrature.

M and P cells perform simultaneous and overlapping analyses of the visual field using different strategies to minimize entropy. This enables higher order visual neurons, similar to those of the dorsal and ventral cortical streams (Ungerleider and Mishkin, 1982; Milner and Goodale, 1995), to combine M and P inputs in different ways; and this could explain why M and P inputs converge in their cortical pathways.

ACKNOWLEDGMENTS

This work was supported by FINEP/FADESP #4.3.90.0082.00 and CNPq #52.1749/94-8. The authors are in debt to Keith Purpura and Ehud Kaplan for making available additional unpublished information of their work. The authors thank Antonio T. Silveira and Hermínio S. Gomes for helpful suggestions. The authors are grateful to Carlos Eduardo G. Rocha-Miranda, Barbara L. Finlay, Elizabeth Yamada, and Jan Kremers for critiquing the manuscript. Cézar A. Saito and Walther A. de Carvalho helped with the illustrations.

REFERENCES

Blasdell, G. G. and Lund, J. S., 1983, Termination of afferent axons in macaque striate cortex. J. Neurosci., 3:1389–1413.

Boycott, B. B. and Dowling, J. E., 1969, Organization of the primate retina: light microscopy. Phil. Trans. R. Soc. Lond. B, 255:109–184.

Boycott, B. B. and Wässle, H., 1991, Morphological classification of bipolar cells of the primate retina. Eur. J. Neurosci., 3:1069–1088.

Bracewell, R., N., 1986, "The Fourier Transform and its Applications. Second edition." McGraw-Hill, New York.

Calkins, D. J., Schein, S. J., Tsukamoto, Y. and Sterling, P., 1995, Ganglion cell circuits in primate fovea. In: "Colour Vision Deficiencies XII", B. Drum, ed., Kluwer Academic Publishers, Dordrecht, pp. 267–274.

Casagrande, V. A., 1994, A third parallel visual pathway to primate area V1. TINS, 17:305–310.

Casagrande, V. A. and Norton, T. T., 1991, Lateral geniculate nucleus: a review of its physiology and function. In: "Vision and Visual Dysfunction, Vol. 4 The Neural Basis of Visual Function", A. G. Leventhal, ed., J. R. Cronly-Dillon, ser. ed., MacMillan Press, Houndmills, pp. 41–84.

Chui, C. K., 1992, "An Introduction to Wavelets", Academic Press, Boston.

Conley, m. and Fitzpatrick, D., 1989, Morphology of retinogeniculate axons in the macaque. Vis. Neurosci., 2:287–296.

Conley, M., Penny, G. R. and Diamond, I. T., 1987, Terminations of individual optic tract fibers in the lateral geniculate nuclei of Galago crassicaudatus and Tupaia belangeri. J. Comp. Neurol., 256:71–87.

Connolly, M. and Van Essen, D., 1984, The representation of the visual field in parvicellular and magnocellular layers of the lateral geniculate nucleus in the macaque monkey. J. Comp. Neurol., 226:554–564.

Creutzfeldt, O., Lee, B. B. and Elepfandt, A., 1979, A quantitative study of chromatic organisation and receptive fields of cells in the lateral geniculate body of the rhesus monkey. Exp. Brain Res., 35:527–545.

Crook, J. M., Lee, B. B., Tigwell, D. A. and Valberg, A., 1987, Thresholds to chromatic spots of cells in the macaque geniculate nucleus as compared to detection sensitivity in man. J. Physiol. (Lond.), 392:193–211.

Crook, J. M., Lange-Malecki, B., Lee, B. B. and Valberg, A., 1988, Visual resolution of macaque retinal ganglion cells. J. Physiol. (Lond.), 396:205–224.

Croner, L. J. and Kaplan, E., 1995, Receptive fields of P and M ganglion cells across the primate retina. Vision Res., 35:7–24.

Dacey, D. M., 1993, The mosaic of midget ganglion cells in the human retina. J. Neurosci., 13:5334–5355.

Dacey, D. M., 1996, Circuitry for color coding in the primate retina. Proc. Natl. Acad. Sci. U.S.A., 93:583–588.

Dacey, D. M. and Brace, S., 1992, A coupled network for parasol but not midget ganglion cells in the primate retina. Vis. Neurosci., 9:279–290.

Dacey, D. M. and Lee, B. B., 1994, The "blue-on" opponent pathway in primate retina originates from a distinct bistratified ganglion cell type. Nature, 367:731–735.

Dacey, D. M. and Petersen, M. R., 1992, Dendritic field size and morphology of midget and parasol ganglion cells of the human retina. Proc. Natl. Acad. Sci. U.S.A., 89:9666–9670.

Daugman, J. G., 1980, Two-dimensional spectral analysis of cortical receptive field profiles. Vision Res., 20:847–856.

Daugman, J. G., 1983, Six formal properties of two-dimensional anisotropic visual filters: structural principles and frequency / orientation selectivity. IEEE Transactions on Systems, Man, and Cybernetics, 13:882–887.

Daugman, J. G., 1984, Spatial visual channels in the Fourier plane. Vision Res., 24:891–910.

Daugman, J. G., 1985, Uncertainty relation for resolution in space, spatial frequency, and orientation optimized by two-dimensional visual cortical filters. J. Opt. Soc. Am. A, 2:1160–1169.

de Monasterio, F. M. and Gouras, P., 1975, Functional properties of ganglion cells of the rhesus monkey retina. J. Physiol. (Lond.), 251:167–195.

de Monasterio, F. M., Gouras, P. and Tolhurst, D. J., 1975a, Trichromatic colour opponency in ganglion cells of the rhesus monkey retina. J. Physiol. (Lond.), 251:197–216.

de Monasterio, F. M., Gouras, P. and Tolhurst, D. J., 1975b, Concealed colour opponency in ganglion cells of the rhesus monkey retina. J. Physiol. (Lond.), 251:217–229.

Derrington, A. M. and Lennie, P., 1984, Spatial and temporal contrast sensitivities of neurons in lateral geniculate nucleus of macaque. J. Physiol. (Lond.), 357:219–240.

Derrington, A. M., Krauskopf, J. and Lennie, P., 1984, Chromatic mechanisms in lateral geniculate nucleus of macaque. J. Physiol. (Lond.), 357:241–265.

DeYoe, E. A. and Van Essen, D. C., 1985, Segregation of efferent connections and receptive field properties in visual area V2 of the macaque. Nature, 317:58–61.

DeYoe, E. A. and Van Essen, D. C., 1988, Concurrent processing streams in monkey visual cortex. TINS, 11:219–226.

Diamond, I. T., Conley, M., Itoh, K. and Fitzpatrick, D., 1985, Laminar organization of geniculocortical projections in *Galago senegalensis* and *Aotus trivirgatus*. J. Comp. Neurol., 242:584–610.

Ding, Y. and Casagrande, V. A., 1997, The distribution and morphology of LGN K pathway axons within the layers and CO blobs of owl monkey V1. Vis. Neurosci., 14:691–704.

Dreher, B., Fukada, Y. and Rodieck, R. W., 1976, Identification, classification and anatomical segregation of cells with X-like and Y-like properties in the lateral geniculate nucleus of Old-World primates. J. Physiol. (Lond.), 258:433–452.

Dogiel, A. S., 1891, Ueber die nervösen Elemente in der Retina des Menschen. Archiv für Mikroskopische Anatomie und Entwicklungsmechanik, 38:317–344.

Felleman, D. J. and Van Essen, D. C., 1991, Distributed hierarchical processing in the primate cerebral cortex. Cerebral Cortex, 1:1–47.

Ferrera, V. P., Nealey, T. A. and Maunsell, J. H. R., 1992, Mixed parvocellular and magnocellular geniculate signals in visual area V4. Nature, 358:756–758.

Ferrera, V. P., Nealey, T. A. and Maunsell, J. H. R., 1994, Responses in macaque visual area V4 following inactivation of the parvocellular and magnocellular LGN pathways. J. Neurosci., 14:2080–2088.

Field, D., 1987, Relations between the statistics of natural images and the response properties of cortical cells. J. Opt. Soc. Am. A, 12:2379–2394.

Field, D. and Tolhurst, D. J., 1986, The structure and symmetry of simple-cell receptive-field profiles in the cat's visual cortex. Proc. R. Soc. Lond. B, 228: 379–400.

Fitzpatrick, D., Itoh, K. and Diamond, I. T., 1983, The laminar organization of the lateral geniculate body and the striate cortex in the squirrel monkey (*Saimiri sciureus*). J. Neurosci., 3:673–702.

Gabor, D., 1946, Theory of communication. J. Inst. Electr. Eng., 93:429–457.

Ghosh, K. K., Goodchild, A. K., Sefton, A. E. and Martin, P. R., 1996, The morphology of retinal ganglion cells in the New World marmoset monkey *Callithrix jacchus*. J. Comp. Neurol., 366:76–92.

Ghosh, K. K., Martin, P. and Grünert, U., 1997, Morphological analysis of the blue cone pathway in the retina of a New World monkey, the marmoset *Callithrix jacchus*. J. Comp. Neurol., 379:211–225.

Gielen, C. C. A. M. and van Gisbergen, J. A. M., 1980, Signal processing by neurones in the visual pathway of the rhesus monkey. In: "Signal Processing - Theories and Applications", M. Kunt and F. de Coulon, eds., North-Holland Publishing, pp. 351–356.

Gielen, C. C. A. M., van Gisbergen, J. A. M. and Vendrik, A. J. H., 1981, Characterization of spatial and temporal properties of monkey LGN Y-cells. Biol. Cybern., 40:157–170.

Gielen, C. C. A. M., van Gisbergen, J. A. M. and Vendrik, A. J. H., 1982, Reconstruction of cone-system contributions to responses of colour-opponent neurones in monkey lateral geniculate. Biol. Cybern., 44:211–221.

Goodchild, A. K., Ghosh, K. K. and Martin, P. R., 1996, A comparison of photoreceptor spatial density and ganglion cell morphology in the retina of human, macaque monkey, cat, and the marmoset *Callithrix jacchus*. J. Comp. Neurol., 366:55–75.

Gouras, P., 1968, Identification of cone mechanisms in monkey ganglion cells. J. Physiol. (Lond.), 199:533–547.

Hendrickson, A. E., Wilson, J. R. and Ogren, M. P., 1978, The neuroanatomical organization of pathways between the dorsal lateral geniculate nucleus and visual cortex in old world and new world monkeys. J. Comp. Neurol., 182:123–136.

Hendry, S. H. C. and Casagrande, V. A., 1996, A common pattern for a third visual channel in the primate LGN. Society for Neuroscience Abstracts, 22:1605.

Hendry, S. H. C. and Yoshioka, T., 1994, A neurochemically distinct third channel in the macaque dorsal lateral geniculate nucleus. Science, 264:575–577.

Hicks, T. P., Lee, B. B. and Vidyasagar, T. R., 1983, The responses of cells in the macaque lateral geniculate nucleus to sinusoidal gratings. J. Physiol. (Lond.), 337:183–200.

Hubel, D. H. and Livingstone, M., 1985, Complex-unoriented cells in a subregion of primate area 18. Nature, 315:325–327.

Hubel, D. H. and Livingstone, M., 1987, Segregation of form, color, and stereopsis in primate area 18. J. Neurosci., 7:3378–3415.

Hubel, D. H. and Livingstone, M., 1987, Color and contrast sensitivity in the lateral geniculate body and primary visual cortex of the macaque monkey. J. Neurosci., 10:2223–2237.

Hubel, D. H. and Wiesel, T. N., 1972, Laminar and columnar distribution of geniculocortical fibers in the macaque monkey. J. Comp. Neurol., 146:421–450.

Jacobs, G. H., 1983, Differences in spectral response properties of LGN cells in male and female squirrel monkeys. Vision Res., 23:461–468.

Jacobs, G. H. and De Valois, R. L., 1965, Chromatic opponent cells in squirrel monkey lateral geniculate nucleus. Nature, 206:487–489.

Jacoby, R. A. and Marshak, D. W., 1995, Diffuse bipolar cell inputs to parasol ganglion cells in macaque retina. Invest. Opthalmol. Vis. Sci., 36:S4.

Jacoby, R. A. and Marshak, D. W., 1996, Inputs to parasol ganglion cells in macaque retina. Invest. Opthalmol. Vis. Sci., 37:S950.

Jasinschi, R.S., 1991, Energy filters, motion uncertainty, and motion sensitive cells in the visual cortex: a mathematical analysis. Biol. Cybern., 65:515–523.

Jones, J. P. and Palmer, L. A., 1987, An evaluation of the two-dimensional Gabor filter hypothesis of simple receptive fields in cat striate cortex. J. Neurophysiol., 58:1233–1258.

Kaas, J. H. and Huerta, M. F., 1988, The subcortical visual system of primates. In: "Comparative Primate Biology, Vol. 4 Neurosciences, H. D. Steklis and J. Erwin, eds., J. Erwin, ser. ed., Alan R. Liss, New York, pp. 327–391.

Kaiser, P. K., Lee, B. B., Martin, P. R. and Valberg, A., 1990, The physiological basis of the minimally distinct border demonstrated in the ganglion cells of the macaque retina. J. Physiol. (Lond.), 422:153–183.

Kaplan, E. and Shapley, R. M., 1982, X and Y cells in the lateral geniculate nucleus of macaque monkey. J. Physiol. (Lond.), 330:125–143.

Kaplan, E. and Shapley, R. M., 1986, The primate retina contains two types of ganglion cells, with high and low contrast sensitivity. Proc. Natl. Acad. Sci. USA, 83:2755–2757.

Kaplan, E., Lee, B. B. and Shapley, R. M., 1990, New views of primate retinal function. Prog. Retinal Res., 9:273–336.

Kremers, J. and Weiss, S., 1997, Receptive feild dimensions of lateral geniculate cells in the common marmoset (*Callithrix jacchus*). Vision Res., 37:2171–2181.

Kremers, J., Lee, B. B. and Kaiser, P. K., 1992, Sensitivity of macaque retinal ganglion cells and human observers to combined luminance and chromatic temporal modulation. J. Opt. Soc. Am. A, 9:1477–1485.

Kremers, J., Lee, B. B., Pokorny, J. and Smith, V. C., 1993, Responses of macaque ganglion cells and human observers to compound periodic waveforms. Vision Res., 33:1997–2011.

Kolb, H., 1970, Organization of the outer plexiform layer of the primate retina: electron microscopy of Golgi-impregnated cells. Phil. Trans. R. Soc. Lond. B, 258:261–283.

Kolb, H., Boycott, B. B. and Dowling, J. E., 1969, A second type of midget bipolar cell in the primate retina. Phil. Trans. R. Soc. Lond. B., 255:177–184.

Kolb, H. and DeKorver, L. 1991, Midget ganglion cells of parafovea of the human retina: a study by electron microscopy and serial-section reconstruction. J. Comp. Neurol. 303:617–636.

Kolb, H., Linberg, K. and Fisher, S. K., 1992, Neurons of the human retina: a Golgi study. J. Comp. Neurol., 318:147–187.

Kouyama, N. and Marshak, D. W., 1992, Bipolar cells specific for blue cones in the macaque retina. J. Neurosci., 12:1233–1252.

Kulikowski, J. J. and Bishop, P. O., 1981, Fourier analysis and spatial representation in the visual cortex. Experientia, 37:160–163.

Kulikowski, J. J., Marcelja, S. and Bishop, P. O., 1982, Theory of spatial position and spatial frequency relations in the receptive fields of simple cells in the cat's striate cortex. Biol. Cybern., 43:187–198.

Lachica, E. A. and Casagrande, V. A., 1988, Development of primate retinogeniculate axon arbors. Vis. Neurosci., 1:103–123.

Lachica, E. A. and Casagrande, V. A., 1992, Direct W-like geniculate projections to the cytochrome oxidase (CO) blobs in primate visual cortex: axon morphology. J. Comp. Neurol., 319:141–158.

Lachica, E. A., Beck, P. D. and Casagrande, V. A., 1992, Parallel pathways in macaque monkey striate cortex: anatomically defined columns in layer III. Proc. Natl. Acad. Sci. U.S.A., 89:3566–3570.

Lee, B. B., 1996, Receptive field structure in the primate retina. Vision Res., 36:631–644.

Lee, B. B., Valberg, A., Tigwell, D. A. and Tryti, J., 1987, An account of responses of spectrally opponent neurons in macaque lateral geniculate nucleus to successive contrast. Proc. R. Soc. Lond. B, 230: 293–314.

Lee, B. B., Martin, P. R. and Valberg, A., 1988, The physiological basis of heterochromatic flicker photometry demonstrated in the ganglion cells of the macaque retina. J. Physiol. (Lond.), 404:323–347.

Lee, B. B., Martin, P. R. and Valberg, A., 1989a, Sensitivity of macaque retinal ganglion cells to chromatic and luminance flicker. J. Physiol. (Lond.), 414:223–243.

Lee, B. B., Martin, P. R. and Valberg, A., 1989b, Amplitude and phase of responses of macaque retinal ganglion cells to flickering stimuli. J. Physiol. (Lond.), 414:245–263.

Lee, B. B., Martin, P. R. and Valberg, A., 1989c, Nonlinear summation of M- an L-cone inputs to phasic retinal ganglion cells of the macaque. J. Neurosci., 9:1433–1442.

Lee, B. L., Pokorny, J., Smith, V. C., Martin, P.R. and Valberg, A., 1990, Luminance and chromatic modulation sensitivity of macaque ganglion cells and human observers. J. Opt. Soc. Am. A, 7:2223–2236.

Lee, B. B., Wehrhahn, C., Westheimer, G. and Kremers, J., 1993, Macaque ganglion cell responses to stimuli that elicit hyperacuity in man: detection of small displacements. J. Neurosci., 13:1001–1009.

Lee, B. B., Pokorny, J., Smith, V.C. and Kremers, J., 1994, Responses to pulses and sinusoids in macaque ganglion cells. Vision Res., 34:3081–3096.

Lee, B. B., Wehrhahn, C., Westheimer, G. and Kremers, J., 1995, The spatial precision of macaque ganglion cell responses in relation to vernier acuity of human observers. Vision Res., 35:2743–2758.

Lee, B. B., Silveira, L. C. L., Yamada, E. S. and Kremers, J., 1996, Parallel pathways in the retina of Old and New World primates. Rev. Brasil. Biol., 56(Supl. 1):323–338.

Lee, B. B., Kremers, J. and Yeh, T., 1997, Receptive field structure of primate retinal ganglion cells. Vis. Neurosci., in press.

Lennie, P., 1980a, Perceptual signs of parallel pathways. Phil. Trans. R. Soc Lond. B, 290:23–37.

Lennie, P., 1980b, Parallel visual pathways. Vision Res., 20:561–594.

Leventhal, A. G., Rodieck, R. W. and Dreher, B., 1981, Retinal ganglion cell classes in the Old World monkey: Morphology and central projections. Science, 213:1139–1142.

Leventhal, A. G., Ault, S. J., Vitek, D. J. and Shou, T., 1989, Extrinsic determinants of retinal ganglion cell development in primates. J. Comp. Neurol., 286:170–189.

Lima, S. M. A., Silveira, L. C. L. and Perry, V. H., 1993, The M-ganglion cell density gradient in New World monkeys. Brazil. J. Med. Biol. Res. 26:961–964.

Lima, S. M. A., Silveira, L. C. L. and Perry, V. H., 1996, Distribution of M retinal ganglion cells in diurnal and nocturnal New World monkeys. J. Comp. Neurol., 368:558–552.

Livingstone, M. and Hubel, D. H., 1982, Thalamic inputs to cytochrome oxidase-rich regions in monkey visual cortex. Proc. Natl. Acad. Sci. U.S.A., 79:6098–6101.

Livingstone, M. and Hubel, D. H., 1983, Specificity of cortico-cortical connections in monkey visual system. Nature, 304:531–534.

Livingstone, M. and Hubel, D. H., 1984a, Anatomy and physiology of a color system in the primate visual cortex. J. Neurosci., 4:309–356.

Livingstone, M. and Hubel, D. H., 1984b, Specificity of intrinsic connections in primate primary visual cortex. J. Neurosci., 4:2830–2835.

Livingstone, M. and Hubel, D. H., 1987, Connections between layer 4B of area 17 and thick cytochrome oxidase stripes of area 18 in the squirrel monkey. J. Neurosci., 7:3371–3377.

Lund, J. S., 1988, Anatomical organization of macaque monkey striate visual cortex. Ann. Rev. Neurosci., 11:53–288.

Lund, J. S. and Boothe, R. G., 1975, Interlaminar connections and pyramidal neuron organization in the visual cortex, area 17, of the macaque monkey. J. Comp. Neurol., 159:305–334.

Lynch III, J. J., Silveira, L. C. L., Perry, V. H. and Merigan, W. H., 1992, Visual effects of damage to P ganglion cells in macaques. Vis. Neurosci., 8:575–583.

MacLennan, B., 1991, Gabor representations of spatiotemporal visual images. Technical Report CS-91-144. Computer Science Department, University of Tennessee, Knoxville.

Marr, D., 1982, "Vision". W. H. Freeman, New York.

Marcelja, S., 1980, Mathematical description of the responses of simple cortical cells. J. Opt. Soc. Am., 70:1297–1300.

Mariani, A. P., 1984, Bipolar cells in monkey retina selective for cones likely to be blue-sensitive. Nature, 308:184–186.

Maunsell, J. H. R., Nealey, T. A. and DePriest, D. D., 1990, Magnocellular and parvocellular contributions to responses in the middle temporal visual area (MT) of the macaque monkey. J. Neurosci., 10:3323–3334.

McLean, J., Raab, S. and Palmer, L. A., 1994, Contribution of linear mechanisms to the specification of local motion by simple cells in areas 17 and 18 of the cat. Vis. Neurosci., 11:271–294.

McLean, J. and Palmer, L. A., 1994, Organization of simple cell responses in the three-dimensional (3-D) frequency domain. Vis. Neurosci., 11:295–306.

Merigan, W. H. and Maunsell, J. H. R., 1993, How parallel are the primate visual pathways? Ann. Rev. Neurosci., 16:369–402.

Merigan, W. H., Katz, L. M. and Maunsell, J. H. R., 1991a, The effects of parvocellular lateral geniculate lesions on the acuity and contrast sensitivity of macaque monkeys. J. Neurosci., 11:994–1101.

Merigan, W. H., Byrne, C. and Maunsell, J. H. R., 1991b, Does primate motion perception depend on the magnocellular pathway? J. Neurosci., 11:3422–3429.

Michael, C. R., 1988, Retinal afferent arborization patterns, dendritic field orientations and the segregation of functions in the lateral geniculate nucleus of the monkey. Proc. Natl. Acad. Sci. U.S.A., 85:4914–4918.

Milner, A. D. and Goodale, M. A., 1995, "The Visual Brain in Action". Oxford University Press, Oxford.

Mollon, J. D., 1989, "Tho' she kneel'd in that Place where they grew". J. Exp. Biol., 146:21–38.

Mollon, J. D., 1991, Uses and evolutionary origins of primate colour vision. In: "Vision and Visual Dysfunction, Vol. 2 Evolution of the Eye and Visual System, J. R. Cronly-Dillon and R. L. Gregory, eds., J. R. Cronly-Dillon, ser. ed., MacMillan Press, Houndmills, pp 306–319.

Mollon, J. D., Estévez, O. and Cavonius, C. R., 1990, The two subsystems of colour vision and their rôles in wavelength discrimination. In: "Vision: Coding and Efficiency", C. Blakemore, ed., Cambridge University Press, Cambridge, pp. 119–131.

Nealey, T. A. and Maunsell, J. H. R., 1994, Magnocellular and parvocellular contributions to the responses of neurons in macaque striate cortex. J. Neurosci., 14:2069–2079.

Norton, T. T. and Casagrande, V. A., 1982, Laminar organization of receptive-field properties in lateral geniculate nucleus of bush baby (Galago crassicaudatus). J. Neurophysiol, 47:715–741.

Padmos, P. and van Norren, D., 1975, Cone systems interaction in single neurons of the lateral geniculate nucleus of the macaque. Vision Res., 15:617–619.

Palmer, L. A., Jones, J. P. and Mullikin, W. H., 1985, Functional organization of simple receptive fields. In: "Models of the Visual Cortex", D. Rose and V. G. Dobson, eds., John Wiley & Sons, Chichester, pp. 273–280.

Palmer, L. A., Jones, J. P. and Stepnoski, R. A., 1991, Striate receptive fields as linear filters: characterization in two dimensions of space. In: "Visual and Visual Dysfunction, Vol. 4 The Neural Basis of Visual Function", A. G. Leventhal, ed., J. R. Cronly-Dillon, ser. ed., Houndmills, Macmillan Press, pp. 246–265.

Perry, V. H. and Cowey, A., 1981, The morphological correlates of X- and Y-like retinal ganglion cells in the retina of monkeys. Exp. Brain Res. 43:226–228.

Perry, V. H., Oehler, R. and Cowey, A., 1984, Retinal ganglion cells that project to the dorsal lateral geniculate nucleus in the macaque monkey. Neuroscience 12:1101–1123.

Pollen, D. A. and Ronner, S. F., 1981, Phase relationships between adjacent simple cells in the visual cortex. Science, 212:1409–1411.

Polyak, S. L., 1941, "The Vertebrate Retina." University of Chicago Press, Chicago.

Press, W. H., Teukolsky, S. A., Vetterling, W. T. and Flannery, B. P., 1992, "Numerical Recipes in C - The Art of Scientific Computing", Second Edition, Cambridge University Press, Cambridge.

Purpura, K., Kaplan, E. and Shapley, R. M., 1988, Background light and contrast gain of primate P and M retinal ganglion cells. Proc. Natl. Acad. Sci. U.S.A., 85:534–4537.

Purpura, K., Kaplan, E., Tranchina, D. and Shapley, R. M., 1990, Light adaptation in the primate retina: analysis of changes in gain and dynamics of monkey retinal ganglion cells. Vis. Neurosci., 4:75–93.

Rodieck, R. W., 1991, Which cells code for color? In: "NATO ASI Series A Life Sciences, Vol. 203 From Pigments to Perception - Advances in Understanding Visual Processes, A. Valberg and B. B. Lee, eds., Plenum Press, New York, pp. 83–93.

Rodieck, R. W., Binmoeller, K. F., Dineen, J., 1985, Parasol and midget ganglion cells of the human retina. J. Comp. Neurol., 233:115–132.

Rodieck, R. W., Brening, R. K. and Watanabe, M., 1993, The origin of parallel visual pathways. In: "Proceedings of the Retina Research Foundation Symposia, Vol. 5 Contrast Sensitivity", R. Shapley and D. Man-Kit Lam, eds., The MIT Press, Cambridge, Massachusetts, pp. 117–144.

Schiller, P. H. and Malpeli, J. G., 1978, Functional specificity of lateral geniculate nucleus laminae of the rhesus monkey. J. Neurophysiol., 41:788–797.

Schiller, P. H., Logothetis, N. K. and Charles, E. R., 1990a, Role of the color-opponent and broad-band channels in vision. Vis. Neurosci., 5:321–346.

Schiller, P. H., Logothetis, N. K. and Charles, E. R., 1990b, Functions of the colour-opponent and broad-band channels of the visual system. Nature (Lond.), 343:68–70.

Shapley, R. M., Kaplan, E. and Soodak, R., 1981, Spatial summation and contrast sensitivity of X and Y cells in the lateral geniculate nucleus. Nature, 292:543–545.

Shapley, R. M., Reid, R. C. and Kaplan, E., 1991, Receptive field structure of P and M cells in the monkey retina. In: "NATO ASI Series A Life Sciences, Vol. 203 From Pigments to Perception - Advances in Understanding Visual Processes", A. Valberg and B. B. Lee, eds., Plenum Press, New York, pp. 95–104.

Sherman, S. M., Wilson, J. R., Kaas, J. H. and Webb, S. W., 1976, X- and Y-cells in the dorsal lateral geniculate nucleus of the owl monkey (Aotus trivirgatus). Science, 192:475–477.

Shipp, S. and Zeki, S., 1985, Segregation of pathways leading from area V2 to areas V4 and V5 of macaque monkey visual cortex. Nature, 315:322–325.

Silveira, L. C. L., 1996, Joint entropy loci of M and P cells: a hypothesis for parallel processing in the primate visual system. Rev. Brasil. Biol., 56(Supl. 1):345–367.

Silveira, L. C. L. and Perry, V. H., 1991, The topography of magnocellular projecting ganglion cells (M-ganglion cells) in the primate retina. Neuroscience, 40:217–237.

Silveira, L. C. L., Yamada, E. S., Perry, V. H. and Picanço-Diniz, C. W., 1994, M and P retinal ganglion cells of diurnal and nocturnal New World monkeys. NeuroReport, 5:2077–2081.

Silveira, L. C. L., Lee, B. B., Yamada, E. S., Kremers, J. and Hunt, D. M., 1997, Post-receptoral mechanisms of colour vision in new world primates. Vision Res., submitted.

Stone, J., 1983, "Parallel Processing In The Visual System - The Classification of Retinal Ganglion Cells and its Impact on the Neurobiology of Vision." Plenun Press, New York.

Ungerleider, L. G. and Mishkin, M., 1982, Two cortical visual systems, pp. 549–586. In: "The Analysis of Visual Behavior", D. J. Ingle, R. J. W. Mansfield and M. S. Goodale (eds.). The MIT Press, Cambridge, Massachusetts.

Van Essen, D. C., Anderson, C. H. and Felleman, D. J., 1992, Information processing in the primate visual system: an integrated systems perspective. Science, 255:419–423.

Valberg, A., Lee, B. B., Kaiser, P. K. and Kremers J, 1992, Responses of macaque ganglion cells to movement of chromatic borders. J. Physiol. (Lond.), 458:579–602.

Wang, Y., Qi, X., Jing, X. and Yu, D., 1988, Extended Gabor function model and simulation of some characteristic curves of receptive field. Scientia Sinica B, 31: 1185–1194.

Wang, Y., Qi, X., Yao, G. and Wang, M., 1993, Neural wave representation in early vision. Science in China B, 36: 677–684.

Wässle, H. and Boycott, B. B., 1991, Functional architecture of the mammalian retina. Physiol. Rev., 71:447–480.

Wässle, H., Grünert, U., Martin, P. R. and Boycott, B. B., 1994, Immunocytochemical characterization and spatial distribution of midget bipolar cells in the macaque monkey retina. Vision Res., 34:561–579.

Watanabe, M. and Rodieck, R. W., 1989, Parasol and midget ganglion cells. J. Comp. Neurol., 289:434–454.

Weber, J. T., Huerta, M. F., Kaas, J. H. and Harting, J. K., 1983, The projections of the lateral geniculate nucleus of the squirrel monkey. Studies of the interlaminar zones and the S layers. J. Comp. Neurol., 213:135–145.

Webster, M. A. and De Valois, R. L., 1985, Relationship between spatial-frequency and orientation tuning of striate-cortex cells. J. Opt. Soc. Am. A, 2:1124–1132.

Wiesel, T. N. and Hubel, D. H., 1966, Spatial and chromatic interactions in the lateral geniculate body of the rhesus monkey. J. Neurophysiol., 29:1115–1156.

White, A. J. R., Wilder, H. D., Goodchild, A. K., Sefton, A. J., and Martin, P. R., 1997, The origin and distribution of cone opponent responses in the lateral geniculate nucleus of the marmoset (*Callithrix jacchus*). Eur. J. Neurosci., in press.

Yamada, E. S., Silveira, L. C. L. and Perry, V. II., 1996a, Morphology, dendritic field size, somal size, density and coverage of M and P retinal ganglion cells of dichromatic *Cebus* monkeys. Vis. Neurosci., 13:1011–1029.

Yamada, E. S., Silveira, L. C. L., Gomes, F. L. and Lee, B. B., 1996b, The retinal ganglion cell classes of New World primates. Rev. Brasil. Biol., 56(Supl. 1):381–396.

Yamada, E. S., Marshak, D. W., Silveira, L. C. L. and Casagrande, V. A., 1997, Morphology of P and M retinal ganglion cells of the bush baby *Galago garnetti*. Vision Res., submitted.

Yamada, E. S., Silveira, L. C. L., Perry, V. H. and Franco, E. C. S., 1998, Morphology and dendritic field size of M and P retinal ganglion cells of the owl monkey. J. Neurosci., submitted.

Yeh, T., Lee, B. B., Kremers, J., Cowing, J. A., Hunt, D. M., Martin, P. R., and Troy, J. B., 1995, Visual responses in the lateral geniculate nucleus of dichromatic and trichromatic marmosets (*Callithrix jacchus*). Journal of Neuroscience, 15:7892–7904.

Yoshioka, T., Levitt, J. B. and Lund, J. S., 1994, Independence and merger of thalamocortical channels within macaque monkey primary visual cortex: anatomy of interlaminar projections. Vis. Neurosci., 11:467–489.

11

TRANSIENT AND PERSISTENT Na$^+$, Ca^{2+}, AND MIXED-CATION CURRENTS IN RETINAL GANGLION CELLS

A. T. Ishida

Section of Neurobiology, Physiology, and Behavior
University of California
Davis, California 95616-8519

1. INTRODUCTION

Cascading photochemical, biochemical, electrophysiological, and synaptic events in non-spiking photoreceptors and interwoven arrays of interneurons drive vertebrate retinal ganglion cells to generate volleys of action potentials that encode properties of the visual field and images within it. The frequency and kinetics of spiking in these cells are determined extrinsically as well as intrinsically, by changes in the balance of excitatory and inhibitory synaptic inputs from bipolar and amacrine to ganglion cells; by the distribution, voltage-sensitivity, and kinetics of post-synaptic responses to these inputs; and by voltage-gated ion channels that enable retinal ganglion cells to spike. This information has not been easy to collect because proximal neurons of the retina are anatomically and electrically more complex than photoreceptors and the distal interneurons; because a greater variety of neurotransmitters mediate synaptic transmission from interneurons in the inner nuclear layer to retinal ganglion cells than from photoreceptors to these interneurons; and because several novel structures and mechanisms have had to be recognized and carefully fleshed-out.

A decade of voltage-clamp studies has now shown that retinal ganglion cells are packed with at least a dozen types of voltage-gated Na$^+$, K$^+$, Ca^{2+}, Cl$^-$, and mixed-cation currents (Table 1); that the voltage-sensitivities of these currents seem suitable for generating, shaping, and regulating spikes (see Ishida 1995); and that a few of these currents may subserve specific functions (e.g., Lukasiewicz and Werblin 1988; Fohlmeister et al. 1990; Kaneda and Kaneko 1991a; Tabata and Ishida 1996; Taschenberger and Grantyn 1995; Zhang et al. 1997). Such functions have not been—and are unlikely to be—simple to establish for several reasons. To begin with, it would hardly be possible to describe the function of these and other currents until the density, voltage-dependence, time-depend-

Development and Organization of the Retina, edited by Chalupa and Finlay.
Plenum Press, New York, 1998.

Table 1. Ion currents identified in voltage-clamped
retinal ganglion cells

Na⁺	low-threshold, persistent[1] high-threshold, transient[2,3,4,5]
Ca²⁺	low-threshold, transient[5,6,7,8,9] high-threshold, ω-CTx-GVIA-sensitive[6,8,9] high-threshold, ω-CTx-MVIIC-sensitive[7,8] high-threshold, ω-CTx-MVIID-sensitive[9] high-threshold, dihydropyridine-sensitive[6,7,8,9,10] conotoxin-agatoxin-dihydropyridine-resistant[11]
K⁺	outwardly rectifying, sustained[2,3,12] outwardly rectifying, rapidly inactivating[2,3,12] outwardly rectifying, slowly inactivating[3,13] slightly rectifying, slowly inactivating[3] Ca²⁺-activated[2,3]
Mixed-cation	I_h[14] leak[15] cGMP-activated[16]
Cl⁻	outwardly rectifying, non-decaying[17]

1: Hidaka and Ishida (1996), 2: Lipton and Tauck (1987), Lasater and Witkovsky (1989), 3: Lukasiewicz and Werblin (1988), 4: Barres et al. (1989), Kaneda and Kaneko (1991a), Skaliora et al. (1993), 5: Ishida (1991), 6: Karschin and Lipton (1989), Guenther et al. (1994), 7: Liu and Lasater (1994a), 8: Bindokas and Ishida (1996), 9: Tabata et al. (1996), 10: Kaneda and Kaneko (1991b), 11: Taschenberger and Grantyn (1995), 12: Skaliora et al. (1995), 13: Sucher and Lipton (1992), 14: Tabata and Ishida (1996), 15: Robinson et al. (1994), see also Tabata and Ishida (1996), 16: Ahmad et al. (1994), 17: Tabata and Ishida (1997). Sensitivities of these currents to modulation by neurotransmitters, neurotransmitter-activated mixed-cation and Cl⁻ currents, and ion fluxes across transporters or exchangers, are not listed.

ence, and modulation of distinct current components are clarified, without contamination by other currents.

This chapter summarizes recent findings about voltage-gated Na⁺, Ca²⁺, and mixed-cation currents in retinal ganglion cells, and addresses some of the problems alluded to above. The aim is two-fold, namely to (a) update measurements of the kinetics, current density, resistance to inactivation, and pharmacological profile of these currents, and (b) draw attention to components of all three of these currents that may transiently augment excitability of retinal ganglion cells at the termination of light- or dark-driven hyperpolarizations. These currents were measured under voltage-clamp, using whole-cell patch-clamp electrodes (see Sakmann and Neher 1995) and step-wise changes in membrane potential from a "holding" or "conditioning" value to a "test" value. Starting from holding potentials at which current had recovered from inactivation (see below), voltage jumps to various test potentials were used to measure the maximum amplitude, rate of current growth, and rate of decay (if any), at those test potentials. The dependence of these rates on membrane potential, and the relative amplitude of current components that differ in voltage-sensitivity and kinetics, were then studied by shifting membrane potential to various conditioning values just before the test polarizations. In all cases, the currents described below are either the difference between currents recorded in the presence and absence of permeant ions, the difference between currents recorded in the presence and absence of blocking agents, or currents after linear-leak subtraction and pharmacological isolation.

Four basic properties of these currents will be referred to, with the following meanings:

    a. *Activation threshold* will denote the test potential at which a whole-cell voltage-clamp current grows to a just-detectable amplitude. The threshold of current activated by depolarization will be taken as the most negative membrane potential at which whole-cell current can be detected. Conversely, the threshold of current activated by hyperpolarization will be taken as the most positive membrane potential at which whole-cell current can be detected.

    b. *Post-peak decay* will denote the decline (if any) in whole-cell current amplitude at a fixed test potential, from the peak amplitude at that voltage, without specifying or implying mechanisms that produce the reduction in current amplitude.

    c. *Steady-state inactivation* will be used more specifically to denote the time- and voltage-dependent loss, at a given conditioning potential, of current available for activation.

    d. *Deactivation* will denote the decline in probability that an ion channel opens or remains open, at membrane potentials where steady-state inactivation is minimal. This is typically monitored upon repolarization from a conditioning potential that activates the current, to a test potential equal to or near the holding potential.

2. Na⁺ CURRENT

It has long been known that the action potentials of retinal ganglion cells can be blocked by tetrodotoxin (TTX; Byzov et al. 1970), that optic nerve and ganglion cell somata bind Na⁺-channel-specific toxins and antibodies directed against Na⁺ channels (Tang et al. 1979; Sarthy et al. 1983; Wollner and Catterall 1986), and that depolarization under voltage clamp activates a regenerative inward Na⁺ current in these cells. This current has been found to have the following properties in retinal ganglion cells of goldfish, salamander, turtle, rat, and cat:

2.1. Activation Threshold

Activation threshold has been reported in most studies to be around −50 or −45 mV (Lipton and Tauck 1987; Lasater and Witkovsky 1989; Ishida 1991; Kaneda and Kaneko 1991a; Skaliora et al. 1993). This is near the value of "spike threshold"—i.e., the membrane potential at which the rate of depolarization abruptly increases during the rising phase of action potentials (see Baylor and Fettiplace 1979; Fohlmeister et al. 1990). However, Na⁺ current activation threshold values as positive as −31 mV have been reported (Barres et al. 1989). At face value, this suggests a large spread in activation threshold values—nearly as large as the difference between voltages that activate the smallest and largest inward Na⁺ currents in single cells. The spread in activation thresholds may actually be larger than previously thought, because we have recently found that TTX-sensitive Na⁺ current activates regeneratively at membrane potentials as negative as −65 mV in goldfish retinal ganglion cells (Hidaka and Ishida 1996). Presence of this low-threshold Na⁺ current in some ganglion cells (and not others) might account for the spread in previously reported threshold values.

2.2. Post-Peak Decay

The rate at which Na⁺ current decays after activation, during a maintained test depolarization, may be limited by the voltage dependence of the rate at which single-channel currents inactivate, or by the voltage-dependence of channel opening probability (see

Aldrich et al. 1983). Retinal ganglion cell Na⁺ currents have been reported to decay exponentially and completely at rates that hasten with depolarization between −40 and 0 mV. At 0 mV, an exponential decay rate constant of 0.5 msec can be fitted to the whole-cell current decline (Barres et al. 1988; Skaliora et al. 1993). We have measured the decay of whole-cell Na⁺ currents in goldfish retinal ganglion cells during test depolarizations between 30 and 300 msec in duration (Hidaka and Ishida 1996). We find that the decay of these currents is not exponential, but can be fit instead by an equation of the form:

$$I_{Na} = A_{fast} \exp(-t/\tau_{fast}) + A_{slow} \exp(-t/\tau_{slow}) + C \tag{1}$$

In reference to their macroscopic gating kinetics, the component fit by the term "τ_{fast}" can be referred to as a "transient" Na⁺ current, while the current fit by the term "C" is clearly "persistent". Na⁺ current decline of this shape was first reported in squid axon (Chandler and Meves 1970).

Between −40 and 0 mV, our measurements of τ_{fast} agree with the time constants reported by Barres et al. (1988) and by Skaliora et al. (1993). At membrane potentials where both transient and persistent Na⁺ current are maximum-amplitude (around −10 mV), the transient current is roughly two orders of magnitude larger than the persistent current in goldfish retinal ganglion cells (Hidaka and Ishida 1996). The persistent Na⁺ current is also no larger than 1% of the peak current, in numerous other tissues (squid axon, node of Ranvier, skeletal and cardiac muscle).

2.3. Steady-State Inactivation

The amount of Na⁺ current available for activation, upon depolarization from a given membrane potential, is expressed as a fraction (h_∞) of the maximum Na⁺ current that can be recorded from a given cell (Hodgkin and Huxley 1952a). This fraction ranges from unity to zero for currents that are completely inactivated by prolonged ($\geq$ 100 msec) conditioning depolarizations. The position of the Boltzmann distribution of these values as a function of voltage (commonly referred to as "h_∞ curves") is commonly given by the membrane potential that reduces the amplitude of Na⁺ current available for activation to 50% of the maximum value. For retinal ganglion cells, this value ("$V_{1/2}$") is around −50 mV (range: −55 mV to −40 mV). Two other values of interest may be read from these h_∞ curves. One is that the most negative conditioning potential that produces full steady-state inactivation of Na⁺ current in salamander, turtle, and cat retinal ganglion cells is −20 mV (Lukasiewicz and Werblin 1988; Lasater and Witkovsky 1989; Kaneda and Kaneko 1991a; Skaliora et al. 1993). The corresponding value for rat ranges from −30 to −10 mV (Lipton and Tauck 1987; Barres et al. 1989). A second result of interest is that the fraction of maximum Na⁺ current available for activation is roughly 0.9 at −70 mV, viz. near the most negative values of resting potential reported for ganglion cells (Wiesel 1959; Baylor and Fettiplace 1979; Coleman and Miller 1989).

We have found that a small fraction of the whole-cell Na⁺ current resists steady-state inactivation in fish retinal ganglion cells (Hidaka and Ishida 1996). In other words, we have found that most of the whole-cell Na⁺ current can be inactivated by conditioning depolarizations, and that $V_{1/2}$ is around −45 mV when measured as in previous studies. At the same time, our results differ from those cited above in that h_∞ does not fall to zero at any membrane potential. Thus, the whole-cell Na⁺ current includes components that are transient and persistent not only in terms of decay kinetics, but also in terms of steady-state inactivation. We have been unable to resolve from whole-cell recordings whether exclusively separate Na⁺ channel populations differ in their susceptibility to steady-state inactivation. This possibility is supported by the preferential staining of central grey matter over white matter by antibodies directed against a sodium channel subunit associated with non-inactivating kinetics (Westenbroek et al. 1989). An alternate possibility is that the gating kinetics of single Na⁺ channels can shift between inactivating and non-inactivating "modes" (Patlak and Ortiz 1985). We have found preliminary evidence consistent with the latter, in that elevation of cytoplasmic adenosine $3',5'$ monophosphate (cAMP) reduces the amplitude of the faster component Na⁺ current component, and augments the amplitude of the more slowly decaying component (Hidaka and Ishida 1995a).

2.4. Deactivation

For at least three reasons, little is known about Na⁺ current deactivation in retinal ganglion cells. First of all, there are no published descriptions of experiments designed to systematically measure it. Second, most of the current traces published for retinal ganglion cells have not included the decay of Na⁺ current at the termination of test depolarizations (i.e., Na⁺ 'tail' currents), or they are obscured by capacitive currents. Third, the few available traces of current recorded at the termination of test depolarizations in different preparations either do not show Na⁺ currents deactivating, or the apparent deactivation rates differ substantially. Cell-attached patch recordings from rat retinal ganglion cells—the Na⁺ current traces least marred by capacitive currents upon repolarization to very negative holding potentials—have shown no closures of open channels corresponding to a tail current (see the ensemble current records in Fig. 3 of Barres et al. 1989). Whole-cell recordings from goldfish retinal ganglion cells have shown that Na⁺ current deactivates rapidly, and that at least in some cells at –84 mV, the decay can be fit by an exponential time constant of less than 1 msec (Ishida, unpublished observations). More slowly-decaying whole-cell current can be seen in whole-cell recordings from rat retinal ganglion cells (see Fig. 4D of Lipton and Tauck 1987).

2.5. Function

The Na⁺ current measurements summarized above include three observations of functional interest. One is a substantial difference in the extent that activation and steady-state inactivation curves overlap -- from 1 mV in the measurements of Barres et al. (1989) to 30 mV or more in the measurements of Lipton and Tauck (1987), Lasater and Witkovsky (1989), Kaneda and Kaneko (1991a), and Skaliora et al. (1993). This overlap not only indicates the membrane potentials at which Na⁺ current can activate and not fully inactivate, but also the relative amplitude of such so-called "window" Na⁺ currents from the h_∞ value at which these curves intersect (cf. Attwell et al. 1982). The overlap of activation and steady-state inactivation curves may be large enough that the window current augments excitability. On the other hand, cells whose activation and inactivation curves do not overlap are likely to spike transiently.

Second, Na⁺ current density ranges between 100 and 300 pA/pF (Lipton and Tauck 1987; Barres et al. 1988; Hidaka and Ishida 1996) when measured in dissociated retinal ganglion cells with whole-cell patch electrodes. Current densities of these magnitudes could depolarize cells at rates equal to or greater than the maximum rates of "rise" seen in spikes recorded from various species (e.g. Baylor and Fettiplace 1979).

Third, these measurements give a rough estimate of the extent to which Na⁺ current inactivates during spike trains. $V_{1/2}$ values reported for retinal ganglion cells (between –50 and –40 mV) implies that 50% of the largest Na^+ current amplitudes recorded in these cells would be available for activation during volleys of action potentials—because these cells repolarize to membrane potentials around –45 mV between successive spikes (e.g. Baylor and Fettiplace 1979; Fohlmeister et al. 1990). It is not surprising to find that Na^+ current is available for repeated activation, because some retinal ganglion cells spike at sustained frequencies that reflect luminance (Sakmann and Creutzfeld 1969; Barlow and Levick 1969).

Because its activation threshold is around –65 mV, persistent Na^+ current would not be activated at (and thus not measurably contribute to) resting potentials more negative than –65 mV (Wiesel 1959; Baylor and Fettiplace 1979; Coleman and Miller 1989; Ishida 1991). However, persistent Na^+ current seems geared for recruitment under a variety of conditions. First, this current could be activated by `rebound' depolarizations at the termination of hyperpolarizations—i.e., depolarizations produced by inward currents that are activated by hyperpolarization (e.g., I_h—see below), or by inward currents that recover from inactivation during these hyperpolarizations (e.g., transient Ca^{2+} current—see below). Second, although we find that persistent Na^+ current amplitude is small in retinal ganglion cell somata under normal conditions (ca. 1 pA/pF), it is increased in amplitude by activation of serotonin receptors, and by elevation of intracellular cAMP (Hidaka and Ishida 1995b). This is consistent with the presence of cAMP-dependent phosphorylation sites in fish Na^+ channels (Emerick and Agnew 1989), and we suppose that (within limits) augmentation of persistent inward Na^+ current amplitude by serotonin will be excitatory. It is not yet known if neurotransmitters modulate voltage-gated Na^+ current in mammalian retinal ganglion cells. Recently, dopamine receptor antagonists have been found to reduce the amplitude of Na^+ current in rat retinal ganglion cells (Ito et al. 1996). Thirdly, because the activation threshold of Na^+ conductance is just slightly more positive than typical resting potentials, the positive slope of the resting leak conductance will tend to be offset by the negative slope due to regeneratively increasing Na^+ conductance (Hotson et al. 1979). Persistent, sub-threshold Na^+ current is so large in other central neurons that it can reduce the size of synaptic currents needed to influence excitability and it can generate repetitive spiking (Llinás and Sugimori 1980; Strafstrom et al. 1982; Magee and Johnston 1995; Stuart and Sakmann 1995; Schwindt and Crill 1995). Whether persistent Na^+ current affects dendritic integration and axonal excitability in retinal ganglion cells remains to be tested.

3. Ca^{2+} CURRENT

In addition to the Na^+ current described above, depolarization can activate Ca^{2+} current in retinal ganglion cells. The following properties of whole-cell Ca^{2+} current have been recognized in goldfish, turtle, rat, and cat retinal ganglion cells:

3.1. Activation Threshold

The total Ca^{2+} current recorded from most single retinal ganglion cells consists of a low-threshold component that can activate at relatively negative test potentials, and high-threshold current that begins to activate only at more positive test potentials. The activation thresholds for these currents differ by between 15 and 30 mV (Karschin and Lipton 1989; Ishida 1991; Liu and Lasater 1994a; Guenther et al. 1994; Bindokas and Ishida 1996). In extracellular solutions containing physiological levels of Ca^{2+} (2–2.5 mM), the most negative test potential that can activate a regenerative, whole-cell, voltage-clamp

Ca^{2+} current is around −65 mV (Ishida 1991; Kaneda and Kaneko 1991b; Bindokas and Ishida 1996). The activation threshold for high-threshold Ca^{2+} current under the same recording conditions is around −45 mV (e.g. Bindokas and Ishida 1996). In extracellular solutions containing elevated divalent concentrations (e.g. 10 mM Ca^{2+} or 10 mM Ba^{2+}), the activation threshold is near −50 mV for low-threshold current, and near −30 mV for high-threshold current (Karschin and Lipton 1989; Guenther et al. 1994; Liu and Lasater 1994a; Zhang et al. 1997) as if both were shifted positively by surface charge effects.

3.2. Decay, Inactivation, and Deactivation

In ruptured-patch whole-cell recordings with the pipet solution Ca^{2+} level buffered to less than 10 nM, low- and high-threshold Ca^{2+} currents decay at different rates and to different extents. Low-threshold Ca^{2+} current inactivates fully at conditioning potentials between −60 and −50 mV (Karschin and Lipton 1989; Guenther et al. 1994; Bindokas and Ishida 1996; Tabata et al. 1996), whereas 50% of the high-threshold Ca^{2+} current resists steady-state inactivation at conditioning potentials as positive as +35 mV (Bindokas and Ishida 1996). Low-threshold Ca^{2+} current decays exponentially and quickly (e.g. τ_{decay} around 10–30 msec at −55 mV), while maximum-amplitude high-threshold Ca^{2+} currents decay around 10-times more slowly (Bindokas and Ishida 1996).

High-threshold Ca^{2+} current reportedly decays faster when intracellular Ca^{2+} levels rise (Kaneda and Kaneko 1991b; Liu and Lasater 1994a). Particularly for spike modeling purposes, it would be of interest to compare Ca^{2+} current decay rates and the rate at which submembrane Ca^{2+} levels decline during sustained depolarizations *in situ*. Fura-2 fluorescence intensity measurements in K^{+}-depolarized ganglion cells indicate that high-threshold Ca^{2+} influx does not decay completely during depolarizations lasting minutes (Bindokas et al. 1994).

Rates of Ca^{2+} current deactivation have not been examined systematically in retinal ganglion cells. However, published records suggest that these rates are rapid at very negative potentials (Karschin and Lipton 1989; Taschenberger and Grantyn 1995; Bindokas and Ishida 1996). Upon repolarization from (a conditioning potential of) −44 mV to (a test potential of) −84 mV, Ca^{2+} tail currents decay exponentially in goldfish retinal ganglion cells, with a decay time constant typically measuring around 1–2 msec (Ishida, unpublished observations). Upon repolarization of these cells from 0 mV to test potentials between −90 and −70 mV, the decay of Ca^{2+} tail currents are fit better by the sum of two exponentials than by single exponentials. Good fits are achieved with one decay time constant that typically ranges between 1 and 3 msec, and a second time constant faster than 1 msec (Bindokas and Ishida, unpublished observations).

3.3. Conotoxin-Sensitivity

The voltage-clamp measurements described above demonstrate conditions under which different Ca^{2+} channels gate. These types of differences (particularly those in activation threshold and susceptibility to steady-state inactivation) have been used to distinguish Ca^{2+} currents in a wide variety of tissues (Hagiwara et al. 1975; Tsien et al. 1988). For reasons not yet known, no two studies have fully agreed on the subtypes of Ca^{2+} current in retinal ganglion cells. As few as one type—and as many as two, three, and four types—of Ca^{2+} channel have been recognized in single retinal ganglion cells (Karschin and Lipton 1989; Kaneda and Kaneko 1991b; Huang et al. 1994; Liu and Lasater 1994a; Rothe and Grantyn 1994; Taschenberger and Grantyn 1995; Bindokas and Ishida 1996;

Tabata et al. 1996; Zhang et al. 1997). Furthermore, the pharmacological properties of both low- and high-threshold Ca^{2+} channels have been found to vary substantially from species to species. High-threshold Ca^{2+} currents differ in susceptibility to block by different "conotoxins"—small-peptide toxins from fish-eating, marine snails of the genus *Conus*—and in the reversibility of block. For example, the conotoxin fraction ω-GVIA (referred to hereafter as "ω-CTx-GVIA") blocks high-threshold Ca^{2+} currents reversibly in some rat retinal ganglion cells (Karschin and Lipton 1989), blocks most of the high-threshold Ca^{2+} current irreversibly in fish retinal ganglion cells (Bindokas and Ishida 1996), and is without effect on a high-threshold Ca^{2+} current blocked by a different conotoxin (ω-CTx-MVIIC) in turtle retinal ganglion cells [Liu and Lipton 1994a; see Jensen (1995) and Tamura et al. (1995) for different types of evidence of a pharmacologically similar Ca^{2+} channel]. Low-threshold Ca^{2+} currents are blocked by ω-CTx-GVIA in rat (Karschin and Lipton 1989; Guenther et al. 1994), by dihydropyridine antagonists in turtle (Liu and Lasater 1994a), and by neither in fish (Bindokas and Ishida 1996).

The molecular basis of these differences has not yet been resolved in retinal ganglion cells. One possibility is that Ca^{2+}-channels in the species mentioned above bear different α_1 subunits. The pharmacological profiles in rat, fish, and turtle, differ as if a substantial fraction of their high-threshold Ca^{2+} channels incorporated α_{1D} subunits (those blocked by ω-CTx-GVIA reversibly: Williams et al. 1992a), α_{1B} subunits (those blocked by ω-CTx-GVIA irreversibly or with very slow reversibility; Williams et al. 1992b), and α_{1A} subunits (those blocked by ω-CTx-MVIIC, but not by ω-CTx-GVIA: Mori et al. 1991; Sather et al. 1993; Zhang et al. 1993), respectively. We checked these possibilities with a pallete of other pharmacological agents that included three toxin fractions from the funnel-web spider *Agelenopsis aperta* (ω-Aga-IA, ω-Aga-IIIA, ω-Aga-IVA), a toxin cloned from *Conus magus* (ω-CTx-MVIID; Monje et al. 1993), and Ni^{2+}. We found that the high-threshold Ca^{2+} current in fish retinal ganglion cells is resistant to block by ω-Aga-IVA, yet is markedly reduced in amplitude by ω-CTx-MVIID, ω-CTx-MVIIC, ω-CTx-GVIA, and Ni^{2+} (Bindokas and Ishida 1996; Tabata et al. 1996). This current thus differs from P-type Ca^{2+} current, Q-type Ca^{2+} current, and current through α_{1A}-subunit-bearing Ca^{2+} channels (Mori et al. 1991; Bertolino and Llinás 1992; Sather et al. 1993; Zhang et al. 1993). This current also differs from amphibian and mammalian N-type Ca^{2+} current, R-type Ca^{2+} current, and current through α_{1E}-subunit-bearing Ca^{2+} channels, as it is only slightly reduced in amplitude by ω-Aga-IIIA, and only partially inactivated by large depolarizations (Bindokas and Ishida 1996; cf. Mintz et al. 1991; Zhang et al. 1993; Soong et al. 1993). Finally, because it is inactivation-resistant, largely (if not entirely) blocked by ω-CTx-MVIIC, and irreversibly blocked by ω-CTx-GVIA, this current differs from that carried through doe-1-subunit-bearing Ca^{2+} channels (Ellinor et al. 1993). On the basis of these results, we have suggested that the conotoxin-sensitive Ca^{2+} current in fish retinal ganglion cells differs from the high-threshold Ca^{2+} currents types denoted N, P, Q, R, and doe-1 (see Bindokas and Ishida 1996; Tabata et al. 1996; Table 2).

3.4. Dihydropyridine-Sensitivity

A minor fraction (10–30% of the maximum amplitude) of the high-threshold Ca^{2+} current in adult retinal ganglion cells is dihydropyridine-sensitive. Three aspects of this dihydropyridine-sensitivity are known. First, BAY-K-8644 can augment the amplitude of Ca^{2+} current that resists block by saturating concentrations of conotoxin (Bindokas and Ishida 1996; Tabata et al. 1996). This result implies that the dihydropyridine-sensitive Ca^{2+} channels are physically distinct from conotoxin-sensitive Ca^{2+} channels (Aosaki and Kasai

Table 2. Ca²⁺ channel types (by functional designation and associated α_1 subunit) whose combined properties (voltage- and/or ligand-sensitivity) differ from those of conotoxin-sensitive Ca²⁺ current in goldfish retinal ganglion cells

Class	α_1 subunit	Voltage-sensitivity, ligands
T		low-threshold[1]
		fully inactivates @ −55 mV[2]
L	α_{1C}	augmented by BAY-K-8644[1,3,4]
		unaffected by ω-CTx-GVIA[3,4]
L	α_{1D}	augmented by BAY-K-8644[5]
		blocked reversibly by ω-CTx-GVIA[5]
N	α_{1B}	blocked irreversibly by ω-CTx-GVIA[6,7]
		$V_{1/2}$ between −70 and −60 mV[1,2,7]
P		blocked by *Agelenopsis* toxins[8]
		blocked by ω-CTx-MVIIC[9,10]
		non-inactivating[10,11]
Q	α_{1A}	not blocked by ω-CTx-GVIA[12,13]
		blocked by ω-CTx-MVIIC[4,9,12]
		blocked partially by ω-Aga-IIIA[4]
		blocked partially by ω-Aga-IVA[4]
		blocked by Ni[10,12] (however, see 13)
		not augmented by BAY-K-8644[4,12,13]
R	α_{1E}	not blocked by ω-CTx-GVIA[10,14]
		blocked by Ni[10,14]
		unaffected by BAY-K-8644[14]
		fully inactivates @ −30 mV[10,14]
doe-1	doe-1	reversibly blocked by ω-CTx-GVIA[15]
		blocked slightly by ω-CTx-MVIIC[15]
		blocked slightly by BAY-K-8644[15]
		fully inactivates @ 0 mV[15]

1: Nowycky et al. (1985), 2: Fox et al. (1987a), 3: Hess et al. (1984), Aosaki and Kasai (1989), Plummer et al. (1989), 4: Sather et al. (1993), 5: Williams et al. (1992a), 6: Hess (1990), 7: Williams et al. (1992b), 8: Bertolino and Llinás (1992), Mintz et al. (1992), 9: Hillyard et al. (1992), 10: Zhang et al. (1993), 11: Usowicz et al. (1992), 12: Stea et al. (1994), 13: Mori et al. (1991), 14: Soong et al. (1993), 15: Ellinor et al. (1993)

1989; Plummer et al. 1989), and more likely to bear α_{1C} subunits than α_{1D} subunits (Birnbaumer et al. 1994). A second effect characteristic of BAY-K-8644 is to slow the deactivation of Ca²⁺ currents—in whole-cell voltage-clamp records, to slow the rate of decay of Ca²⁺ tail currents—upon repolarization to negative membrane potentials. No obvious slowing of deactivation has been seen in retinal ganglion cells of any species upon repolarization to the very negative holding potentials typically used (−90 mV). However, BAY-K-8644 markedly slows the decay of retinal ganglion cell Ca²⁺ tail currents at less negative membrane potentials (e.g. −50 mV; see Tabata et al. 1996), as in photoreceptors (Wilkinson and Barnes 1996), bipolar cells (Kaneko et al. 1989), other central neurons (Plummer et al. 1989), peripheral neurons (Fox et al. 1987b), and cardiac myocytes (Hess et al. 1984). A third result reported as evidence that retinal ganglion cells possess dihydropyridine-sensitive Ca²⁺ channels is the reduction of whole-cell current amplitude by dihydropyridine Ca^{2+} channel "antagonists" (e.g., nifedipine, nimodipine, nisoldipine, and nicardipine). 10–15 μM nifedipine, for example, reduces high-threshold Ca^{2+} current in goldfish, salamander, and rat retinal ganglion cell somata by 10–30% of the control amplitude (Taschenberger and Grantyn 1995; Bindokas and Ishida 1996; Zhang et al. 1997). This sort of effect would be expected if ganglion cell Ca^{2+} channels were sensitive to BAY-K-8644 and other dihydropyridine Ca^{2+} channel "agonists". Moreover, nifedipine and ω-Aga-IIIA reduce total Ca^{2+} current in ganglion cells by similar amounts (Bindokas and Ishida 1996), as might be expected from block of L-type Ca^{2+} channels by ω-Aga-IIIA (Mintz et al. 1991).

At high concentrations, both dihydropyridine agonists and antagonists can produce side effects. At concentrations exceeding 5 μM, BAY-K-8644 has been found to reduce high-threshold Ca^{2+} current amplitude in cat and rat retinal ganglion cells (Kaneda and Kaneko 1991b; Taschenberger and Grantyn 1995). At concentrations exceeding 10 μM, nifedipine can reduce the amplitude of low-threshold Ca^{2+} current, conotoxin-sensitive Ca^{2+} current, and certain K^+ currents (see Akaike et al. 1989; Jones and Jacobs 1990; Grissmer et al. 1994). Thus, the fractions of whole-cell Ca^{2+} current suppressed by conotoxin and high concentrations of a dihydropyridine antagonist may exhibit order-dependence.

3.5. Function

The transience and voltage-sensitivity of low-threshold Ca^{2+} current enables certain central neurons to generate bursts of action potentials at the termination of hyperpolarizations (e.g. Llinás and Yarom 1981). However, a comparison of voltage-clamp data and spike records suggests that this might not be so in at least some retinal ganglion cells of at least three species. For example, cat and salamander ganglion cells generate transient flurries of spikes at the offset of hyperpolarizing stimuli (see Fig. 3 of Saito 1983; Fig. 6 of Belgum et al. 1984), yet neither species has been reported to possess a transient, low-threshold Ca^{2+} current (Lukasiewicz and Werblin 1988; Kaneda and Kaneko 1991b). Turtle ganglion cells possess a transient, low-threshold Ca^{2+} current (Liu and Lasater 1994a), yet bursts of spikes can be triggered by the offset of lights that reduce excitatory input to these cells but do not hyperpolarize them (Marchiafava 1976; Baylor and Fettiplace 1979). These three pairs of observations are consistent with at least one of several alternative possibilities: (a) that excitability at the termination of hyperpolarizations is augmented by other voltage-gated ion currents (Tabata and Ishida 1996), (b) that the release of excitatory synaptic transmitters onto ganglion cells is transient (Kaneko et al. 1989; von Gersdorff al. 1996), and (c) that responses to excitatory transmitters desensitize (Downing and Kaneko 1992; Lukasiewicz et al. 1995).

Immunoreactivity, voltage-clamp currents, and intracellular Ca^{2+} concentration increases provide evidence for high-threshold Ca^{2+} channels in proximal dendrites, somata, and axon terminals of retinal ganglion cells (e.g. Ahlijanian et al. 1990; Karschin and Lipton 1989; Ishida et al. 1991; Tabata and Ishida 1994; Taschenberger and Grantyn 1995). Three functions of high-threshold Ca^{2+} channels have been studied. The first—to support neurotransmitter release at axon terminals—entails activation of multiple types of Ca^{2+} current, including one that is conotoxin-, agatoxin-, and dihydropyridine-resistant (Taschenberger and Grantyn 1995). Whether high-threshold Ca^{2+} current supports transmitter release from somata (Huang and Neher 1996) or dendrites (Sakai et al. 1986) is not known. A second function—to foster processing of spatio-temporal information—is implied by the block of directional selectivity by conotoxin ω-CTx-MVIIC in rabbit retinal ganglion cells (Jensen 1995). How, and in what cells, this striking effect is mediated remain to be established. A third function—to foster excitability—is implied by the reduction of spike frequency by GABA$_B$ receptor agonists (Zhang et al. 1997) which reduce high-threshold Ca^{2+} current amplitudes (Bindokas and Ishida 1991; Zhang et al. 1997). Because high-threshold Ca^{2+} currents contribute to rises in intracellular Ca^{2+} concentration (Bindokas et al. 1994), because retinal ganglion cells possess Ca^{2+}-activated K$^+$ channels (Lipton and Tauck 1987), and because Co^{2+} broadens spikes and suppresses repetitive spiking (Fohlmeister et al. 1989; Liu and Lasater 1994a), GABA$_B$ receptor activation conceivably slows the rate of repolarization during the falling phase of invididual action potentials enough to exacerbate Na$^+$ current inactivation. The extent to which these steps contribute to inhibition, if any, remain to be measured in detail. Moreover, while this means of inhibition may be conspicuous in some ganglion cells, it may be small in cells that lack spike after-hyperpolarizations (Taschenberger and Grantyn 1995).

4. I$_H$

Inward current can be activated in retinal gangion cells not only by shifts in membrane potential from values near resting potential to more positive values (as described above), but also by shifts in membrane potential from values near resting potential to more negative values (viz., by hyperpolarization). Hyperpolarization-activated inward current has been measured in goldfish retinal ganglion cells (Tabata and Ishida 1996), and deduced on the basis of voltage-recordings from rat optic nerve (Eng et al. 1990). Hyperpolarization-activated current has presented the following properties.

4.1. Identification

The hyperpolarization-activated inward current in goldfish retinal ganglion cells resembles the current known as I$_h$, I$_q$, and I$_f$ in various preparations (see Brown et al. 1990). In terms of ion selectivity and susceptibility to block by pharmacological agents, this current differs from hyperpolarization-activated K$^+$ current (I$_K$: see Hagiwara and Takahashi 1974; Standen and Stanfield 1978) and hyperpolarization-activated Cl$^-$ current (I$_{Cl}$: see Chesnoy-Marchais 1982; Wilson and Gleason 1991). I$_{Cl}$ differs from both I$_h$ and I$_K$ in that is carried by anions, and in that it is resistant to block by extracellular Cs$^+$. Hyperpolarization-activated inward cationic current can be identified as I$_h$ rather than I$_K$ from its resistance to block by Ba^{2+} (at 1 mM) and by tetraethylammonium (as much as 30 mM); Cs$^+$ blocks I$_h$, but also blocks I$_K$. The reversal potential of I$_h$ depends on extracellular concentrations of Na$^+$ and K$^+$, whereas the reversal potential of I$_K$ depends on extracellular K$^+$ concentration. These results imply that I$_h$ is normally carried by Na$^+$ and K$^+$, while K$^+$ normally carries I$_K$.

In retinal ganglion cells, as in other tissues, I$_h$ appears to be carried somewhat more readily by K$^+$ ions than by Na$^+$ ions. However, the relative permeability to Na$^+$ and K$^+$ is difficult to specify from whole-cell records in two respects. First, the Na$^+$ permeability of channels that pass I$_h$ cannot be measured in bi-ionic conditions because I$_h$ ceases to flow in the absence of extracellular K$^+$ (Tabata and Ishida 1996; see also Wollmuth and Hille 1992; Frace et al. 1992). Second, the relative permeability of these channels to Na$^+$ and K$^+$ are difficult to assess in solutions containing 'physiological' concentrations of extracellular Na$^+$ (ca. 150 mM), because I$_h$ then reverses at voltages that activate other K$^+$ currents (Tabata and Ishida 1996).

4.2. Activation and Inactivation

In some respects, I_h looks like the outwardly-rectifying K^+ current modeled by Hodgkin and Huxley (1952b) with an inverted voltage-sensitivity: the rate at which I_h grows in amplitude increases at increasingly negative test potentials, the steady-state amplitude of I_h increases non-linearly with voltage, and I_h shows no post-peak decline in amplitude during test hyperpolarizations maintained for as long as 2 sec (Tabata and Ishida 1996).

I_h activates at membrane potentials more negative than -70 mV. Typically, I_h grows exponentially in amplitude at a fixed test potential, and apparent activation rate speeds up with more negative test potentials (e.g., time constants measuring 349 msec at -85 mV, versus 73 msec at -105 mV). Because I_h also deactivates at membrane potentials more positive than -75 mV, it is not surprising that Cs^{2+} does not measurably change resting potential or holding current (see Figures 4 and 5 of Tabata and Ishida 1996). If I_h activated at membrane potentials as positive as it does in photoreceptors (-50 mV), I_h would depolarize resting retinal ganglion cells to levels that inactivate substantial amounts of Na^+ current.

Even at test potentials as negative as -100 mV, I_h hardly exceeds 100 pA in amplitude (see Figs. 2F and 4E in Tabata and Ishida 1996). At membrane potentials where I_h is likely to activate in situ—between resting potential (-70 mV) and the K^+ equilibrium potential (-100 mV)—the current density of I_h is typically not more than 3 pA/pF (Tabata and Ishida 1996). I_h is thus strikingly smaller than the largest Na^+, Ca^{2+}, and K^+ currents that can be recorded in retinal ganglion cells: Na^+ currents reach 100–300 pA/pF (Barres et al. 1989; Hidaka and Ishida 1996); high-threshold Ca^{2+} currents in 2–2.5 mM external Ca^{2+} reach 20–80 pA/pF (Bindokas and Ishida 1996), and K^+ current amplitudes can exceed those of maximal Na^+ currents. I_h is also smaller in amplitude than inwardly-rectifying K^+ current in other retinal neurons (e.g., horizontal cells) even in terms of chord conductance (i.e., at test potentials differing from the respective reversal potential by comparable amounts). On the other hand, the current density of I_h is not so small in three different respects. First, it is similar in amplitude to that of I_h recorded in cones and bipolar cells. Second, its amplitude equals (or exceeds) that of persistent Na^+ current in retinal ganglion cells, and equals (or is slightly smaller than) that of low-threshold Ca^{2+} currents in these cells (Bindokas and Ishida 1996; Hidaka and Ishida 1996). Third, it is the only hyperpolarization-activated ion current found so far in retinal ganglion cells (Tabata and Ishida 1996). That is, block of I_h by Cs^+ or by removal of extracellular K^+ leaves a relatively small, linear, leak-type conductance (viz., currents that increase ohmically in amplitude with voltage, with no time-dependent change in amplitude at any voltage; see Fig. 2A of Tabata and Ishida 1996). This implies that I_h is the only voltage-gated current available to produce the delayed depolarization of ganglion cells—i.e., the decline in the amplitude of hyperpolarizations that develops—during injections of constant-amplitude, hyperpolarizing current (see Fig. 2 of Eng et al. 1990; Fig. 2 of Skaliora et al. 1993; Fig. 5 of Tabata and Ishida 1996).

4.3. Deactivation

In addition to producing voltage-rectification when activated in retinal ganglion cells, we wondered whether I_h might contribute to light- or dark-modulated spike trains as it deactivated in retinal ganglion cells. We found that I_h tail currents decay exponentially, and can be fit by time constants of 100 msec at test potentials between -75 mV and -55 mV (Tabata and Ishida 1996). The deactivation of I_h is thus roughly 50–100 times slower than that of voltage-gated Na^+ and Ca^{2+} currents (see above). It is also more than three-times slower than the deactivation of currents gated by GABA and glycine (see below), and it is roughly three-times slower than the post-peak decay of low-threshold Ca^{2+} current (cf. Bindokas and Ishida 1996). This result suggests that, at the termination of synaptic events that hyperpolarize retinal ganglion cells, I_h will decay so slowly that it continues to depolarize ganglion cells (toward spike threshold) after other currents deactivate or inactivate.

4.4. Function

We tested this possibility in two ways. First, we took advantage of the absence of other hyperpolarization-activated currents in retinal ganglion cells, and modeled the contribution of I_h to membrane potential changes elicited by changes in synaptic inputs—including receptive field center stimulation of off-center ganglion cells, receptive field surround stimulation of on-center ganglion cells, and hyperpolarizations generated in a push-pull manner. For these calculations, cell capacitance, leak conductance (after block by Cs^+), I_h activation, and I_h deactivation, were measured from individual cells (Tabata and Ishida 1996); glutamate- and GABA-activated currents were assumed to deactivate instantaneously because excitatory postsynaptic current decay rates (Taylor et al. 1995), the slow corner frequency of GABA-current noise power spectra (Ishida and Cohen 1988), GABAergic IPSC decay rates (Protti et al. 1997), and glycinergic IPSC decay rates (Mittman and Copenhagen 1985; Protti et al. 1997) are faster than the deactivation time constant of I_h; and the amplitude of glutamate- and GABA-activated currents were varied over an arbitrary but wide range (relative to the amplitude of I_h). These calculations indicate that although I_h is relatively small in amplitude, it can depolarize retinal ganglion cells to around −55 mV (see Fig. 7 of Tabata and Ishida 1996), and thus well to the activation thresholds of Na^+ and Ca^{2+} currents (Bindokas and Ishida 1996; Hidaka and Ishida 1996).

Secondly, we tested whether Cs^+ can block action potentials triggered by the termination of hyperpolarizations, because Cs^+ blocks I_h without effects on resting potential. Cs^+ blocked these spikes, at concentrations (3 mM) that block I_h (Tabata and Ishida 1996).

Because I_h is a non-synaptic current, cells equipped with it should display voltage rectification after the onset of any hyperpolarizing light configuration, and transient bursts of spikes should be triggered by terminating those lights. "Off" responses recorded from carp, catfish, frog, rabbit and cat retinal ganglion cells to four different stimuli are consistent with this expectation: spot-shaped receptive field center illumination, annular illumination of receptive field surrounds, oriented light bars, and broad-field illumination (Wiesel 1959; Tomita et al. 1961; Murakami and Shimoda 1977; Nelson et al. 1978; Saito 1983; Sakai and Naka 1990; Bloomfield 1994).

4.5. Distribution

I_h has been detected in retinal ganglion cells of some species, but not others. Neither the basis nor functional consequence of this difference is presently known. One might wonder if I_h were simply not expressed in certain species, or if other currents assume the function of I_h. A list (and Table 3) of hyperpolarization-activated currents found in retinal neurons suggests that, with one exception, the answers to these questions are both 'No'.

The presence in photoreceptors of current having the voltage-sensitivity and reversal potential of I_h was originally deduced in turtle cones (Baylor et al. 1974) and toad rods (Fain et al. 1978). I_h has been found in zebrafish cones (Fan and Yazulla 1996), salamander rods (Attwell and Wilson 1980; Bader et al. 1982; Hestrin 1987; Wollmuth and Hille 1992), salamander cones (Barnes and Hille 1989), *Xenopus* rods (Akopian and Witkovsky 1996a), lizard cones (Maricq and Korenbrot 1990), guinea pig rods (Dementis and

Table 3. Inwardly-rectifying mixed-cation,
Cl⁻, and K⁺ currents in retinal neurons

I_h	I_{Cl}	I_K
Rod		
salamander[1]		
Xenopus[2]		
guinea pig[3]		
Cone	**Cone**	
zebrafish[4]	chicken[8]	
salamander[5]		
lizard[6]		
monkey[7]		
		Horizontal cell
		goldfish[9]
		catfish[10]
		turtle[11]
		rabbit[12]
		cat[13]
Bipolar cell		
goldfish[14]		
bass[15]		
rat[16]		
Retinal ganglion cell		
goldfish[17]		
rat[18]		

1: Attwell and Wilson (1980), Bader et al. (1982), Hestrin (1987), Wollmuth and Hille (1992), 2: Akopian and Witkovsky (1996a), 3: Demontis and Cervetto (1994), 4: Fan and Yazulla (1996), 5: Attwell et al. (1982), Barnes and Hille (1989), 6: Maricq and Korenbrot (1990), 7: Yagi and MacLeish (1994), 8: Wilson and Gleason (1991), 9: Tachibana (1983), Shingai and Quandt (1986), Yagi and Kaneko (1988), 10: Schwartz (1987), Takahashi and Copenhagen (1995), Dong and Werblin (1995), 11: Golard et al. (1992), 12: Lohrke and Hofmann (1994), 13: Ueda et al. (1992), 14: Kaneko and Tachibana (1985), 15: Lasater (1988), 16: Karschin and Wässle (1990), 17: Tabata and Ishida (1996), 18: Eng et al. (1990)

Cervetto 1994), and monkey cones (Yagi and MacLeish 1994). Cl⁻ current with a voltage-sensitivity, reversal potential, and gating kinetics resembling that of I_h has been found in chicken cones (Wilson and Gleason 1991).

I_h has not been detected in any horizontal cells. Inward rectification is due to a K⁺ conductance in horizontal cells of goldfish (Tachibana 1983; Shingai and Quandt 1986; Yagi and Kaneko 1988), catfish (Schwartz 1987; Takahashi and Copenhagen 1995; Dong and Werblin 1995), turtle (Golard et al. 1992), rabbit (Lohrke and Hofmann 1994), cat (Ueda et al. 1992), and perhaps skate (Malchow et al. 1990). Inward rectification can be activated in salamander horizontal cells, but not at test potentials less negative than −100 mV (Gilbertson et al. 1991; Kamermans and Werblin 1992).

I_h has been found in bipolar cells of goldfish (Kaneko and Tachibana 1985), bass (Lasater 1988), and rat (Karschin and Wässle 1990). A Cs⁺-sensitive, slowly activating, inwardly rectifying, inward current has been found in skate (Malchow et al. 1991) and mouse (Kaneko et al. 1989) bipolar cells. A bit of inward rectification can be seen in current-voltage measurements from axolotl and salamander bipolar cells (Figs. 5A and 10A of Tessier-Lavigne et al. 1988; Fig. 3B of Lasansky 1992), but neither its ionic basis nor whether it is solely voltage-gated *in situ* are known. I$_h$ is either miniscule or absent in rabbit bipolar cells (Gillette and Dacheux 1995).

I$_h$ has been found in retinal ganglion cells of goldfish (Tabata and Ishida 1996), and appears to be present rat optic nerve (Eng et al. 1990). Inward rectification appears to be absent in salamander (Lukasiewicz and Werblin 1988; Coleman and Miller 1989) and turtle (Lasater and Witkovsky 1989; see also Fig. 2 of Marchiafava 1976) retinal ganglion cells.

Three patterns in the distribution of I$_h$ emerge from these results. First, I$_h$ is either present in photoreceptor, bipolar, and ganglion cells (as in teleost and rodent), *or* I$_h$ is present in photoreceptors but absent in ganglion cells (as in salamander and turtle). Second, I$_h$ is the only hyperpolarization-activated inward current in cells that possess it. For that matter, the results summarized here imply that if hyperpolarization activates any inwardly rectifying current in a retinal neuron, then that current is of a single type—either I$_h$, an anomalously rectifying K$^+$ current, or an anomalously rectifying Cl$^-$ current. (Note that inwardly rectifying K$^+$ current could not functionally substitute for I$_h$ in retinal ganglion cells under all conditions, because these currents begin to flow inwardly at membrane potentials differing by roughly 30 mV, and because their reversal potentials would fall 80 mV apart on opposite sides of resting potential.) Third, I$_h$ is absent in one cell class in all species examined so far (horizontal cells). These results imply that I$_h$ is available for activation in only certain cells and at different membrane potentials (due to differences in activation threshold). However, the current density of I$_h$ seems high enough to contribute to light-evoked responses in only some cells or under some conditions (see Attwell et al. 1982; Demontis and Cervetto 1994; Akopian and Witkovsky 1996a; Tabata and Ishida 1996).

5. SUMMARY

The results discussed in this chapter can be summarized by nine observations:

1. Na$^+$, Ca^{2+} and I$_h$ currents constitute most (if not all) of the voltage-gated inward currents that activate in goldfish retinal ganglion cells at membrane potentials between −110 mV and −45 mV. At resting potential (around −70 mV), none of these currents are activated to a measurable extent. However, Na$^+$ and Ca^{2+} currents can be activated by depolarization from resting potential; I$_h$ is activated by hyperpolarization from resting potential; and pharmacological blockade of Na$^+$, Ca^{2+}, and I$_h$ currents leaves an ohmic and time-independent `leak' conductance. We have recently found a voltage-activated, outwardly rectifying Cl$^-$ current in retinal ganglion cells, but have not yet resolved its activation threshold (Tabata and Ishida 1997).

2. The voltage sensitivity and kinetics of gating (activation, inactivation, and deactivation) of these currents endow single retinal ganglion cells with as many as three types of *transient* inward current, and three types of *persistent* inward current. Transient Na$^+$ current activates upon depolarization to membrane potentials more positive than −45 mV; transient Ca^{2+} current is activated by depolarizations to membrane potentials more positive than −65 mV; I$_h$ deactivates (and can therefore be treated as a `transient' inward current) at membrane potentials more positive than −80 mV. Persistent Na$^+$ current activates (and does not inactivate) at membrane potentials more positive than −65 mV; persistent Ca^{2+} current activates (and resists complete inactivation) at membrane potentials more positive than −45 mV; I$_h$ activates (and does not inactivate) at membrane potentials more negative than −70 mV. The activation ranges of these currents (i.e., the mem-

Table 4. Activation of transient and persistent
cation currents in goldfish retinal ganglion cells

Profile	Current	E_{hold} or $E_{condition}$ or $E_{activation}$	Recorded at E_m more:
Transient	Na^+	E_{hold} between -100 and -35 mV	positive than -55 mV
	Ca^{2+}	E_{hold} between -100 and -65 mV	positive than -65 mV
	I_h	E_{activ} more negative than -75 mV	positive than -80 mV
Persistent	Na^+	$E_{condition}$ as positive as $+35$ mV	positive than -65 mV
	Ca^{2+}	$E_{condition}$ as positive as $+35$ mV	positive than -45 mV
	I_h	E_{hold} between -80 and -40 mV	negative than -75 mV

brane potential over which they can be recorded) are summarized in Table 4. Note that currents can be activated by test polarizations *from* holding potentials within the range listed, or *after* termination of conditioning polarizations to membrane potentials within the range listed.

3. Of the current components listed in Table 4, transient Na^+ current has been found in all retinal ganglion cells described to date (those of goldfish, salamander, turtle, rat, and cat); low-threshold Ca^{2+} current has been described in goldfish, turtle, and rat retinal ganglion cells; and high-threshold Ca^{2+} current has been found in all retinal ganglion cells (and shown to include a dihydropyridine-sensitive component in goldfish, turtle, rat, and cat). Of the other currents listed, persistent Na^+ current might explain the hyperpolarization of mudpuppy and rat optic nerves by tetrodotoxin (Tang et al. 1979; Stys et al. 1993), and it has recently been found in dopaminergic amacrine cells of mouse retina (Feigenspann et al. 1997). I_h has been inferred from the Cs^+-sensitive, Na^+-dependent voltage rectification in rat optic nerve (e.g., Eng et al. 1990). Finally, most of the inactivation-resistant Ca^{2+} current that we find (Bindokas and Ishida 1996; Tabata et al. 1996) exhibits a sensitivity to pharmacological treatments (including Ni^{2+} ions and ω-toxins from *Conus geographus*, *Conus magus*, and *Agelenopsis aperta*) that has not reported in any other tissue, species, or expression system that we are aware of.

4. Voltage-gated Na^+ and Ca^{2+} currents differ *diametrically* in kinetics and voltage-sensitivity: Low-threshold Na^+ current is persistent, while low-threshold Ca^{2+} current inactivates rapidly and completely. High-threshold Na^+ current inactivates rapidly and completely, while at least half of the high-threshold Ca^{2+} current is inactivation-resistant. A further contrast is that fully inactivating Ca^{2+} current is one of the smallest types of current that can be recorded from ganglion cells, whereas fully inactivating Na^+ current is one of the largest currents that can be recorded from ganglion cells. A possible explanation for this latter difference may be that (a) the transient Na^+ current must be large enough to depolarize cells rapidly and prominently (at least to beyond 0 mV) even if cells do not repolarize to resting potential between successive spikes in a train (see Baylor and Fettiplace 1979; Taschenberger and Grantyn 1995), whereas (b) transient Ca^{2+} current (like I_h) seems geared to best depolarize ganglion cells whose membrane capacitance is small, only upon depolarization from membrane potentials more negative than resting potential, and only at membrane potentials where they would not be outweighed by voltage-gated outward currents (viz. at voltages more negative than -50 mV). Persistent Na^+ current may be small in amplitude for similar reasons, but these reasons are not so unique that the function of

persistent Na$^+$ current could not be subserved by a persistent (and low-threshold) Ca^{2+} current. The resistance of high-threshold Ca^{2+} current (Bindokas and Ishida 1996) and of Ca^{2+}-activated K$^+$ current (Lipton and Tauck 1987; Lukasiewicz and Werblin 1988; Lasater and Witkovsky 1989) to voltage-induced inactivation seem matched for recruitment during prolonged spike trains. However, the relative contributions of different high-threshold Ca^{2+} currents to Ca^{2+} and K$^+$ channel gating, and to spiking, remain to be quantified.

5. The three cation currents that activate at membrane potentials more negative than −50 mV (persistent Na$^+$ current, low-threshold Ca^{2+} current, and I$_h$ tail currents) are similar in amplitude, and in no case exceed a few hundred pA (Ishida 1991; Bindokas and Ishida 1996; Tabata and Ishida 1996; Hidaka and Ishida 1996). The possibility that these currents contribute to the generation of action potentials in retinal ganglion cells might seem remote, given that these currents are 1–2 orders of magnitude *smaller* than the largest transient Na$^+$ and outwardly rectifying K$^+$ currents that can be activated in retinal ganglion cells, and that spike trains triggered by injection of depolarizing current into resting cells can be modeled without them (Fohlmeister et al. 1990).

However, three observations suggest that these currents (persistent Na$^+$ current, low-threshold Ca^{2+} current, and I$_h$ tail currents) can contribute to retinal ganglion cell excitability at the termination of hyperpolarizations, i.e., under conditions different than those considered previously. First, the amplitude of these currents are equal to or larger than the amounts of exogenous current required to elicit spikes in retinal ganglion cells (Lipton and Tauck 1987; Lukasiewicz and Werblin 1988; Fohlmeister et al. 1990; Mobbs et al. 1992; Skaliora et al. 1993; Boos et al. 1993; Tabata and Ishida 1996; see also Baylor and Fettiplace 1979). Second, retinal ganglion cell capacitances are relatively small (8–30 pF per soma) and input resistance is large (1 GΩ, when measured in the absence of synaptic input, and in either K$^+$ or Cs$^+$-based solutions at voltages that do not activate K$^+$ currents). We have calculated that even one of these currents (I$_h$) can depolarize retinal ganglion cells to Na$^+$ and Ca^{2+} current activation threshold, at the termination of hyperpolarizations (Tabata and Ishida 1996). Third, we have found that pharmacological suppression of I$_h$ blocks spikes recorded at the termination of hyperpolarizations (Tabata and Ishida 1996).

6. Hyperpolarizations from resting potential (toward K$^+$ equilibrium potential, viz. from −70 mV toward −100 mV) *prime* currents that can transiently depolarize ganglion cells: I$_h$ deactivates at resting potential, and will activate during hyperpolarizations. Low-threshold Ca^{2+} current will substantially inactivate at resting potential, and will recover from inactivation during hyperpolarizations. Transient Na$^+$ current will partially inactivate at resting potential, and will also recover from inactivation during hyperpolarization.

7. At the termination of large hyperpolarizations, I$_h$, Ca^{2+} current, and Na$^+$ current will activate in a sequential *cascade*—I$_h$ will speed repolarization toward the activation thresholds of transient Ca^{2+} current and of persistent Na$^+$ current; any or all three of these currents will tend to depolarize cells to the activation threshold of persistent Ca^{2+} current and transient Na$^+$ current.

8. I$_h$ and low-threshold Ca^{2+} current will tend to produce depolarizations that are, in effect, "*excitatory non-synaptic potentials*". These types of currents could depolarize ganglion cells beyond resting potential and toward spike threshold, because the reversal potentials of I$_h$ and low-threshold Ca^{2+} current are more

positive than resting potential. These depolarizations will be transient, because I_h will deactivate and low-threshold Ca^{2+} current will inactivate. The transience of these depolarizations can produce transience in spiking at the termination of hyperpolarizations—as loss of even one of these currents by other means (Cs^+ block of I_h) can block spike bursts initiated at the termination of hyperpolarizing current pulses (Tabata and Ishida 1996). This transience serves to *"high-pass filter"* retinal ganglion cell responses to light, as it augments responses to the moment at which hyperpolarizing light stimuli terminate (and not to the ambient light level thereafter).

9. Due to recovery from partial steady-state inactivation, larger Na^+ currents can be activated in ganglion cells by depolarizations from -100 mV than from -65 mV. This can account for classical *anode-break* excitation (Hodgkin and Huxley 1952b), and should contribute to anode-break responses of retinal ganglion cells. However, membrane conductance (and therefore membrane potential) will not change during recovery from inactivation alone. The "off" responses of certain retinal ganglion cells—the bursts of action potentials that follow termination of hyperpolarizations in these cells—differ from anode-break excitation in two ways. First, these hyperpolarizations "sag" before spikes are triggered by their termination—for example, see Fig. 1 of Tomita et al. (1961), Fig. 3 of Saito (1983), Fig. 2A of Skaliora et al. (1993), Fig. 2B of Murakami and Shimoda (1977), Fig. 7 of Sakai and Naka (1990), and Fig. 5C of Tabata and Ishida (1996). Second, the spike bursts are blocked by Cs^+ (Tabata and Ishida 1996), whereas anode-break excitation does not entail activation of a Cs^+-sensitive inward current.

6. NEXT

Patch-clamp measurements have demonstrated the presence of more than ten different types of voltage-gated ion channel in retinal ganglion cells [see Ishida (1995) and Table 1]. Most of what is known about ion channels in retinal ganglion cells has been measured in isolated somata or somatic membrane patches. This has permitted voltage-sensitivity, kinetics, ion selectivity, and pharmacology to be studied, while minimizing the possibility of spurious currents due to insufficient space-clamp (see, for example, White et al. 1995).

Because only one of the currents identified in retinal ganglion cells so far (I_h) is activated by hyperpolarization, it has been possible to model and test a contribution of this current to excitability in these cells (Tabata and Ishida 1996). By contrast, at least a half dozen types of current have been found to be activated by depolarization in individual retinal ganglion cells of every species examined to date (Ishida 1995). Although the voltage-sensitivities of activation and inactivation, charge carriers, and block by pharmacological agents, have been measured for many of these currents, the contribution of most of these currents to spiking patterns in intact cells remain to be unraveled. Two major efforts will now be required even to develop a single-compartment model of spike firing in retinal ganglion cells:

First, gating kinetics and current densities will be required to develop quantitative, single-compartment models of spiking (cf. Hodgkin and Huxley 1952b). It is now known that Na^+, K^+, and Ca^{2+} ions can carry transient inward and outward currents, persistent inward and outward currents, and both low-threshold and high-threshold components, in retinal ganglion cells. However, the relative amplitude and precise timing of all of these currents are not yet known for any single cell-type.

Second, the malleability of these currents under several conditions will have to be tested, because recent studies have shown that the amplitude and kinetics of voltage-gated Na^+, Ca^{2+}, K^+, and Cl^- currents can be modulated by neurotransmitters (glutamate, dopamine, serotonin, and GABA), elevated intracellular Ca^{2+}, activation of protein kinase A or protein kinase C, and sustained depolarization (Lipton and Tauck 1987; Lukasiewicz and Werblin 1988; Bindokas and Ishida 1991, 1996; Kaneda and Kaneko 1991a,b; Liu and Lasater 1994a,b; Guenther et al. 1994; Rothe et al. 1994; Hidaka and Ishida 1995a,b; Akopian and Witkovsky 1996b; Zhang et al. 1997; Tabata and Ishida 1997). Once the rate, extent, and channel-specificity of these effects are known, re-calculating spike-shape and spike trains will be necessary to either confirm or reformulate ideas about the function of specific currents *in situ*.

When these hurdles are surpassed, powerful means will be in hand to analyze and mechanistically describe the generation and control of action potentials in intact retinal ganglion cells.

ACKNOWLEDGMENTS

The author thanks Vytautas Bindokas, Toshihide Tabata, Soh Hidaka, Gloria Partida and Ming-Hsing Cheng for measurements that form the core of this review, and also gratefully acknowledges support by NIH grant EY 08120.

REFERENCES

Ahlijanian M.K., R.E. Westenbroek, and W.A. Catterall (1990) Subunit structure and localization of dihydropyridine-sensitive calcium channels in mammalian brain, spinal cord, and retina. Neuron *4*:819–832.

Ahmad I., T. Leinders-Zufall, J.D. Kocsis, G.M. Shepherd, F. Zufall, and C.J. Barnstable (1994) Retinal ganglion cells express a cGMP-gated cation conductance activatable by nitric oxide donors. Neuron *12*:155–165.

Akaike N., P.G. Kostyuk, and Y.V. Osipchuk (1989) Dihydropyridine-sensitive low-threshold calcium channels in isolated rat hypothalamic neurones. J. Physiol. *412*:181–195.

Akopian A., and P. Witkovsky (1996a) D2 dopamine receptor-mediated inhibition of a hyperpolarization-activated current in rod photoreceptors. J. Neurophysiol. *76*:1828–1835.

Akopian A., and P. Witkovsky (1996b) Activation of metabotropic glutamate receptors decreases a high-threshold calcium current in spiking nuerons of the *Xenopus* retina. Visual Neurosci. *13*: 549–557.

Aldrich R.W., D.P. Corey, and C.F. Stevens (1983) A reinterpretation of mammalian sodium channel gating based on single channel recording. Nature *306*:436–441.

Aosaki, T., and H. Kasai (1989) Characterization of two kinds of high-voltage activated calcium channels in chick sensory neurones. Pflügers Arch. *414*:150–156.

Attwell D., I. Cohen, D. Eisner, M. Ohba, and C. Ojeda (1979) The steady state TTX-sensitive ("window") sodium current in cardiac Purkinje fibres. Pflügers Arch. *379*:137–142.

Attwell D., F.S. Werblin, and M. Wilson (1982) The properties of single cones isolated from the tiger salamander retina. J. Physiol. *328*:259–283.

Attwell D., and M. Wilson (1980) Behaviour of the rod network in the tiger salamander retina mediated by membrane properties of individual rods. J. Physiol. *309*:287–315.

Bader C.R., D. Bertrand, and E.A. Schwartz (1982) Voltage-activated and calcium-activated currents studied in solitary rod inner segments from the salamander retina. J. Physiol. *331*:253–284.

Barlow H.B., and W.R. Levick (1969) Changes in the maintained discharge with adaptation level in the cat retina. J. Physiol. (Lond.) *202*:699–718.

Barnes S., and B. Hille (1989) Ionic channels of the inner segment of tiger salamander cone photoreceptors. J. Gen. Physiol. *94*:719–743.

Barres B.A., L.L.Y. Chun, and D.P. Corey (1989) Glial and neuronal forms of the voltage-dependent sodium channel: Characteristics and cell-type distribution. Neuron 2:1375–1386.

Barres B.A., B.E. Silverstein, D.P. Corey, and L.L.Y. Chun (1988) Immunological, morphological, and electrophysiological variation among retinal ganglion cells purified by panning. Neuron 1:791–803.

Baylor D.A., and R. Fettiplace (1979) Synaptic drive and impulse generation in ganglion cells of turtle retina. J. Physiol. Lond. 288:107–127.

Baylor D.A., A.L. Hodgkin, and T.D. Lamb (1974) The electrical response of turtle cones to flashes and steps of light. J. Physiol. Lond. 242:685–727.

Belgum J.H., D.R. Dvorak, and J.S. McReynolds (1984) Strychnine blocks transient but not sustained inhibition in mudpuppy retinal ganglion cells. J. Physiol. Lond. 354:273–286.

Bertolino M., and R. Llinás (1992) The central role of voltage-activated and receptor-operated calcium channels in neuronal cells. Ann. Rev. Pharmacol. Toxicol. 32:399–421.

Bezanilla F., and C.M. Armstrong (1977) Inactivation of the sodium channel. I. Sodium current experiments. J. Gen. Physiol. 70:549–566.

Bindokas, V.P., and A.T. Ishida (1991) (-)Baclofen and γ-aminobutyric acid inhibit calcium currents in isolated retinal ganglion cells. Proc. Natl. Acad. Sci. USA 88:10759–10763.

Bindokas, V.P., and A.T. Ishida (1996) Conotoxin-sensitive and conotoxin-resistant Ca^{2+} currents in fish retinal ganglion cells. J. Neurobiol. 29:429–444.

Bindokas, V.P., M. Yoshikawa, and A.T. Ishida (1994) Na^+-Ca^{2+} exchanger-like immunoreactivity and regulation of intracellular Ca^{2+} levels in fish retinal ganglion cells. Journal of Neurophysiology 72:47–55.

Birnbaumer, L., K.P. Campbell, W.A. Catterall, M.M. Harpold, F. Hofmann, W.A. Horne, Y. Mori, A. Schwartz, T.P. Snutch, T. Tanabe, and R.W. Tsien (1994). The naming of voltage-gated calcium channels. Neuron 13:505–506.

Bloomfield, S.A. (1994) Orientation-sensitive amacrine and ganglion cells in the rabbit retina. J. Neurophysiol. 71:1672–1691.

Boos R., H. Schneider, and H. Wässle (1993) Voltage- and transmitter-gated currents of AII amacrine cells in a slice preparation of the rat retina. J. Neurosci. 13:2874–2888.

Brown D.A., B.H. Gahwiler, W.H. Griffith, and J.V. Halliwell (1990) Membrane currents in hippocampal neurons. Progr. Brain Res. 83:141–160

Byzov A.L., N.A. Polishchuk, and G.M. Zenkin (1970) On the transmission of signals in vertebrate retina in the presence and absence of impulses. Neirofiziol. 2:536–543.

Chandler W.K., and H. Meves (1970) Slow changes in membrane permeability and long-lasting action potentials in axons perfused with fluoride solutions. J. Physiol. Lond. 211:707–728.

Chesnoy-Marchais, D. (1982) A Cl^- conductance activated by hyperpolarization in Aplysia neurons. Nature 299:359–361.

Chiu, S.Y., J.M. Ritchie, R.B. Bogart, and D. Stagg (1979) A quantitative description of membrane currents in rabbit myelinated nerve. J. Physiol. 292:149–166.

Coleman, P.A. and R.F. Miller (1989) Measurements of passive membrane parameters with whole-cell recording from neurons in the intact amphibian retina. J. Neurophysiol. 61:218–230.

Demontis, G.C. and L. Cervetto (1995) Hyperpolarization activated current (I_h) and is modulation in mammalian retinal rods. Biophys. J. 68:23.

Diamond, J.S., and D.R. Copenhagen (1993) The contribution of NMDA and non-NMDA receptors to the light-evoked input-output characteristics of retinal ganglion cells. Neuron 11: 725–738.

Dong C.-J., and F.S. Werblin (1995) Inwardly rectifying potassium conductance can accelerate the hyperpolarizing response in retinal horizontal cells. J. Neurophysiol. 74:2258–2265.

Downing, J.E.G., and A. Kaneko (1992) Cat retinal ganglion cells show transient responses to acetylcholine and sustained responses to L-glutamate. Neurosci. Lett. 137:114–118.

Ellinor, P.T., J.-F. Zhang, A.D. Randall, M. Zhou, T.L. Schwarz, R.W. Tsien, and W.A. Horne (1993) Functional expression of a rapidly inactivating neuronal calcium channel. Nature 363:455–458.

Emerick, M.C., and W.S. Agnew (1989) Identification of phosphorylation sites for cAMP dependent protein kinase on the voltage-gated sodium channel from Electrophorus electricus. Biochem. 28:8367–8380.

Eng, D.L., T.R. Gordon, J.D. Kocsis, and S.G. Waxman (1990) Current-clamp analysis of a time-dependent rectification in rat optic nerve. J. Physiol. 421:185–202.

Fain G.L., F.N. Quandt, B.L. Bastian, and H.M. Gerschenfeld (1978) Contribution of a caesium-sensitive conductance increase to the rod photoresponse. Nature 272:446–449.

Fan S.-F., and S. Yazulla (1997) Electrogenic hyperpolarization-elicited chloride transporter current in blue cones of zebrafish retinal slices. J. Neurophysiol. 77:1447–1459.

Feigenspan, A., S. Gustincich, and E. Raviola (1997) Spontaneous activity of solitary mouse dopaminergic amacrine cells. Investig. Ophthal. Vis. Sci. 38:s709.

Fohlmeister, J.F., P.A. Coleman, and R.F. Miller (1990) Modeling the repetitive firing of retinal ganglion cells. Brain Res. *510*:343–345.

Fox, A.P., M.C. Nowycky, and R.W. Tsien (1987a) Kinetic and pharmacological properties distinguishing three types of calcium currents in chick sensory neurones. J. Physiol. *394*: 149–172.

Fox, A.P., M.C. Nowycky, and R.W. Tsien (1987b) Single-channel recordings of three types of calcium channels in chick sensory neurones. J. Physiol. *394*: 173–200.

Frace, A.M., F. Maruoka, and A. Noma (1992) External K⁺ increases Na⁺ conductance of the hyperpolarization-activated current in rabbit cardiac pacemaker cells. Pflügers Arch. *421*:97–99.

Gilbertson, T.A., S. Borges, and M. Wilson (1991) The effects of glycine and GABA on isolated horizontal cells from the salamander retina. J. Neurophysiol. *66*:2002–2013.

Gillette, M.A. and R.F. Dacheux (1995) GABA- and glycine-activated currents in the rod bipolar cell of the rabbit retina. J. Neurophysiol. *74*:856–875.

Golard, A., P. Witkovsky, and D. Tranchina (1992) Membrane currents of horizontal cells isolated from turtle retina. J. Neurophysiol. *68*:351–361.

Grissmer, S., A.N. Nguyen, J. Aiyar, D.C. Hanson, R.J. Mather, G.A. Gutman, M.J. Karmilowicz, D.D. Auperin, K.G. Chandy (1994) Pharmacological characterization of five cloned voltage-gated K⁺ channels, types Kv1.1, 1.2, 1.3, 1.5, and 3.1, stably expressed in mammalian cell lines. Molec. Pharmacol. *45*:1227–1234.

Guenther, E., T. Rothe, H. Taschenberger, and R. Grantyn (1994) Separation of calcium currents in retinal ganglion cells from postnatal rat. Brain. Res. *633*:223–235.

Hagiwara, S., S. Ozawa, and O. Sand (1975) Voltage clamp analysis of two inward current mechanisms in the egg cell membrane of a starfish. J. Gen. Physiol. *65*:617–644.

Hagiwara, S. and K. Takahashi (1974) The anomalous rectification and cation selectivity of the membrane of a starfish egg cell. J. Membr. Biol. *18*:61–80.

Hess, P. (1990) Calcium channels in vertebrate cells. Annu. Rev. Neurosci. 13: 337–356.

Hess, P., J.B. Lansman, and R. W. Tsien (1984) Different modes of Ca channel gating behaviour favoured by dihydropyridine Ca agonists and antagonists. Nature *311*:538–544.

Hestrin, S. (1987) The properties and function of inward rectification in rod photoreceptors of the tiger salamander. J. Physiol. *390*:319–333.

Hidaka, S. and A.T. Ishida (1995a) Intracellular modulation of sodium channels in goldfish retinal ganglion cells. Invest. Ophthalmol. Vis. Sci. *36*:623.

Hidaka, S. and A.T. Ishida (1995b) Modulation of voltage-gated Na⁺ currents in retinal ganglion cells by serotonin receptor activation. Soc. Neurosci. Abstr. *21*:1035.

Hidaka, S. and A.T. Ishida (1996) Non-inactivating, tetrodotoxin-sensitive sodium current in goldfish retinal ganglion cells. Invest. Ophthalmol. Vis. Sci. *37*: 1154

Hillyard, D.R., V.D. Monje, I.M. Mintz, B.P. Bean, L. Nadasdi, J. Ramachandran, G. Miljanich, A. Azimi-Zoonooz, J.R. McIntosh, L.J. Cruz, J.S. Imperial, and B.M. Olivera (1992) A new *Conus* peptide ligand for mammalian presynaptic Ca²⁺ channels. Neuron *9*:69–77.

Hodgkin, A.L., and A.F. Huxley (1952a) The dual effect of membrane potential on sodium conductance in the giant axon of *Loligo*. J. Physiol. Lond. *116*:497–506.

Hodgkin, A.L., and A.F. Huxley (1952b) A quantitative description of membrane current and its application to conduction and excitation in nerve. J. Physiol. Lond. *117*:500–544.

Hotson, J.R., D.A. Prince, and P.A. Schwartzkroin (1979) Anomalous inward rectification in hippocampal neurons. J. Neurophysiol. *42*:889–895.

Huang, L.-Y.M. and E. Neher (1996) Ca²⁺-dependent exocytosis in the somata of dorsal root ganglion neurons. Neuron *17*:135–145.

Huang, S.-J., D.W. Robinson, R.P. Scobey, and L.M. Chalupa (1994) Voltage-gated calcium channels in developing cat retinal ganglion cells. Soc. Neurosci. Abstr. *20*:900.

Ishida, A.T. (1991) Regenerative sodium and calcium currents in goldfish retinal ganglion cell somata. Vision Res. *31*:477–485.

Ishida, A.T. (1995) Ion channel components of retinal ganglion cells. Prog. Retinal and Eye Res. *15*:261–280.

Ishida, A.T., V.P. Bindokas, and R. Nuccitelli (1991) Calcium ion levels in resting and depolarized goldfish retinal ganglion cell somata and growth cones. J. Neurophysiol. *65*:968–979.

Ishida, A.T., and B.N. Cohen (1988) GABA-activated whole-cell currents in isolated retinal ganglion cells. J. Neurophysiol. *60*:381–396.

Ito, K., Y. Nishimura, Y. Uji, K. Nakano, T. Yamamoto (1996) Effects of haloperidol on transient sodium current in the retinal ganglion cell of young rats. Soc. Neurosci. Abstr. *22*:58.

Jensen, R.J. (1995) Effects of Ca²⁺ channel blockers on directional selectivity of rabbit retinal ganglion cells. J. Neurophysiol. *74*:12–23.

Jones, S.W., and L.S. Jacobs (1990). Dihydropyridine actions on calcium currents of frog sympathetic neurons. J. Neurosci. *10*:2261–2267.

Kamermans, M., and F.S. Werblin (1992) GABA-mediated positive autofeedback loop controls horizontal cell kinetics in tiger salamander retina. J. Neurosci. *12*:2451–2463.

Kaneda, M., and A. Kaneko (1991a) Voltage-gated sodium currents in isolated retinal ganglion cells of the cat: relation between the inactivation kinetics and the cell type. Neurosci. Res. *11*: 261–275.

Kaneda, M., and A. Kaneko (1991b). Voltage-gated calcium currents in isolated retinal ganglion cells of the cat. Jpn. J. Physiol. *41*:35–48.

Kaneko, A., L.H. Pinto, and M. Tachibana (1989) Transient calcium current of retinal bipolar cells of the mouse. J. Physiol. *410*:613–629.

Kaneko, A. and Tachibana M. (1985) A voltage-clamp analysis of membrane currents in solitary bipolar cells dissociated from Carassius auratus. J. Physiol. 358: 131–152.

Karschin, A., and Lipton, S. A. (1989). Calcium channels in solitary retinal ganglion cells from post-natal rat. J. Physiol. *418*:379–396.

Karschin, A., and Wässle H. (1990) Voltage- and transmitter-gated currents in isolated rod bipolar cells of rat retina. J. Neurophysiol. *63*:860–876.

Lasansky, A. (1992) Properties of depolarizing bipolar cell responses to central illumination in salamander retinal slices. Brain Res. *576*:181–196.

Lasater, E.M. (1988) Membrane currents of retinal bipolar cells in culture. J. Neurophysiol. *60*:1460–1480.

Lasater, E.M., and P. Witkovsky (1989) Membrane currents of spiking cells isolated from turtle retina. J. Comp. Physiol. *167A*:11–21.

Lipton, S.A., and D.L. Tauck (1987) Voltage-dependent conductances of solitary ganglion cells dissociated from the rat retina. J. Physiol. *385*:361–391.

Liu, Y., and E.M. Lasater (1994a) Calcium currents in turtle retinal ganglion cells. I. The properties of T- and L-type currents. J. Neurophysiol. *71*:733–742.

Liu, Y., and E.M. Lasater (1994b) Calcium currents in turtle retinal ganglion cells. II. Dopamine modulation via a cyclic AMP-dependent mechanism. J. Neurophysiol. *71*: 743–752.

Llinás, R., and M. Sugimori (1980) Electrophysiological properties of in vitro Purkinje cell somata in mammalian cerebellar slices. J. Physiol. *305*:171–195.

Llinás, R., and Y. Yarom (1981) Properties and distribution of ionic conductances generating electroresponsiveness of mammalian inferior olivary neurones in vitro. J. Physiol. *315*:569–584.

Lohrke, S., and H.D. Hofmann (1994) Voltage-gated currents of rabbit A- and B-type horizontal cells in retinal monolayer cultures. Visual Neurosci. *11*:369–378.

Lukasiewicz, P.D., J.E. Lawrence, and T.L. Valentino (1995) Desensitizing glutamate receptors shape excitatory synaptic inputs to tiger salamander retinal ganglion cells. J. Neurosci. *15*:6189–6199.

Lukasiewicz, P.D., F.S. and Werblin (1988) A slowly inactivating potassium current truncates spike activity in ganglion cells of the tiger salamander retina. J. Neurosci. *8*:4470–4481.

Magee, J.C., and D. Johnston (1995) Characterization of single voltage-gated Na^+ and Ca^{2+} channels in apical dendrites of rat CA1 pyramidal neurons. J. Physiol. *481*:67–90.

Malchow R.P., R.L. Chappell, and H. Ripps (1991) Voltage- and ligand-gated conductances of bipolar cells from the skate retina. Biol. Bull. *181*:323–324.

Malchow R.P., H. Qian, H. Ripps, and J.E. Dowling (1990) Structural and functional properties of two types of horizontal cell in the skate retina. J. Gen. Physiol. *95*:177–198.

Marchiafava, P.L. (1976) Centrifugal actions on amacrine and ganglion cells in the retina of the turtle. J. Physiol. *255*:137–155.

Maricq, A.V., and J.I. Korenbrot (1990) Inward rectification in the inner segment of single retinal cone photoreceptors. J. Neurophysiol. *64*:1917–1928.

Mintz, I.M., M.E. Adams, and B.P. Bean (1992) P-type calcium channels in rat central and peripheral neurons. Neuron *9*:85–95.

Mintz, I.M., V.J. Venema, M.E. Adams, and B.P. Bean (1991) Inhibition of N- and L-type Ca^{2+} channels by the spider venom toxin ω-Aga-IIIA. Proc. Natl. Acad. Sci. USA *66*:6628–6631.

Mittman, S., and D.R. Copenhagen (1985) Glycine mimics transient inhibitory synaptic input to whole-cell patch clamped retinal ganglion cells. Invest. Ophthalmol. Vis. Sci. *26*:312.

Mittman, S., W.R. Taylor and D.R. Copenhagen (1990) Concomitant activation of two types of glutamate receptor mediates excitation of salamander retinal ganglion cells. J. Physiol. *428*:175–197.

Mobbs, P., K. Everett, and A. Cook (1992) Signal shaping by voltage-gated currents in retinal ganglion cells. Brain Res. *574*:217–223.

Monje, V.D., J.A. Haack, S.R. Naisbitt, G. Miljanich, J. Ramachandran, L. Nasdasdi, B.M. Olivera, D.R. Hillyard, and W.R. Gray (1993) A new *Conus* peptide ligand for Ca channel subtypes. Neuropharmacol. *32*:1141–1149.

Mori, Y., T. Friedrich, M.-S. Kim, A. Mikami, J. Nakai, P. Ruth, E. Bosse, F. Hofmann, V. Flockerzi, T. Furuichi, K. Mikoshiba, K. Imoto, T. Tanabe, and S. Numa (1991) Primary structure and functional expression from complementary DNA of a brain calcium channel. Nature *350*:398–402.

Murakami, M. and Y. Shimoda (1977) Identification of amacrine and ganglion cells in the carp retina. J. Physiol. Lond. *264*:801–818.

Naka, K.-I., H.M. Sakai (1991) The messages in optic nerve fibers and their interpretation. Brain Res. Rev. *16*:135–149.

Nelson, R., E.V. Famiglietti, Jr., and H. Kolb H. (1978) Intracellular staining reveals different levels of stratification for on- and off-center ganglion cells in cat retina. J. Neurophysiol. *41*:472–483.

Nowycky, M.C., A.P. Fox, and R.W. Tsien (1985). Three types of neuronal calcium channel with different calcium agonist sensitivity. Nature *316*:440–443.

Patlak, J.B., and M. Ortiz (1985) Slow currents through single sodium channels of the adult rat heart. J. Gen. Physiol. *86*:89–104.

Plummer, M.R., D.E. Logothetis, and P. Hess (1989) Elementary properties and pharmacological sensitivities of calcium channels in mammalian peripheral neurones. Neuron *2*:1453–1463.

Protti, D.A., H.M. Gerschenfeld, and I. Llano (1997) GABAergic and glycinergic IPSCs in ganglion cells of rat retinal slices. J. Neurosci. *17*:6075–6085.

Robinson, D.W., S.-J. Huang, R.P. Scobey, and L.M. Chalupa (1994) Whole-cell and single channel properties of a linear membrane conductance in cat rgcs. Soc. Neurosci. Abstr. *20*:1528.

Rothe, T., V. Bigl, and R. Grantyn (1994) Potentiating and depressant effects of metabotropic glutamate receptor agonists on high-voltage-activated calcium currents in cultured retinal ganglion neurons from postnatal mice. Pflügers Arch. *426*:161–170.

Rothe, T., and R. Grantyn (1994) Retinal ganglion neurons express a toxin-resistant developmentally regulated novel type of high-voltage-activated calcium channel. J. Neurophysiol. *72*:2542–2546.

Saito, H.-A. (1983) Morphology of physiologically identified X-, Y-, and W-tyoe retinal ganglion cells of the cat. J. Comp. Neurol. *221*:279–288.

Sakai, H.M. and K.-I. Naka (1990) Dissection of the neuron network in the catfish inner retina. IV. Bidirectional interactions between amacrine and ganglion cells. J. Neurophysiol. *63*:105–119.

Sakai, H.M., K.-I. Naka, and J.E. Dowling (1986) Ganglion cell dendrites are presynaptic in catfish retina. Nature *319*:495–497.

Sakmann, B., and O.D. Creutzfeld (1969) Scotopic and mesopic light adaptation in the cat's retina. Pflügers Arch. *313*:168–185.

Sakmann, B., and E. Neher (1995) *Single-Channel Recording*, 2nd ed. Plenum, NY.

Sarthy, P.V., B.M. Curtis, and W.A. Catterall (1983) Retrograde labeling, enrichment, and characterization of retinal ganglion cells from the neonatal rat. J. Neurosci. *3*:2532–2544.

Sather, W.A., T. Tanabe, J.-F. Zhang, Y. Mori, M.E. Adams, and R.W. Tsien (1993) Distinctive biophysical and pharmacological properties of class A (BI) calcium channel α₁ subunits. Neuron *11*:291–303.

Schwartz, E.A. (1987) Depolarization without calcium can release γ-aminobutyric acid from a retinal neuron. Science *238*:350–355.

Schwindt, P.C., and W.E. Crill (1995) Amplification of synaptic current by persistent sodium conductance in apical dendrite of neocortical neurons. J. Neurophysiol. *74*:2220–2224.

Shingai, R., and F.N. Quandt (1986) Single inward rectifier channels in horizontal cells. Brain Res. *369*:65–74.

Skaliora, I., D.W. Robinson, R.P. Scobey, and L.M. Chalupa (1995) Properties of K⁺ conductances in cat retinal ganglion cells during the period of activity-mediated refinements in retinofugal pathways. Eur. J. Neurosci. *7*:1558–1568.

Skaliora, I., R.P. Scobey, and L.M. Chalupa (1993) Prenatal development of excitability in cat retinal ganglion cells: Action potentials and sodium currents. J. Neurosci. *13*:313–323.

Slaughter, M.M., and Bai, S.-H. (1989) Differential effects of baclofen on sustained and transient cells in the mudpuppy retina. J. Neurophysiol. *61*:374–381.

Soong, T.W., A. Stea, C.D. Hodson, S.J. Dubel, S.R. Vincent, and T.P. Snutch (1993) Structure and functional expression of a member of the low voltage-activated calcium channel family. Science *260*:1133–1136.

Standen, N.B., and P.R. Stanfield (1978) A potential and time-dependent blockade of inward rectification in frog skeletal muscle fibres by barium and strontium ions. J. Physiol. *280*:169–191.

Stea, A., W.J. Tomlinson, T.W. Soong, E. Bourinet, S.J. Dubel, S.R. Vincent, and T.P. Snutch (1994) Localization and functional properties of a rat brain α₁ₐ calcium channel reflect similarities to neuronal Q- and P-type channels. Proc. Natl. Acad. Sci. USA *91*:10576–10580.

Strafstrom, C.E., P.C. Schwindt, and W.E. Crill (1982) Negative slope conductance due to a persistent subthreshold sodium current in cat neocortical neurons in vitro. Brain Res. *236*:221–226.

Stuart, G., and B. Sakmann (1995) Amplification of EPSPs by axosomatic sodium channels in neocortical pyramidal neurons. Neuron *15*:1065–1076.

Stys, P.K., H. Sontheimer, B.R. Ransom, and S.G. Waxman (1993) Noninactivating, tetrodotoxin-sensitive Na^+ conductance in rat optic nerve axons. Proc. Natl. Acad. Sci. USA *90*:6976–6980.

Sucher, N.J., and S.A. Lipton (1992) A slowly inactivating K^+ current in retinal ganglion cells from postnatal rat. Vis. Neurosci. *8*:171–176.

Tabata, T., and A.T. Ishida (1994) Whole-cell patch-clamp recordings from tissue-printed retinal ganglion cells. Soc. Neurosci. Abstr. *20*:218.

Tabata, T., and A.T. Ishida (1996) Transient and sustained depolarization of retinal ganglion cells by I_h. J. Neurophysiol. *75*:1932–1944.

Tabata, T., and A.T. Ishida (1997) Intracellular zinc may sustain PKC-mediated reduction of outwardly rectifying Cl^- current in retinal ganglion cells. Soc. Neurosci. Abstr. *23*: in press.

Tabata, T., B.M. Olivera, and A.T. Ishida (1996) ω-Conotoxin-MVIID blocks an ω-conotoxin-GVIA-sensitive, high-threshold Ca^{2+} current in fish retinal ganglion cells. Neuropharmacology *35*:633–636.

Tachibana, M. (1983) Ionic currents of solitary horizontal cells isolated from goldfish retina. J. Physiol. *345*:329–351.

Takahashi, K.-I., and D.R. Copenhagen (1995) Intracellular alkalization enhances inward rectifier K^+ current in retinal horizontal cells of catfish. Zool. Sci. *12*:29–34.

Tamura, N., K. Yokotani, Y. Okuma, M. Okada, H. Ueno, and Y. Osumi (1995) Properties of the voltage-gated calcium channels mediating dopamine and acetylcholine release from the isolated rat retina. Brain. Res. *676*:363–370.

Tang, C.M., G.R. Strichartz, and R.K. Orkand (1979) Sodium channels in axons and glial cells of the optic nerve of *Necturus maculosa*. J. Gen. Physiol. *74*:629–642.

Taschenberger, H., and R. Grantyn (1995) Several types of Ca^{2+} channels mediate glutamatergic synaptic responses to activation of single Thy-1-immunolabeled rat retinal ganglion neurons. J. Neurosci. *15*:2240–2254.

Taylor, W.R., E. Chen, and D.R. Copenhagen (1995) Characterization of spontaneous excitatory synaptic currents in salamander retinal ganglion cells. J. Physiol. *486*:207–221.

Tessier-Lavigne, M., D. Attwell, P. Mobbs, and M. Wilson (1988) Membrane currents in retinal bipolar cells of the axolotl. J. Gen. Physiol. *91*:49–72.

Tomita, T., M. Murakami, Y. Hashimoto, and Y. Sasaki (1961) Electrical activity of single neurons in the frog's retina. In: The Visual System: Neurophysiology and Psychophysics. Eds: R. Jung and H. Kornhuber (Springer-Verlager, Berlin); pp. 24–30.

Tsien, R.W., D. Lipscombe, D.V. Madison, K.R. Bley, and A.P. Fox (1988) Multiple types of neuronal calcium channels and their selective modulation. TINS *11*:431–438.

Ueda, Y, A. Kaneko, and M. Kaneda (1992) Voltage-dependent ionic currents in solitary horizontal cells isolated from cat retina. J. Neurophysiol. *68*:1143–1150.

Usowicz, M., M. Sugimori, B. Cherksey, and R. Llinás (1992) P-type calcium channels in the somata and dendrites of adult cerebellar Purkinje cells. Neuron *9*:1185–1199.

von Gersdorff, H., E. Vardi, G.G. Matthews, and P. Sterling (1996) Evidence that vesicles on the synaptic ribbon of retinal bipolar neurons can be rapidly released. Neuron *6*:1221–1227.

Westenbroek, R.E., D.K. Merrick, and W.A. Catterall (1989) Differential subcellular localization of the R_I and R_{II} Na^+ channel subtypes in central neurons. Neuron *3*:695–704.

White, J.A., N.S. Sekar, and A.R. Kay (1995) Errors in persistent inward currents generated by space-clamp errors: A modeling study. J. Neurophysiol. *73*:2369–2377.

Wiesel, T. (1959) Recording inhibition and excitation in the cat's retinal ganglion cells with intracellular electrodes. Nature *183*:264–265.

Wilkinson, M.F., and S. Barnes (1996) The dihydropyridine-sensitive calcium channel subtype in cone photoreceptors. J. Gen. Physiol. *107*:621–630.

Williams, M.E., D.H. Feldman, A.F. McCue, R. Brenner, G. Velicelebi, S.B. Ellis, and M.M. Harpold (1992a) Structure and functional expression of α_1, α_2, and β subunits of a novel human neuronal calcium channel subtype. Neuron *8*:71–84.

Williams, M.E., D.H. Feldman, P.F. Brust, D.H. Feldman, S. Patthi, S. Simerson, A. Maroufi, A.F. McCue, G. Velicelebi, S.B. Ellis, and M.M. Harpold (1992b) Structure and functional expression of an ω-conotoxin-sensitive human N-type calcium channel. Science *257*:389–395.

Wilson, M., and E. Gleason (1991) An unusual voltage-gated anion channel found in the cone cells of the chicken retina. Vis. Neurosci. *6*:19–23.

Wollmuth, L.P., and B. Hille (1992) Ionic selectivity of I_h channels of rod photoreceptors in tiger salamanders. J. Gen. Physiol. *100*:749–765.

Wollner, D.A., and W.A. Catterall (1986) Localization of sodium channels in axon hillocks and initial segments of retinal ganglion cells. Proc. Natl. Acad. Sci. USA *83*:8424–8428.

Yagi, T., and A. Kaneko A (1988) The axon terminal of goldfish retinal horizontal cells: A low membrane conductance measured in solitary preparations and its implications to the signal conduction from the soma. J. Neurophysiol. *59*:482–494.

Yagi, T., and P.R. MacLeish (1994) Ionic conductances of monkey solitary cone inner segments. J. Neurophysiol. *71*:656–665.

Zhang, J., W. Shen, and M.M. Slaughter (1997) Two metabotropic γ-aminobutyric acid receptors differentially modulate calcium currents in retinal ganglion cells. J. Gen. Physiol. *110*:45–58.

Zhang, J.-F., A.D. Randall, P.T. Ellinor, W.A. Horne, W.A. Sather, T. Tanabe, T.L. Schwarz and R.W. Tsien (1993) Distinctive pharmacology and kinetics of cloned neuronal Ca²⁺ channels and their possible counterparts in mammalian CNS neurons. Neuropharmacol. *32*:1075–1088.

12

SYNAPTIC TRANSMISSION BETWEEN RETINAL NEURONS

Martin Wilson[*]

Section of Neurobiology, Physiology and Behavior
Division of Biological Sciences
University of California
Davis, California 95616

RETINAL SYNAPSES

The retina is a uniquely favorable place in which to hunt for the design principles of the brain. Aside from its well-known virtues of being compact, accessible and having well understood inputs and outputs, the retina possesses the paramount virtue that we understand what it is for. The business of elucidating design principles is really one of reverse engineering, in which forming a list of components and their properties is the essential first step. In pursuit of this goal, the last fifteen years have seen a great deal of progress in understanding the properties of individual cells in the retina and the way in which their voltage-gated channels allow signals to be shaped and transformed. Parallel with this have been discoveries about the transmitter-gated channels present on neurons in the retina. These important advances have, however, tended to obscure a major gap in our understanding of the way in which transmitter is actually released at synapses and how it is experienced by those postsynaptic receptors.

Many of the questions for which we need answers center on Ca^{++}, its economy at synapses and its exact relationship to transmitter release and synaptic plasticity. Still other questions have to do with the quantal nature of transmission. The beautiful picture of synaptic transmission presented by Bernard Katz almost 3 decades ago (Katz, 1969) and presented in all neuroscience textbooks has recently come under sceptical reinterpretation from several perspectives (see for example, Edwards et al., 1990), so that issues that were generally thought to be settled are now reopened. None of these questions is peculiar to the retina; our understanding of central synapses in general is very poor, a fact that has

[*] Correspondence: Martin Wilson, Section of Neurobiology, Physiology, and Behavior, Division of Biological Sciences, University of California, Davis, CA 95616. Fax: 916-752-1449; Phone: 916-752-7250; e-mail: mcwilson@ucdavis.edu

Development and Organization of the Retina, edited by Chalupa and Finlay.
Plenum Press, New York, 1998.

been brought into sharp focus by the intense interest in the synaptic phenomenon of Long Term Potentiation in the hippocampus.

The two synapses about which most is known, the vertebrate neuromuscular junction and the squid giant synapse, may not be good models for central synapses because they are specialized for very unusual tasks. The properties in which they excel are the faithful and rapid generation of a post-synaptic action potential for every pre-synaptic action potential, capabilities that are probably inappropriate to most central synapses and certainly inappropriate to retinal neurons, many of which signal without action potentials. While we should be cautious in assuming that the two canonical synapses are similar to more usual synapses in the brain, it should also be remembered that central synapses are themselves unlikely to be uniformly similar.

At the EM level it is possible to discern anatomical peculiarities that characterize synapses between particular types of neuron and it would be surprising if these anatomical embellishments were without functional significance. A clear of example of anatomical differences between synapses in the retina would be the distinction to be drawn between ribbon synapses existing in photoreceptors and bipolar cells and the more usual presynaptic structures found elsewhere in the retina. Recent work on transmitter release from bipolar cells indicates that presynaptic ribbons may indeed be associated with unusual release properties at these synapses (Rieke and Schwartz, 1996; Lagnado et al., 1996). In general, we may expect that synapses vary considerably in all important regards since evolution has had hundreds of millions of years in which to sculpt the functional properties of these crucial components of brain hardware.

Ultimately, it will be necessary to examine how synapses work in the real retina with its real geometry and real milieu of ions, modulators and the like. Unfortunately though, many of the questions to which we would most like an answer cannot yet be addressed in the real retina and some other way has to be found to begin answering them. The reasons for postponing a frontal assault on the major questions are technical and stem largely from the fact that retinal neurons are richly interconnected so that the properties of any one synapse are hard to separate from the effects of others. Our strategy has been to adopt a two step approach: to look first at a highly simplified preparation where the properties of individual synapses can be easily examined, with the expectation that, having found these answers, we might then see to what extent these answers hold true in a preparation more nearly approximating the real retina.

Simplified Retinal Synapses

The preparation we have examined uses synapses between retinal neurons cultured at low density. After 9 or 10 days in culture many of the cells originally dissociated from 8 day old chick embryo retina are multipolar as shown in Figure 1 and have amacrine-like properties (Huba and Hoffman, 1990, 1991). Since these cells bind the antibody HPC1 (Gleason et al., 1993), which is thought to be specific for amacrine cells (Barnstable et al., 1985; Akagawa, 1990), it seems likely that they are indeed amacrine cells. Pairs of these amacrine cells may be found in our cultures and where cells appear to touch they frequently make functional synapses. Usually, functional synapses are formed in both directions so that defining which cell is pre- and which post-synaptic is an arbitrary matter. Along with chemical synapses, and in fact preceding their formation, it is often possible to see electrical coupling between pairs of cells (Gleason et al., 1993). In our experiments on chemical synaptic transmission, pairs of cells that are electrically coupled are not useful since electrical coupling prevents a clean separation of pre and post-synaptic currents. Regrettably then, we have been obliged to ignore

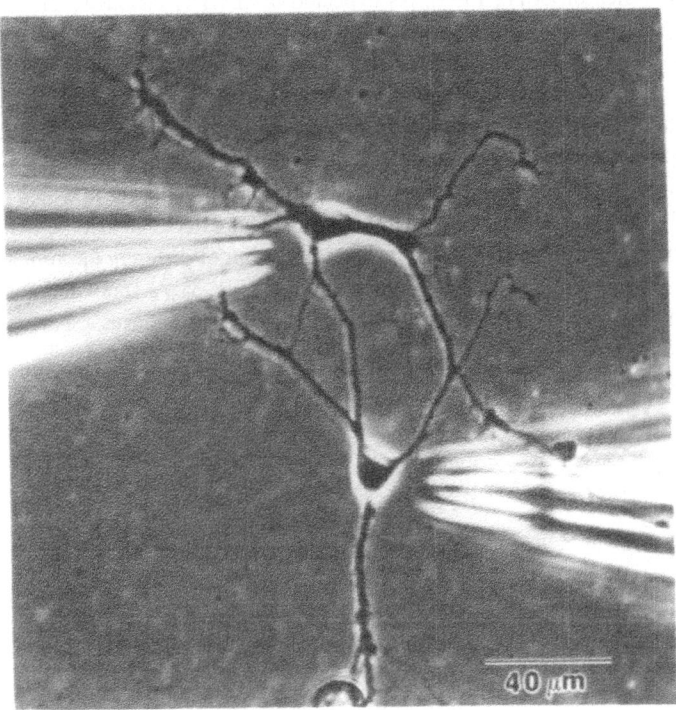

Figure 1. A pair of amacrine cells growing in culture on a plastic substrate. Cultured in the absence of glial cells, neurons never extend long processes, thereby permitting good voltage-clamp of the entire cell membrane. As shown here, pairs of cells often appear to touch and at these points of contact they form synapses. A pair of patch pipettes, slightly out of focus, may be seen positioned above the cell bodies of these two cells prior to voltage-clamping them.

the large fraction of cell pairs that show electrical coupling in addition to chemical transmission. Solitary amacrine cells, on the other hand, have been useful in our work. Solitary cells frequently form synapses back on to themselves and for some kinds of experiments these autapses are a convenient stand-in for synapses whose properties they closely resemble.

In addition to the complicating wealth of connections within the real retina, there is another factor creating a technical barrier to the examination of synapses. This is the difficulty of adequately voltage-clamping long neuronal processes. This problem is not necessarily solved by culturing cells since amacrine neurons invariably send out long processes that are impossible to voltage-clamp, when grown in the presence of glial cells. To discourage amacrine cells from producing long processes we have employed culture conditions unfavorable to the growth of glial cells. A second trick by which we have promoted good voltage-clamp is to increase the input resistance of cells by blocking those channels in which we have no interest, in particular Na^+ and K^+ channels.

THE GENERAL PROPERTIES OF TRANSMISSION BETWEEN AMACRINE CELLS

Presynaptic depolarization generates postsynaptic currents. These postsynaptic currents are always mediated by $GABA_A$ channels and are completely blocked by low concentrations of bicuculline (Gleason et al., 1993). As expected of GABA$_A$ channels, Cl$^-$ ions carry the majority of the current and manipulation of the Cl$^-$ gradient allows us to see these postsynaptic currents as either inward or outward. There are several salient features of transmission elicited by presynaptic steps of voltage. The first is that, as shown in Figure 2, voltage steps negative to about –45mV are ineffective in generating postsynaptic currents. Positive to this voltage, responses become larger and saturate in their maximum amplitude but although amplitude saturates, bigger voltage steps generally elicit longer responses in which postsynaptic current clearly outlasts the presynaptic voltage step. This effect can also be easily seen in a series of presynaptic voltage steps to the same voltage but of different durations. The longer steps elicit responses that are disproportionately longer (Fig. 3). At small depolarizations it is very clear that transmission is quantal and that postsynaptic responses are actually the summation of miniature currents generated by a single quantum of transmitter. For larger presy-

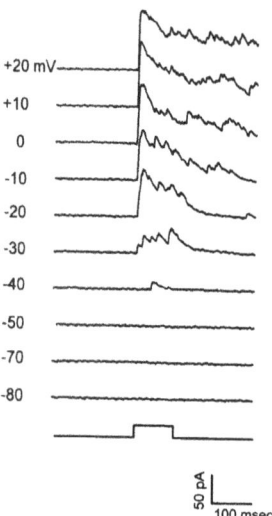

Figure 2. Depolarization of one cell in a pair of amacrine cells elicits postsynaptic currents in its partner (from Gleason et al., 1993). These individual current records show postsynaptic currents generated by 100 msec voltage steps of the presynaptic cell from its holding voltage of –70 mV to the voltage indicated to the left of each current record. The postsynaptic cell was voltage-clamped at 0 mV so that postsynaptic currents were always outwards. Presynaptic voltages negative to –40 mV elicit no response in the postsynaptic cell but at –40 mV a single discrete current (a mini) is seen. Larger presynaptic depolarizations elicit bigger postsynaptic currents that outlast the duration of the presynaptic step.

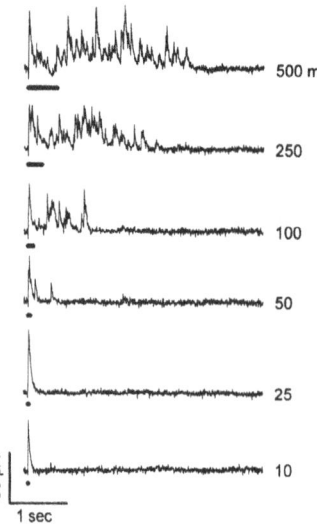

Figure 3. Postsynaptic currents recorded when the presynaptic cell is stepped from –70 mV to 0 mV for durations indicated to the right of the current record (from Gleason et al., 1994). As voltage steps become longer, postsynaptic responses become disproportionately extended as a consequence of long-lasting elevation in the Ca^{++} concentration at synapses.

naptic depolarizations it is harder to make out individual minis since so many are piled on top of each other. As we will see it is nevertheless possible, using the mathematical concept of deconvolution, to work out how minis are distributed in time, even though this is not readily apparent from inspection of the postsynaptic currents themselves.

ENTRY OF CALCIUM INTO TERMINALS PROMOTES TRANSMITTER RELEASE

As it is at other synapses, external Ca^{++} is required in order for presynaptic depolarizations to elicit postsynaptic currents. Because of its crucial role in triggering transmitter release we have examined how Ca^{++} enters the cell and how it is subsequently removed. Examination of individual amacrine cells reveals that they have a Ca^{++} current that begins to turn on at about –45 mV (Fig. 4A) and, relative to other Ca^{++} currents, is slowly activat-

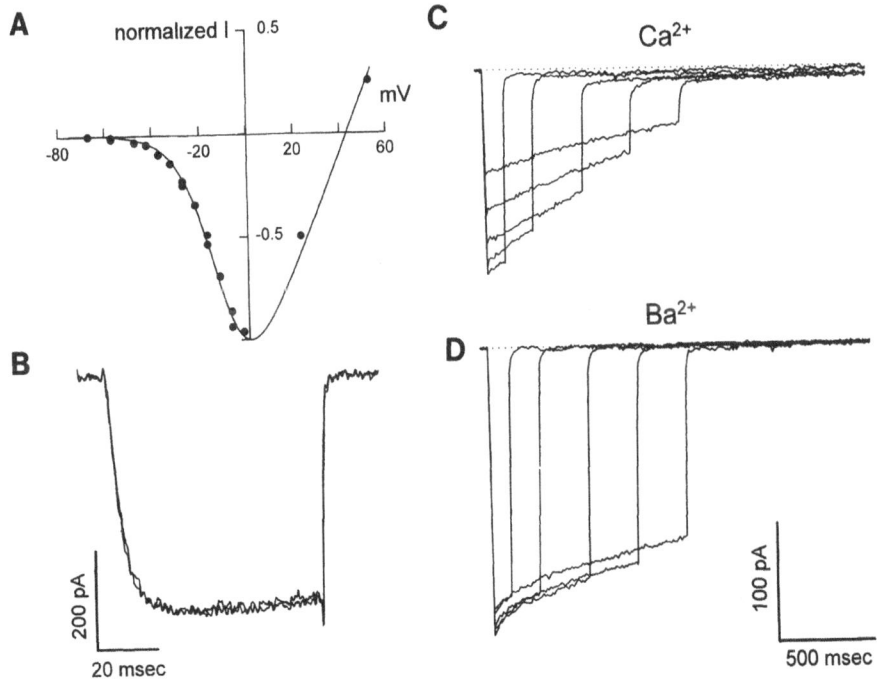

Figure 4. The properties of presynaptic Ca^{++} currents (from Gleason et al., 1994). **A.** Normalized peak Ca^{++} currents elicited from an isolated amacrine cell shown as a function of voltage. At about –40 mV Ca^{++} current begins to activate and reaches a peak at approximately 0 mV, positive to which it declines. The apparent reversal of the current at about +40 mV suggests that some ions other than Ca^{++} must also be passing through these Ca^{++} channels. **B.** Two superimposed current records showing Ca^{++} currents elicited by a voltage step to 0 mV preceded by a 2 sec step to either –90 or –60 mV. In either case the currents are the same magnitude, consistent with the conclusion that these presynaptic Ca^{++} currents most resemble L-type current. **C & D.** Superimposed currents elicited by a series of 5 steps from –70 to 0 mV for increasing durations. Twelve sec rest periods were left between voltage steps but in C it is clear that from one step to the next, the peak current that can be elicited has declined. This decline is attributable to Ca^{++}-dependent inactivation since, as shown in **D**, when Ba^{++} is the charge carrier this effect is not seen. In Ba^{++} there is nevertheless still some inactivation of the current occurring during the voltage step and this inactivation component we attribute to voltage-dependent inactivation. With Ca^{++} as the charge carrier, but not with Ba^{++}, a small persistent inward current due to Na^{+}–Ca^{++} exchange may be seen after the Ca^{++} current has deactivated.

ing and slowly inactivating (Fig. 4B). The voltage region at which the Ca^{++} current begins to activate corresponds roughly to the voltage at which transmission begins (Fig. 12), leading us to suppose that it is in fact this Ca^{++} current, rather than some different, hard-to-see current found only at synapses, that is responsible for synaptic transmission. Two forms of inactivation occur in this Ca^{++} current. A small but measurable inactivation is voltage-dependent and another more severe and long-lasting component of inactivation depends on Ca^{++} entry (Fig. 4C). When external Ba^{++} is substituted for Ca^{++}, the Ca^{++}-dependent inactivation is largely removed, even though Ba^{++} is a better charge carrier through these channels (Fig. 4D).

Very likely, the Ca^{++} current described in Figure 4 is a mixture of at least two currents. The dihydropyridine agents nifedipine and nimodipine that block L-type Ca^{++} channels generally block about 60% of the Ca^{++} currents seen in cultured amacrine cells (Gleason et al., 1994). Higher concentrations of these agents fail to block all the current and the fraction blocked varies considerably between individual cells, leading us to propose that, although the majority of Ca^{++} current can be described as L-type, there is some other component that constitutes a variable fraction of the Ca^{++} channels in these cells. While the identity of this other kind of Ca^{++} channel is presently unknown, in the important properties of kinetics and activation range this unknown current closely resembles the L-type component. Both kinds of Ca^{++} channel seem to be involved in synaptic transmission since nifedipine blocks a majority of synaptic transmission, but not all of it (Gleason et al., 1994).

REMOVAL OF CALCIUM FROM SYNAPSES

A clue to the way in which Ca^{++} is removed from amacrine cells can be seen in Figure 4C. In the presence of external Ca^{++}, though not external Ba^{++}, divalent cation currents activated by depolarization are followed by a small, long-lasting inward current after the membrane is stepped back to its hyperpolarized holding voltage. These long-lasting inward currents are clearly dependent on the influx of Ca^{++} since their magnitude correlates with the magnitude of the preceding Ca^{++} current. Cells loaded with the Ca^{++} buffer, BAPTA, fail to show these long-lasting currents, suggesting that the currents are activated by an increase in cytoplasmic Ca^{++} (Gleason et al., 1995). Calcium-gated currents such as Ca^{++}-gated Cl^- or Ca^{++}-gated K^+ currents are obvious candidates for the identity of this current, but manipulation of K^+ and Cl^- gradients has little effect on the long-lasting current (Gleason et al., 1995). Changes in external Ca^{++} concentration, in contrast, have large and curious effects. Removal of all external Ca^{++} of course abolishes the long-lasting current but strangely, increasing Ca^{++} concentration (Fig. 5) also diminishes the magnitude of the current, and extends its timecourse. The interpretation of this effect is undoubtedly that the long-lasting current is not a Ca^{++}-gated current but rather is a Na^+–Ca^{++} exchange current in which the influx of Na^+ down its electrochemical gradient is coupled to the expulsion of Ca^{++} ions against their electrochemical gradient. Increasing the Ca^{++} gradient in such a system might well have the effect of slowing the rate of exchange, thereby reducing the maximum current, consistent with Figure 5.

Where Na^+–Ca^{++} exchange has been characterized in other systems, its requirement for Na^+ has been shown to be very specific, so that Li^+ ions, in particular, which can permeate readily through Na^+ channels, are unable to substitute for Na^+ in the exchanger (Blaustein, 1977). In amacrine cells we similarly find that Li^+ is not able to substitute for Na^+ (Fig. 6).

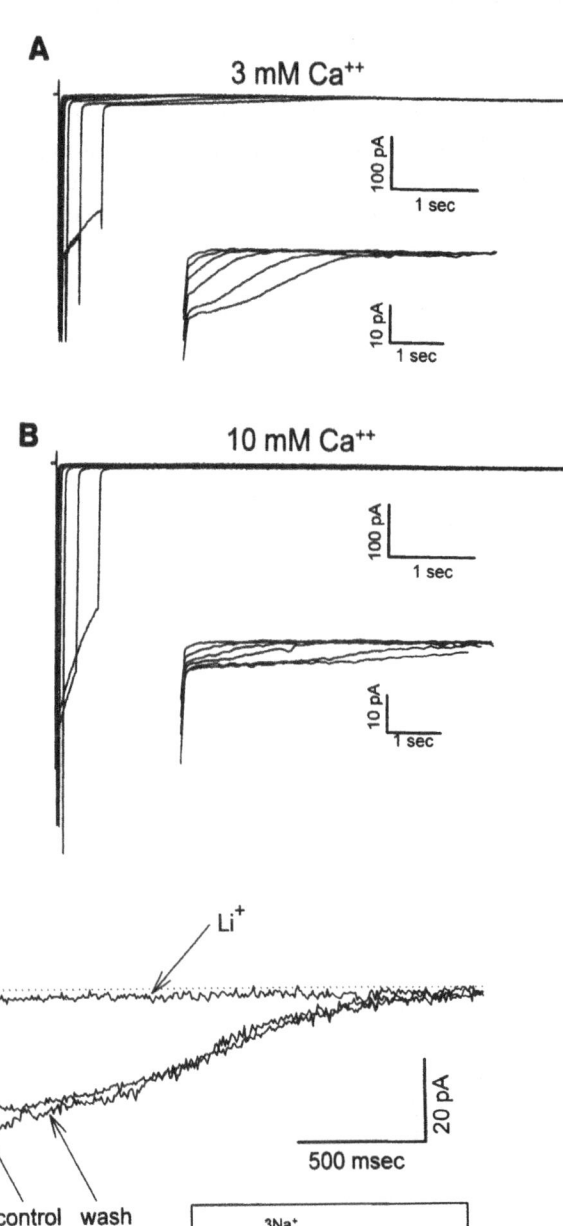

A

3 mM Ca^{++}

100 pA | 1 sec

10 pA | 1 sec

B

10 mM Ca^{++}

100 pA | 1 sec

10 pA | 1 sec

Figure 5. Calcium currents elicited by voltage steps of increasing duration to 0 mV in a protocol similar to that of Figure 4C (from Gleason et al., 1995). Two sets of currents have been elicited, one in 3 mM Ca^{++} the other in 10 mM Ca^{++}. In 10 mM Ca^{++}, Ca^{++} currents are larger but the persistent tail currents following these are smaller. This is best seen in the insets where tail currents have been aligned to the end of their preceding Ca^{++} currents. Though smaller in 10 mM Ca^{++} these Na$^+$–Ca^{++} exchange currents are longer lasting.

Li$^+$

20 pA

500 msec

control   wash

3Na$^+$

outside

inside

Ca^{2+}

Figure 6. Substitution of Li$^+$ for Na$^+$ in the external medium suppresses the Na$^+$–Ca^{++} exchange current (from Gleason et al., 1995). In this experiment an amacrine cell was stepped from −70 to 0 mV for 500 ms, indicated by the step, in the presence of external Na$^+$ or, alternatively, external Li$^+$. In either condition the current elicited during the step is the same amplitude but too big to display at this gain. In the presence of Na$^+$, either before the application of Li$^+$ (control) or after the application of Li$^+$ (wash) the persistent tail current due to Na$^+$–Ca^{++} exchange has a maximum amplitude of about 30 pA. When Li$^+$ substitutes for Na$^+$ though this inward tail current is almost entirely suppressed. Inset: the presumed stoichiometry of this exchanger results in the net entry of one positive charge for every Ca^{++} expelled.

Two forms of Na^+–Ca^{++} exchange are known. The form best characterized in cardiac muscle has a usual stoichiometry of three Na^+ ions entering for every Ca^{++} ion leaving the cell (Reeves and Hale, 1984), and it is the inequality in charges moved during every turn of the cycle that constitutes the small current generated by this exchanger. Another form of Na^+–Ca^{++} exchange has been described in rod photoreceptor outer segments and is more complicated since K^+ ions also move down their concentration gradient to help power the expulsion of Ca^{++} (Cervetto et al., 1989). Sudden changes in K^+ gradient seem to have no effect on exchange current in amacrine cells (Gleason et al., 1995), leading us to think that their exchanger probably conforms to the cardiac type with respect to its stoichiometry, though this is not yet certain.

How much of the Ca^{++} entering via voltage activated Ca^{++} channels is expelled by Na^+–Ca^{++} exchange? In principle we should be able to answer this question by comparing the charge entering as Ca^{++} current with the charge entering as exchange current. A key assumption here is, of course, the exchange stoichiometry about which we are not completely certain. Nevertheless, if we presume the stoichiometry of three Na^+s entering to one Ca^{++} leaving we expect the charge moved during a Ca^{++} current to be exactly twice that moved during the exchange current, if every Ca^{++} entering leaves the cell via the exchanger. In some cells this expectation is very nearly met but there is a high degree of variability between cells and, on average, it seems that about 60% of Ca^{++} entering is removed by the exchanger.

Some insight into the fate of the remainder of the Ca^{++} derives from experiments like that shown in Figure 7. In this experiment we have used a voltage step to load a cell with

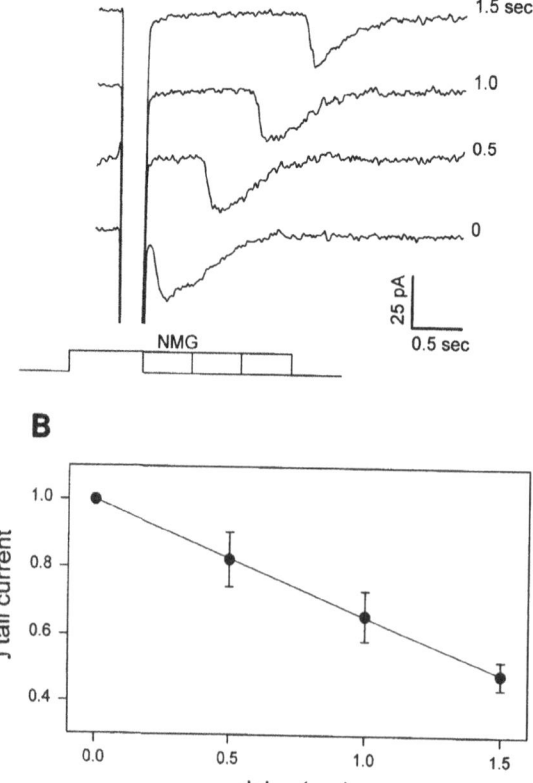

Figure 7. Most, though not all, Ca^{++} entering an amacrine cell during a voltage step leaves via the exchanger (from Gleason et al., 1995). **A.** Four superimposed currents showing a Ca^{++} current elicited by a voltage step to 0 mV (indicated by black bar) followed by Na^+–Ca^{++} exchange current. In these four trials Na^+ was removed from the external medium just before the voltage step and is substituted with N-methyl-d-glucamine (NMG), a cation that will not support Na^+–Ca^{++} exchange. At variable times following the voltage step, indicated to the right of the four current traces, Na^+ is restored to the external medium. When Na^+ is added back it allows Na^+–Ca^{++} exchange to start and thereby remove the available excess Ca^{++}. As the delay in adding back external Na^+ is increased, the amount of Ca^{++} available to the exchanger is reduced, as evidenced by the amount of charge moved (the time integral of the exchange current). **B.** The amount of charge moved by the exchanger falls approximately linearly as a function of the delay in adding back Na^+. This linear function must represent the kinetics of a process other than Na^+–Ca^{++} exchange responsible for removal of Ca^{++}.

Ca^{++} during which the exchanger has been unable to work owing to an absence of external Na$^+$. By waiting a variable length of time before adding back the Na$^+$, and thereby allowing the exchanger to operate again, we are able to look at the kinetics of that process of Ca^{++} removal that does not involve the exchanger. The longer external Na$^+$ is withheld, the smaller is the fraction of Ca^{++} that can subsequently be removed by the exchanger. About 30% of a Ca^{++} load is made unavailable to the exchanger every second. It is not clear what happens to this Ca^{++} but the most likely interpretations are either that mitochondria take it up or else it diffuses into cytoplasmic compartments from which it is only slowly released. In neither case though could Ca^{++} be permanently removed and ultimately this Ca^{++}, like all excess Ca^{++}, must be expelled by the exchanger or else removed by an ATP-requiring Ca^{++} pump, the only other known mechanism by which Ca^{++} is expelled from cells.

HOW MUCH CALCIUM IS REQUIRED FOR TRANSMITTER RELEASE?

In experiments such as those shown in Figures 2 and 3, it is apparent that transmitter release can outlast the opening of Ca^{++} channels by a considerable margin. Putting this together with what we have learned about the Na$^+$–Ca^{++} exchanger in amacrine cells we might propose an explanation for this phenomenon along the following lines. During depolarization Ca^{++} channels open and Ca^{++} enters the cytoplasm increasing free Ca^{++} sufficiently to cause transmitter release. An elevated Ca^{++} concentration outlasts the opening of the Ca^{++} channels and is restored to resting values largely through the action of the Na$^+$–Ca^{++} exchanger.

This straightforward narrative has two unexpected implications. The first of these is that Na$^+$–Ca^{++} exchange controls the timecourse of transmitter release. The truth of this statement can readily be demonstrated by removing external Na$^+$ which, as shown in Figure 8, brings about a huge prolongation in postsynaptic current (Gleason et al., 1994). The second implication is that relatively small cytoplasmic concentrations of Ca^{++} must be sufficient to promote transmitter release. The standard view of transmitter release is that high concentrations, at least tens of micromolar (Augustine et al., 1991), are required for transmission. It is inconceivable that such high concentrations of Ca^{++} can exist over the hundreds of milliseconds of release seen in experiments such as that shown in Figure 3. Exactly what concentration of free Ca^{++} is required to promote transmitter release in amacrine cells is somewhat uncertain, but in preliminary experiments using buffered Ca^{++} solutions to internally perfuse amacrine cells (Frerking et al., 1997) we have found that activities as low as 50 nM Ca^{++} are sufficient to promote a slow but measurable rate of release (Fig. 9). Long-lasting transmitter release, though less dramatic than that seen here, has been observed in hippocampal neurons where it has been termed asynchronous release (Goda and Stevens, 1994). In those cells, as in amacrine cells, its significance is not yet understood.

WHAT ACCOUNTS FOR THE VARIABILITY IN MINI SIZE?

Minis seen in amacrine cells are more or less stereotyped in their waveform but show striking variability in peak amplitude, having a coefficient of variation (standard deviation/mean) of around 60% (Fig. 10). An obvious possible explanation for this is that small minis originate on the tips of distant dendrites and are small because of the attenu-

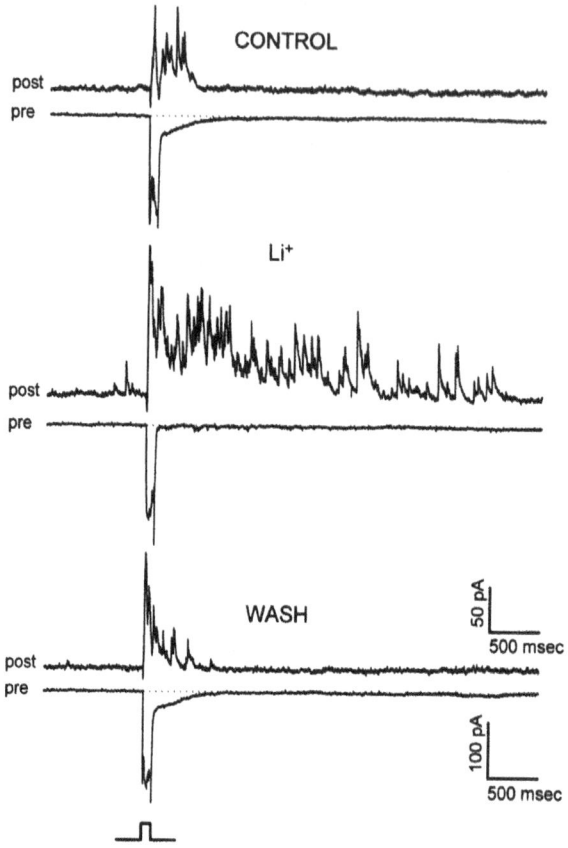

Figure 8. Simultaneous pre and post-synaptic recordings from a pair of amacrine cells. In the control and wash experiments Na$^+$ was present in the external medium (Gleason et al., 1994). Under this condition a presynaptic voltage step of 100 msec from −70 to 0 mV elicits a Ca^{++} current in the presynaptic cell, contaminated with autaptic currents, as well as a longer lasting tail current due to the Na$^+$–Ca^{++} exchanger. In the postsynaptic cell, currents comprising superimposed minis are visible during the Ca^{++} current and for a while after its termination. Characteristically, the duration of postsynaptic current corresponds to the duration of the exchange current, implying that once the exchanger has removed presynaptic Ca^{++}, transmission ceases. When Li$^+$ is substituted for external Na$^+$ no exchange current is visible presynaptically following the Ca^{++} current and postsynaptically the current is very long-lasting since without the action of the exchanger Ca^{++} persists in the presynaptic terminals.

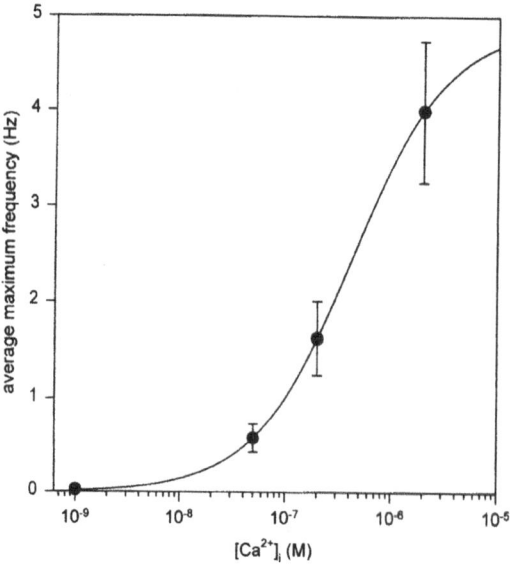

Figure 9. Dependence of release rate on cytoplasmic Ca^{++} (from Frerking et al., 1997). Forty nine amacrine cells were internally perfused with buffered Ca^{++} solutions to derive the relationship shown here. Data points show the average maximum frequency of minis at autapses. While the form of this relationship probably reflects the influence of several processes, such as redocking of vesicles, and not just the last step of vesicle fusion, the most significant point is that, even at only 50 nM Ca^{++}, release rate is not zero.

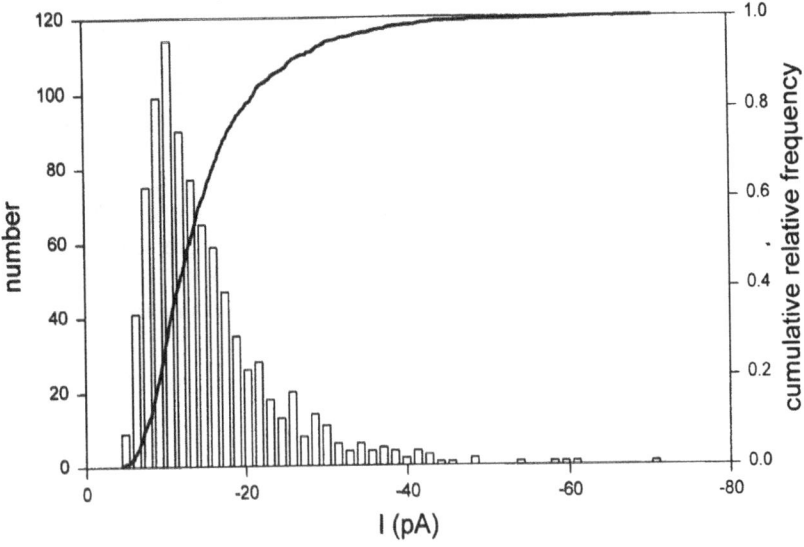

Figure 10. The distribution of mini amplitudes as seen at autapses in a single, isolated amacrine cell (from Frerking et al., 1995). The data, comprising 896 mini amplitudes, is displayed in 2 ways: as a histogram and as a cumulative relative frequency, a form of display that avoids the arbitrary choice of bin width. These plots clearly show the large variations in peak size of minis as well as the pronounced skew in the distribution of values.

ation imposed by the cable properties of the dendrite. This explanation predicts that small minis should also be slow; a situation that is not seen experimentally (Gleason et al., 1993). Furthermore, calculations based on the geometry of amacrine cells shows that cable attenuation can account for only a negligible fraction of the variability seen (Frerking et al., 1995). Two alternative kinds of explanation with rather different implications are nevertheless possible.

The first kind of explanation is that individual release sites each have their own size of mini that remains invariant from one quantum to the next, so that the variation seen in Figure 10 would have to be attributed to differences between individual release sites. An inescapable corollary of this explanation is that a quantum of transmitter would have to saturate its postsynaptic receptors in order for minis at any one release site to have no variance (Frerking and Wilson, 1996). At anything less than saturation, minis would show fluctuations in peak size just on the basis of the probabilistic behavior of channels, even if every quantum comprised the same number of transmitter molecules (Faber et al., 1992). We might note that this stochastic variance is expected to be substantial since the mean number of channels opened at the peak of a mini is small, probably only about 15. An implication of this general explanation is that individual release sites might well have their strength set by a mechanism that strives to adjust this to some optimal value.

A quite different explanation for mini variance views the variability in mini sizes as a source of noise or unreliability in synaptic transmission. This second view conceives of each release site being associated with a variable mini: in other words at any particular release site mini size varies from one quantum to the next, perhaps because different quanta comprise different numbers of transmitter molecules.

Which of these two views is right? An obvious way to settle this issue would be to examine the properties of a single site. So far this has not been possible in any preparation although some attempts have come very close (e.g. Liu and Tsien, 1995). Our approach has been rather different and takes advantage of an unexpected observation. The unexpected discovery was that as shown in Figure 11, if two connected amacrine cells are examined at high gain with one of them slightly depolarized, it is possible to see that some fraction of the minis in the postsynaptic cell are coincident with minis seen in the presynaptic cell. That there should be minis in the presynaptic cell is not so surprising since we already know that amacrine cells very often have autapses. The fact, however, that both pre- and postsynaptic minis occur at the same time requires some explanation.

Electrical coupling cannot account for this since direct measurements show simultaneous minis in pairs of cells with no apparent electrical coupling. The explanation instead has to be that a single quantum of transmitter is experienced simultaneously by both

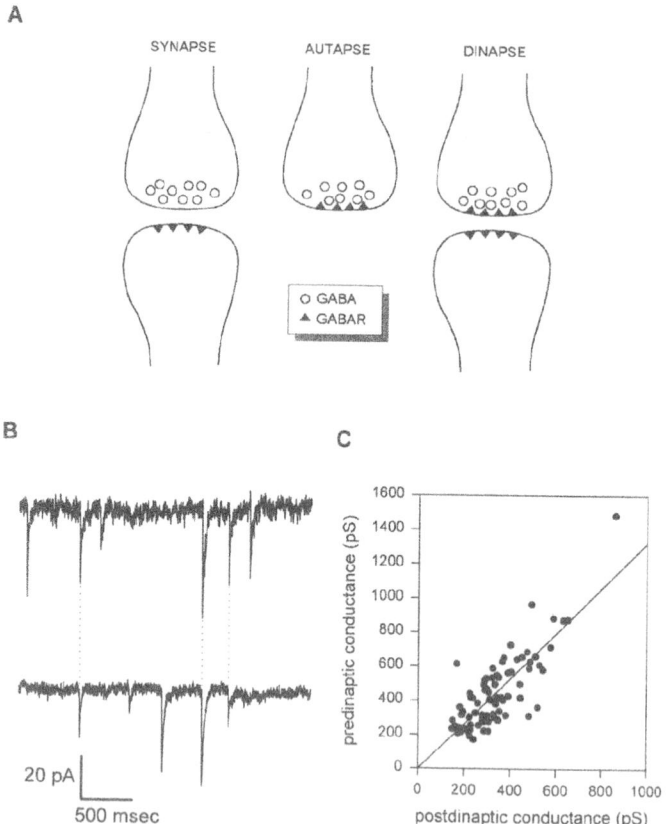

Figure 11. Dinapses as a physiological tool (Frerking et al., 1995). **A.** A schematic representation of the three forms of transmission seen in our amacrine cells. Synapses represent transmission from one cell to another in which transmitter is released from the presynaptic cell and received by the postsynaptic cell. At autapses, transmitter is released and received by the same cell. Dinapses represent a hybrid between synapses and autapses so that transmitter released by one cell causes a signal in both that cell and its postdinaptic partner. The presence of dinapses is inferred from records like those in **B** in which high gain current recordings are made from a pair of cells one of which is slightly depolarized so that it releases quanta at a low rate. Some of the minis seen in the current records from the two cells are coincident in time and must result from a single quantum being sensed by both neurons. When the amplitudes of simultaneous minis in the two cells are plotted against each other it is seen that there is a high degree of correlation, suggesting that mini amplitude is determined largely by variability in the amount of transmitter from one quantum to the next.

pre-and postsynaptic cells. The connection that mediates such a phenomenon is really a conjunction of a synapse and an autapse, as shown in Figure 11A, to which we have given the name "dinapse" (Frerking et al., 1995). Dinapses have been seen in hippocampal cultures (Vautrin et al., 1994) and there is indirect evidence that they exist also in the spinal cord (Liu et al., 1994).

The usefulness of dinapses for our purposes is that they should allow two independent samples of the same quantum, thereby providing us with a way of examining the causes of variability in mini size. If there is quantum to quantum variation at a single release site, with some quanta comprising more transmitter than other quanta, one would expect strong correlation between the amplitudes of the simultaneous minis seen in the two dinaptic partners. On the other hand, if quanta completely saturate postsynaptic receptors and all variations in mini size are due to differences between sites, correlation between dinaptic mini sizes would not necessarily be expected.

As shown in Figure 11C, a strong correlation exists in dinaptic mini amplitudes, thereby favoring the idea that quantum to quantum variation in the amount of transmitter contributes significantly to the variability of minis. Using some statistical manipulation it is possible to derive from the strength of this correlation, the fraction of variance in mini size that could be attributed to variations in transmitter between quanta. We calculate (Frerking et al., 1995) that as much as 75% of mini amplitude variance could be attributed to variations in transmitter concentration with the rest of the variance of unknown origin, though probably a large fraction of this remainder is due to stochastic variations in the number of channels that open. Further support for the nonsaturation of receptors in these amacrine cells comes from experiments in which we have applied diazapam, an agent thought to increase binding affinity of GABA to the $GABA_A$ receptor. If it were true that every GABA receptor bound GABA during a mini (i.e. receptor saturation), increasing the binding affinity of the receptor for its ligand would not result in bigger minis. Experimentally we find that minis do reversibly become larger in the presence of diazapam (Frerking et al., 1995).

If, as we propose here, minis vary because quanta are not all the same size, what accounts for the amount of transmitter in the quantum? A lot of processes must influence the amount of transmitter in the quantum, many of which are very poorly understood. For example, the rate at which vesicles fill with transmitter and the mechanism controlling the final concentration is unknown. Vesicles fill with transmitter through the action of exchangers that utilize the proton gradient across the vesicle membrane as the motive force. For acetylcholine-filled vesicles in *Xenopus* neuromuscular junction it seems that filling does not reach thermodynamic equilibrium (Song et al., 1997), though whether that is the case in GABA-filled vesicles in amacrine cells is not known. Despite these uncertainties, a very simple idea can be shown to fit the data well. At synapses where careful measurements have been made, the diameters of synaptic vesicles at presynaptic locations have been found to vary according to a Gaussian distribution (Palay and Chan-Palay, 1974: Bekkers et al., 1990). The number of molecules of transmitter within a vesicle then, because it is determined by vesicle volume, should depend on the third power of vesicle radius. Moreover, since two GABA molecules are usually required to open a $GABA_A$ channel, we would suppose that the number of postsynaptic channels open should depend on the square of this volume, in other words, the sixth power of the radius. Presuming that all vesicles are filled to the same concentration, we can calculate the expected distribution in mini amplitudes from the fact that the original distribution of diameters is Gaussian and has a particular coefficient of variation (around 12.5%). The expected shape of mini amplitude distributions is remarkably close to that seen experimentally and, in particular, it reproduces the strong skew seen in the distribution seen in Figure 10 (Frerking et al., 1995).

THE KINETICS OF RELEASE

As shown in Figure 12, small depolarizations of the presynaptic cell elicit low but steady rates of release that can be measured directly. Large depolarizations elicit rates of release that are not directly measurable since individual minis cannot be readily discriminated (Fig. 2). Furthermore, it is evident that for big depolarizations, the rate of release is not constant but changes with time elapsed since the onset of the depolarization. To examine the kinetics of release at these large depolarizations it is necessary to use a less direct approach..

Postsynaptic currents produced by strong depolarizations are built up from minis piled on top of each other according to some function of time that represents the instantaneous release rate. The postsynaptic current can be thought of as the result of a mathematical convolution operation of two functions (Bracewell, 1978), one the timecourse of a mini and the other, the instantaneous release rate function that we would like to derive. In order to extract this release rate function we have to work backwards, so to speak, by deconvolving the timecourse of a mini with the postsynaptic current. In practice this is trivial to implement using one of the many mathematics and engineering software packages commercially available.

Because minis and postsynaptic currents are variable, as we have seen, a good estimate of release rate requires that we work with the mean mini timecourse and the mean postsynaptic current derived from averaging responses to a number of trials (Fig. 13A,B). The result of such a deconvolution is shown in Figure 13C,D. Shortly after the beginning of the depolarizing step there is an initially high rate of release that lasts for a few milliseconds, followed by a much lower continuous rate of release that outlasts the depolarizing voltage step. It looks as though there is roughly synchronous release of vesicles initially, followed by a much lower rate of release. Direct measurements of the Ca^{++} current show

A

B

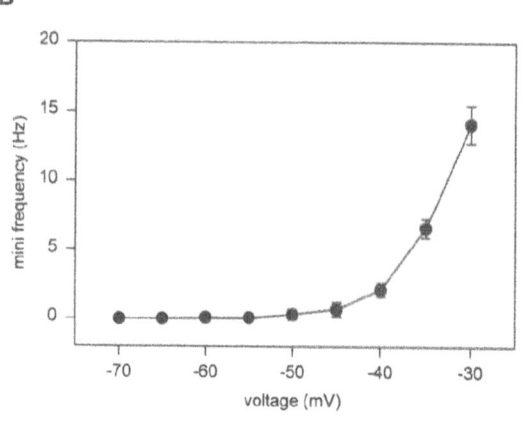

Figure 12. Mini frequency as a function of membrane voltage (from Frerking et al., 1997). A. Autaptic minis are shown for a holding voltage of –45 mV. B. For small depolarizations, where the rate of observed minis is constant, iterated depolarizations lasting 1.3 sec have been used to determine the release rate as a function of potential.

that it does not diminish significantly during the period in which release rate falls so profoundly. The explanation for this drop in release rate must therefore lie elsewhere.

To explore the release rate in greater depth it would be helpful to normalize the release rate function with respect to individual release sites. In order to do this it is necessary, of course, to estimate the number of release sites present on the presynaptic cell. A method for doing this is based on the identification of quantal peaks (Korn and Faber, 1991) like those originally described by del Castillo and Katz (1954) but the conditions necessary for this method to work are stringent and in the case of amacrine cells, the method would fail because the huge variance seen in mini sizes would smudge out any quantal peaks that might be present. A different approach to finding the number of release sites is to solve for N, the number of release sites, in the pair of simultaneous equations we can write below.

$$I_{e,peak} = N p\, I_{m,peak}$$

$$\mathrm{var}\, I_{e,peak} = I_{m,peak}^2\, N p\, (1-p) + N p\, \mathrm{var}\, I_{m,peak}$$

The first of these two equations describes the mean size of the peak response, $I_{e,peak}$, elicited by a big presynaptic depolarizing step, using the approximation that the peak post-

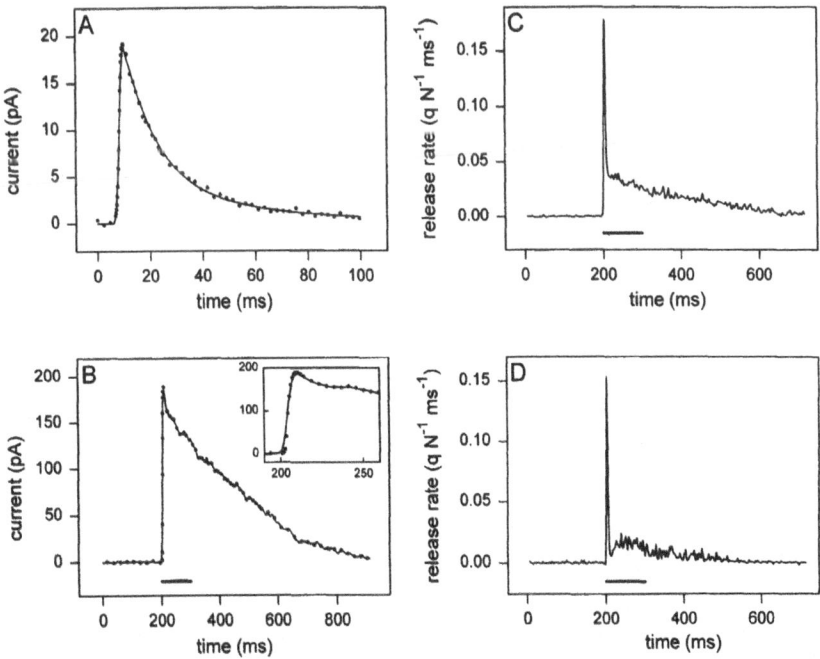

Figure 13. Estimation of release rate at synapses as a function of time (from Borges et al., 1995). Experiments used to derive the data shown here consisted of stepping the presynaptic cell from −70 to 0 mV for 100 ms. The postsynaptic current averaged from 18 trials is shown in **B** with an inset showing the rising phase of this current. Individual minis seen in the same postsynaptic cell have been averaged together after aligning them at their peaks as shown in **A**. The two waveforms shown in **A** & **B** are deconvolved to yield the instantaneous quantal release rate shown in panel **C**. Panel **D** shows a similar deconvolution done for another pair of cells. The release rates shown in **C** & **D** demonstrate that initially, for a period of a few msec, release rate is high but is followed by a much lower rate that is maintained even after the presynaptic membrane has been returned to its holding voltage.

synaptic current is due to synchronized minis piled on top of each other. The expected value is equal to the number of release sites, N, multiplied by the probability of release at a site, p, (assuming they are all the same), multiplied by the mean amplitude of a mini, $I_{m,peak}$. The second equation says the variance in the height of the evoked peak current, var $I_{e,peak}$, is the sum of two terms, the first of which is the binomial term describing variance resulting from the different numbers of quanta that could be released from trial to trial. The second term arises from the extra variance contributed by the variance in mini size, (which we assume is identical at all sites). Solving for N and p gives the result that p, the probability of release at all sites, is very close to one, and for most cells N, the number of release sites, is around ten. Finding that p is one during the peak evoked current implies that every release site more or less synchronously releases a quantum but following that event, even though the Ca^{++} influx through Ca^{++} channels is continuing unabated, the rate of release drops to much lower values, about 22 quanta/second.

A number of schemes are consistent with this kind of result. The one we proposed has the virtue of parsimony and can be described as a scheme of "fire and reload" (Borges et al., 1995). In this scheme, every release site has one vesicle primed to be released, just waiting for a suitable Ca^{++} signal. When this Ca^{++} signal arrives, release is very rapid, but following this event some much slower process is required to get the next vesicle ready for release: the "loading" process. Although simple and attractive this scheme does not fit very easily with the observation that at presynaptic sites at central synapses more than one vesicle can usually be seen docked at the plasma membrane (Gleason and Wilson, 1989). To be consistent with our scheme we would have to argue that only one of these apparently equivalent vesicles is sitting at the right location to be released, with the others forming the reserves. This is possible but at present there is no evidence that directly supports it. A slightly different interpretation would be that although there are a number of positions from which a vesicle might be released, the release of one vesicle somehow temporarily inhibits the release of other vesicles. This view has been articulated by others (Korn et al., 1994) but it is quite unclear how such an inhibitory message could be propagated between vesicles.

ACKNOWLEDGMENTS

I thank Salvador Borges for help in the preparation of this chapter. This work was supported by NEI, EY04112 to MW.

REFERENCES

Akagawa K, Takada M, Hayashi H, Uyemura K., 1990, Calcium- and voltage-dependent potassium channel in the rat retinal amacrine cells identified *in vitro* using a cell type-specific monoclonal antibody. Brain Res 518: 1–5.

Augustine, G.J., Adler, E.M., and Charlton, M.P., 1991, The calcium signal for transmitter secretion from presynaptic nerve terminals. Ann. N.Y. Acad. Sci 635: 365–381.

Barnstable, C.J., Hofstein R., and Akagawa K., 1985, A marker of early amacrine cell development in rat retina. Dev. Brain Res. 20: 286–290.

Blaustein, M.P., 1977, Effects of internal and external cations and of ATP on sodium-calcium and calcium-calcium exchange in squid axons. Biophys. J. 20: 79–110.

Bekkers, J.M., Richerson, G.B., and Stevens, C.F., 1990, Origin of variability in quantal size in cultured hippocampal neurons and hippocampal slices. Proc. Natl. Acad. Sci. USA 87: 5359–5362.

Borges, S., Gleason, E., Turelli, M. and Wilson, M., 1995, The kinetics of quantal transmitter release from retinal amacrine cells. Proc. Natl. Acad. Sci. USA. 92: 6896–6900.

Bracewell, R.N., 1978 The Fourier Transform and Its Application. McGraw-Hill, New York.

Cervetto, L., Lagnado, L., Perry. R.J., Robinson, D.W., and McNaughton, P.A., 1989, Extrusion of calcium from rod outer segments is driven by both sodium and potassium gradients. Nature 337: 740–743.

del Castillo, J., and Katz, B., 1954, Quantal components of the end-plate potential. J. Physiol. (Lond) 124:560–573.

Edwards, F.A., Konnerth, A., Sakmann, B., 1990, Quantal analysis of inhibitory synaptic transmission in the dentate gyrus of rat hippocampal slices: a patch-clamp study. J. Physiol. (Lond) 430:213–49.

Faber, D.S., Young, W.S., Legendre, P. and Korn, H., 1992, Intrinsic quantal variability due to stochastic properties of receptor-transmitter interactions. Science 258: 1494–1498.

Frerking, M., Borges, S. and Wilson M., 1995, Variation in GABA mini amplitude is the consequence of variation in transmitter concentration. Neuron 15: 885–895.

Frerking, M. and Wilson M., 1996, Effects of variance in mini amplitude on stimulus-evoked release: a comparison of two models. Biophys. J. 70: 2078–2091.

Frerking, M., Borges, S. and Wilson M., 1997. Are some minis multiquantal? J. Neurophysiol. 78: 1293–1304.

Frerking, M. and Wilson, M., 1996, Saturation of postsynaptic receptors at central synapses? Current Opinion in Neurobiology 6: 395–403.

Gleason, E., Borges, S. and Wilson, M., 1993, Synaptic transmission between pairs of retinal anacrine cells in culture. The Journal of Neuroscience 13: 2359- 2370.

Gleason, E., Borges, S., and Wilson, M., 1994, Control of transmitter release from retinal amacrine cells by Ca^{2+} influx and efflux. Neuron 13: 1109–1117.

Gleason, E., Borges, S. and Wilson M., 1995, Electrogenic Na-Ca exchange clears Ca^{2+} loads from retinal amacrine cells in culture. J. Neurosci. 15: 3612–3621.

Gleason, E. and Wilson, M., 1989, Development of synapses between chick retinal neurons in dispersed culture. J. Comp. Neurol 287: 213–224.

Goda, Y. and Stevens, C.F., 1994, Two components of transmitter release at a central synapse. Proc. Natl. Acad. Sci. USA. 91: 12942–12946.

Katz, B., 1969, The release of neural transmitter substances. Liverpool University Press, Liverpool.

Korn, H. and Faber, D.S., 1991, Quantal analysis and synaptic efficacy in the CNS. TINS 14: 439–445.

Korn, H , Sur, C., Charpier, S., Legendre, P. and Faber, D.S. 1994, The one-vesicle hypothesis and multivesicular release. In Molecular and Cellular Mechanisms of Neurotransmitter Release. L. Stjärne, P. Greengard, S. Grillner, T. Hökfelt, and D. Ottoson, editors. Raven Press, Ltd., New York. 301–322.

Huba, R. and Hofmann, H.-D., 1990, Identification of GABAergic amacrine cell-like neurons developing in chick retinal monolayer cultures. Neurosci. Lett. 117: 37–42.

Huba, R. and Hofmann, H.-D., 1991, Transmitter-gated currents of GABAergic amacrine-like cells in chick retinal cultures. Visual Neurosci. 6: 303–314.

Lagnado, L., Gomis, A. and Job, C., 1996, Continuous vesicle cycling in the synaptic terminal of retinal bipolar cells. Neuron 17: 957–67.

Liu, G., and Tsien, R.W., 1995, Properties of synaptic transmission at single hippocampal synaptic boutons. Nature 375: 404–408.

Liu, H., Wang, H., Sheng, M., Jan, L.Y., Jan, Y.N. and Basbaum, A.I., 1994, Evidence for presynaptic N-methyl-D-aspartate autoreceptors in the spinal cord dorsal horn. Proc. Natl. Acad. Sci. USA. 91: 8383–7.

McNaughton, P.A., 1991, Fundamental properties of the Na-Ca exchange. An Overview. Annals of the New York Academy of Sciences 639: 2–9.

Palay, S. and Chan-Palay, V., 1974, Cerebellar Cortex: Cytology and Organization Springer, New York.

Reeves, J.P. and Hale, C.C., 1984, The stoichiometry of the cardiac sodium-calcium exchange system. J. Biol. Chem. 259: 7733–7739.

Rieke, F. and Schwartz, E.A., 1996, Asynchronous transmitter release: control of exocytosis and endocytosis at the salamander rod synapse. J. Physiol. (Lond), 493:1–8.

Song, H.J., Ming, G.L., Fon, E. and Bellocchio, E., 1997, Expression of a putative vesicular acetylcholine transporter facilitates quantal transmitter packaging. Neuron 18:815–826.

Vautrin, J., Schaffner, A., and Barker, J., 1994, Fast presynaptic $GABA_A$ receptor mediated Cl^- conductance in cultured rat hippocampal neurones. J. Physiol. (Lond) 479: 53–63.

13

SCALING THE RETINA, MICRO AND MACRO

Barbara L. Finlay and Randolph L. Snow

Developmental Neuroscience Group
Cornell University
Ithaca, New York 14853

1. ABSTRACT

The scaling of retinal cell morphology following experimental manipulation and of retinal cell numbers across a wide range of species is reviewed. Morphological changes in retinal ganglion, amacrine and bipolar cells following manipulations to change retinal area or convergence are reviewed to demonstrate cell scaling at the single cell level. The results of these studies are compared to a cross species analysis of changes in cell size and number to demonstrate cell scaling at the level of ensembles of neurons. The convergence of these two levels of scaling is discussed in light of current knowledge of normal retinal development.

2. INTRODUCTION

Vertebrate eyes come in a wide range of sizes. The axial length of the smallest vertebrate eyes, such as the functioning eyes of newly mobile teleosts, is appropriately measured in microns (Easter, Nicola and Burrill 1998), and the largest axial lengths, such as those found in whales and elephants, are best measured in centimeters (Hughes 1977). Vertebrate neurons, however, have nothing like this range of size. While the lengths of axons and some dendrites may vary from microns to centimeters, the diameter of cell bodies and their processes has a much smaller range, limited by diffusion and short-range intracellular transport mechanisms. The essential problem is how to build an organ that scales over a large range using as building blocks elements that may scale only in piecemeal fashion.

Different organizational features of the eye have different scaling requirements: for some features, the relative dimensions of elements are critical, but for others, the absolute size of an element must be maintained over a range of eye sizes. For example, the gross dimensions of the eye—such as axial length, lens thickness and retinal area—scale relatively, such that the schematic eyes of related vertebrates of different sizes can be virtually superimposed (for example, Remtulla and Hallet 1985). In contrast, retinal thickness has

Development and Organization of the Retina, edited by Chalupa and Finlay.
Plenum Press, New York, 1998.

an absolute limit: light must be transmitted through it; nutrients must be able to reach the inner layers; and non-spiking potentials must propagate across its depth. Furthermore, the requirements of retinal construction itself must be considered: components of the brain, including the retina, scale in a very predictable way based on order of neurogenesis (Finlay and Darlington 1995). The outcome of this developmental constraint appears basically unrelated to preservation of eye function at various eye sizes, but must be accommodated in eye structure.

Scaling of the eye and retina in vertebrates is not only a problem that must be solved in evolutionary time but one that must also be adjusted in the lifetimes of individuals. Coordination of cytogenesis, neurogenesis, and simple expansion of the eye must necessarily occur in those vertebrate like teleosts whose eye growth is indeterminate (Fernald 1989). However, eyes grow in substantial amounts during the development and early function of most mammals and birds. This growth occurs well after neurogenesis has ceased and a version of scaling problem must then be solved with a fixed number of neurons.

We propose to examine evolutionary and developmental scaling of eyes and retinas as a natural class of phenomena. We will review a series of our own experiments in which we experimentally manipulate the rate of growth of the eye and the number of retinal ganglion cells in the developing chick to determine how the morphology of various cell classes changes with varying growth and convergence conditions. Then, we will examine the scaling of numbers and distributions of cells across mammals, to determine what variation in convergence and growth must be accommodated in the evolution of the eye. Finally, we compare the results of these experimental manipulations to observed cross-species variation in cell morphology consequent to evolutionary variation in relative and absolute cell densities. These investigations point at structural and developmental constraints common to both developmental and evolutionary scaling.

3. DEVELOPING RETINAS: REGULATION OF RETINAL GANGLION, AMACRINE, AND BIPOLAR CELL ARBORIZATION IN THE GROWING CHICK RETINA

Generally, the retina has two tasks: 1. to indicate the presence and location of contrast differences in the visual array, and 2. to begin the analysis of the cross-retinal pattern of visual stimulation. These tasks are reflected in the radial and tangential organization of the retina, respectively. Photoreceptors, both rods and cones, must transmit stimulus information through bipolar cells and, optionally, through amacrine cells to reach retinal ganglion cells, which transform the graded information received into a spike discharge that can be communicated to the rest of the brain. Photoreceptor type, retinal location, and particular species of animal are all factors that set the convergence and divergence of the radial pathway of information flow. In the tangential organization of the retina, horizontal cells, amacrine cells, and retinal ganglion cells spread their processes widely to allow the comparison of information from neighboring and more distant photoreceptors. Because both the radial and tangential pathways must converge on the single output cell of the retina, the retinal ganglion cell, retinal development may be seen as a push/pull of these two forces.

The chick is a convenient animal in which to observe and manipulate retinal development during early function. The hatchling chick will peck at food items, avoid looming objects, and begin the visual learning of species discriminations—its retina is quite differentiated. However, the chick eye at hatching is only about 80% of its adult size, so the chick is still in the process of solving its scaling problem (Troilo et al. 1996). At this point, both the generation of and death of neuronal elements in the retina is well over, so the existing cells must be reconformed to fit the expanding area. The spatial resolution of information transmitted through the retina is maintained or improved during this period (Troilo 1992). How is the ensemble of cells jointly regulated in their connections to maintain both acuity and pattern vision?

Simple observation of changes in cell morphology during this period would be informative, but the non-uniformity of the normal retina confounds a complete answer to the question. Retinal growth is non-uniform across its surface; the periphery stretches more than the center (Reichenbach et al. 1991; Robinson 1991). Convergence from photoreceptors to inner nuclear cells (bipolars and amacrines) to retinal ganglion cells also is non-uniform over the retinal surface: convergence is greater in the periphery than center. Both of these factors are likely to affect the arborization of retinal neurons. Therefore, we designed two experimental manipulations to separate retinal growth from varying convergence.

3.1. Eye Growth, Retinal Stretch, and Emmetropization

The chicken eye (Wallman et al. 1987), as well as a number of mammalian eyes [monkey (Wiesel and Raviola 1979), tree shrew (Norton 1990), and marmoset (Troilo et al. 1993)], has been shown to use visual experience to match the axial length of the eye to the power of its optics (cornea and lens); that is, to assure that the visual image is focused on the retina—not in front of or behind it. This process is called emmetropization. In the chicken, a substantial component of this process occurs directly in the eye and retina itself, not requiring the brain (Troilo et al. 1987). If the retina receives a high-contrast image, maximally activating the photoreceptors and other neurons of the retina, then optics and axial length must match, and the growth of the eye is checked. However, if the image is blurred, the eye continues to grow, in an apparent attempt to find a match between optics and axial length. If the blur has been caused experimentally by a diffusing lens instead of by actual defocus, this manipulation will anomalously increase the axial length of the eye at a rate faster than normal growth, producing a condition called "experimental myopia". The axial length of the eye can be made to increase by as much as a third during the first 4–6 weeks post-hatch by this procedure, increasing both the diameter of the eye and the area of the retina (Troilo et al. 1996). Although other tissues of the eye may add cells during this period of growth, the retina does not add neurons. Therefore, the retina is essentially stretched, and existing neurons must grow or reconform their processes within the stretched retina. Like a balloon with areas of greater and lesser wall thickness, this stretch is not uniform. Those parts of the retina that are already thicker (the area centralis) stretch little, while the thinner, more elastic parts of the retina (the periphery in general) will stretch the most (Kelling et al. 1989; Reichenbach et al. 1991; Robinson et al. 1989). Therefore, experimental myopia can be used to examine how retinal neurons alter to fit increased area, particularly in the retinal periphery. Since the numbers of neurons are fixed, there is no change in convergence between the different layers of the retina.

3.2. Experimental Manipulations, Retinal Neurons Studied, and Visualization Procedures

We used either experimental myopia, or depletion of retinal ganglion cells to examine the control of arborization across several classes of retinal neurons (Table 1). Experimental myopia was produced by putting a diffusing lens in front of one eye for the first three weeks posthatch, which produced an increase in retinal area of approximately one

Table 1. Strategy and sources for comparison across manipulations of retinal size and retinal cell convergence

Manipulation	Cell type		
	Retinal ganglion cell	Bipolar	Amacrine
Retinal enlargement via form deprivation	Troilo et al. 1996	Finlay et al. 1997	Snow et al. 1997 Teakle et al. 1993
Variation of retinal convergence via partial-optic nerve section	Troilo et al. 1996	Finlay et al. 1997	Xiong et al. 1997

third. Alternatively, a patchy depletion of retinal ganglion cells was produced by a partial crush of the optic nerve behind the orbit at hatching, causing cell loss in the retinal ganglion cell layer averaging about 50%. After three weeks of posthatch growth, retinas from both manipulations were examined. Therefore, retinal neurons grew during comparable periods in either a situation of increased retinal stretch, where the density of all retinal neurons was reduced, or during a selective reduction of the density of retinal ganglion cells, causing increased convergence of inner nuclear cells on the remaining retinal ganglion cells. We were able to make use of a previous study in chickens by Teakle et al. 1993 on the effects of increased retinal stretch on dopaminergic amacrine cells to fill out this examination (Table 1).

In wholemounts we examined the small bushy retinal ganglion cells of the chicken following application of DiI crystals to the cut ends of optic nerves in fixed tissues obtained from both conditions. Small bushy cells are the majority retinal ganglion cell population, whose reasonably compact arbors are advantageous for this technique (Snow et al. 1994; Thanos et al. 1992). We also measured the arborization of dopaminergic amacrine cells in flatmount, as visualized by tyrosine hydroxylase immunohistochemistry, after retinal ganglion cell depletion. These results are compared to those of Teakle et al. after retinal expansion (1993). Using the Golgi technique in sectioned tissue (Sherry and Yazulla 1993), we examined a collection of various types of bipolar cells and a class of amacrine cells of medium arbor size with a diffuse stratification. Examination of the diameter or tangential area of arbors allowed us to study how retinal coverage (tangential integration) was preserved: analysis of the number and density of branches, the total length of all processes and the pattern of stratification of the cell's processes in the inner plexiform layer allowed us to study how radial transmission of information was maintained.

3.3. Tangentially Oriented Cells (Retinal Ganglion and Amacrine Cells) Differ from Radially Oriented Cells (Bipolars) in Their Response to Retinal Expansion

Retinal ganglion and amacrine cells reflect the amount of retinal expansion by increasing the area over which they arborize without increasing their total number of branches, which reduces process density (Figure 1) (Troilo et al. 1995); see also (Bloomfield and Hitchcock 1991; Mastronarde et al. 1984; Teakle et al. 1993). Diffuse amacrine cells respond in an exactly comparable manner (Figure 1) (Snow et al. 1997). This preserves relative coverage and relative internal geometry, but reduces absolute coverage of the retinal surface. The mechanism of this growth is likely interstitial addition of membrane between branches, since addition of more branches would alter the internal geometry of these cells' arborization and process density (Maslim et al. 1986). In the same

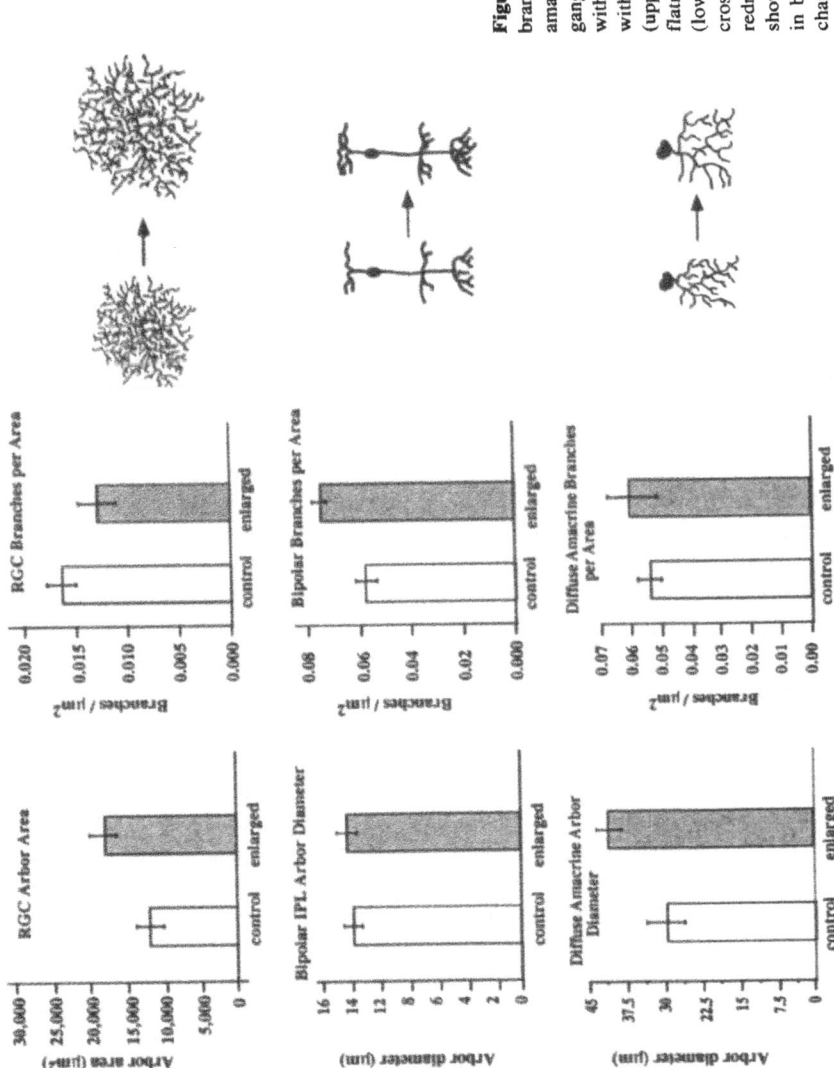

Figure 1. Comparison of arbor area or diameter and branch density for retinal ganglion, bipolar and diffuse amacrine cells following retinal enlargement. Retinal ganglion cells significantly increase their arbor area without significantly changing numbers of branches within the arbor therefore arbor density is decreased (upper graphs: redrawn from Troilo et al. 1996—DiI flatmounts). Diffuse amacrine cells respond similarly (lower graphs: redrawn from Snow et al. 1997—Golgi cross sections). However, bipolar cells (middle graphs: redrawn from Finlay et al. 1997—Golgi cross sections) show no increase in diameter but a significant increase in branch density. A schematic representation of each change is shown in the last column.

conditions, bipolar cells do not change their total tangential diameter and arbor area, thus failing to maintain coverage (Finlay et al. 1997) (Figure 1). On the other hand, bipolar cells do increase the number of branches within an arborization (Figure 1). A priori, it would seem that all cell types should enlarge symmetrically when subjected to a mechanical force such as stretch, but consideration of the information-processing requirements of the retina and the physical constraints of synaptic transmission suggest reasons why that should not be so.

When retinal ganglion cells are enlarged during retinal growth in the goldfish, interstitial addition of membrane increases both the length and diameter of the dendritic processes (Bloomfield and Hitchcock 1991). Since synapses do not change in area with increasing process size, more synapses must be added to depolarize the larger cell; therefore, the total number of synapses per cell goes up (Hitchcock 1993)—the cable properties of the cell remain constant. In our experimental manipulation, as the retinal ganglion cell or amacrine cell is enlarged via interstitial growth, each synaptic contact of the bipolar cell is effectively weakened by being moved further from the soma. In addition, the same change in process diameter described by Bloomfield and Hitchcock (1991) may also occur. In order to maintain function, bipolar cells must either increase the strength of individual synapses in some fashion, or increase the number of synapses. Purves (1988) has described a remarkable consistency in the size of synapses across species and conditions, and since increase in the number of synaptic contacts is a common observation in various cases of experience-dependent plasticity, we suggest that an increase in the number of synapses is the most likely solution. We hypothesize the increased local branching of bipolar cells is the morphological correlate of increased synaptic number. In this fashion, the bipolar cell supplies the increased number of synapses required to stimulate an enlarged amacrine or retinal ganglion cell in response to change in their cable properties, preserving information transmission.

Is it important that coverage of the inner plexiform layer by bipolar cell neurites is lost as the retina expands? If it is the case that the main function of bipolar cells is to faithfully transmit local spatial information, and other cells cover the function of horizontal integration of information, then there may be no need to increase the size of bipolar arbors, provided that the transmission of excitation is not compromised.

3.3.1. Push–Pull Relationship of Stratification and Process Density in Bipolar Cells.
A quite unexpected relationship emerged between the number of layers of stratification of bipolar cells and their total branchedness: in bipolar cells, the total amount of processes appeared to be conserved, both in normal development and in our experimental manipulation (Figure 2). Bipolar cells may arborize in up to four strata in the inner plexiform layer in chickens. In the normal animal, if a cell has three layers of arborization, each of those arbors has just a little more than a third of the extent of arborization of a monostratified cell. In the experimentally manipulated retinas, this effect is even stronger, and moreover, the mean numbers of layers of stratification is significantly reduced. Normal animals average 2.1 layers of stratification, but after retinal expansion and the resulting increase in branch density described above, experimental animals average only 1.7 strata (Figure 2). The total amount of arborization is conserved—it appears that the increase in branch numbers within a stratum occurs at the expense of total stratification. Since strata in the IPL normally correspond to either functional divisions (such as on- or off-center cells) or the type of postsynaptic cell contacted (amacrine or retinal ganglion cell class), this result suggests that each bipolar cell in expanded retinas distributes its information to fewer cell classes.

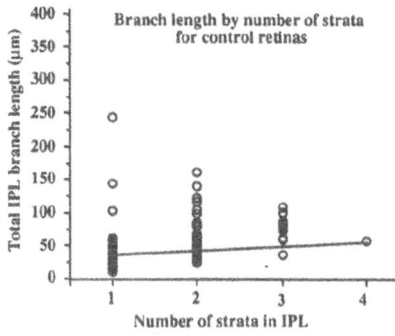

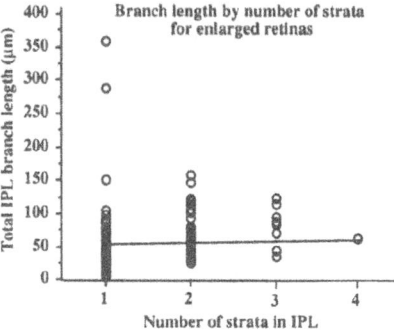

Figure 2. Changes in bipolar arbor stratification in response to retinal enlargement. These graphs show branch length divided out by strata for control (upper graph) and enlarged (middle graph) retinas (redrawn from Finlay et al. 1997). Note, as the number of strata increase the length is divided equally between strata. A significant decrease in number of strata was observed for cells from enlarged retinas (lower graph: redrawn from Finlay et al. 1997). A schematic representation of these changes is shown in the last panel.

3.4. Tangentially Oriented Cells May Differ from Radially Oriented Cells in Their Response to Retinal Ganglion Cell Depletion

When the growing retina is partially depleted of ganglion cells (increasing the number of inner nuclear layer cells compared to ganglion cells) the remaining retinal ganglion cells sprout new branches, increasing both their tangential extent and process density (Troilo et al. 1995). This suggests some trophic effect of either bipolar or amacrine cell synaptic availability on process growth in retinal ganglion cells. Symmetrically and oppositely, dopaminergic cells respond to reduced total postsynaptic availability by reducing their total number of processes (Figure 3) (Xiong et al. 1997). However, there was no hint of a decrease in the total number of processes or consistent change in any feature of bipolar cellular organization following depletion. So, while amacrine and retinal ganglion cells appear to respond in a fairly robust fashion to alterations in the total pool of synapses in the inner plexiform layer, bipolar cells retain the size and conformation of their arbors.

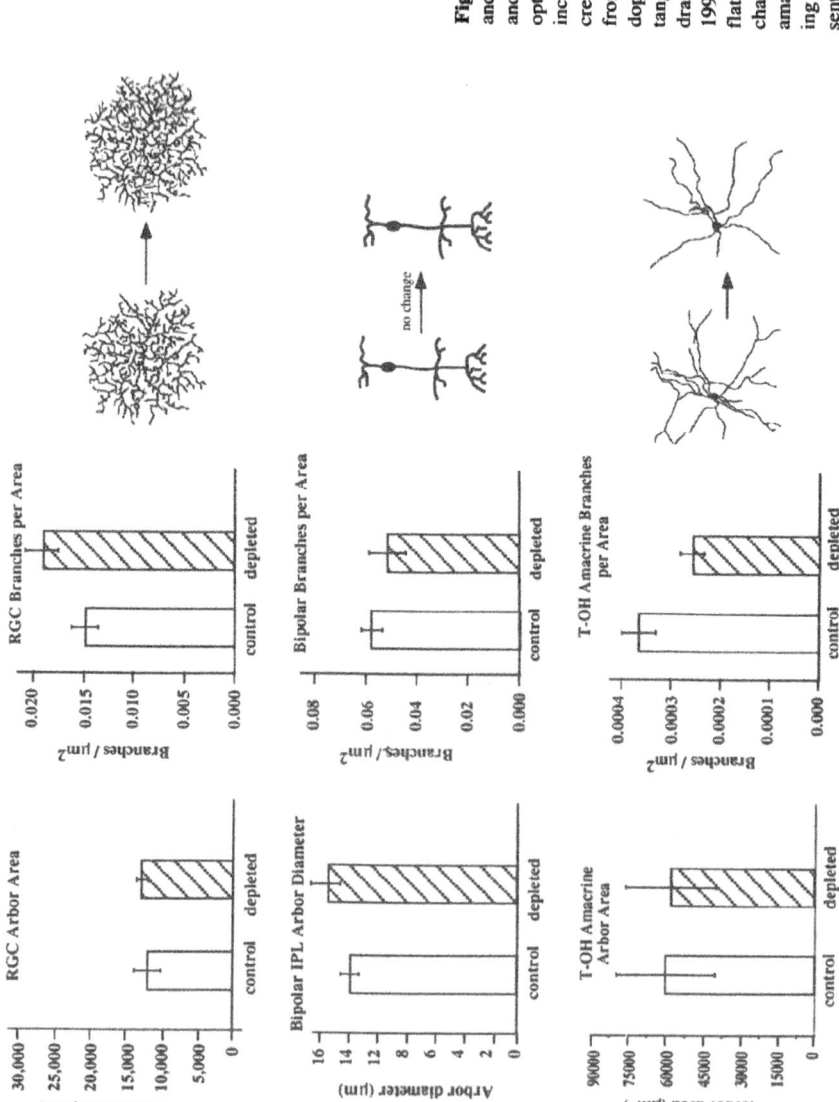

Figure 3. Comparison of arbor area or diameter and branch density for retinal ganglion, bipolar and dopaminergic amacrine cells following partial optic nerve section. Retinal ganglion cells slightly increase their arbor area with a significant increase in arbor density (upper graphs: redrawn from Troilo et al. 1996). However, bipolar and dopaminergic amacrine cells show no change in tangential extent (middle and lower graphs: redrawn from Finlay et al. 1997 and Xiong et al. 1997—tyrosine hydroxylase immunoreactivity in flatmounts). Additionally, bipolar cells show no change in arbor density while dopaminergic amacrine cells show a decrease in density following partial optic nerve section. A schematic representation of each change is shown in the last column.

3.5. Radial versus Tangential Cell Deployment and Growth Strategies in the Retina

Even though mechanical stretch prompts the addition of interstitial membrane in retinal ganglion and amacrine cells, bipolar cells appear to be virtually unresponsive to stretch. We found little evidence in bipolar cells for the "seeking", terminal branching kind of growth that is seen in the horizontally extended cell classes when synaptic sites become available, although there is a mechanism that promotes very local branching, perhaps prompted by failing synaptic efficacy. These dichotomies in the sprouting behavior of bipolars versus retinal ganglion and amacrine cells are echoed in other observations of retinal development. If retinal clones are labeled with retroviruses, radially oriented cells (photoreceptors, bipolar cells and Müller cells) remain more tightly linked to their radial position of origin, while retinal ganglion, amacrine, and horizontal cells may migrate substantially away (Reese et al. 1995). Similarly, the relative numbers of radially oriented cells in a column tend to be conserved across the retina while horizontally oriented cells vary (Reichenbach et al. 1994). The conservative manner of bipolar cell growth may serve a central function in development and at maturity. By the simple mechanism of restricted growth potential in early development for one class of cell, the retinal mosaic is kept intact, with faithful vertical transmission of high-acuity spatial information in the retina.

4. EVOLVING RETINAS AND EYES: DEVELOPMENTAL STRUCTURE IN EVOLUTIONARY TRENDS

In this section we will shift the focus somewhat. Fundamentally, we will be arguing that because of pronounced regularities in the relative numbers of neurons in brain components as the brain enlarges, the retina (as part of the brain) will be presented with predictable developmental problems to solve in mapping one population of neurons onto the next. These evolutionary problems are not unlike the experimental manipulations of the developing retina described above. This developmental work gives us some structure to dissect the role that absolute and relative neuronal densities play in the cellular organization of the retina across different species. In the following section we will describe what some of those predictable relationships are, and then describe some preliminary observations coming from the first phase of a large scale collaboration with the laboratory of Luiz Carlos L. Silveira concerning scaling of the primate visual system. Finally, we will compare the structure of bipolar cells and retinal ganglion cells across species to see if their structures differ in the fashion predicted from the experimental manipulations of population convergence.

4.1. Regularities of Brain Evolution in the Case of the Retina

In a survey of patterns of neurogenesis in seven mammals, ranging from possum to macaque, extreme conservation of sequence was observed (Finlay and Darlington 1995). While the total duration of neurogenesis could vary by almost a factor of 10, it was possible to make a simple, two-factor model that would predict with extreme accuracy the day that a given structure was generated in a particular animal: the fit of data to model was 98.8%. The conservation of the order of neurogenesis in turn translated into highly predictable patterns of brain enlargement in animals with long durations of neurogenesis and

correspondingly large brains: late generated structures or cell groups became disproportionately large in large brained animals.

This relationship seems reasonably non-controversial when comparing cell groups
like spinal cord motor-neurons to such structures as the isocortex and the cerebellum.
However, it is not so obvious that this should hold for structures like the retina, where the
distribution and number of cell types would seem to be under strong functional limitations
and environmental adaptations, as contrasted with developmental constraint. The predictions made by this hypothesis are quite clear, however, and easy to test: across mammals,
as brain size increases, the first generated cells of the retina, retinal ganglion cells, should
increase the least in number with brain size; the next generated cells, cones, with a slightly
higher slope; bipolar and amacrine cells intermediate; and rods with the steepest slope.
The last prediction is of most interest, since an adaptation model must clearly predict that
the number of rods should be relatively higher in nocturnal animals, for example.

A survey of numbers of retinal ganglion cells, cones, and rods in published work on
rodents and primates supports the predicted change in numbers of each cell group surprisingly well (Figure 4). These observations are taken from: (Curcio et al. 1987; Fischer and
Kirby 1991; Henderson 1985; Perry et al. 1983; Rhoades et al. 1979; Troilo et al. 1993;
Wikler and Rakic 1990; Wikler et al. 1990; Williams et al. 1996). Retinal ganglion cell
numbers scale with the shallowest slope, cone numbers with an intermediate slope, and
rod numbers with the steepest slope. In this still-limited data set, it is the definitively non-
nocturnal humans that have the most rods. Comparing rats to humans, humans have about
one million more retinal ganglion cells, 4 million more cones, and 70 million more rods.

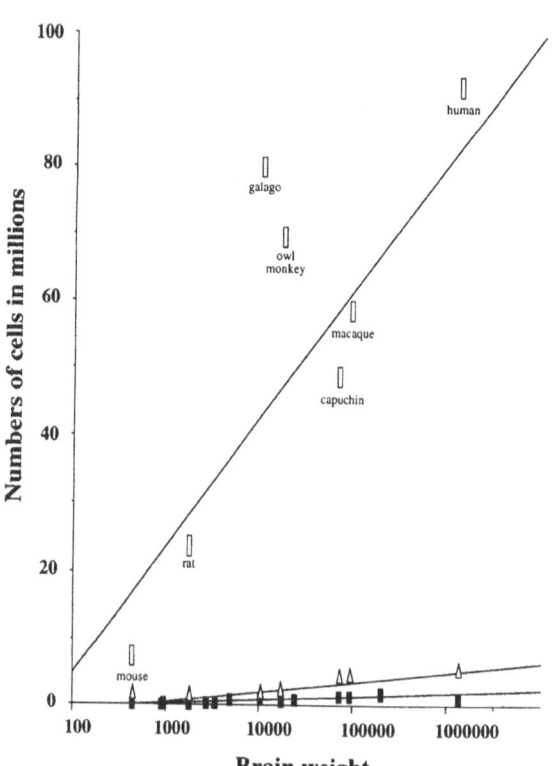

Figure 4. Number of cells (retinal ganglion cells, rods, or cones) versus the
brain weight for species from mice to
men. Note the dramatic increase in number of rods (rectangles) versus cones (triangles) or retinal ganglion cells (squares).
Scatter plots are fit with simple regression
lines, $r^2 = 0.649$ for rods, 0.799 for cones
and 0.762 for retinal ganglion cells. Species for which rod, cone and retinal ganglion cell numbers are graphed and
labeled; references are as in text.

However, it is clear the adaptation argument gets some support as well: the nocturnal owl monkey and galago clearly have many more rods than would be predicted by allometric scaling, and we are interested in determining just what about normal neurogenesis is altered to produce this adaptation.

4.2. An Aside on the Primate Fovea

Most animals have a central concentration of cells, such as an area centralis or visual streak, and this concentration is usually most pronounced for retinal ganglion cells and cones, and less pronounced for other inner nuclear cell classes and rods. In most primates, this central concentration is yet more specialized, with cones found at their maximal packing density, rods absent, and the cell bodies of the photoreceptors and all other retinal neurons displaced to the outside of the region of the maximal packing density of cones. It is quite clear what the virtue of maximizing acuity is for an animal like a primate that is diurnal and does a great deal of close-range manipulation. It is less clear why the area of maximal acuity should be so limited in the extent of its visual angle. The constraint on the absolute numbers of cones that can be generated gives a clue to the limited size of the primate fovea. An animal with a pronounced central specialization like a fovea covering a small visual angle must be prepared to allocate more neural resources to the control of eye movements, and to extracting information from glance-to-glance comparisons. Since central neural resources do not have this same limit on number (the isocortex scales steeply in neuron number with brain size), a computationally-intensive solution to a limit on peripheral acuity is viable.

4.3. Scaling of Retinal Ganglion and Bipolar Cells across Species

In the animals surveyed above (Figure 4), retinal ganglion cells become distributed in a lesser and more variable density in a larger eye, and are a lower and lower fraction of the number of bipolar and amacrine cells potentially serving as input to them. Bipolar cells, therefore, confront a reduced population of retinal ganglion cell targets and a somewhat expanded eye. According to our experimental manipulations of development, we should predict that ganglion cells should be highly responsive to lowered density and eye size, and bipolar cells relatively impervious. More detailed predictions about the nature of arbor structure could also be made for both cell groups, but would require control of the location of retina sampled, and the amount of convergence. The general prediction holds true: across species as diverse in size as rats and humans, 7 to 9 types of bipolar cells are typically described, with a consistent axonal coverage of 18–50 μm (Euler and Wässle 1995; Kolb et al. 1981; Massey and Mills 1996; Wässle et al. 1991) In comparison, the somal diameter alone of retinal ganglion cells has this range (10–60 μm), and the arbor diameter of these cells, often hard to trace because of extreme size, can very from less than 20 μm to greater than 1000 μm (Amthor et al. 1983; Brown 1965; Dawson et al. 1989; Kolb et al. 1981; Peichl et al. 1987; Schall et al. 1987; Wässle et al. 1990; Wässle et al. 1981; Amthor et al. 1983; Brown 1965; Dawson et al. 1989; Kolb et al. 1981; Peichl et al. 1987; Schall et al. 1987; Wässle et al. 1990; Wässle et al. 1981; Thanos et al. 1991). The same advantages in this strategy that applied to scaling the eye in development apply also to scaling the eye in evolution: restricted growth of the bipolar cell retains the faithful transmission of high-acuity spatial information, while the generous expansion of tangentially-directed cells, like retinal ganglion cells, retains the potential for lateral interactions regardless of eye size or convergence ratios.

The challenge of work that relates developmental mechanisms to evolutionary paths is to determine if there might be some relatively limited set of developmental rules that, when executed in different contexts, produce all the functional diversity and specificity of the cell classes and functions seen in the retina. When may various features of the eyes of different species be understood as the necessary outcomes of conserved developmental mechanisms? When are developmental mechanisms altered to produce differing retinal structures across species? Such a taxonomy will allow us to account not only for the development and variation of any particular eye, but also the range of possible structure for all vertebrate eyes.

ACKNOWLEDGMENTS

Supported by NIH19245 and NSF INT 9604599 to BLF.

REFERENCES

Amthor, F. R., Oyster, C. W., and Takahashi, E. S. (1983). "Quantitative morphology of rabbit retinal ganglion cells." Proceedings of the Royal Society of London B, 217, 341–355.

Bloomfield, S. A., and Hitchcock, P. F. (1991). "The dendritic arbors of large-field ganglion cells show scaled growth during expansion of the goldfish retina: a study of morphometric and electrotonic properties." Journal of Neuroscience, 11, 910–917.

Brown, J. E. (1965). "Dendritic fields of retinal ganglion cells of the rat." Journal of Neurophysiology, 28, 1091–1110.

Curcio, C. A., Sloan, K. R., Packer, O., Hendrickson, A. E., and Kalina, R. E. (1987). "Distribution of cones in human and monkey retina: Individual variability and radial asymmetry." Science, 236, 579–582.

Dawson, W. W., Hawthorne, M. N., Parmer, R., Hope, G. M., and Hueter, R. (1989). "Very large neurons of the inner retina of humans and other mammals." Retina, 9, 69–74.

Easter, S. S., Nicola, G. N., and Burrill, J. D. (1998) "Development of vision and the pre-visual system." In: Development and Organization of the Retina: From Molecules to Function. L. M. Chalupa and B. L. Finlay, eds. Plenum Press, New York, 1998

Euler, T., and Wässle, H. (1995). "Immunocytochemical identification of cone bipolar cells in the rat retina." Journal of Comparative Neurology, 361(3), 461–478.

Fernald, R. D. (1989). "Seeing through a growing eye." Perspectives in Neural Systems and Behavior, T. J. Carew and D. B. Kelly, eds., Liss, New York, 151–174.

Finlay, B. L., Crowley, J. C., Rubin, B. D., and Snow, R. L. (1997). " Developmental regulation of the spread and stratification of ganglion cell processes." Manuscript submitted

Finlay, B. L., and Darlington, R. B. (1995). "Linked regularities in the development and evolution of mammalian brains." Science, 268(5217), 1578–1584.

Fischer, Q. S., and Kirby, M. A. (1991). "Number and distribution of retinal ganglion cells in anubis baboons (Paio anubis)." Brain, Behavior and Evolution, 37, 189–204.

Henderson, Z. (1985). "Distribution of ganglion cells in the retina of adult pigmented ferret." Brain Research, 358, 221–228.

Hitchcock, P. F. (1993). "Mature, growing ganglion cells acquire new synapses in the retina of the goldfish." Visual Neuroscience, 10, 219–224.

Hughes, A. (1977). "The topography of vision in mammals of contrasting life style: comparative optics and retinal organization." Handbook of Sensory Physiology, F. Crescitelli, ed., Springer-Verlag, Berlin, 613–756.

Kelling, S. T., Sengelaub, D. R., Wikler, K. C., and Finlay, B. L. (1989). "Differential elasticity of the immature retina: A contribution to the development of the area centralis?" Visual Neuroscience., 2, 117–120.

Kolb, H., Nelson, R., and Mariani, A. (1981). "Amacrine cells, bipolar cells and ganglion cells of the cat retina: a Golgi study." Vis. Res., 21, 1081–1114.

Maslim, J., Webster, M., and Stone, J. (1986). "Stages in the structural differentiation of retinal ganglion cells." Journal of Comparative Neurology, 254, 382–402.

Massey, S. C., and Mills, S. L. (1996). "Calbindin-immunoreactive cone bipolar cell type in the rabbit retina." Journal of Comparative Neurology, 366(1), 15–33.

Mastronarde, D. N., Thibeault, M. A., and Dubin, M. W. (1984). "Non-uniform postnatal growth of the cat retina." Journal of Comparative Neurology, 228, 598–608.

Norton, T. (1990). "Experimental myopia in tree shrews." Myopia and the Control of Eye Growth, G. Bock and K. Widdows, eds., Wiley, Chichester, 178–199.

Peichl, L., Buhl, E. H., and Boycott, B. B. (1987). "Alpha ganglion cells in the rabbit retina." Journal of Comparative Neurology, 263, 25–41.

Perry, V. H., Henderson, Z., and Linden, R. (1983). "Postnatal changes in retinal ganglion cell and optic axon populations in the pigmented rat." Journal of Comparative Neurology, 219, 356–368.

Purves, D. (1988). Body and Brain: A Trophic Theory of Neural Connections, Harvard University Press, Cambridge, Ma.

Reese, B. E., Harvey, A. R., and Tan, S. S. (1995). "Radial and tangential dispersion patterns in the mouse retina are cell-class specific." Proceedings of the National Academy of Science, USA, 92(7), 2494–2498.

Reichenbach, A., Eberhardt, W., Scheibe, R., Deich, C., Seifert, B., Reichelt, W., Dahnert, K., and Rodenbeck, M. (1991). "Development of the rabbit retina IV. Tissue tensility and elasticity in dependence on topographic specializations." Experimental Eye Research, 53, 241–251.

Reichenbach, A., Ziegert, M., Schnitzer, J., Pritzhohmeier, S., Schaaf, P., Schober, W., and Schneider, H. (1994). "Development of the rabbit retina .5. the question of 'columnar units'." Brain Research and Developmental Brain Research, 79(1), 72–84.

Remtulla, S., and Hallet, P. E. (1985). "A schematic eye for the mouse and comparisons with the rat." Vision Research, 25, 21–31.

Rhoades, R. W., Hsu, L., and Parfett, G. (1979). "An electronmicroscopic analysis of the optic nerve in the golden hamster." Journal of Comparative Neurology, 186, 491–504.

Robinson, S. R. (1991). "Development of the mammalian retina." Neuroanatomy of the Visual Pathways and their Development, B. Dreher and S. R. Robinson, eds., Macmillan, London, 69–128.

Robinson, S. R., Dreher, B., and McCall, M. J. (1989). "Nonuniform retinal expansion during the formation of the rabbit's visual streak: Implications for the ontogeny of mammalian retinal topography." Visual Neuroscience, 2, 201–219.

Schall, J. D., Perry, V. H., and Leventhal, A. G. (1987). "Ganglion cell dendritic structure and retinal topography in the rat." Journal of Comparative Neurology, 257, 160–165.

Sherry, D. M., and Yazulla, S. (1993). "Goldfish bipolar cells and axon terminal patterns—a Golgi study." Journal of Comparative Neurology, 329(2), 188–200.

Snow, R. L., Kirszrot, J., and Finlay, B. L. (1997). "Changes in amacrine cell arborization as a result of induced ocular enlargement in the chick." Manuscript submitted

Snow, R. L., Robson, J. A., and Stelzner, D. J., (1994). "Dendritic differentiation of retinal ganglion cells in chick." Society of Neuroscience Abstract, 20.

Taylor, L. (1997) "The effect of eye size on looming behavior in primates." Scandinavian Journal of Primatology, 34:248–274.

Teakle, E. M., Wildsoet, C. F., and Vaney, D. I. (1993). "The spatial organization of tyrosine hydroxylase-immunoreactive amacrine cells in the chicken retina and the consequences of myopia." Vision Research, 33(17), 2383–2396.

Thanos, S., Vanselow, J., and Mey, J. (1992). "Ganglion cells in the juvenile chick retina and their ability to regenerate axons in vitro." Experimental Eye Research, 54, 1992.

Troilo, D. (1992). "Neonatal eye growth and emmetropization: a literature review." Eye, 6, 154–160.

Troilo, D., Gottlieb, M. D., and Wallman, J. (1987). "Visual deprivation causes myopia in chicks with optic nerve section." Current Eye Research, 6(8), 993–999.

Troilo, D., Howland, H. C., and Judge, S. J. (1993). "Visual optics and retinal cone topography in the common marmoset (Callithrix-Jacchus)." Vision Research, 33(10), 1301–1310.

Troilo, D. B., Xiong, M., Crowley, J. C., and Finlay, B. L. (1996). "Factors controlling the dendritic arborization of retinal ganglion cells." Visual Neuroscience, 13, 721–734.

Wallman, J., Gottlieb, M. D., Rajaram, V., and Fugate-Wentzek, L. (1987). "Retinal control of myopia and eye growth." Science, 237, 73–77.

Wässle, H., Grünert, U., Rohrenbeck, J., and Boycott, B. B. (1990). "Retinal ganglion cell density and cortical magnification factor in the primate." Vision Research, 30, 1897–1912.

Wässle, H., Peichl, L., and Boycott, B. B. (1981). "Morphology and topography of on and off-alpha cells in the cat retina." Proceedings of the Royal Society of London. (Biology), 212, 157–175.

Wässle, H., Yamashita, M., Greferath, U., Grunert, U., and Muller, F. (1991). "The rod bipolar cell of the mammalian retina." Visual Neuroscience, 7, 99–112.

Wiesel, T. N., and Raviola, E. (1979). "Increase in axial length of the macaque monkey eye after corneal opacification." Investigative Ophthalmology and Visual Science, 18, 1232–1236.

Wikler, K. C., and Rakic, P. (1990). "Distribution of photoreceptor subtypes in the retina of diurnal and nocturnal primates." Journal of Neuroscience, 10, 3390–3402.

Wikler, K. C., Williams, R., and Rakic, P. (1990). "Photoreceptor mosaic: number and distribution of rods and cones in the rhesus monkey retina." Journal of Comparative Neurology, 297, 499–508.

Williams, R. W., Strom, R. C., Rice, D. S., and Goldowitz, D. (1996). "Genetic and environmental control of variation in retinal ganglion cell number in mice." Journal of Neuroscience, 16(22), 7193–7205.

Xiong, M., Kelly, M.G., Troilo, D.B., and Finlay, B.L. (1997). "Developmental regulation of arborization of dopaminergic amacrine cells." Manuscript submitted.

14

RETINAL GANGLION CELL AXONAL TRANSPORT

Moving down the Road to Functional Connections

Kenneth L. Moya

CNRS URA 2210 and INSERM U334, SHFJ, DRM, CEA
Orsay, France

1. INTRODUCTION

Retinal ganglion cells form the essential link between the eye and the brain, conveying visual information transduced at the retina for further visual processing in higher centers. In visual animals, this pathway is marked by a high order of functional and anatomical organization. Developmental studies are one approach aimed at understanding how global connectivity and the specificity of the primary visual projection is established. Since the formation of retinofugal pathways follows the general program of developmental events in the brain, developmental studies of this pathway also shed light on the formation of the neuronal network in general. A highly simplified view of the major developmental events in the nervous system would start with the generation of neuronal precursors, followed by neuronal differentiation which in the case of long projection neurons involves rapid axon outgrowth over considerable distances, the formation of synaptic relationships, and finally the efficient processing of neuronal signals along the neural network. The method described below is particularly adapted for studies of neuronal protein changes during the latter three stages of development.

During rapid axon outgrowth, nerve fibers are often fasciculated into bundles and the advance of the axons is led by the dynamic growing tip, the growth cone. Important cell biological processes at this stage likely include adhesive interactions between axons and guidance mechanisms at the growth cone which may use cues on other cells or in the extracellular matrix. During synaptogenesis, recognition signals must be transduced to tell the growing axons that they have arrived in the appropriate target region and the dynamics of

Development and Organization of the Retina, edited by Chalupa and Finlay.
Plenum Press, New York, 1998.

axon growth must change so that the fibers defasciculate and send out branches to initiate synaptic contacts. The last stage requires that appropriate synaptic contacts be stabilized, that extraneous contacts be eliminated and that the cellular machinery for the regulated release of neurotransmitters be put in place.

The hamster offers one important advantage for developmental studies in that it is born at embryonic day 15.5 (E15.5) while mice and rats are born at about E20–21. Thus, the neonate hamster allows access to a nervous system that is less developed compared to other laboratory rodents. The hamster visual projection is also well characterized in terms of morphogenetic events. The earliest retinal ganglion cell axons leave the eye at about day E11.5 and the first axons reach the contralateral superior colliculus (SC) on E13–14. On the day of birth (P0), many retinal axons are still rapidly elongating to the lateral geniculate nucleus (LGN) and SC, and the axons retain the capacity to regenerate. By P5, retinal axons have shifted their mode of growth, sending out multiple branches and forming terminal arbors in the neuropil of the LGN and SC. At the time of eye-opening on P14, considerable synaptogenesis has occurred, mature arbor types can be recognized and the pathway is capable of mediating visual behavior (Bhide and Frost, 1991; Campbell, So and Lieberman, 1984; Frost, So and Schneider, 1979; Jhaveri, Edwards and Schneider, 1991; Jhaveri et al., 1996; Schneider et al., 1985).

One question regarding the development of this system is what are the proteins synthesized by the retinal ganglion cells at these different stages and more precisely, what are the molecular changes associated with the shift from axon elongation to synaptogenesis to the establishment of mature connections?

2. PROTEIN SYNTHESIS AND AXONAL TRANSPORT

Neurons are highly polarized cells in which the cell body can be separated from the axon ending by up to 1 meter in the case of human lower spinal motor neurons whose cell bodies are in the base of the spinal cord and whose axons innervate the muscles of the toes. While in the case of adult hamster retinal ganglion cell axons, the distance from the back of the eye to midtectum is about 2.2 cm and the distance is about one half this length in an E15/P0 hamster, this polarity nonetheless imposes specific constraints on the retinal ganglion cells since gene transcription and protein translation are in almost all cases restricted to the cell body. In the context of retinal pathway development, the most important site of action is along the axon and at the axon ending, i.e., the growth cone or synaptic terminals, since it is here that adhesion molecules, axon guidance cues, cell recognition signals and synaptic specializations will exert their effects.

Protein translation takes place in the rough endoplasmic reticulum, and then proteins pass through the Golgi where they can be subjected to posttranslational modifications such as glycosylation. Neuronal proteins that are destined for the plasma membrane are packaged in membranous organelles and then transported down axons in the rapid phase of axonal transport by microtubule associated ATP-dependent motor proteins, kinesins (see Hirokawa, 1993, for review). While the separation of the site of protein synthesis from the site of action for axonal proteins imposes specific energy requirements for protein trafficking and a lag time between changes in gene expression and the arrival of a new cohort of proteins at the axon ending, we can use this separation to our advantage in in vivo metabolic labeling experiments. We have only to provide radiolabeled amino acids to the region of the cell body and then recover the newly synthesized, radiolabeled proteins from the distal axon region after an appropriate interval.

2.1. In Vivo Metabolic Labeling

A flow chart of the in vivo metabolic labeling experimental design is presented in Figure 1. Radiolabeled amino acids (35S-methionine/cysteine) are intraocularly injected into the vitreal chamber of deeply anesthetized hamsters. The amino acids are internalized by the cells in the retina and incorporated into proteins. After various survival times, the animals are sacrificed and retinal target areas (LGN and SC) are dissected out. The tissue is homogenized and an aliquot can be counted to determine the amount of radioactivity transported to the target tissue. The solubilized proteins in the rest of the sample are then separated by high resolution 2-D gel electrophoresis and the radiolabeled proteins visualized by fluorography or autoradiography (Moya et al., 1988, 1994b).

In experiments in which we keep the age constant (adult hamsters) but vary the survival time (1hrs to 4 weeks) we can determine a global profile of axonal transport (Fig. 1, bottom). The low levels of radioactivity observed at 1 and 2 hours represent the intrinsic noise in the system due to leakage from the eye and the circulation of radiolabeled amino acids in brain capillaries. What is important to note from this set of data is the arrival of a wave of radiolabeled proteins at 4 hours which corresponds to the arrival of fast axonally transported proteins (Hammerschlag and Stone, 1982). A second wave of radioactivity begins to appear in the SC at 1 week and then increases over a week or two. This second wave represents the arrival of proteins in slow axonal transport which include tubulin, actin, neurofilaments and microtubule associated proteins (Nixon, 1992).

At a finer level of analysis we wanted to identify the various proteins that are constituents of axonal transport and, in particular, characterize the proteins destined for the plasma membrane of the axon and growth cone or synaptic ending. The 2-D gel or blot of

Figure 1. Upper: Flow diagram of the experimental steps for in vivo metabolic labeling. **Lower:** Curve of radioactivity transported to the contralateral SC in adult hamsters after various post intraocular injection intervals in counts per mintue per whole SC. The low signal at 1 and 2 hours post injection represents the intrinsic noise in this in vivo system. The front of the wave corresponding to fast axonal transport appears in the SC at 4 hours (arrow). Radioactivity accumulates gradually and between 1 and 2 weeks post injection a second wave of radioactivity arrives in the SC corresponding to pro- teins in slow axonal transport (arrow).

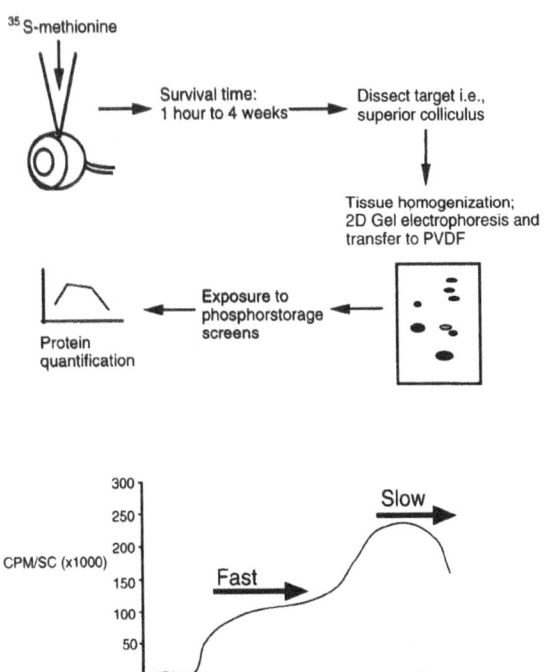

LGN or SC proteins contains all of the proteins in that tissue, including proteins synthesized locally by neuronal and non-neuronal cells, proteins from blood vessels as well as proteins in axons of passage, and all of these proteins are revealed by general protein stains such as Coomassie blue or silver staining. However, the radiolabeled proteins visualized on the fluorographs or on the autoradiograms are those proteins which were synthesized in the retinal ganglion cell neurons and transported to the target tissue.

2.2. Developmental Studies

Since proteins destined for the axonal plasma membrane are transported in the rapid phase of axonal transport, in our developmental studies we kept the survival time constant (4 hours) and varied the age of the animals (Moya et al., 1988, 1992). Figure 2 (upper) shows the pattern of proteins synthesized in retinal ganglion cells and transported to the axon endings at P4 when retinal fibers are shifting from rapid elongation to terminal arborization. Inspection of the fluorograms suggests a number of developmental protein changes. There appears to be a decrease in GAP-43 and NCAM during maturation of the retinal pathway while proteins such as SNAP-25 and a 64kD protein increase. Other proteins may undergo more complex developmental changes.

Semiquantitation of the changes in the proteins rapidly transported down retinal axons to the SC or LGN showed that proteins could be classified into at least 3 categories based on their developmental profiles of synthesis and axonal transport (Fig. 2, lower). GAP-43, NCAM and L1 were synthesized and transported at highest levels relative to other proteins, in neonatal hamster retinal ganglion cells, when many axons are still elongating. Several proteins (APP and a 67kD protein) were at relatively low levels in the neonate, then increased in their synthesis and transport during arborization and then declined. A third group of proteins (SNAP-25 and a 64kD protein) showed an increase during maturation of the retinal projection and remained high in adult animals.

3. ELONGATION PROTEINS

3.1. GAP-43

GAP-43 (Growth Associated Protein, 43kD), also known as B-50, F1, pp46 and neuromodulin, is a protein kinase C substrate phosphoprotein (for reviews see Skene, 1989 and Benowitz and Routtenberg, 1997). This protein is enriched in growth cones and undergoes major upregulation during axon growth and regeneration. GAP-43 is attached along the inner face of the plasma membrane by a fatty acid anchor where the protein may modulate or participate in the transduction of signals across the growth cone membrane. In adult brain, GAP-43 is abundant in regions which maintain a known capacity for synaptic plasticity.

The levels of GAP-43 synthesis and axonal transport in vivo were highest in P2 hamster retinal axons and then decreased during development. Neonatal hamster GAP-43 from growth cones is phosphorylated under conditions that stimulate protein kinase C activity (Fig. 3A, B). GAP-43 immunocytochemistry of embryonic hamster retinal explants show abundant localization of the protein in growth cones, with labeling extending into the filopodia (Fig. 3C). GAP-43 is also distributed along the retinal ganglion cell axons. In the developing optic tract, GAP-43 is distributed along retinal axon fascicles in late embryonic and early neonatal hamsters over the LGN and continuing

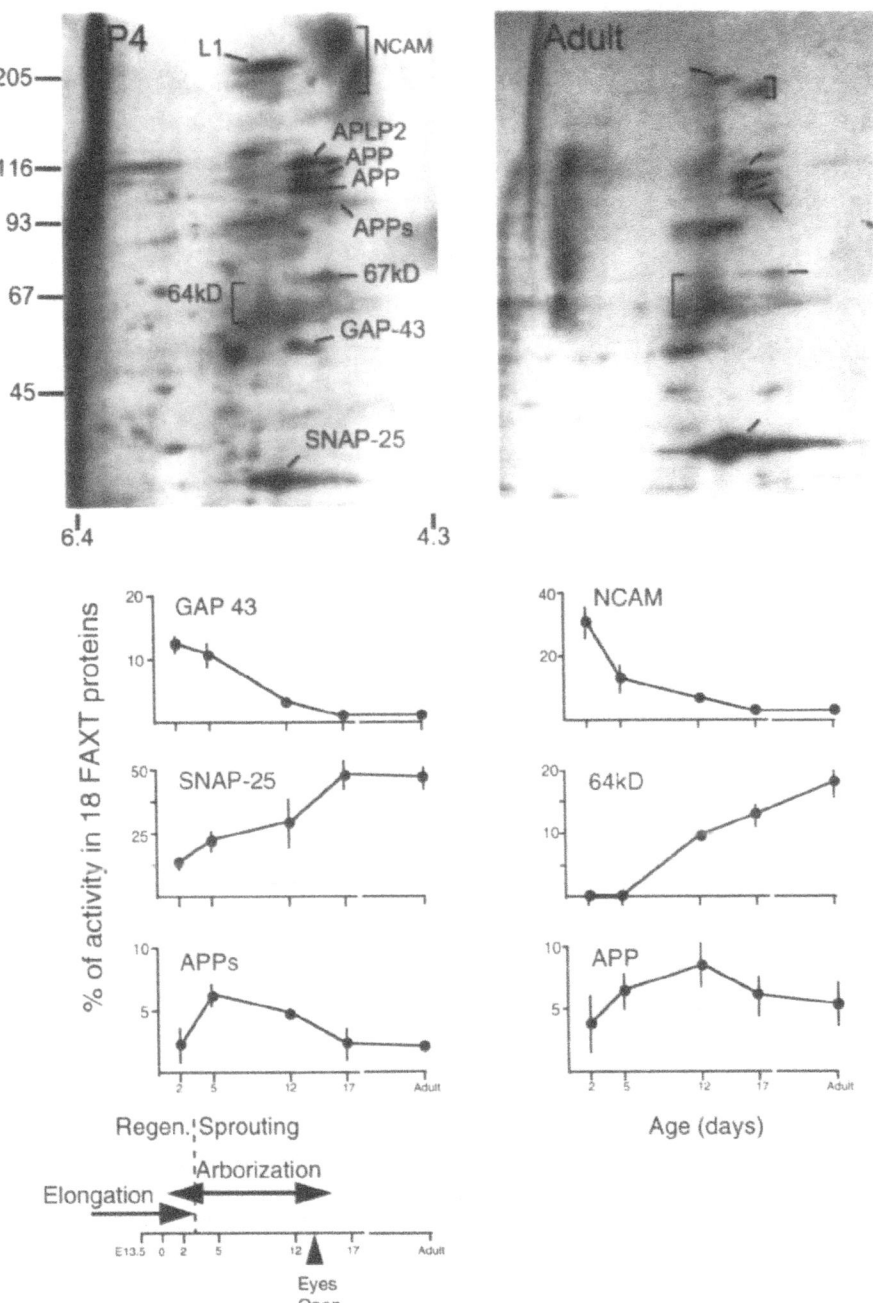

Figure 2. Upper: Fluorograms of radiolabeled proteins synthesized in retinal ganglion cells and rapidly fast transported to the contralateral SC 4 hours after intraocular injection in P4 and adult hamsters. The numbers along the left indicate the apparent molecular weight of proteins in kiloDaltons (kD). The numbers along the bottom indicate the isoelectric gradient (pI). Identified proteins (see text) are indicated. **Lower:** Developmental profiles of fast axonally transported proteins 4 hours post eye injection. The radioactivity in each protein is normalized to the total radioactivity in 18 proteins transported to the contralateral SC in P2, P5, P12, P17 and adult hamsters. At the bottom is a schematic summary of the major morphogenetic events in the hamster retinofugal projection. Top row: proteins whose timecourse of synthesis and transport correlated with retinal axon elongation. Middle row: proteins whose timecourse increased during development and remained high. Bottom row: proteins whose developmental profile correlated with synaptogenesis. Modified with permission from Moya et al., 1988.

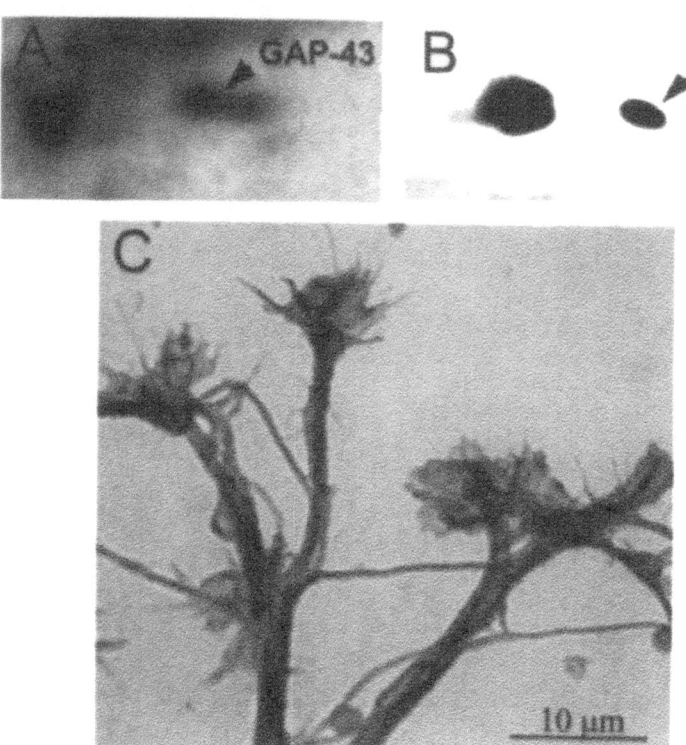

Figure 3. GAP-43 is a growth cone phosphoprotein. **A.** The region of the P4 gel in Figure 2 is enlarged to show in vivo metabolically labeled GAP-43. **B.** A growth cone fraction prepared from neonatal hamster brain was incubated with radiolabeled ATP under conditions to stimulate protein kinase C-mediated phosphorylation. Prominent among the growth cone phosphoproteins is GAP-43, arrowhead). **C.** E14 hamster retinal explants were cultured for 48 hrs, fixed and processed for GAP-43 immunocytochemistry. Note the localization of GAP-43 along retinal axons and in growth cones including the lamellapodia and filopodia.

into the superficial layers of the SC (Moya et al., 1989). The staining of fiber bundles diminished during the first postnatal week and GAP-43 immunoreactivity was more obvious in the neuropil of the LGN and SC. By P12, two days before eye opening, the neuropil staining had diminished and little or no GAP-43 immunoreactivity could be observed in optic axon bundles.

The immunolocalization studies closely parallel metabolic labeling results and together show that high levels of GAP-43 are present in retinal axons during their elongation and initial target contact. In this regard it is interesting to note that in GAP-43 knockout mice, retinal ganglion cell axons are severely impaired in pathfinding, especially at the optic chiasm which has been taken as further evidence that the protein is a modulator of signal transduction at the growth cone, perhaps through coupling to a G-protein-receptor signal transduction (Strittmatter et al., 1995). The decrease in GAP-43 synthesis and transport in hamster retinal axons and the protein's shift to the neuropil coincide with the arborization of retinal terminals and a diminished demand for axon pathfinding. Finally, levels of GAP-43 are at barely detectable levels in the mature hamster retinal projection which has little capacity for plastic remodeling.

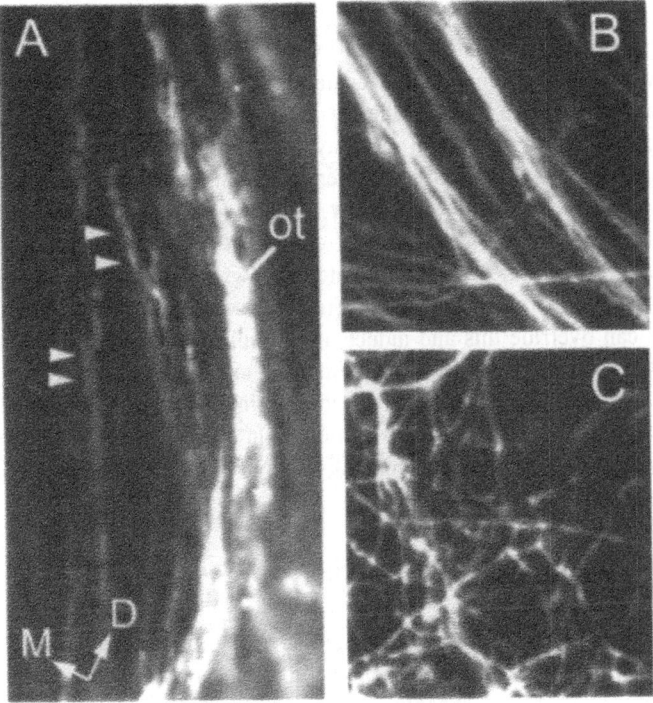

Figure 4. A. L1 immunohistochemistry shows the presence of this adhesion molecule along fiber bundles in the superficial optic tract (ot) over the P0 LGN. M=medial; D=dorsal. **B.** L1 decorates retinal axon fascicles of E14 hamster retinal explants grown on poly-lysine/laminin coated coverslips for 3 days. **C.** E14 hamster retina 3 days in culture grown on a confluent layer of astrocytes from P3 hamster cortex express L1 along their axon. Note, however, that the pattern of growth is markedly different; retinal axons are no longer predominantly fasciculated and instead form a meshwork of crossing single fibers.

3.2. Adhesion Proteins

Two proteins whose developmental profiles show a correlation with retinal axon elgonation are NCAM and L1 (Fig. 2). These two transmembrane glycoproteins are part of the IgG superfamily and participate in adhesive interactions between cells during development and appear to be involved in synaptic modifications during LTP in the adult hippocampus (Rutishauser, 1984; Hortsch, 1996; Lüthi et al., 1994).

In adult retinal axons, in vivo metabolic labeling studies have shown NCAM and L1 to be glycosylated and sulfated in retinal ganglion soma before being axonally transported (Moya et al., 1994a, b). Protein sulfation of NCAM and L1 can conceivably alter the charge of these molecules, thus changing the adhesive properties. It remains to be determined, however, if different levels of sulfation contributes to changes in cell adhesion during development.

L1 is localized to fibers in the optic tract over the LGN and in the SC during the period of axon elongation when retinal fibers are tightly fasciculated (Fig. 4A). During the time when the retinal axons defasciculate and arborize into the neuropil of the LGN and SC, L1 immunoreactivity is decreased in the optic tract and in adult animals the tract is devoid of L1 immunoreactivity (Moya, Confaloni, Lyckman and Jhaveri, in preparation).

While L1 has demonstrated adhesive properties and its expression and localization correlate with the elongation of axons in bundles, preliminary evidence suggests that the presence of L1 along axons is not sufficient for axon fasciculation. Embryonic hamster retinal explants grown on poly-lysine/laminin coated coverslips show a typical pattern of L1-labeled retinal axon growth with bundles of axon emanating from the explant (Fig. 4B). However, when explants are grown on a layer of confluent astrocytes, the retinal axon growth revealed by L1 staining is very different in that axons are now oriented in all directions, crossing over one another with few fascicles (Fig. 4C). The immunocytochemistry shows that the retinal axons express L1 consistent with their embryonic age, but the presence of L1 is not sufficient for fasciculated axon growth on astrocytes. Thus, it is as if the explants followed an intrinsic progam (i.e., L1 expression) but extrinsic cue(s) from the astrocytes can override this and induce a morphology of a later age. This, then, suggests that there is a hierarchy of signals that regulate axon growth.

In future studies, it will be of interest to examine the function of the L1 and NCAM adhesion molecules. One promising possibility might involve the selective blockade of L1 and NCAM trafficking to retinal axon endings in vivo by the mannosidase inhibitor, deoymannojirimycin (McFarlane et al., 1997).

4. SYNAPTOGENESIS-ASSOCIATED PROTEINS

The metabolic labeling studies revealed several proteins that have developmental profiles in which the peak of synthesis and axonal transport coincided with the period of retinal axon arborization and synaptogenesis (Fig. 2). These synaptogenesis-associated proteins include the amyloid precursor protein (APP) and an acidic 67kD polypeptide (Moya et al., 1992, 1994a).

4.1. APP in the Retinofugal Pathway

APP is a membrane glycoprotein heavily expressed in brain but also found in other tissues. The 695 amino acid form of APP is predominantly neuron-specific while other isoforms arising by alternative splicing of the mRNA transcribed from the single APP gene are expressed in other cell types in the brain and other organs. APP has been the focus of considerable attention since a fragment of the protein, the βA4 peptide, is found in the amyloid deposits in senile plaques seen in the brains of patients with Alzheimer's disease (for review, see Selkoe, 1994).

APP expression undergoes considerable changes in the developing brain (Neve et al., 1988; O'Hara et al., 1989) and in vitro studies suggest that APP may function as an adhesion protein in cell–cell and/or cell–substrate interactions and promotes neurite outgrowth (Breen et al., 1991; Schubert et al., 1989; Milward et al., 1992). In our developmental fast axonal transport studies we observed two isoforms of APP (110kD and 100kD) which first increased after birth and then decreased after eye opening (Moya et al., 1994a). The 110kD isoform is full-length transmembrane APP while the 100kD APP is lacking the C-terminal and is enriched in the soluble fraction. The 100kD soluble APP isoform corresponds to the proteolytically cleaved and secreted N-terminal ectodomain and the results suggest that the secretion of APP from retinal axon terminals is maturation-dependent and may require the formation of functional contacts and/or physiological activity (see Nitsch et al., 1993).

It is important to note that another full-length transmembrane APP isoform (120kD) and an isoform of the related Amyloid Precursor-Like Protein 2 (APLP2 140kD), have developmental profiles which show a general decline with development from the earliest ages examined (Lyckman et al., 1995; Moya et al., 1994a). APLP2 has also been associated with axogenesis in the olfactory system (Thinakaran et al., 1995). Since the respective antibodies available do not distinguish between the 'elongation-' and 'synaptogenesis-' associated forms of these proteins, immunolocalization studies offer a composite picture of the changes in these proteins. In embryonic hamster retinal explants, APP is distributed along the retinal axons and is present in the growth cones (Fig. 5C). In the developing brain, APP is localized to fibers in the optic tract for the first few days after birth (Fig. 5A). Similar developmental distributions for the related Amyloid Precursor-Like Protein 2 (APLP2) have also been observed in the hamster retinal projection (Confaloni et al., 1995). This axonal localization at young ages may reflect the earlier elongation forms of these proteins and would be consistent with a role for these proteins in cell-cell or cell-substrate adhesion as reported for APP (Breen et al., 1991; Schubert et al., 1989).

We have performed two types of experiments to test whether APP has a direct role in axon growth. In one set of experiments, we used an oligonucleotide antisense to block APP synthesis in embryonic rat cortical neurons in vitro (Allinquant et al., 1995). APP synthesis was diminished by about 30% after the antisense treatment and this was accompanied by a reduction in neurite (axon and dendrite) outgrowth. The APP synthesis inhibition was reversible as was the morphological effect. In a second set of experiments, we interferred with APP at the level of the protein on the surface of retinal axons in vitro (M. Young and K.L. Moya, unpublished observations). Monoclonal antibody 22C11 which recognizes both APP and APLP2 from hamster (Moya et al., 1994a; Lyckman et al., 1995) was administered to embryonic hamster retinal explants. We noted a marked and dose-dependent effect of the blocking antibody supporting a direct role for APP and APLP2 in the growth of retinal axons.

After axon elongation, APP and APLP2 immunoreactivity shifts to the neuropil of the LGN and SC during the second postnatal week. In adult hamsters, no APP or APLP2 could be detected in the optic tract while puncta of immunoreactivity were found over and around large cell bodies in the LGN (Fig 5B), which is consistent with the reported synaptic localization for APP (Schubert et al., 1991). Although, relative to other proteins, the

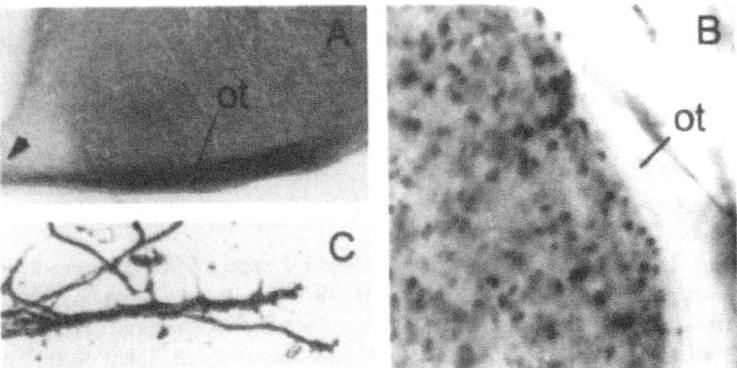

Figure 5. APP is localized to developing retinal axons in vivo and in vitro. **A.** Fibers in the optic tract (ot) along the ventral surface of the hypothalamus and coursing from the optic chiasm (arrowhead) are immunoreactive for APP in P1 hamsters. **B.** E14 retinal explants processed for APP immuncytochemistry show that APP is distributed along retinal axons and is present in their growth cones. **C.** In the adult hamster LGN, the superficial optic tract (ot) is devoid of APP immunoreactivity, while in the neuropil of the LGN, APP is localized in and around cell bodies.

synthesis and transport of APP and APLP are greatly diminished in adult retinal ganglion cells, detectable levels of these proteins reach adult hamster synaptic endings. Our metabolic labeling studies show that both proteins are sulfated and glycosylated (Moya et al., 1994a, b; Lyckman et al., 1995; Lyckman et al., in preparation). Furthermore, these proteins have an extremely short half-life in synaptic endings in vivo; these proteins are eliminated from retinal synapses with a half-life estimated on the order of 2–3 hours (Lyckman et al., 1995).

4.2. The 67kD Acidic Protein

The identity of the 67kD, pI 4.7 protein remains unknown. Preliminary data suggest that the hamster 67kD protein is similar to a myristoylated phosphoprotein in rat cortical neurons (Moya, unpublished observations; Perrone-Bizzozero et al., 1989). These characteristics and the 2-D electrophoretic migration properties closely resemble the proto-oncogene, pp60src. pp60src is found in growth cones, is developmentally regulated in brain and is involved in cell-substrate interactions (Maness et al., 1988; Steedman and Landreth, 1989; Hynes, 1982). While further studies will be necessary to determine if the 67kD hamster protein is related to pp60src, these results raise the intriguing possibility that the 67kD protein may participate in the growth cone-mediated extension of axon arbors in the LGN and SC, and influence the establishment of synaptic relations through mechanisms dependent on pp60src (see Maness et al., 1988).

5. PROTEINS ABUNDANT IN THE MATURE RETINAL PROJECTION

Once the retinal projection has been formed and synapses established, the function of retinal axons is the reliable transfer of information for visual processing. At this stage, one might expect that the pattern of proteins transported to retinal axon endings would include proteins necessary for correct synaptic function such as constituents of the synaptic vesicle cycle, presynaptic ion channels and receptors, and proteins that structurally stabilize the synapse. Thus far, most of the proteins identified in axonal transport studies of mature hamster retinal terminals fall into the first category. A few possible candidates for proteins that stabilize the synaptic membrane have been identified; while little has yet emerged regarding the transport of ion channel or receptor proteins, this will likely soon change as more sensitive methods are used to visualize 2-D separated radiolabeled proteins such as phosphorimaging.

5.1. Proteins of the Synaptic Vesicle Cycle

5.1.1. Synaptic Core Complex Proteins. SNAP-25 and a 64kD protein smeared in the molecular weight dimension (64–82kD) were observed in early studies as proteins that were not detectable in elongating retinal axons and which first appeared during arborization (Fig 2 and Moya et al., 1988). These proteins gave the strongest signal on fluorograms of rapidly transported proteins to adult SC. SNAP-25 is a 25kD synaptosomal associated protein linked to the inner face of the synaptic plasma membrane by palmitic acid bridge. In 2-D Western blot experiments, monoclonal antibodies specific for syntaxin also recognized the 64kD (T. Smirnova and K.L. Moya, unpublished results). One interpretation is that the 64kD (64–82kD) rapidly transported 'protein' consists of a multimeric, SDS-resistant complex with syntaxin (34kD) SNAP-25 (25kD) and VAMP (18kD) (see Hayashi et al., 1994). We have also recently been able to identify VAMP as a protein in rapid axonal transport in adult hamster retinal axons (A.W. Lyckman and K.L. Moya, unpublished results).

According to a current model of the synaptic vesicle cycle, these three proteins play a crucial role in the early steps of vesicular release of neurotransmitters (Südhof, 1995). During the docking step, synaptic vesicles approach the active zone of the synaptic plasma membrane where transmembrane syntaxin is situated. During the priming phase, SNAP-25 is recruited to complex with syntaxin forming a high affintiy site for VAMP (vesicle associated membrane protein) on the synaptic vesicle. This synaptic core complex draws the vesicle towards the plasma membrane allowing other proteins to act at the site so the end result is synaptic vesicle fusion and neurotransmitter release.

Further studies will be necessary to definitively identify the constituents of the 64kD protein. Thus far, however, the in vivo metabolic labeling studies suggest that these proteins of the synaptic core complex are synthesized in retinal ganglion cells and conveyed to the synaptic ending by rapid axonal transport, even though their mode of attachment to membranes (transmembrane or fatty acid tail) and their membrane localization (plasma membrane or synaptic vesicle membrane) differs. In addition, developmental studies should determine if VAMP has a developmental profile similar to SNAP-25 and the 64kD protein. If this turns out to be the case, then this would provide strong in vivo evidence that very diverse gene products which participate in a common function are developmentally regulated in parallel.

5.1.2. Synaptic Vesicle Protein 2 (SV2). SV2 is a integral synaptic vesicle membrane protein ubiquitous to all synapses for which it has been examined (Buckley and Kelly, 1982; Okada et al., 1994). In a recent study we examined changes in SV2 in developing retinal axons (Confaloni et al., 1997). In the adult hamster, SV2 was not evident in retinal axons in the optic tract over the LGN nor were SV2-labeled retinal fascicles in the SC, consistent with its localization to synaptic vesicles in axon terminals. In the adult LGN, SV2 immunoreactivity revealed a gradient, with the outer shell of the LGN being more darkly stained, and distinct immunoreactive profiles were revealed (Fig. 6A–C). These

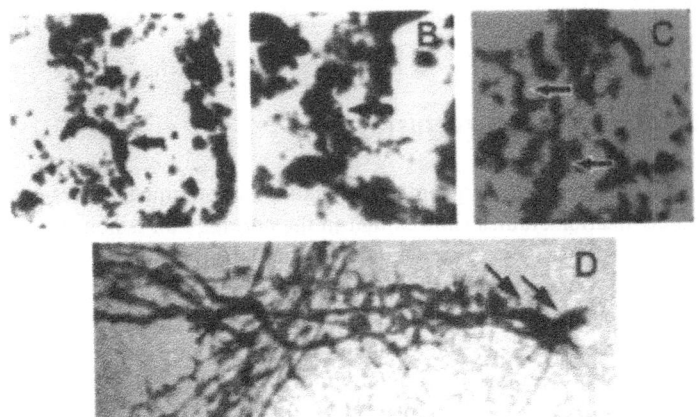

Figure 6. SV2 in adult retinal terminals and embryonic retinal axons. **A–C.** The adult hamster LGN contains SV2-positive terminal profiles that resemble R1, R2, and R3 terminal types, respectively, as described by Erzurumlu et al. (1988). **D.** SV2 is present along E14 hamster retinal axons and growth cones in vitro (arrows).

profiles resembled the three previously identifed retinal axon terminal types (Erzurumlu et al., 1988). Adult eye enucleation experiments demonstrated the anterograde degenerative loss of these SV2-labeled retinal terminals in the LGN and also exquisitely revealed the segregation of the ipsi- and contralateral retinal projections to the adult LGN.

What was entirely unexpected in this study was the localization of SV2 in developing retinal axons. SV2 labeled retinal axons (bundles and single fibers) in the optic tract over the LGN and in the superficial layers of the SC in hamsters from P0 to late P1. This was unexpected because at these ages there is no evidence for mature synapses nor for the presence of synaptic vesicles. Furthermore, previous developmental studies of different synaptic vesicle proteins had never shown a prominent axonal localization in rat hippocampus and neocortex (Stettler et al., 1994, 1996). Immunocytochemistry of embryonic retinal explants showed that SV2 was present along the axons and extended into the growth cone (Fig. 6D). This raises the question as to whether SV2 has an additional function in elongating axons.

SV2 has a molecular structure which suggests that the protein transports molecules into or out of synaptic vesicles, although its ligand has not been identified (Bajjaleih et al., 1992; Feaney et al., 1992). Based on the predicted topology of the SV2 glycoprotein, the glycodomains are oriented within the lumen of the synaptic vesicle (Bajjaleih et al., 1992), and thus when the vesicle fuses with the plasmalemma, the carbohydrate moeities are oriented toward the extracellular surface. In developing axons and growth cones then, SV2 could serve as a cell surface glycoprotein, perhaps participating in adhesive interactions, or alternatively as a growth cone surface transporter.

7. TARGET INTERACTIONS AND REGULATORY SIGNALS

The studies of normal retinal axon development described above identified a number of changes in protein synthesis, axonal transport and localization during different stages of axon growth. Some of the striking changes in neuronal proteins occurred at the time when developing axons reach their targets, cease elongating and commence elaborating synaptic contacts in brain nuclei. This raised the question of whether some of these changes might be triggered by signals generated by target interactions. To begin to investigate this we prevented optic axons from encountering their normal targets in the brain and induced the formation of abnormal connections.

7.1. Changes in Target Structures

In normal hamsters, retinal ganglion cell axons form terminal arbors in the LGN and SC. Immunolocalization studies showed that in the normal, non-lesioned hamster, the lateral posterior nculeus (LP) was moderately immunoreactive for GAP-43 (Moya et al., 1989). This structure is adjacent to the LGN and normally does not receive a retinal projection. However, neonatal lesions of the SC can result in the formation of dense patches of hyperinnervation in the LGN and an abnormal retinal projection to the LP. We traced the abnormal retinal projections using anterograde techniques and oberved that the GAP-43 immunoreactivity was excluded precisely from the territory of the retinal projections (Moya et al., 1990). Additional experiments showed that much of the GAP-43 in the LP was expressed in local interneurons. In these experiments, the retinal axons displaced GAP-43-rich local terminals while colonizing synaptic territory in the LP.

7.2. Alterations in Retinal Ganglion Cells

We examined the extent to which the developmental regulation of rapidly transported membrane proteins in the mammalian optic pathway is influenced by interactions between the advancing nerve endings and the target cells they encounter. We found that the synthesis and rapid axonal transport of several proteins destined for the nerve terminal membrane are altered when retinal axons form abnormal connections (Moya et al., 1992).

One of the proteins that showed a significant change in its synthesis and axonal transport after our experimental manipulations is the neural cell adhesion molecule, NCAM, which is involved in neuronal adhesion, neurite adhesion and axonal guidance. During normal development, the decline in NCAM just after birth may explain the defasciculation of the axons and the subsequent elaboration of terminal arbors in target regions. In the experimental animals, however, the abnormally elevated levels of the protein at later ages may reflect a continued growth of retinal axons in a fasciculated state when they do not encounter their normal target cells.

The acidic 67kDa protein (see above), which normally increases during terminal arborization and then decreases, also remained significantly elevated in animals with early SC lesions. Preliminary biochemical studies described above suggest a number of similarities between the 67kD hamster protein and the tryrosine kinase, pp60src. pp60src is known to be involved in cell-substrate interactions through extracellular matrix receptors, and the misrouted fibers may require a prolonged period of growth through the extracellular matrix as they search for available synaptic territory.

The anatomical studies showing that the formation of abnormal neural connections can alter the distribution of a molecular marker in the postsynaptic territory. The metabolic labeling experiments in the early lesioned hamsters show that various target manipulations can selectively alter the expression of a subset of proteins but not others in retinal ganglion cell axons. Together these results suggest that there may be a multiplicity of signals governing the pattern of proteins in developing axons.

8. SUMMARY

In vivo metabolic labeling studies show that elongating retinal ganglion cells express a specific set of proteins that includes GAP-43 and adhesion proteins L1 and NCAM along with certain isoforms of APP and APLP2. Not only are the proteins synthesized in retinal ganglion cells and transported down their axons, these proteins are localized along the axons and can be detected in growth cones. This phase of axon growth requires molecular mechanisms for growth cone motility, axon adhesion for fasciculated growth and guidance cues in order for the growing fibers to successfully reach the target. Blocking experiments suggest that APP and/or APLP2 are necessary for hamster retinal axon growth, at least in vitro, while the expression of L1 does not appear to be sufficient for retinal axon fasciculation in the presence of other extracellular cues. Further functional tests will be necessary to determine if NCAM and L1, for example, are required for retinal axon growth.

When retinal axons arborize in their target nuclei, they change their pattern of protein synthesis and axonal transport. Isoforms of APP, and an acidic 67kD protein are upregulated while elongation-associated proteins decline. Cellular interactions at this stage involve target recognition, axon branching, the initial steps for putting in place the molecular machinery for synaptic transmission, and the stabilization or elimination of synapses. Although the precise signals for the change from elongation to arborization are not yet known, experimental manipulations showed that the expression of some retinal ganglion cell proteins depends on target interactions.

Adult animals depend on the efficient and reliable processing of visual information. The pattern of proteins synthesized in mature retinal ganglion cells and transported to terminals is dominated by synaptic proteins, some of which form part of the synaptic vesicle docking complex. Other proteins may help stabilize synapses and the metabolic labeling studies show that certain proteins have a rapid turnover in the retinal terminals in vivo.

In future studies, it will be of general cell biological interest to determine if the turnover of the different proteins changes during development. A second open question of particular relevance to the retinofugal projection is what is the local signal that stops the retinal axons when they have reached the correct targets (i.e., the LGN and SC) so that the pattern of retinal ganglion cell protein expression can change. And a third area of inquiry centers on testing which retinal ganglion cell proteins are essential for the correct formation of retinal synapses in the brain.

REFERENCES

Bajjaleih, S. M., Peterson, K., Shinghal, R. and Scheller, R.H., 1992, SV2, a synaptic vesicle protein homologous to bacterial transporters. Science 257: 1271–1273.

Benowitz, L.I. and Routtenberg, A., 1997, GAP-43: an intrinsic determinant of neuronal development and plasticity. Trends Neurosci. 20:84–91.

Bhide, P. G. and Frost, D.O., 1991, Stages of growth of hamster retinofugal axons: implications for developing pathways with multiple targets. J. Neurosci. 11: 485–504.

Breen, K.C., Bruce, M. and Anderton, B.H., 1991, Beta amyloid precursor protein mediates neuronal cell–cell and cell–surface adhesion. J. Neurosci. Res. 28: 90–100.

Brugge, J.S., Cotton, P.C., Queral, A.E., Barrett, J.N., Nonner, D. and Keane, R.W., 1985, Neruons express high levels of a structurally modified, activated form of pp60c-src. Nature 316: 554–557.

Buckley, K. and Kelly, R.B., 1985, Identification of a transmembrane glycoprotein specific for secretory vesicles of neural and endocrine cells. J. Cell Biol. 100: 1284–1294.

Campbell, G., So, K.-F. and Lieberman, A.R., 1984, Normal post-natal development of retinogeniculate axons and terminals and identification of inappropriately-located transient synapses. Neurosci. 13:743–759.

Chen D. F., Jhaveri, S. and Schneider G.E., 1995, Intrinsic changes in developing retinal neurons result in regenerative failure of their axons. P. N. A. S. 92: 7287–7291.

Confaloni A., Lyckman , A. W., Thinikaran, G., Sisodia, S.S. and Moya, K. L., 1995, APP and APLP2 localization in the developing hamster primary visual pathway. Neurosci. Abstr. 21: 1307.

Confaloni, A., Lyckman, A.W. and Moya, K.L., 1997, Developmental shift of synaptic vesicle protein 2 (SV2) from axons to terminals in the primary visual projection of the hamster. Neuroscience 77: 1225–1236.

Erzurumlu, R. S., Jhaveri, S. and Schneider, G.E., 1988, Distribution of morphologically different retinal axon terminals in the hamster dorsal lateral geniculate nucleus. Brain Res. 461: 175–181.

Feany, M. B., Lee, S., Edwards, R.H. and Buckley, K.M., 1992, The synaptic vesicle protein SV2 is a novel type of transmembrane transporter. Cell 70: 861–867.

Frost, D.O., So, K.-F. and Schneider, G.E., 1979, Postnatal development of retinal projections in Syrian hamsters: a study using autoradiographic and anterograde degeneration techniques. Neurosci. 4: 1649–1677.

Hammerschlag, R. and Stone, G.C., 1982, Membrane delivery by fast axonal transport. Trends in Neurosci. 5:12.

Hayashi, T., MacMahon, H., Yamasaki, S., Binz, T., Südhof, T.C. and Nieman, H., 1995, Synaptic vesicle membrane fusion complex: action of clostidial neurotoxins on assembly. EMBO J. 13: 5051–5061.

Hirokawa, N., 1993, Axonal transport and the cytoskeleton. Curr. Op. Neurobiol. 3: 724–731.

Hortsch, M., 1996, The L1 family of neural cell adhesion molecules: old proteins performing new tricks. Neuron 17: 587–593.

Hynes, R.O., 1982, Phosphorylation of vinculin by pp60src: what it might mean. Cell 28: 437–438.

Jhaveri, S., Edwards, M.A. and Schneider, G.E., 1991, Initial stages of retinofugal axon development in the hamster: evidence for two distinct modes of growth. Exp. Brain Res. 87: 371–382.

Jhaveri, S., Erzurumlu, R.S. and Schneider, G.E., 1996, The optic tract in embryonic hamsters: fasciculation, defasciculation, and other rearrangements of retinal axons. Visual Neurosci. 13: 359–374.

Loewy, A., Liu, W.-S., Baitinger, C., and Willard, M.B., 1991, The major ^{35}S-methionine-labeled rapidly transported protein (Superprotein) is identical to SNAP-25, a protein of synaptic terminals. J. Neurosci. 11: 3412–3421.

Lüthi, A., Laurent, J.P., Figurov, A., Muller D. and Schachner, M., 1994, Hippocampal long-term potentiation and neural cell adhesion molecules L1 and NCAM. Nature 372: 777–779.

Lyckman, A.W., Confaloni, A.M., Thinikaran, G., Sisodia S.S., DiGiamberardino, L., and Moya, K.L., 1995, In vivo turnover of synaptic and cytoskeletal proteins rapidly transported to adult CNS nerve terminals. Neuroscience Abstracts, 21: 50.

Maness, P.F., Aubry, M., Shores, C.G., Frame, L. and Pfenninger, K.H., 1988, c-src gene product in developing rat brain is enriched in nerve growth cone membranes. P.N.A.S. 85: 5001–5005.

McFarlane, I., Breen, K.C., DiGiamberardino, L. and Moya, K.L., 1997, Inhibition of N-glycan processing alters axonal transport in vivo. Neuroscience Abstracts, in press.

Milward, E.A., Papadopoulos, R., Fuller, S.J., Moir, R.D., Small, D., Beyreuther, K. and Masters, C.L., 1992, The amyloid protein precursor of Alzheimer's disease is a mediator of the effects of nerve growth factor on neurite outgrowth. Neuron 9: 129–137.

Moya, K. L., Jhaveri, S., Benowitz, L.I. and Schneider, G.E., 1988, Changes in rapidly transported proteins in developing hamster retinofugal axons. J. Neurosci. 8: 4445–4454.

Moya, K. L., Jhaveri, S., Schneider, G.E. and Benowitz, L.I., 1989, Immunohistochemical localization of GAP-43 in the developing hamster retinofugal pathway. J. Comp. Neurol. 288: 51–58.

Moya, K.L., Benowitz, L.I. and Schneider, G.E., 1990, Abnormal retinal projections suppress GAP-43 in the diencephalon. Brain Res. 527: 259–265.

Moya, K.L., Benowitz, L.I., Sabel, B.A. and Schneider, G.E., 1992, Changes in rapidly transported proteins associated with development of abnormal projections in the diencephalon. Brain Res. 586: 265–272.

Moya, K. L., Benowitz, L.I., Schneider, G.E. and Allinquant, B., 1994a, The amyloid precursor protein is developmentally regulated and correlated with synaptogenesis. Dev. Biol. 161: 597–603.

Moya, K.L., Confaloni, A. and Allinquant, B., 1994b, In vivo neuronal synthesis and axonal transport of KPI-containing forms of the amyloid precursor protein. J. Neurochem. 63: 1971–1974.

Neve, R.L., Finch E.A. and Dawes, L.R., 1988, Expression of the Alzheimer amyloid precursor gene transcripts in the human brain. Neuron 1: 669–677.

Nitsch, R.M., Farber, S.A., Growdon, J.H. and Wurtman, R.J., 1993, Release of amyloid β-protein precursor derivatives by electrical depolarization of rat hippocampal slices. P.N.A.S. 90: 5191–5193.

Nixon, R.A., 1992, Slow axonal transport. Curr. Op. Cell Biol. 4: 8–14.

O'Hara, B.F., Fisher, S., Oster-Granite, M.L., Gearhart, J.D. and Reeves, R.H., 1989, Developmental expression of the amyloid precursor protein, growth-associated protein 43 and somatostatin in normal and trisomy 16 mice. Dev. Brain Res. 49: 300–304.

Okada, M., Erickson, A. and Hendrickson, A., 1994, Light and electron-microscopic analysis of synaptic development in Macaca monkey retina as detected in immunocytochemical labeling for the synaptic vesicle protein SV2. J. Comp. Neurol. 339: 535–558.

Perrone-Bizzozero, N.I., Benowitz, L.I., Apostilides, P.J., Franck, E.R., Finkelstein, S.P. and Bizzozero, O.A., 1989, Protein fatty acid acylation in developing cortical neurons. J. Neurochem. 52: 1149–1155.

Rutishauser, U., 1984, Developmental biology of a neural cell adhesion molecule. Nature 310: 549–554.

Schneider, G.E., Jhaveri, S., Edwards, M.A. and So, K.-F., 1985, Regeneration, re-routing, and redistribution of axons after early lesions: changes with age, and functional impact. In J. C. Eccles and M.R. Dimitrijevic, (eds.): Recent Achievements in Restorative Neurology, Vol 1. Upper Motor Function and Dysfunction. Basel: Karger, pp. 291–310.

Schubert, D., Jin, L.-W., Saitoh, T. and Cole, G., 1989, The regulation of amyloid β protein precursor secretion and its modulatory role in cell adhesion. Neuron 3: 689–694.

Schubert, W., Prior, R., Weidemann, A., Dircksen, H., Multhaup, G., Masters, C.L. and Beyreuther, K., 1991, Localization of Alzheimer βA4 amyloid precursor protein at central and peripheral synaptic sites. Brain Res. 563: 184–194.

Selkoe, D., 1994, Normal and abnormal biology of the β-amyloid precursor protein. Annu. Rev. Neurosci. 17: 489–517.

Skene, J.H.P., 1989, Axonal growth-associated proteins. Annu. Rev. Neurosci. 12: 127–156.

Steedman, J.G. and Landreth, G.E., 1989, Expression of pp60c-src in adult and developing rat central nervous system. Dev. Brain Res. 45:161–167.

Stettler, O., Moya, K.L., Zahraoui, A. and Tavitian, B., 1994, Developmental changes in the localisation of the synaptic vesicle protein rab3A in rat brain. Neurosci. 62: 587–600.

Stettler, O., Tavitian, B. and Moya, K.L., 1996, Differential synaptic vesicle protein expression in the barrel field of developing cortex. J. Comp. Neurol. 375: 321–332.

Strittmatter, S.M., Frankhauser, C., Huang, P.L., Mashimo, H. and Fishman, M.C., 1995, Neuronal pathfinding is abnormal in mice lacking the neuronal growth cone protein GAP-43. Cell 80: 445–452.

Thinikaran, G., Kitt, C.A., Roskams, A.J.I., Slunt, H.H., Masliah, E., von Koch, C., Ginsberg, S.D., Ronnett, G.V., Reed, R.R., Price, D.L. and Sisodia, S.S., 1995, Distribution of an APP homolog, APLP2, in the mouse olfactory system: a potential role for APLP2 in axogenesis. J. Neurosci. 15: 6314–6326.

15

DEVELOPMENTAL CHANGES IN THE SPONTANEOUS BURSTING PATTERNS OF ON AND OFF RETINAL GANGLION CELLS

R. O. L. Wong,[1,*] E. D. Miller,[1] W. T. Wong,[1] C. R. Shields,[2] and K. L. Myhr[1]

[1]Department of Anatomy and Neurobiology
[2]Department of Ophthalmology
Washington University School of Medicine
660 S. Euclid, St. Louis, Missouri 63110

1. INTRODUCTION

Connections in the adult visual system are highly precise. In mammals, the axonal projections of retinal ganglion cells from the two eyes are segregated into distinct layers within the dorsal lateral geniculate nucleus (dLGN)[1-3]. In addition, each geniculate neuron receives input from only one type of ganglion cell. In the ferret and cat, the major classes of ganglion cells that project to the dLGN, the alpha and beta ganglion cells[4,5], each comprise subpopulations of cells which have receptive field centers that respond either to the onset or to the offset of a light stimulus[6]. In the ferret, On and Off retinal inputs are organized into sublaminae within each eye-specific layer[7,8]. No On and Off sublaminae are present in the cat dLGN, but individual geniculate neurons are contacted only by On or Off ganglion cells[9,10].

Early in development, however, dLGN neurons are binocularly driven[11] and the axonal terminals of the ganglion cells of the left and right eyes are not segregated into distinct laminae. Monocular innervation and the confinement of retinal inputs into eye-specific layers occur before the retina is sensitive to light[12-15]. In the cat, the formation of the eye specific layers is prevented when tetrodotoxin, a sodium channel blocker, is chronically delivered to the dLGN[16,17]. This result suggests that neuronal activity, presumably from the retina, is necessary for the segregation of eye-input in the dLGN.

Like the formation of eye-specific layers, the appearance of On and Off sublaminae in the ferret dLGN also occurs before photoreceptors mature[7] and is dependent on neuro-

* Phone: (314) 362-4941; Fax: (314) 747-1150.

Development and Organization of the Retina, edited by Chalupa and Finlay.
Plenum Press, New York, 1998.

nal activity[15]. When an NMDA receptor antagonist, MK-801 or APV (2-amino-5-phos-phono-pentanoic acid), is infused into the region of the ferret dLGN throughout the period when On and Off sublaminae normally develop in vivo (between the second and third postnatal weeks), these sublaminae fail to form[15]. Examination of individual axonal arbors of the ganglion cells at the dLGN further showed that in the absence of postsynaptic acti-vation, the terminal arbors of individual ganglion cells arborize diffusely within each eye-specific layer[15].

What is the nature of the ganglion cell activity during the period when the retina is in-sensitive to light stimuli? The first direct evidence that the immature retina generates spe-cific temporal patterns of activity came from in vivo extracellular recordings of the embryonic rat retina[18]. These recordings revealed that the immature retina spontaneously generates a pattern of rhythmic bursting activity, which is temporally synchronized between neighboring cells. By using a multielectrode array to monitor the in vitro activity of 50–100 cells in the ganglion cell layer simultaneously, we subsequently showed that the neonatal ferret retina also generates a similar pattern of synchronized bursting activity[19,20]. The bursts of action potentials of neighboring cells are correlated by propagating waves of excitation. This pattern of activity persists until just before eye-opening (at P30), at which time the spontaneous activity of retinal cells is no longer rhythmic, nor is it synchronized by waves[20].

It was suggested from the multielectrode recordings that the correlated bursting ac-tivity can provide cues that are suitable for instructing the segregation of eye-inputs in the dLGN[19]. This is based on the Hebbian model[21], which proposes that synchronized activity leads to the strengthening of connections between coactive cells. Thus, synchronized bursts of activity within an eye would ensure that cells from the same retina would "wire together". Conversely, cells from the two eyes are unlikely to burst simultaneously, and thus are unlikely to maintain connections with the same target neuron[22–25].

In the ferret, On and Off sublaminae form in the dLGN between the second and third postnatal weeks, after eye-specific layers are established, but during the period when reti-nal waves are still present. Because On and Off ganglion cells are spatially organized into overlapping mosaics[26,27], it would appear that their activity would always be correlated with each other when a wave propagates through the area. If so, then based on the Heb-bian rule, it would seem unlikely that their activity patterns would lead to the segregation of their inputs at the dLGN. Because the class of cells recorded by the multielectrode ar-ray cannot be identified, optical recording techniques, using calcium indicator dyes, were employed in order to determine the bursting patterns of identified On and Off ganglion cells at different ages. We demonstrate here that the patterns of spontaneous activity of On and Off beta ganglion cells are initially indistinguishable but become distinct during the period when On and Off sublaminae emerge in the dLGN. We will address how, and in particular, which retinal circuits can account for the observed age-related changes in activ-ity patterns. Finally, we will consider whether the bursting patterns of On and Off gan-glion cells contain the appropriate cues that would ensure that their inputs onto dLGN neurons are segregated at maturity.

2. BURSTING RHYTHMS OF ON AND OFF GANGLION CELLS

2.1. Identification of Bursting Patterns of On and Off Ganglion Cells

The spontaneous bursting activity of retinal ganglion cells can be observed using optical recording techniques which monitor intracellular calcium concentrations ($[Ca^{2+}]_i$) over time. Retinae were dissected from the eyecup and incubated in fura-2AM (Molecular

Probes) in HEPES-buffered Ames medium, pH 7.4 (Sigma Chemicals) for an hour, initially at room temperature and then at 30°C for half an hour[28]. Upon loading of the calcium indicator dye into cells, the retinae were placed in a temperature-controlled recording chamber and maintained in oxygenated Ames medium (27–30°C) for the duration of the recording. Consecutive pairs of images were acquired during 340 nm and 380 nm excitation using a low light level camera (SIT, Hamamatsu) under computer control (Image-1FL, Universal Imaging Inc). Intracellular calcium levels were estimated using a calibration paradigm which converts the ratio of the fluorescence intensity measured at 340 nm and 380 nm to a known calcium concentration[28].

In order to identify the recorded cells, spontaneously active cells were intracellularly dye-filled with 1% Lucifer yellow in 0.1 M LiCl and 3% Neurobiotin in 0.1 M Tris-buffer, pH 7.6. Putative On and Off ganglion cells were classified according to their dendritic morphology and stratification in the inner plexiform layer, IPL[6]. Figure 1a,b shows an example of a putative On and a putative Off beta ganglion cell which were injected with Lucifer yellow after calcium imaging. On cells were classifed as those whose arbors stratified in the inner two-fifths of the IPL whereas cells whose arbors stratified in the outer three-fifths of the IPL were classified as Off cells. When many neighboring cells were injected subsequently, the On and Off types of beta cells were found to tile the recorded region as described previously[28] (Figure 1c). Like beta ganglion cells in the cat retina[26,27], the dendritic overlap of ferret beta ganglion cells of the same subclass (On or Off) was minimal, compared to the overlap of the dendritic territories of the different subclasses (Figure 1c). Based on the dendritic stratification and the mosaic arrangement of the cells, the activity patterns of putative On and Off beta cells in the recorded region can be compared at different developmental time points.

2.2. Bursting Rhythms of On and Off Ganglion Cells Alter with Age

Putative On and Off beta cells can be identified beginning from P9, when differential dendritic stratification in the IPL becomes discernable[28]. At this age (P9–10), the bursting patterns of On and Off beta ganglion cells were indistinguishable (Figure 2). In particular, the burst frequencies of On and Off beta cells were the same, at around 30–50 bursts/hour. However, during the second and third postnatal weeks, On and Off beta cells developed distinct bursting patterns. On beta cells burst much less frequently (at least 3 or 4 times less often) than Off beta cells, which showed a dramatic increase in their bursting rate with age (Figure 2). However, despite the changes in the burst frequencies of these cells, the bursts of On cells remain correlated with corresponding bursts in neighboring Off cells.

3. MECHANISMS UNDERLYING DEVELOPMENTAL CHANGES IN ON AND OFF BURSTING RHYTHMS

3.1. Differential Modulation of On and Off Bursting Activity by GABAergic Amacrine Cells

Amacrine cells have been demonstrated to participate in the synchronized bursting activity of the neonatal ferret retina[29,30]. But, not all amacrine cells appear to be involved in this pattern of activity. We have begun to examine whether spontaneously active amacrine

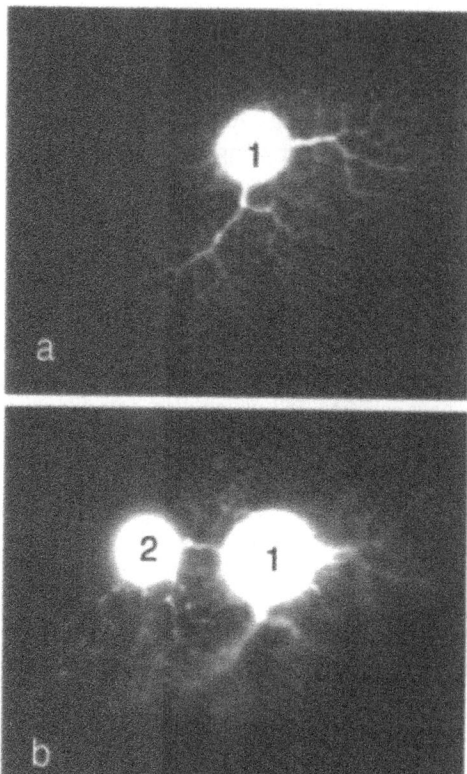

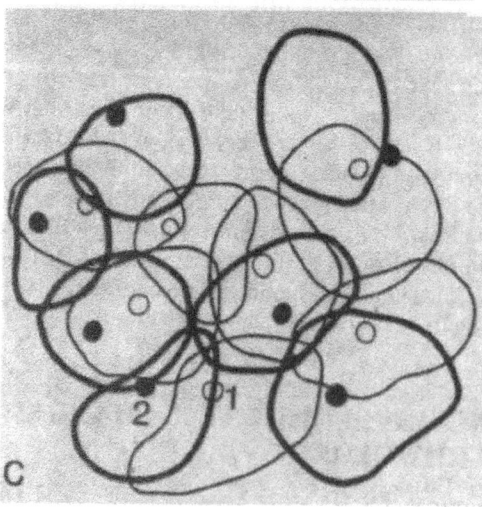

Figure 1. (a and b) Lucifer yellow filled putative On (1) and Off (2, in focus) beta ganglion cells of a P20 ferret retina, classified according to the level of their dendritic stratification. The dendritic arbors of the putative On and Off cells stratify at different levels of the inner plexiform layer. (c) Cells '1' and '2' form part of two mosaics of beta cells, which were reconstructed after dye-filling of many of the neighboring cells. The dendritic territories of presumed On cells is represented by the thin boundaries and the On somata by the open circles, whereas the territories of the presumed Off cells are denoted by the thick outlines and their somata by the filled circles. Note that the dendritic territories of On and Off cells overlap extensively whereas the territories of cells belonging to the same subpopulation have a reduced extent of overlap.

cells share a common morphology or transmitter content, in order to understand what network or circuits form the substrate for the synchronized bursting activity. Figure 3 shows that many different morphological subtypes of displaced amacrine cells exhibit spontaneous rhythmic bursting activity. However, these cells may share a common transmitter because all amacrine cells contain either gamma-aminobutyric acid (GABA) or glycine[31].

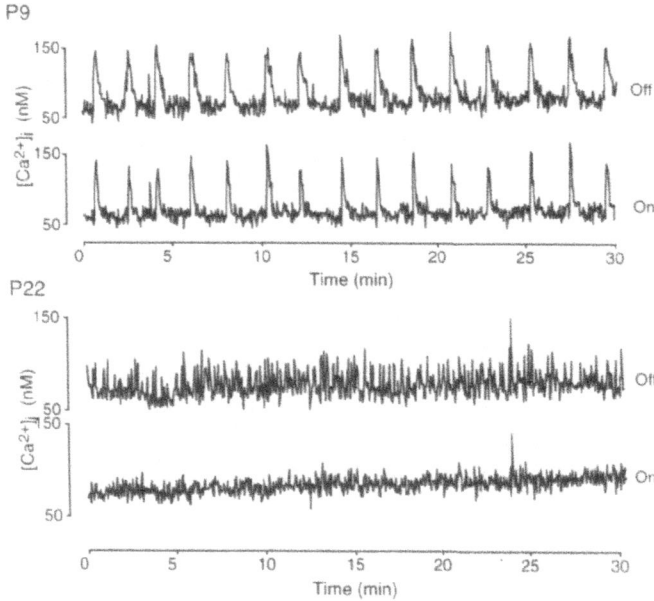

Figure 2. Examples of the bursting patterns of putative On and Off beta ganglion cells at P9 and at P22.

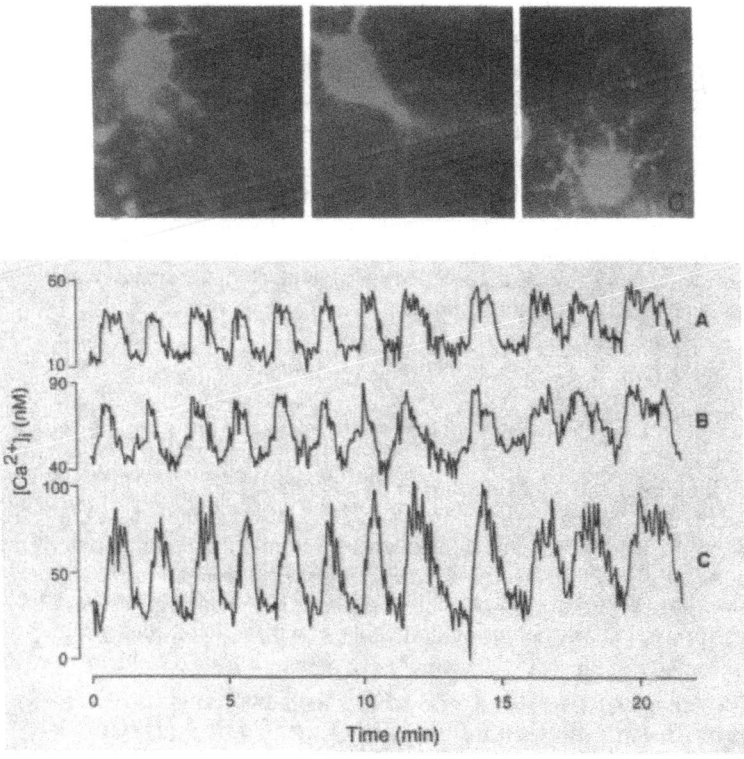

Figure 3. (A–C) Examples of lucifer yellow filled displaced amacrine cells from a P15 ferret retina. These cells displayed rhythmic bursting activity, which is shown in the plots below.

We have explored the possibility that GABAergic and/or glycinergic amacrine cells participate in the wave activity. Our immunohistochemical studies demonstrate that GABA and glycine are present in conventionally placed and displaced amacrine cells in the ferret retina by embryonic day 38 (GABA[32]; glycine-unpublished observations). In addition, immunohistochemistry for synapse-associated proteins suggests the presence of synapses in the ferret IPL by birth[32], a finding that corroborates a previous electron microscopy study demonstrating the presence of conventional synapses in the ferret IPL by P4[33].

Immunocytochemistry for GABAa receptor subunits also revealed the presence of GABAa receptors in the ferret IPL by birth[32]. Whole-cell recordings from neonatal ganglion cells show that these cells have functional GABAa receptors and calcium imaging studies also demonstrate the presence of functional GABA and glycine receptors[34,35]. Taken together, these observations suggest that GABAergic and glycinergic amacrine cells could be interacting synaptically with ganglion cells in the ferret retina as early as at the time of birth. In order to examine whether the spontaneous bursting activity of the ganglion cells depends on interactions with GABAergic or glycinergic amacrine cells, we compared the activity patterns of neonatal ferret ganglion cells in the presence and absence of antagonists to GABAa and glycine receptors[34].

Bicuculline methobromide (100–150 μM; Sigma Chemicals), a GABAa receptor antagonist, reduced or abolished the bursting activity of all ganglion cells in the first ten postnatal days (Figure 4, P10). This effect was reversible upon washout of the antagonist. SR95531 and picrotoxin, agents which also antagonize the effect of GABA, produced the same effect as bicuculline on the bursting activity. Thus, by blocking GABAa receptors, the spontaneous bursting activity of the ganglion cells in the early neonates was diminished, suggesting that GABA promotes the bursting activity of the cells at this stage. By contrast, after P18, bicuculline, SR95531 and picrotoxin caused an increase in bursting activity in the ganglion cells. Examination of the effects of bicuculline on the bursting activity of identified On and Off ganglion cells between P18 and P24, demonstrated that bicuculline caused On cells to exhibit a marked increase in their burst frequency (Figure 4, P24). The effects of bicuculline on the burst frequency of Off cells was less dramatic. Interestingly, the bursts of On and Off cells remained synchronized even in the presence of GABAa receptor antagonists (Figure 4, P24).

By contrast, when strychnine (2–5 μM), a glycine receptor antagonist, was applied, there was no detectable effect on the bursting activity at all ages studied (P0 to P24; data not shown). Only at higher concentrations (25–100 μM), did strychnine cause a change in the bursting activity; at the younger ages (up to P10), strychnine at these higher concentrations abolished the bursting activity, whereas in the older neonates (by about P18), it caused a dramatic increase in bursting activity of both On and Off ganglion cells.

These results suggest that initially, during the first and second postnatal weeks, GABA is not inhibitory and appears to promote the bursting activity of the ganglion cells. As the action of GABA becomes inhibitory at around P18, the activity of On cells in particular appear to be heavily suppressed, compared to Off cells. The role of glycine at the various ages, however, is less clear and warrants further investigation. The apparent ineffectiveness of low concentrations of strychnine, compared to bicuculline, may be due to a reduced sensitivity of the glycine receptor to strychnine during development.

3.2. Contribution of Bipolar Cells to On and Off Activity

During the period when On and Off ganglion cells develop distinct bursting patterns, ribbon synapses can be detected in the maturing IPL[33]. Immunostaining at different post-

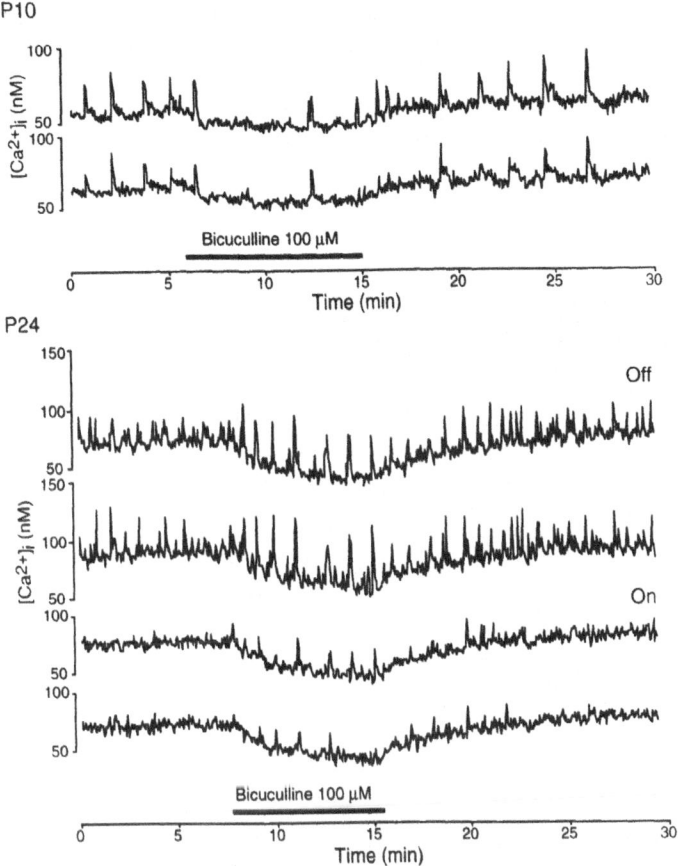

Figure 4. Bicuculline, a GABAa receptor antagonist, reduces the bursting activity of all classes of ganglion cells at P10. At P24, bicuculline causes an increase in activity instead, with On cells cells showing a marked increase in their burst frequency in the presence of this antagonist.

natal ages for calbindin (which labels putative cone bipolar cells) and for protein kinase C (PKC) (found in rod bipolar cells), demonstrated that putative cone and rod bipolar cells are present by the end of the second postnatal week. Examples of immunolabelling for calbindin and PKC of a P23 retina are shown in Figure 5a, b. To ascertain when bipolar cells form functional connections with their postsynaptic targets, the ganglion cells and amacrine cells, retinal slices were obtained and bipolar cells were stimulated by puffs of potassium chloride at the outer plexiform layer (OPL), while recording optically from the neighboring cells. These recordings demonstrate that in mid-peripheral retina, functional bipolar connections emerge by around P13. By the third postnatal week, responses to stimulation of the bipolar cell dendrites in the OPL consistently evoked responses in ganglion cells and amacrine cells. An example of responses evoked in the ganglion cell layer and inner nuclear layer by a puff of potassium at the OPL is shown in Figure 5c. In order to demonstrate that the rise in $[Ca^{2+}]_i$ in the ganglion cells and amacrine cells resulted from depolarization of bipolar cells, the stimulations were repeated in the presence of the non-NMDA receptor antagonist, NBQX (1,2,3,4,-Tetrahydro-6-nitro-2,3,-dioxo-benz[f]quinoxaline-7-sulfonamide; RBI) and the NMDA receptor antagonist, APV (RBI). Responses which were abolished in the presence of these antagonists imply that the evoked responses were glutamate-mediated, whereas responses that persisted during superfusion of these antagonists suggested that the cells were most likely depolarized by the diffusion of potassium from the puffer pipette. Taken together, the anatomical and physiological results suggest that functional glutamatergic synapses between bipolar cells and ganglion cells or amacrine cells emerge by the end of the second postnatal week.

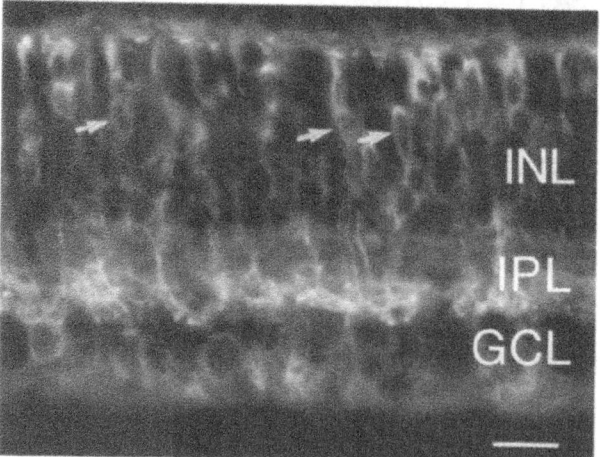

Figure 5. A cross-section of a P23 ferret retina double-immunolabelled for calbindin (a) and protein kinase C (b). Putative cone bipolar cells, revealed by calbindin labelling (examples, arrows, a) and rod bipolar cells, immunoreactive for PKC (examples, arrows, b) are present at this age. Inner nuclear layer (INL); inner plexiform layer (IPL); ganglion cell layer (GCL). Scale bar = 20 μm. (c) Cross section of a P22 ferret retina loaded with fura-2AM, and imaged optically during stimulation of a region in the outer plexiform layer (asterisk, OPL) with a puffer pipette containing 1M KCL. Increases in intracellular calcium levels are highlighted by the black pixels and are found in the INL and GCL. The activated cells are observed after digital subtraction of an image during the KCL puff and a background image prior to the puff. Shown here is the superimposition of the subtracted image (the black pixels) and the background staining with fura-2.

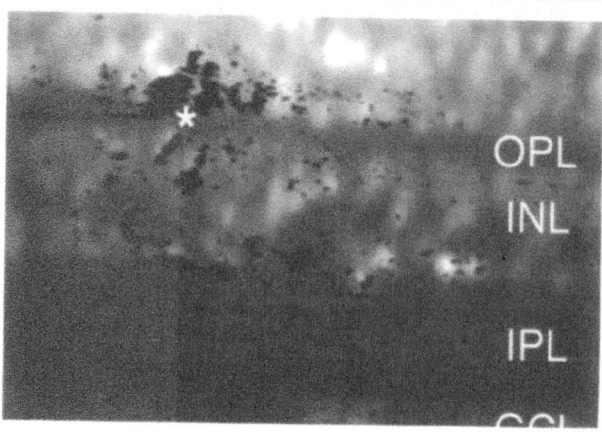

In order to determine whether the spontaneous bursting activity involved interactions with bipolar cells during the period when ribbon synapses are formed, the effects of NBQX and APV on the bursting activity was examined in retinae between P15 and P21. Previous recordings at early neonatal ages (before P10) suggested that ionotropic glutamate receptor antagonists had no effect on the bursting activity of the ganglion cells[29]. Our preliminary recordings indicate that at later ages, at P21, NBQX and APV reduced or abolished all bursting activity (Figure 6). In the mature retina, On-bipolar cells are hyperpolarized by the glutamate analog, DL-2-amino-4-phosphonobutyrate acid (APB)[36]. To further assess whether glutamate-mediated activity significantly contributes to the acquisition of different bursting rhythms of the On and Off cells, we examined the effects of APB on the bursting activity. At P21, APB (Calbiochem) depressed the bursting activity of the ganglion cells. On-beta ganglion cell activity, which is enhanced in the presence of bicuculline, was abolished in the presence of APB. Off beta ganglion cells still burst in the presence of APB, but the frequency (and amplitude) of the bursts appeared to be diminished (Figure 7). Taken together, these preliminary results indicate that glutamate-mediated interactions are not involved in the bursting activity of the ganglion cells initially, but become important as bipolar cells mature and form functional connections during the third postnatal week.

4. DISCUSSION

4.1. A Changing Retinal Circuitry Underlies the Development of Distinct Patterns of On and Off Bursting Activity

Our in vitro recordings from the neonatal ferret retina revealed that the immature ganglion cells spontaneously undergo rhythmic bursting activity prior to eye opening (P32). Initially, the bursting rhythms of all classes of ganglion cells are indistinguishable,

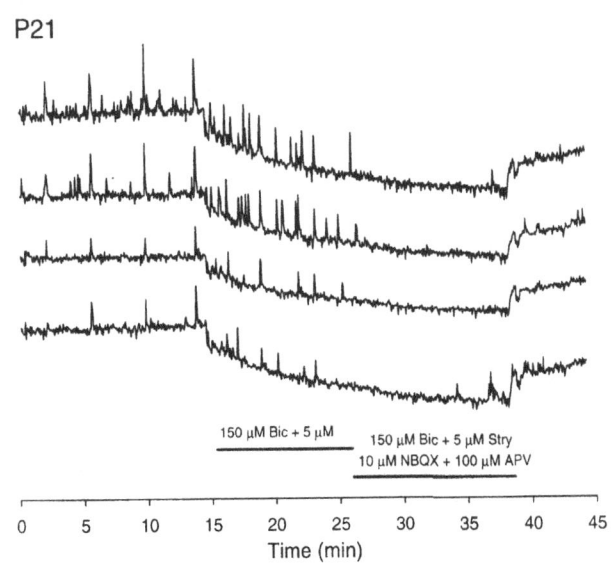

Figure 6. By P21, bicuculline (and strychnine) causes an increase in bursting activity which is abolished in the presence of NBQX and APV.

P21

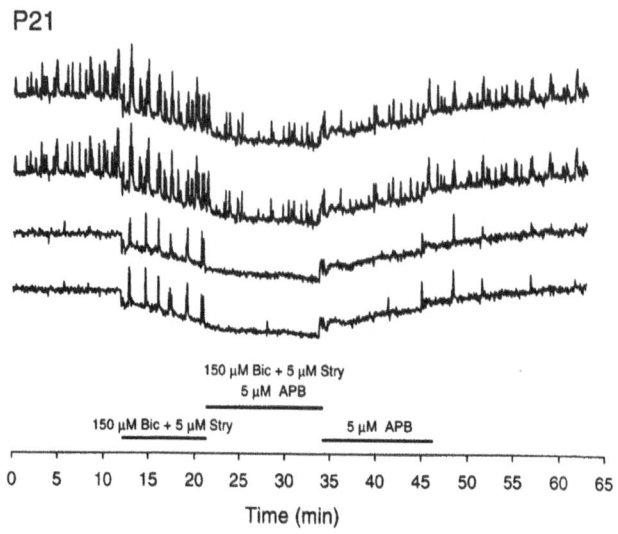

Figure 7. The metabotropic glutamate receptor blocker, APB, is also effective in depressing the activity of On and Off ganglion cells by P21. On cells are silenced by APB, whereas Off cells continue to burst in the presence of this agonist, but at a reduced frequency and amplitude.

but by the third postnatal week, distinct rhythms emerge. Specifically, putative On and Off beta ganglion cells develop different temporal patterns of spontaneous activity during the period in which On and Off sublaminae form in the dLGN; On beta cells participate in only a small fraction of the waves during this period.

We have concentrated on examining the contributions of amacrine cells and bipolar cells to changes in the On and Off ganglion cell activity patterns. Our results to date suggest that interactions with amacrine cells and bipolar cells contribute to the differences in the bursting activity of On and Off ganglion cells. GABAergic inhibition appears to be greater for On cells compared to Off cells, which in part, results in the relatively suppressed bursting activity of the On cells compared to the Off cells. In addition, glutamate-mediated activity, presumably arising from interactions with bipolar cells, is important during the period when On and Off cells adopt different bursting rhythms.

The action of GABA and glutamate are clearly age-dependent. Prior to the emergence of On and Off rhythms, GABA is not inhibitory, and appears to promote the occurrence of bursting activity. In addition, the bursting activity does not depend on the activation of ionotropic glutamate receptors because antagonists to these receptors do not block the activity. This is not surprising, because bipolar cells have yet to differentiate and form functional connections, processes which occur only just prior to the appearance of On and Off distinct rhythms during the third postnatal week.

Since GABAergic amacrine cells form direct connections with ganglion cells, the modulation of ganglion cell bursting activity could arise directly via these connections. In addition, GABAergic amacrine cells could affect the bursting activity indirectly by modulating the release of glutamate from bipolar cells. Whole-cell patch recordings from neonatal ferret bipolar cells show that they are sensitive to exogenously applied GABA (Figure 8). The GABA-evoked current in the neonatal bipolar cells is largely mediated by GABAa receptors. Thus, On ganglion cells may become much less active with age partly because the release of glutamate from On bipolar cell terminals is inhibited by the maturing GABAergic inputs.

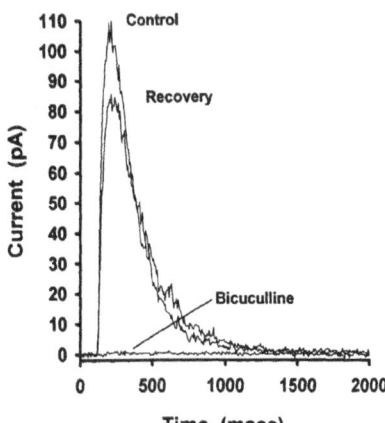

Figure 8. Responses of a P19 bipolar cell under whole cell patch clamp to puffs of GABA (200 μM in the pipette) in the inner plexiform layer. The responses of this cell to GABA was completely abolished in the presence of bicuculline (200 μM). Each trace is an average of 2 or 3 responses. The cell was held at 0 mV (see 35 for details).

Another possible explanation for the age-related decrease in the spontaneous bursting activity of On ganglion cells is that in the absence of light stimulation, On bipolar cells are hyperpolarized by glutamate acting on APB-sensitive receptors on the dendrites of these interneurons. By contrast, Off bipolar cells are depolarized in the absence of light stimulation. Thus, the activity patterns of On and Off ganglion cells might simply reflect the activity of the On and Off bipolar cells to which they are connected, respectively. Future work will be aimed at further addressing which cellular interactions produce the patterns of spontaneous bursting activity both in the early stages of development and when On and Off ganglion cells adopt different temporal patterns of activity. Our interest in determining the cellular interactions that underlie these unique patterns of spontaneous activity are based on the hope that it will be possible to manipulate the temporal patterns of activity in vivo, and therefore examine more directly how these patterns of activity affect the formation of visual connections during development.

4.2. Developmental Role for Changing Patterns of On and Off Bursting Patterns?

In order to fully understand how On and Off retinogeniculate connections are established, several aspects of the development of this visual pathway need to be explored further. In the cat, previous studies have demonstrated that in the absence of retinal activity resulting from chronic injections of tetrodotoxin (TTX) into kitten eyes, neurons in the mature dLGN show mixed On and Off visually-evoked responses[37]. This result may be because the inputs of On and Off cells failed to segregate in the absence of retinal activity, or because new connections developed during the period of activity blockade. Likewise, in the ferret, the failure of sublaminae (presumed On or Off) to form in the absence of postsynaptic activity[15] may have arisen due to the sprouting of the axonal terminals of On and Off cells, rather than the failure of diffuse axonal arbors to segregate. In future studies, it would be helpful to determine whether individual dLGN neurons initially receive converging synaptic inputs from On and Off ganglion cells. If so, these observations would provide strong evidence that segregation of On and Off ganglion inputs is necessary in order to establish separate On and Off channels within the dLGN. Such experiments are important, although at present are limited technically because it is not possible to record On or Off responses prior to the formation of their sublaminae distribution in their dLGN, nor is it possible, as yet, to label specifically the terminals of On or Off ganglion cells. However, even if geniculate neurons do not receive mixed On and Off inputs initially, it is evident from the mentioned studies[15,37], that the maintenance of separate On and Off channels in the dLGN requires continual synaptic interactions between the ganglion cells and the dLGN neurons.

Our current finding that On and Off ganglion cells exhibit distinct rhythms at the time when the On and Off sublaminae develop in the dLGN suggests that this change in the rhythms of ganglion cell bursting activity could be important for the activity-dependent establishment of their connectivity patterns. The different burst frequencies of On and Off cells within an eye results in a decrease in the degree to which their activity is correlated during the period when On and Off sublamine form in the dLGN[28]. Our preliminary modelling studies suggest that this decrease in the correlations may be sufficient to provide the basis by which activity-dependent segregation of On and Off inputs can proceed or is maintained[38]. However, whether these changes in the activity patterns of On and Off ganglion cells are instructive or permissive in vivo remains to be tested experimentally.

It should be noted that while a decrease in the correlations in the activity patterns of On and Off cells could lead to the functional segregation of their inputs, these patterns do not contain cues that would be instructive for the formation of the sublaminae themselves. How layers in the dLGN, dedicated to different channels of visual information, take shape during development remains a mystery. While this is an intriguing problem, perhaps what is more important to bear in mind is that individual target (dLGN) cells receive connections only from a distinct functional class of presynaptic (ganglion) cell. It is the segregation of inputs at the level of individual geniculate neurons that ultimately ensures that visual information reaches the visual cortex in distinct channels, to be recombined later to provide a greater diversity of responses. Activity in the immature visual system is undoubtably important to the developmental process employed to attain functionally distinct pathways from the retina to the dLGN. What remains to be determined is what specific role the patterns of activity from the immature retina play in this process and how it achieves that role during development.

ACKNOWLEDGMENTS

Supported by NIH EY 10699 and a Lucille P. Markey award (Washington University) to R.O.L. Wong, and in part by an Esther A. and Joseph Klingenstein Fellowship and an Alfred P. Sloan Fellowship. We wish to thank Gee Kow Wong for his help with the immunohistochemistry, Dennis M. Oakley for his expert and valued technical support. Whole-cell recordings were performed in collaboration with Dr. Peter Lukasiewicz.

REFERENCES

1. Stone, J. (1983) Parallel processing in the visual system. New York, Plenum Press.
2. Sherman, S.M. (1985) Functional organization of the W-, X-, and Y-cell pathways in the cat: A review and hypothesis. Prog. in Psychobiol. and Physiol. Pscyh. *11*: 233–313.
3. Garraghty, P.E. and M. Sur (1993) Competitive interactions influencing the development of retinal axonal arbors in cat lateral geniculate nucleus. Physiol. Reviews. *73*: 529–45.
4. Boycott, B.B. and H. Wässle (1974) The morphological types of ganglion cells of the domestic cat's retina. J. Physiol. Lond. *240*: 397–419.
5. Wingate, R.J., T. Fitzgibbon and I.D. Thompson (1992) Lucifer yellow, retrograde tracers, and fractal analysis characterise adult ferret retinal ganglion cells. J. Comp. Neurol. *323*: 449–474.

6. Nelson, R., E.V. Famiglietti and H. Kolb (1978) Intracellular staining reveals different levels of stratification for on-centre and off-centre ganglion cells in the cat retina. J. Neurophysiol. *41*: 472–483.
7. Linden, D.C., R.W. Guillery and J. Cucchiaro (1981) The dorsal lateral geniculate nucleus of the normal ferret and its postnatal development. J. Comp. Neurol. *203*: 189–211.
8. Stryker, M.P. and K. Zahs (1983) On and Off sublaminae in the lateral geniculate nucleus of the ferret. J. Neurosci. *3*: 1943–1951.
9. Hubel, D.H. and T.N. Wiesel (1961) Integrative activity in the cat's lateral geniculate body. J. Physiol. Lond. *155*: 385–398.
10. Cleland, B.G., M.W. Dubin and W.R. Levick (1971) Simultaneous recording of input and output of lateral geniculate neurons. Nature New Biol. *231*: 191–192.
11. Shatz, C.J. and P.A. Kirkwood (1984) Prenatal development of functional connections in the cat's retinogeniculate pathway. J. Neurosci. *4*: 1378–1397.
12. Cucchiaro, J. and R.W. Guillery (1984) The development of the retinogeniculate pathways in normal and albino ferrets. Proc. Roy. Soc. Lond. (B) *223*: 141–164.
13. Sherman, S.M. (1985) Development of retinal projections to the cat's lateral geniculate nucleus. TINS *60*: 350–355.
14. Sretavan, D.W. and C.J. Shatz (1986) Prenatal development of retinal ganglion cell axons: Segregation into eye-specific layers. J. Neurosci. *6*: 234–251.
15. Hahm, J.O., R.B. Langdon and M. Sur (1991) Disruption of retinogeniculate afferent segregation by antagonists to NMDA receptors. Nature *351*: 568–570.
16. Shatz, C.J. and M.P. Stryker (1988) Prenatal tetrodotoxin infusion blocks segregation of retinogeniculate afferents. Science *242*: 87–89.
17. Sretavan, D.W., C.J. Shatz and M.P. Stryker (1988) Modification of retinal ganglion cell axon morphology by prenatal infusion of tetrodotoxin. Nature *336*: 468–471.
18. Galli, L. and L. Maffei (1988) Spontaneous impulse activity of rat retinal ganglion cells in prenatal life. Science *242*: 90–91.
19. Meister, M., R.O.L. Wong, D.A. Baylor and C.J. Shatz (1991) Synchronous bursts of action potentials in ganglion cells of the developing mammalian retina. Science *252*: 939–943.
20. Wong, R.O.L., M. Meister and C.J. Shatz (1993) Transient period of correlated bursting activity during development of the mammalian retina. Neuron *11*: 923–938.
21. Hebb, D.O. (1949) The organization of behaviour. New York, Wiley.
22. Willshaw, D. J. and C. von der Malsburg (1976) How patterned neural connections can be set up by self-organization. Proc. Roy. Soc. Lond. (B) *194*: 431–445.
23. Miller, K.D. (1996) Synaptic economics: Competition and cooperation in synaptic plasticity. Neuron *17*: 371–374.
24. Goodman, C.S. and C.J. Shatz (1993) Developmental mechanisms that generate precise patterns of neuronal connecivity. Cell/Neuron *72/10* (suppl): 77–98.
25. Katz, L.C. and C.J. Shatz (1996) Synaptic activity and the construction of cortical circuits. Science *274*: 1133–1138.
26. Wässle, H., B.B. Boycott and R.B. Illing (1981) Morphology and mosaic of On and Off-beta cells in the cat retina and some functional considerations. Proc. Roy. Soc. Lond. (B) *212*: 177–195.
27. Wässle, H., L. Peichl and B.B. Boycott (1981) Dendritic territories of cat retinal ganglion cells. Nature *292*: 344–345.
28. Wong, R.O.L. and D.M. Oakley (1996) Changing patterns of spontaneous bursting activity of On and Off retinal ganglion cells during development. Neuron *16*: 1087–1095.
29. Wong, R.O.L., A. Chernjavsky, S.J Smith and C.J. Shatz (1995) Early functional neural networks in the developing retina. Nature *374*: 716–718.
30. Feller, M.B., D.P. Wellis, D. Stellwagen, F.S. Werblin, and C.J. Shatz (1996) Requirement for cholinergic synaptic transmission in the propagation of spontaneous retinal waves. Science *272*: 1182–1187.
31. Vaney, D.I. (1990) The mosaic of amacrine cells in the mammalian retina. In: Progress in Retinal Research, N.N. Osborne and G. Chader (eds), Oxford. U.K., Pergamon, vol 9, pp 49–100.
32. Karne, A., G.-K. Wong, D.M. Oakley and R.O.L. Wong (1997) Immunocytochemical localization of GABA, GABA$_A$ receptors and synapse associated proteins in the developing and adult ferret retina. Vis. Neurosci. (in press).
33. Greiner, J.V. and T.A. Weidman (1981) Histogenesis of the ferret retina. Exp. Eye Res. *33*, 315–332.
34. Fischer, K.F., P.D. Lukasiewicz and R.O.L. Wong (1997) Age-dependent and cell-class specific modulation of retinal ganglionce cell bursting activity by GABA. (in revision, Neuron).
35. Lukasiewicz, P.D. and R.O.L. Wong (1997) GABA$_C$ receptors on ferret retinal bipolar cells; a diversity of subtypes in mammals? Vis. Neurosci. (in press).

36. Slaughter, M.M. and R.F. Miller (1981) 2-Amino-4-phosphonobutyric acid: a new pharmacological tool for retinal research. Science *211*: 182–185.

37. Dubin, M.W., L.A. Stark and S.M. Archer (1986) A role for action potential activity in the development of neuronal connections in the kitten retinogeniculate pathway. J. Neurosci. *6*, 1021–1036.

38. Lee, C.W. and R.O.L. Wong (1996) Developmental patterns of On/Off retinal ganglion cell activity lead to segregation of their afferents under a Hebbian synaptic rule. Soc. Neurosci. Abst. *22*, p. 1202.

16

THE DISEASED RETINAL GANGLION CELL AND ITS PROPENSITY TO SURVIVE AND REGENERATE AN AXON WITH FUNCTIONAL SIGNIFICANCE

Rita Naskar, Christiane Köbbert, and Solon Thanos

Department of Experimental Ophthalmology
School of Medicine
Westfälische-Wilhelms University of Münster
Germany

1. INTRODUCTION

The conceptual approach to understanding regeneration of lesioned nerve fibers has long been based on developmental neurobiology. Ideally, regeneration would recapitulate development, and many paradigms to study development do, in fact, overlap with repair. Regeneration, in fact, has to be subdivided into the questions of cell survival and reinitiation of axon growth, of target finding, and of the formation of functional connections. Although the retina is an extension of the central nervous system (CNS), and lacks the ability to regenerate in mature stages, it represents an opportune system for the study of regeneration as retinal neurites can regenerate under experimentally selected conditions. Such conditions are created either by replacement of the optic nerve with a peripheral nerve graft (Aguayo, 1985) or by explantation and culture of retinal stripes *in vitro* (Thanos *et al.*, 1989). The regeneration of neurites is then the result of enhanced post-injury survival of cell bodies and facilitation of growth cone movement which seems strongly influenced by interactions with components of their immediate environment as well as their targets.

The retinal ganglion cell (RGC) is a neuronal cell type which fulfills the requisites to study the cascade of de- and regeneration in the CNS, and this is facilitated by the fact that the retina is the only part of the CNS that can be studied as a flatmount. The retina also has the advantage of representing an anatomically well-defined system which is readily accessible for morphological, biochemical and functional studies. Mammalian retinae contain a variety of ganglion cells exhibiting different forms, functional properties and central projections (Enroth-Cugell and Robson, 1974; Cleland and Levick, 1974). The developmental processes by which ganglion cells differentiate into distinct cell types remain

Development and Organization of the Retina, edited by Chalupa and Finlay.
Plenum Press, New York, 1998.

unknown (Leventhal *et al.,* 1988), but are undoubtedly influenced by intrinsic and genetic factors. Accordingly, the developmental factors responsible for morphological differentiation must be responsible for the development of many of the functional properties of the RGCs and their afferent connections. The development of any complex pathway in the CNS requires coordination of a number of distinct processes including cytogenesis, cellular migration, dendritic differentiation, axonal growth, synaptogenesis and cell death. The scope of this coordination, furthermore, must include interactions between a number of widely separated cell populations, and in many systems must involve temporal as well as spatial gradients (Webster and Rowe, 1991). An important issue emphasized by a number of recent studies is the role of glial cells, in particular with regard to guidance, promotion and -in the adult CNS- inhibition of neurite growth (Schwab and Selkoe, 1995). A knowledge of the sequence of events beginning with the birth of a retinal ganglion cell, its functional diversity and its struggle to survive an injury are vital in order to gain an insight into the 'psychology' of this cell.

2. ORIGIN OF THE RETINAL GANGLION CELL

Understanding the development and differentiation of individual retinal cells provides a link between developmental events such as the establishment of retinal topography, synapse formation and cell death of excessively formed neurons.

Ganglion cells are the progeny of the germinal epithelium that forms the wall of the neural tube. Once the lateral tip of the optic stalk has invaginated, the germinal epithelium of the stalk forms the inner wall of the eye cup, from which the neural retina develops. The layer of dividing cells in the developing neural retina has been termed the cytoblast layer, and two zones of cell division have been distinguished within it (Robinson *et al.,* 1985). Ganglion cells are the first to develop (Sidman 1961; Walsh *et al.,* 1983) as progeny of the major zone of proliferation within the cytoblast layer. The youngest identified RGCs are bipolar in shape, resembling intermitotic forms of germinal cells (Hinds and Hinds, 1974; Dütting *et al.,* 1983). Their somata are in the cytoblast layer, and from each soma one process extends radially to the inner, and one to the outer surface of the retina (Maslim *et al.,* 1986). The inner process forms an axon that extends towards the optic disc identifying the cell as a ganglion cell. More mature RGCs become detached from the surface of the retina, and their somata move to the ganglion cell layer and develop dendrites that branch profusely. It is useful to describe the morphological differentiation of RGCs as occurring in three stages: a stage of *axonal growth and migration*; a stage of *active growth of dendritic trees*; an extended stage of *passive or interstitial growth*. The differentiation of the major classes of ganglion cells occurs in the second of these stages (Maslim *et al.,* 1986). Several major aspects of ganglion cell differentiation seem to be determined by intrinsic, presumably genetic factors. In particular, the relative size of a ganglion cell's dendritic tree and the pattern of its branches (two main identifiers of cell class) emerge while the dendrites are actively growing and before they have received substantial synaptic input (Maslim *et al.,* 1986). On the other hand, extrinsic determinants of ganglion cell morphogenesis can be identified, such as the overall dimension of dendrites which appears to be influenced by cell density (e.g., Kirby & Chalupa, 1986).

During development of the retina as in other regions of the vertebrate nervous system, far more neurons are generated than are ultimately needed with almost one half of them undergoing programmed cell death shortly before establishing meaningful contacts within their targets. This phenomenon, termed programmed cell death (recently replaced by the term apoptosis), is vital to our discussion and understanding of the reasons why adult, mammalian CNS neurons, unlike the peripheral nervous system (PNS) and other organs, lack the intrinsic ability to regenerate after an injury. In lower vertebrates, the high degree of regenerative capacity may be related to the fact that the retina in these animals continually enlarges with increasing age and body size. This requires the generation of new retinal neurons in adult life (Johns and Easter, 1977), and also the development of new synaptic connections within the retina as well as continual reorganization of retinotectal connections (Meyer, 1978). Consideration of the adult retina as a regenerating organ leads us to confront some fundamental problems of ontogenetic development as well as nervous system repair (Grafstein, 1986).

The present article is mainly concerned with the intrinsic and extrinsic factors that play a role in the fate of an injured mammalian retinal ganglion cell. These factors will be considered in the context of neuronal responses to injury and are mainly derived from our knowledge of neuronal differentiation during ontogenetic development.

3. FUNCTIONAL SIGNIFICANCE OF APOPTOSIS: A POSSIBLE MECHANISM FOR IMMUNOLOGICAL SURVEILLANCE OF SURVIVORS?

Two general types of morphological changes accompanying cell degeneration can be distinguished and are termed *necrosis* and *apoptosis*, respectively. The Greek term *necrosis* is relatively general and is applicable to a number of devastating toxic events with 'sudden cell death' and no predictable sequence of events. Apoptosis is also a Greek term referring, in its original use, to the naturally occurring seasonal loss of flowers, and involves a progressive contraction of cell volume and widespread chromatic condensation, but with the initial preservation of the integrity of cytoplasmic organelles. The affected cells separate into membrane-bound fragments that are rapidly phagocytosed by adjacent cells (Oppenheim, 1991). Apoptosis is characteristic of normal tissue turnover, embryonic cell death, and metamorphosis. Thus, apoptosis refers to a sequence of events with a predictable hierarchy.

Why did naturally occurring cell death evolve as an apparently integral aspect of neuronal development? One suggestion is that the overproduction and subsequent loss of neurons is a genetically efficient way to bring about a numerical matching between neurons, targets and afferents (Oppenheim, 1985). Cell groups initially develop independently of each other but eventually must become interconnected in a quantitatively invariant fashion in order to assure optimal function. An alternative or additional adaptive role for cell death is to eliminate errors or inaccuracies in axonal projections or terminal contacts (Oppenheim, 1985). Once excessive numbers of neurons are produced by mechanisms intrinsic to precursor cells, they are available for adaptive use, including numerical matching, error correction and other functions, which may co-exist within and between different neuronal populations. Thus, cell death should be viewed as a process that can serve different purposes depending on the specific developmental requirements existing in a given species or in a particular anatomical region.

Although many of the fundamental questions concerning apoptosis remain unsolved, the widespread interest in this phenomenon has aided the understanding of the function and regulation of neuronal death and related regressive events (Oppenheim, 1985). The physiological reduction of initially overproduced neurons is accepted as a functionally significant regulatory process of retinal morphogenesis (Kohno *et al.*, 1982; Linden and Barrabas, 1986). As in other CNS regions, cells undergoing apoptosis condense, their nuclei collapse, and the dying cells become engulfed and degraded by neighboring cells. The entire process takes less than an hour in nematodes. The same speed of degeneration must be assumed in neural tissue of mammals too.

As microglia represent the intraretinal form of phagocytotic cells, it is very likely that they are related to developmental events, such as network formation, synaptic organization and apoptosis which reduces the initial population of retinal neurons by almost 35%. The retina as part of the CNS has unique properties in comparison to other organs. There is no replacement of lost neurons due to the inability of the mature retina to produce new neurons. This accounts for irreversible loss of neuronal populations affected by deleterious diseases. The retina is relatively isolated by virtue of the inner and outer blood-retina-barriers (BRB). This means that only selected groups of small substances have restricted access to the retina. This is the case in the adult, but apparently not in the developing retina where the BRB develops relatively late, after the excess of neurons has begun to decrease (Pearson *et al.*, 1983). Immigration of microglial cells is observed prior to the onset of neuronal death, and continues throughout its major phase to postnatal life, coinciding with maturation of the BRB. The observation that microglial cells reside within the embryonic tissue before the bulk of neuronal death does not contradict their role in apoptosis, but further implicates their regulatory role in earlier regressive events as well. The fact that the speed of cell removal is higher during the phase of rapid cell decay also suggests the intimate relationship between apoptosis and activity of phagocytotic cells (Pearson *et al.*, 1983). With the production of retinal debris during apoptosis, a regular system of clearance must be established to protect healthy cells from any damage initiated by release of death products.

Apoptosis may exert one of its major functions in building up the uniform, regularly distributed microglial system. Theoretically, the requisites needed to create a functionally specific, permanent surveillance of neurons which survive apoptosis are: (1) the uniform death of every second or third embryonic neuron, (2) stoichiometric attraction of equivalent numbers of microglia from the peripheral circulation and (3) contact inhibition between neighboring microglia. The first prerequisite would facilitate the production of a uniform distribution of immunonocompetent cells. The quantitatively high number of dying neurons implies that each surviving neuron is surrounded by at least one immunocompetent cell. In addition, the presence of numerous microglial cells together with homocytic contact inhibition among the microglia would create a microglial network.

The emphasis given to microglia in relation to apoptosis is warranted in view of the association of these cells with the immune system. Expression of markers typical for immunocompetent cells, including the MHC class I and II antigens (Suzumura *et al.*, 1987), Fc and C3 receptors (Perry and Gordon, for review) assign them a close association with immune functions. Other glial cells, such as astrocytes, have been reported to induce immune-like responses *in vitro* (Graeber *et al.*, 1989).

4. THE MOLECULAR MECHANISMS OF GANGLION CELL DEGENERATION

The molecular mechanisms involved in apoptotic cell death are currently being studied in many laboratories; so far the only gene identified with inhibitory effects on apoptotic death is the oncogene *bcl-2* (Rabacchi *et al.*, 1994). *In vitro* evidence that this gene could be involved in neuronal apoptosis comes from overexpression of this gene in neurons deprived of trophic factors (Garcia *et al.*, 1992; Allsopp *et al.*, 1993). Various hypotheses have been proposed concerning the signals that trigger this type of cell death. Substantial evidence points to the lack of neurotrophic factors. The absence of NGF in sympathetic neuronal cultures induces apoptotic cell death (Johnson *et al.*, 1986). These workers inferred the presence of a target-derived neurotrophic factor that would normally have suppressed an intrinsic suicide program. It has also been proposed that neurotrophic factors act post-translationally by inhibiting the activity of these genes (Edwards *et al.*, 1991; Rukenstein *et al.*, 1991). Also consistent with this is the observation that NGF, and other neurotrophic factors, are critical for the survival of axotomized adult RGCs both *in vitro* (Johnson *et al.*, 1986; Thanos *et al.*, 1989) and *in vivo* (Mey and Thanos, 1993; Mansour-Rabay *et al.*, 1992). Other authors point to calcium as playing a critical role in the triggering of apoptosis (Nicotera *et al.*, 1992), by activating a hypothetical calcium-dependent endonuclease, or by inducing its upregulation. A recent hypothesis formulated by Cheng and Mattson (1991) proposes that the mechanisms by which trophic factors exert their protective effects on injured neurons could be through a stabilization of calcium homeostasis.

Porciatti and coworkers (1996) used adult transgenic mice overexpressing the Bcl-2 protein, as a model for preventing injury-induced cell death. Several months after axotomy, the majority of retinal ganglion cells survived and exhibited normal visual responses. In contrast, in control wild type mice, the vast majority of axotomized RGCs degenerated, and the physiological responses were abolished. These results suggest that strategies aimed at increasing Bcl-2 expression, or mimicking its function, might effectively counteract trauma-induced cell death in the central nervous system.

5. THE NEURON–GLIAL INTERACTION

Apoptosis also comes into play in pathological situations such as after axotomy in the CNS (Rabacchi *et al.*, 1994). Damage to the CNS of mammals often leads to permanent loss of neurologic function. By supporting neuronal growth and metabolism, glial cells may determine, in part, the degree of recovery of CNS injury (Aguayo *et al.*, 1981; Reier *et al.*, 1983). Reactive astroglia and microglia, and invading blood-borne macrophages appear acutely following an insult to the CNS (Rio-Hortega, 1932; Bignami and Dahl, 1976; Giulian and Robertson, 1990) and influence neuron survival, axonal regeneration and recovery of neurologic function.

Inherited neurodegenerative diseases are characterized by progressive disappearance of neurons due to intrinsic metabolic deficits or to malfunctioning within the neuronal network. Another cause of neuronal degeneration is the axotomy-induced, traumatic cell death (TCD). It consists of a series of changes in the morphological appearance of the perikarya, resulting in the decay of neurons. TCD is not considered to be a physiological event within the mature nervous system, unless external violence like axotomy, or extreme pathological conditions like stroke, ischemia, infections or toxic agents force the neurons to break down and die. This leads to Wallerian degeneration, disconnecting the cell from its target and the factors provided by this target. Macrophages are activated to remove the lesion-derived debris. Thus, different populations of glia may control recovery of injured CNS by complex and, perhaps, conflicting actions.The simultaneous activation of astrocytes and microglia indicates their interaction during vitreoretinopathic diseases and TCD. Microglial cells might activate astrocytes with mitogens (Giulian *et al.*, 1990), whereas astrocytes may reciprocally activate microglial cells with IL-3 (Frei *et al.*, 1987).

5.1. Microglia Cells

The ubiquitous presence of microglia throughout the brain of different species suggests a crucial role for these cells in various developmental and pathological events. *In vitro*, microglia have been demonstrated to secrete many substances that regulate cellular responses. The substances cover a broad spectrum ranging from neurotrophic factors such as NGF and bFGF (Mallat *et al.*, 1989) to small neurotoxic molecules. Several lines of evidence suggest that mononuclear phagocytes actually engage in neuron-killing behavior. Reactive microglia are at times found in areas of damaged tissue prior to the death of the neurons (Giulian and Robertson, 1990), and neurotoxic factors have been isolated from damaged CNS (Giulian, 1990, 1992). Finally, drug suppression of reactive mononuclear phagocytes reduces motor neuron death and improves functional recovery after ischemic injury to the rabbit spinal cord (Giulian and Robertson, 1990).

One of the first microglial toxins to be identified was superoxide anion (Giulian and Baker, 1986; Colton and Gilbert, 1987). More recent work (Thery *et al.*, 1991) has confirmed the microglial release of this free radical and its involvement in neuron killing when inflammatory cells are in direct contact with neurons. There is common agreement that the association of oxidative stress with excessive activation of glutamate receptors is a final common pathway for cell vulnerabilty in the brain. Microglia also release cytokines (IL-1, IL-3 and IL-4) and tumor necrosis factor (TNF-α) (Frei *et al.*, 1989). Production of TNF-α induces microglial cells to become toxic to tumor cells. TNF-α mRNA expression is inhibited in human fetal microglial cells by antibodies specific to IL-6, IL-10 and TGF-β, showing that these cytokines can modify the beneficial and harmful effects of TNF-α within the developing brain. Microglia-derived interleukines have been shown to affect microglia and other glial cell types (Merril, 1992). IL-1 is a pluripotent and multifunctional molecule that induces the proliferation of cultured astrocytes, indicating that this substance is an astroglial mitogen (Giulian *et al.*, 1993). The regulation of microglial proliferation and their transformation into phagocytes seems to be controlled by cytokines of the colony stimulating group (Giulian and Ingeman, 1988). Colony stimulating factors, in particular, may be instrumental in controlling steady states of microglia in the injured CNS.

In addition to cytokines, microglia release toxic metabolites, including proteases, amino acids such as glutamate (Piani *et al.*, 1991), which mediate their toxic effects via NMDA-receptors. One experimental approach to assess the role of microglia during neuronal degeneration is to interfere with their activity (Srimal and Nathan, 1990). Blockade of their activation in the retina of adult rats made it possible to investigate the relationship of these cells to axotomy-induced ganglion cell degeneration (Thanos, 1991a, b; Thanos *et al.*, 1993, 1994). The macrophage/microglia inhibiting factor (MIF) (Auriault *et al.*, 1983), being a tuftsin-derived tripeptide sequence of the Fc-chain of human immunoglobulin, acts on cells of monocytic origin (Fridkin and Najjar, 1989). MIF alters the morphology of microglia (Thanos *et al.*, 1993), reduces their neurophagic activity, and results in higher viability and efficiency of regeneration of axotomized RGCs (Thanos *et al.*, 1993). Based on the assumption that retinal proteases, which are probably produced by microglia, are directly involved in the cascade of regressive events initiated by optic nerve transection, Thanos (1991) monitored whether blockade of proteolytic activities can rescue lesioned neurons from degeneration. Both *in vivo* and *in vitro* experiments showed that substantial numbers of RGCs can be rescued and induced to express regenerative capacities. Thus, protease inhibitors may interact and inactivate microglial neurophagic activities, implying that microglia either become activated to secrete proteases or they directly attack and phagocytose the RGCs. Protease inhibitors could block intracellular, lysosomal proteases (Thanos, 1991) which digest already degraded and endocytosed material.

Leukocyte-derived cytokines like IL-4 which is released from Th2 cells (Chao et al., 1993) have been found to protect co-cultured neurons from the neurotoxicity of brain microglia. Although these experiments were performed with non-retinal microglia, they allow a certain degree of extrapolation to the possibility of blocking retinal microglia by using inhibiting cytokines. Indeed, transection of the optic nerve and application of either human or rat IL-4 into the vitreous resulted in a remarkable delay of ganglion cell death, similar to that seen after the injection of MIF. Simultaneous grafting of a piece of peripheral nerve (PN), to facilitate regrowth of axons, was accompanied by vigorous growth of nearly half the neurons axotomized at the time of the grafting. It now becomes apparent that cytokine-mediated communication with the circulating immune system plays a key role in both the mechanisms of microglial activation and the mode of cell destruction (Thanos et al., in preparation).

5.2. Astroglia

The mechanisms that link neurons, microglia and astroglia are doubtlessly complex. An injury to the CNS usually leads to the formation of a glial scar, in which astrocytes proliferate with the expression of glial-fibrillary-acidic protein (GFAP). Since neurites cannot grow through the glial scar, it is assumed that this is a physical barrier for axonal regeneration. Unopposed effects of neurotoxins from microglia during the acute phase of inflammation might limit the function or survival of neurons; this neuron killing action could, in turn, be balanced by growth factors released from astroglia at a later stage of wound repair (Reier et al., 1983; Giulian et al., 1992). Activated astroglia release proteins (>10 kDa) that promote neuronal growth, thus challenging the theory that these cells impede regeneration of the mammalian CNS. Co-culture experiments (Giulian et al., 1993) showed that astroglia promote neuronal growth. Retinal astroglia provide a favorable "carpet" (Fig. 1) for axonal regeneration in culture (Vanselow, 1990). This soluble biologic activity, when recovered from astroglial conditioned medium, supported the survival of large, robust neurons with marked neuritic outgrowth. Transection of the facial nerve leads to an activation of protoplasmic astrocytes within the facial nucleus (Graeber and Kreutzberg, 1988). This could play a role in maintaining the synaptic insulation of regenerating motor neurons thereby preventing neuronophagia as long as the neurons themselves are able to survive. If neurons are destined to die, the astroglial reaction takes a different course and eventually results in a glial scar (Streit and Kreutzberg, 1988). Basic fibroblast growth factor (bFGF) is a member of a large family of structurally related polypeptide growth factors known to enhance the survival and neurite outgrowth of both peripheral and central neurons in culture (Morrison et al., Ferrari et al., 1989), and acts as a neurotrophic factor for a broad range of CNS neuronal populations (Matsuda et al., 1990). It has been proposed that bFGF derived from glia may promote neuronal survival and neurite extension (Hatten et al., 1988). Evidence for bFGF receptors on neurons has been obtained from autoradiographic studies in vivo (Ferguson et al., 1990). The possibility that endogenous bFGF acts as an autocrine factor on astroglia was further supported by experiments that tested the mitogenic effects of exogenous bFGF on glial cells. The astroglial mitogen IL-1 turned out to be the most potent stimulator of bFGF release from astroglial cells. The lack of effect of other lymphokines, such as IL-6, suggests that the IL-1 induced increase in astroglial bFGF release is mediated by specific IL-1 receptors. Since astroglial bFGF release is regulated by a variety of lymphokines and trophic factors, this suggests that conditions under which an imbalance of these molecules occur, such as in CNS trauma or disease, may affect bFGF production.

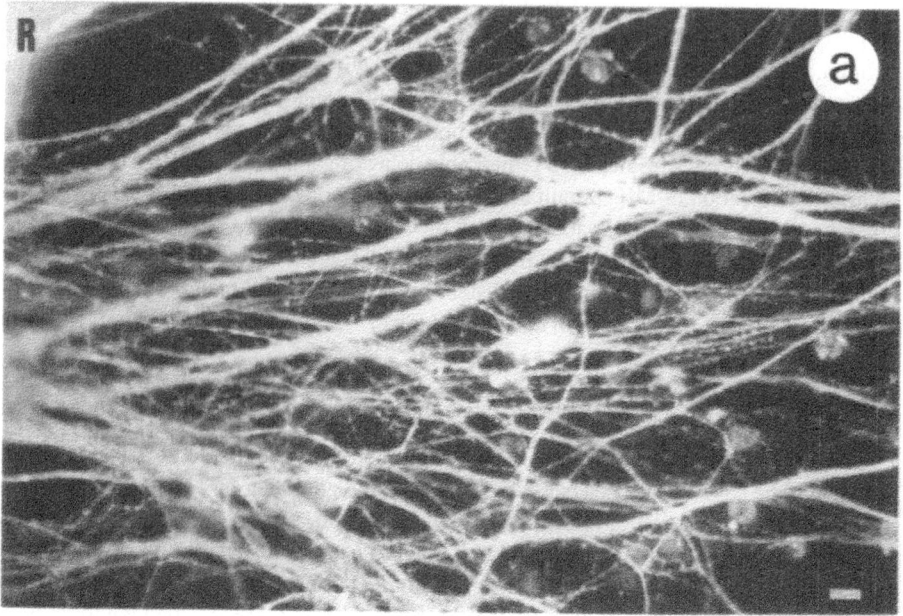

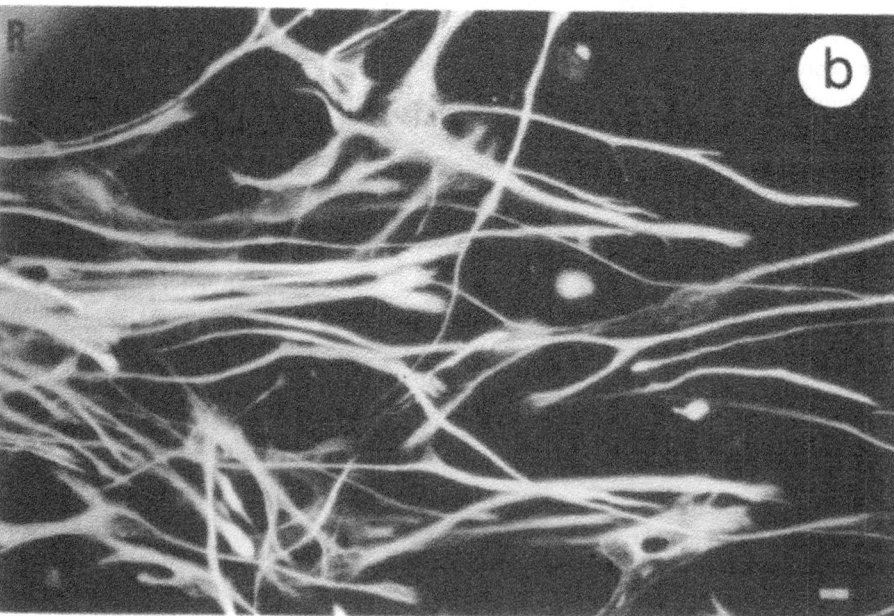

Figure 1. Axonal regrowth *in vitro*. Retinal explants from adult rats were cultured over 7 days in a chemically defined medium without any neurotrophic substitutes. Under these conditions GFAP-stainable astrocytes grow out and develop as a "carpet" (A). Ganglion cell axons which are stained with the Neurofilament antibody SMI-31 grow on the top of the glial cells. Retinal astrocytes do not seem to express inhibitory or repulsive effects on regenerative growth of ganglion cell axons (B). Scale bar =10 μm.

5.3. Oligodendroglia

Oligodendrocytes, which represent the myelinating glia in the maturing CNS, are inhibitory for axon growth *in vitro* and *in vivo*, despite the provision of trophic factors such as NGF and BDNF. Inhibition of CNS inhibitors results in axonal regrowth *in vitro* and *in vivo* (Caroni and Schwab, 1988; Schnell *et al.*, 1994). Interestingly, oligodendrocytes and myelin from the CNS of adult fish appear to be much less inhibitory for axonal regrowth (Vanselow *et al.*, 1990; Bastmeyer *et al.*, 1991), explaining why fish neurons are able to regenerate spontaneously. This suggests that during evolution myelinating cells in the CNS have changed their properties from permissive or supportive to inhibitory.

Ultimately, glial-dependent recovery of neuronal function would rely upon the location and numbers of reactive cells associated with the damaged tissue. The regulation of glial effects upon neurons is further complicated by microglia-astroglia interactions (Giulian *et al.*, 1993). As already mentioned, microglia release astroglia-promoting growth factors including IL-1. Such astroglial growth factors, in turn, influence astroglial production of growth factors such as NGF (Gadient *et al.*, 1990). Moreover, astroglia stimulate microglial growth by release of various protein mitogens (Giulian *et al.*, 1990; Malipiero *et al.*, 1990). In this way, opposing actions of secreted factors from glia might regulate neuronal survival well beyond the period of initial tissue insult (Giulian *et al.*, 1993).

6. PATHFINDING, TARGET RECOGNITION, AND SYNAPTOGENESIS OF REGENERATING GANGLION CELL NEURITES

The first hurdle an injured retinal ganglion cell has to overcome is survival, and this is controlled by the different populations of glia and involves complex and conflicting actions. In addition, intrinsic neuronal determinants of axonal growth seem to influence regeneration: in contrast to the mammalian CNS neurons, fish and frog retinal ganglion cells show a strikingly robust and reproducible growth *in vitro* and *in vivo* (Fawcett, 1992). The re-establishment of interrupted connections between nerve cells in the injured CNS should entail the reiteration of events responsible for the formation of neural circuits in embryonic and early postnatal life. Indeed, the survival of damaged neurons, the extension of axons towards appropriate targets, and the eventual formation of functional synapses are essential steps common to both development and regeneration (Aguayo *et al.*, 1991).

Axon growth and regeneration are cooperative processes; the speed and extent of axon growth are influenced both by the properties of the environment surrounding the axon growth cone, and the properties of the neuron itself (Fawcett, 1992). Although the majority of CNS neurons in the adult animal die after disconnection from their target, a few are able to survive and regrow their axons. The reasons for this differential response to injury remain unknown. Some of the surviving neurons regain or retain the ability to regrow their axons. For instance, both the application of antibodies against myelin-associated inhibitors of neurite growth (Caroni and Schwab, 1988; Schnell *et al.*, 1994) and the transplantation of peripheral nerves replacing lesioned CNS fiber tracts (Aguayo *et al.*, 1985) allow considerable regrowth of injured CNS axons.

The experimental model which has best demonstrated that adult CNS axons are capable of regeneration is the replacement of the inhibiting CNS-environment with a permissive one, best represented by an autologous peripheral nerve graft. CNS neurons of different systems (such as retinal ganglion cells, cortical or brainstem neurons) can regrow

their axons over long distances in the presence of nerve grafts and form synapses with target neurons (Keirstead *et al.*, 1989). The first evidence of reformation of synapses came
from experiments in which adult rat retinal ganglion cells, which had been axotomized at
the optic nerve stump, regrew their axons through transplanted sciatic nerves to approach
the superior colliculus (SC) after mechanical insertion of the distal ends of the nerve grafts
(Keirstead *et al.*, 1989).

In general, regenerative axonal growth is less effective than developmental growth.
In the favorable environment of a peripheral nerve, almost all axons regenerate readily—although some studies have shown a small decrease in speed of regeneration with
age, as well as a decrease in terminal sprouting (Fawcett, 1992). What determines how
vigorously axons grow? Neurons express a growth program, presumably involving the sequential switching of genes, which controls the rate and extent of growth. During development, expression of molecules, such as tubulin, microtubule associated proteins (MAPs),
growth associated proteins (e.g. GAP-43), and neurofilament and adhesion molecules
change in a similar sequence in many neurons (Fawcett for review and references therein).
GAP-43 is rapidly transported by the retinal ganglion cells of neonatal rabbits (Skene and
Willard, 1981), but its relative amount declines drastically with development. It could not
be induced by axotomy of adult optic nerves, but it was induced after axotomy of the adult
hypoglossal nerve which transported very low levels of GAP-43 before axotomy. Thus the
failure of mammalian CNS neurons to express GAP genes may underlie the failure of
CNS axons to regenerate after axon injury (Skene and Willard, 1981).

To explain why older neurons grow their axons less vigorously than young ones,
molecules that are present during embryogenesis and absent or different during regeneration are clearly relevant. It is also possible that some genes are permanently inactivated,
although very few differences between genes expressed during initial development and
those expressed during regeneration have been described. In the cytoskeleton, the pattern
of expression of tubulin and neurofilament genes is very similar in embryonic and regenerative growth. For example, there are at least six alpha-tubulin genes in the rat, one of
which is expressed during embryonic axon growth, and is reactivated during regeneration.
However, MAP1 is expressed in embryos and not reactivated in regeneration, and the different MAPs may play a crucial role in controlling the assembly and stabilization of mi-
crotubules just behind the growth cone. Neurons become less good at growing axons the
older they get, but the provision of trophic factors can certainly help: retinal ganglion cells
grow more vigorously in the presence of fibroblast growth factor (Bähr *et al.*, 1989).

Fish and amphibians in contrast to mammals demonstrate a remarkable ability to
regrow damaged fibers of central nervous tissue. Transection of the goldfish optic nerve
initiates significant increases in RGC biosynthesis, intra-axonal transport, and in the glycosylation of proteins during regeneration (Giulian *et al.*, 1980). Major synthetic events
are the enhanced production of structural proteins, which persist throughout the regrowth
process. Other synthetic events include the production of minor proteins of high molecular
weight which are not detected within the intact visual system (Giulian *et al.*, 1980).

Within PN grafts, the axons of CNS neurons regrow without branching for distances
of 3–4 cm (Aguayo, 1985; Vidal-Sanz *et al.*, 1987). The growth cones of these axons extend in close apposition to Schwann cells and their basal lamina. This anatomical relationship suggests that such contacts mediate neuronal interactions with these substrates
(Aguayo *et al.*, 1991). The mechanisms whereby the regrowing axons interact with some
of the cellular substrate components of the PNS have been investigated *in vitro*. Two major classes of molecules influence cell growth and survival: growth factors and adhesive
molecules. Although there are major overlaps in structure and function, the adhesion

molecules have been classified functionally into cell adhesion molecules (e.g. N-CAM, N-Cadherin) and extracellular matrix (ECM) proteins (e.g. laminin). Cell adhesion molecules often act through homophilic binding between neurons and cells such as Schwann cells, while ECM proteins act through heterophilic interactions with receptors called integrins whose function is specified by different alpha and beta subunits. Secreted forms of CAMs and neural cadherin could affect neuronal function by modulating cascades of second-messenger systems (Aubert *et al.*, 1995). The use of antibodies known to disrupt the function of integrins has stimulated studies on the role of these molecules in regeneration and development (Aguayo, 1991). Laminin, which is transiently expressed in the ECM of the developing vertebrate CNS (Liesi, 1985), also facilitates the regeneration of axons in the injured PNS (Sandrock and Matthew, 1987). Laminin has been shown to cause a sustained increase in growth cone velocity, and to provide clear directional cues, through the activation of protein kinase C intracellular signalling mechanisms (Kuhn *et al.*, 1995). Furthermore, interactions between 1-integrin, a laminin receptor and the cytoskeleton have been reported to influence growth cone guidance (Schmidt *et al.*, 1995). Interestingly, integrins on developing neurons undergo a functional down-regulation that partly involves post-translational modification of these receptors influenced by the connection of the neurons with their targets (Cohen *et al.*, 1989). Also, integrin function becomes apparent following axotomy of peripheral nerves in adult animals (Toyota *et al.*, 1990). Thus, integrins and their ECM ligands which are involved in aspects of neural development, may be recapitulated during regeneration.

7. TARGET INDUCIBLE CELL STABILIZATION

Following injury, the regenerated axons of retinal ganglion cells must regain their communicative properties with their natural targets. In adult rats and hamsters, autologous PN grafts attached to the ocular stump of transected optic nerves (ON) were used as bridges to guide RGC axons to the SC (Vidal-Sanz *et al.*, 1987; Carter *et al.*, 1989a, Thanos *et al.*, 1996). After different time intervals, RGCs were labelled intravitreally with tracer substances and the region of the SC near the end of the PN graft was examined by light and electron microscopy. This revealed that the RGC axons had grown into the superficial layers of the SC for up to 500 μm at 2 months (Carter *et al.*, 1989a) and 1000 μm at 8–10 months. The regenerated RGC axons had formed normal-appearing terminals and synapses. The establishment of synaptic contacts is an essential step in the process of regeneration, indicating that the growth cones of regenerating neurites are secure; before this point they could have regressed to the unstable form. Denervation of the target cells prior to the time of axonal arrival is essential for meaningful synaptogenesis. (Thanos *et al.*, 1996). Thanos and coworkers (1993) also showed that the injection of the neuroprotective factor MIF at the time of grafting permit around 17% of the RGCs to survive and regenerate. However, regenerating ganglion cells are reconnected with a number of postsynaptic neurons, and single postsynaptic neurons receive input from more than one ganglion cell, as shown by extracellular single cell recordings (Sauve *et al.*, 1995). These principles of convergence and divergence in connectivity may partially compensate for the substantial loss of ganglion cells.

Once regenerating retinal axons connect with neurons in the major visual center, the superior colliculus, the growing neurites are capable of restoring the visual circuitry by forming functional synaptic contacts. The minimum number of axons that mediate functional restoration has not yet been determined, but a few thousand fibers (Thanos *et al.*,

1997) or perhaps even hundreds of axons seem sufficient. Although the topographic arrangement of the reinnervating axons may be crude and different from the normal topography, it appears capable of restoring higher visual functions such as the ability to discriminate between light and darkness (Thanos *et al.*, 1996) and to differentiate between simple linear patterns like alternating black and white stripes (Thanos *et al.*, 1997). This is in agreement with previous morphological studies (Carter *et al.*, 1989) and extracellular recordings from the SC (Keirstead *et al.*, 1989; Sauve *et al.*, 1995). In addition, the ascending projections of the reconnected thalamic and midbrain neurons recover their physiological activity, the correlates of which are the transmission of visual evoked potentials (VEPs) and the ability of the animals to discriminate patterns they had been trained to recognize. The fact that the positive VEP responses completely disappear after transection of the PN graft indicates that visual recovery was due to regeneration of RGCs. Thus, the regenerated retino-tectal projections exhibit two important functional attributes: RGCs regenerating axons remain responsive to light and the synapses they form transmit impulses to SC neurons (Aguayo *et al.*, 1991).

During development of the vertebrate visual system, an orderly projection of ganglion cells from the retina onto the SC is established. Mechanisms that might govern this process include the coordinated action of guidance and corresponding receptor molecules that are specifically distributed on axons and their targets. In birds and mammals, information for axonal guidance and targeting appears to be confined to the time when the retino-collicular projection is being formed. Using an *in vitro* model Wizenmann and coworkers (1993) showed that putative guidance activities for temporal and nasal axons, which are not detectable in the normal adult CNS, appear after optic nerve transection in adult rats. Both embryonic and adult retinal axons are able to respond to these guiding cues, although the guidance activities detectable in the deafferented adult rat SC might be different from those found during development. These findings suggest that the reestablishment of an ordered projection after lesions in the adult mammalian visual system is possible.

In adult rats, the retinocollicular pathway is topographically organized in a precise manner that is established after an embryonic and postnatal period of refinement and programmed cell death (O'Leary, 1992). There is no direct indication of similar mechanisms in the regenerating system, although some neuronal decline has been found between the fourth and sixth months after grafting (Thanos and Mey, 1995). A retinotopic organization of the restored pathway occurs, but may not be necessary for simple visual performances since several pretectal and midbrain relay nuclei are involved in the acquisition of spatial discrimination in rats (Legg, 1988). However, the main center involved in visual search is the SC (Heywood and Cowey, 1987), and most of the fibers terminate within the SC (Thanos, 1992). This predominance of termination does not preclude the possibility that VEP responses are due to the retinocollicular projection, since thalamic nuclei are the major contributors to ascending innervation of the cortex. In contrast to the normal condition, the new projection is solely ipsilateral, and may differ from the contralateral one, although uncrossed fibers are also involved in visual discrimination in normal rats (Cowey and Franzini, 1979) and may attain a more prominent compensatory role in the absence of contralateral axons. The fact that induced projections display a retinotopic arrangement has also been demonstrated in hamsters, where retinal neurons terminate and segregate retinotopically within auditory thalamic areas (Frost and Metin, 1985) These results indicate that at least a rough topography is determined by the fiber-fiber interactions (Carter *et al.*, 1994; Wizenmann *et al.*, 1993).

All types of ganglion cells contribute proportionally to regeneration, and display dendritic coverage factors which guarantee the lack of visual field scotomas (Thanos *et al.*, 1997; Thanos and Mey, 1995). These results are in accordance with other studies showing functional recovery by small numbers of regrowing neurites, e.g. in the spinal cord (Bregman *et al.*, 1995) and in retinal axons spared optic nerve injury (Sautter and Sabel, 1993), or treated with the interleukin-2 dimerising glutaminase (Eitan *et al.*, 1994). In addition, intracranial pathways can be restored with neuronal grafts with limited numbers of cells, as shown in the basal ganglia system (Björklund and Stenevi, 1984). Also, neurotrophin-3 was found to be capable of inducing sprouting and restoring corticospinal function with less than 10% of the axons having regenerated (Schnell *et al.*, 1994).

Thus, functional restitution after optic nerve transection requires not only cell survival and regrowth of sufficient numbers of axons but also: 1. Ingrowth of these axons into the SC. 2. Meaningful reconnection of the axons with postsynaptic elements. 3. Contribution of cells from all areas of the retina to axonal growth. 4. Preservation of the intraretinal synaptic circuitry. One of the major therapeutic strategies may therefore be to create functional contingents from a small population of injury-surviving neurons by external treatment, thus facilitating regenerative reconnection with their natural targets.

8. COMPARISON WITH OTHER NEURONS WITHIN THE BRAIN AND SPINAL CORD

In addition to optic nerve transection, other models of traumatic injury to the nervous system provide a separate class of paradigms to investigate the reaction of neurons and microglia to injury. The common feature of these models is physical damage to neurons/axons, arising from some mechanical intervention.

Facial nerve transection is a favorable model, which involves cutting the facial nerve and the retrograde, non-lethal injury of cell bodies within the facial motor nucleus (Kreutzberg, 1968). Kreutzberg and his colleagues found that following facial nerve transection, resident microglia proliferate and begin to express marker molecules, such as vimentin (Graeber *et al.*, 1988), MHC-I and -II antigens (Streit *et al.*, 1988) and the MUC-101 and MUC 102 epitopes (Gehrmann and Kreutzberg, 1991) at different time intervals following the lesion. In addition, the microglia upregulate the expression of cytokines (Frei *et al.*, 1988). The activated microglia are involved in 'synaptic stripping' of the injured axons leading to changes in the electrophysiological properties of these neurons. In addition, transection of the facial nerve causes a rapid increase of glial fibrillary acidic protein in reactive astrocytes (Graeber and Kreutzberg, 1988).

By comparing the response of brain macrophages and resident microglial cells in cortical lesions, Milligan and colleagues (1991) reported that microglia in the developing brain were not involved in the resolution of cellular debris. By contrast, microglia and not macrophages were the predominant responders in the adult brain. These data suggest that distinct populations of phagocytotic cells respond to lesions during development and in the adult.

The rat corticospinal tract is also an ideal model for studying axonal regeneration as it has a well-defined anatomy and is easily accessible to surgical manipulation. In addition, injuries to the corticospinal tract and the development of potential therapeutic strategies are of enormous clinical relevance. The distribution of microglia, macrophages, T-lymphocytes and astrocytes was characterized by Popovich and coworkers (1997) after a spinal cord lesion in rats. By the use of OX42 and ED1 antibodies, peak microglial activation was observed within the lesion epicenter, prior to the bulk of monocyte influx and macrophage activation. Reactive astrocytes were more prominent at later survival times, thus showing that trauma induced CNS inflammation, occurs rapidly at the site of injury and involves the activation of resident and recruited immune cells (Popovich *et al.*, 1997). The relative contribution of a particular macrophage population to the local inflammatory reaction could determine whether a regenerative or a destructive cascade of events is initiated after an injury. A combination of the microglial and astroglial functions could prepare the injured tissue for regeneration. A lipid transport protein, apolipoprotein E, is known to be upregulated on astrocytes and oligodendrocytes after injury and may help deliver lipid droplets to nearby cells, facilitating membrane reconstruction.

9. CONCLUSIONS

Since Frank Nissl's (1894) first description of the microglial cell, and the ascription of 'leucocyte-like' features to it and the pioneering experiments of Ramon y Cajal at the beginning of this century, numerous experimental models have contributed to the understanding of CNS trauma. Several factors such as the rarity of neurogenesis in the adult brain, the presence of inhibitory factors and the lack of trophic factors are responsible for the inability of the adult mammalian CNS to recover from an injury. However, adult CNS neurons do possess the ability to survive an injury if they are provided with the appropriate environment, to the extent that they can extend axons which make functional synaptic contacts with their natural targets. The interaction between central nervous system neurons, their immediate neighbors, the glia, and the factors secreted by these cells influence cell survival, axonal guidance and regrowth, as well as neuronal connectivity during development and regeneration (Fig. 2). This suggests that a better knowledge of develop-

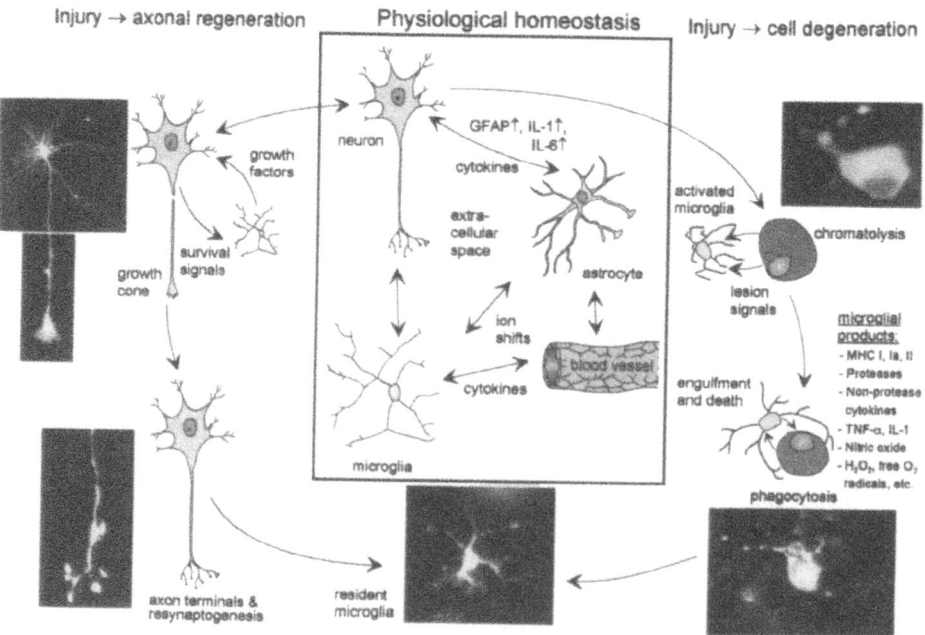

Figure 2. Schematic summary of the relationship between the nerve cell and its neighboring glial and vascular cells.

mental events may enable us to understand more accurately the challenges of axonal regrowth in the adult mammalian CNS. Combining strategies that promote survival of neurons, the regrowth of their axons, and target recognition may contribute to the reduction of neuronal deficits after trauma.

ACKNOWLEDGMENTS

The authors thank Jens Vanselow who provided Figure 1. and Peter Heiduschka for help in providing Figure 2. The technical assistance of Monika Wild is acknowledged. The work was supported by the IRIP (Institute for Research in Paraplegia), the DFG (Deutsche Forschungsgemeinschaft), BMBF (Bundesministerium für Wissenschaft und Bildung), ISRT (International Spinal Cord Trust) and the European Union in the frame of BIOMED-I. R.N. was a recipient of a Friedrich-Ebert Fellowship.

REFERENCES

Aguayo, A.J., David, S. and Bray, G.M. 1981. Influences of glial environment on the elongation of axons after injury: transplantation studies in adult rodents. *J.Exp.Biol.* 95: 231–240.

Aguayo, A.J. 1985. Axonal regeneration from injured neurons in the adult mammalian central nervous system. In: *Synaptic plasticity* (Cotman, C.W. ed), pp 457–484. New York: Guilford Press.

Aguayo, A.J., Rasminsky, M., Bray, G.M., Carbonetto, S., McKerracher.L, Villegas-Perez.M, Vidal-Sanz, M. and Carter, D.A. 1991. Degenerative and regenerative responses of injured neurons in the central nervous system of adult animals. *Phil.Trans.R.Soc.Lond.* 331: 337–343.

Allsopp, T.E., Wyatt, S., Paterson, H.F., Davies, A.M. 1993. The protooncogene bcl-2 can selectively rescue neurotrophic factor-dependent neurons from apoptosis. *Cell.* 73: 295–307.

Aubert, I., Ridet, J-L. and Gage, F.H. 1995. Regeneration in the adult mammalian CNS: guided by development. *Curr.Opin.Neurobiol.* 5: 625–635.

Auriault, C., Joseph, M., Tartar, A. and Capron, A. 1983. Characterisation and synthesis of a macrophage inhibitory peptide from the second constant domain of human immunoglobulin G. *FEBS.Lett.* 153(1): 11–15.

Bähr, M., Vanselow, J. and Thanos, S. 1989. Ability of adult rat ganglion cells to regrow axons *in vitro* can be influenced by fibroblast growth factor and gangliosides. *Neurosci.Lett.* 96: 197–201.

Bastmeyer, M., Beckmann, M., Schwab, M.E. and Stürmer, C.A.O. 1991. Growth of regenerating goldfish axons is inhibited by rat oligodendrocytes and CNS myelin but not by goldfish optic nerve tract oligodendrocyte-like cells and fish myelin. *J.Neurosci.* 16: 621–628.

Bignami, A. and Dahl, D. 1979. The astroglial response to stabbing. Immunofluorescence studies with antibodies to astrocyte-specific protein (GFA) in mammalian and submammalian vertebrates. *Neuropathol. Appl. Neurobiol.* 2: 99–110.

Björklund, A. and Stenevi, U. 1984. Intracerebral neural implants: neuronal replacement and reconstruction of damaged circuitries. *Ann.Rev.Neurosci.* 7: 279–308.

Bregmann, B.S., Kunkel-Bagden, E., Schnell, L., Dal, H.N., Gao, D. and Schwab, M.E. 1995. Recovery from spinal cord injury mediated by antibodies to neurite growth inhibitors. *Nature.* 378: 498–501.

Caroni, P. and Schwab, M.E. 1988. Antibody against myelin-associated inhibitor of neurite growth neutralises nonpermissive substrate properties of CNS white matter. *Neuron.* 1: 85–96.

Carter, D., Bray, G.M. and Aguayo, A.J. 1989a. Regenerated retinal ganglion cell axons can form well-differentiated synapses in the superior colliculus of adult hamsters. *J.Neurosci.* 9: 4042–4050.

Carter, D., Bray, G.M. and Aguayo, A.J. 1989b. Extension and persistence of regenerated retinal ganglion cell axons in the superior colliculus of adult hamsters. *Soc.Neurosci.Abstr.* 15: 872.

Carter, D.A., Bray, G.M. and Aguayo, A.J. 1994. Long-term growth and remodelling of regenerated retino-collicular connections in adult hamsters. *J.Neurosci.* 14(2): 590–598.

Chao, C.C., Molitor, T.W. and Hu, S. 1993. Neuroprotective role of IL-4 against activated microglia. *J.Immunol.* 151(3): 1473–1481.

Cheng, B. and Mattson, M.P. 1991. NGF and bFGF protect rat hippocampal and human cortical neurons against hypoglycemic damage by stablising calcium homeostasis. *Neuron* 7: 1031–1041.

Cleland, B.G. and Levick, W.R. 1974. Properties of rarely encountered types of ganglion cells in the cat's retina and an overall classification. *J.Physiol.* 240: 457–492.

Cohen, J., Nurcombe, V., Jeffrey, P. and Edgar, D. 1989. Developmental loss of functional laminin receptors on retinal ganglion cells is regulated by their target tissue, the optic tectum. *Development.* 107: 381–387.

Colton, C.A. and Gilbert, D.L. 1987. Production of superoxide anion by a CNS macrophage, the microglia. *FEBS.Lett.* 223: 284–288.

Cowey, A. and Franzini, C. 1979. The retinal origin of uncrossed optic nerve fibers in rats and their role in visual discrimination. *Exp.Brain Res.* 35: 443–455.

Dütting, R., Gierer, A. and Hansmann, G. 1983. Self-renewal of stem cells and differentiation of nerve cells in the developing chick retina. *Dev.Brain.Res.* 10: 21–32.

Edwards, S.N., Buckmaster, A.E. and Tolkovsky, A.M. 1991. The death programme in cultured sympathetic neurons can be suppressed at the postranslational level by nerve growth factor, cyclic AMP, and depolarisation. *J.Neurochem.* 57: 2140–2143.

Eitan, S., Solomon, A., Lavie, V., Yoles, E., Hirschberg, D.L., Belkin, M. and Schwartz, M. 1994. Recovery of visual response of injured rat optic nerves treated with transglutaminase. *Science.* 264: 1764–1768.

Enroth-Cugell,C. and Robson, J.G. 1966. The contrast sensitivity of retinal ganglion cells of the cat. *J.Physiol.* 187: 517–552.

Fawcett, J.W. 1992. Intrinsic neuronal determinants of regeneration. *TINS* 15:5–8.

Ferguson, I.A., Schweitzer, J.B. and Johnson, E.M. 1990. Basic fibroblast growth factor: receptor-mediated internalisation, metabolism, and anterograde axonal transport in retinal ganglion cells. *J. Neurosci.* 10: 2176–2189.

Ferrari, G., Minozzi, M.C. Toffano, G. Leaon, A. Skaper, S.D. 1989. Basic fibroblast growth factor promotes the survival and development of mesencephalic neurons in culture. *Dev.Biol.* 133: 140–147.

Frei, K., Siepl, C., Bodmer, S., Scherdel, C. and Fontana, A. 1987. Antigen presentation and tumor cytotoxicity by interferon-gamma treated microglial cells. *Eur.J.Immunol.* 17: 1271–1278.

Fridkin, M. and Najjar, V.A. 1989. Tuftsin: Its chemistry, biology and clinical potential. *Crit. Rev. Biochem. Mol. Biol.* 24: 1–40.

Frost, O.D. and Metin, C. 1985. Induction of functional retinal projections to the somatosensory system. *Nature.* 317: 162–164.

Gadient, R.A., Cron, K.C. and Otten, U. 1990. Interleukin-1 and tumor necrosis factor-alpha synergistically stimulate nerve growth factor (NGF) release from cultured rat astrocytes. *Neurosci.Lett.* 117: 335–340.

Garcia, I., Martinou, I., Tsujimoto, Y. and Martinou, J.C. 1992. Prevention of programmed cell death of sympathetic neurons by the bcl-2 protoonocogene. *Nature.* 258: 302–304.

Gehrmann, J.and Kreutzberg, G.W. 1991. Characterisation of two new monoclonal antibodies directed against rat microglia. *J.Comp.Neurol.* 313: 409–430.

Giulian, D., Ruisseaux, H.D. and Cowburn, D.1980. Biosynthesis and intraxonal transport of proteins during neuronal regeneration. *J.Biol.Chem.* 255: 6494–6501.

Giulian, D and Baker, T.J. 1986. Characterisation of ameboid microglia isolated from developing mammalian brain. *J.Neurosci.* 6: 2163–2178.

Giulian, D. and Ingemann, J.E. 1988. Colony-stimulating factors as promoters of ameboid microglia. *J.Neurosci.* 8: 4707–4717.

Giulian, D. and Robertson, C. 1990. Inhibition of mononuclear phagocytes reduces ischemic injury in the spinal cord. *Ann.Neurol.* 27: 33–42.

Giulian, D., Vaca, K. and Noonan, C. 1990. Secretion of neurotoxins by mononuclear phagocytes infected with HIV-1. *Science.* 250: 1593–1596.

Giulian, D. and Noonan, C. 1992. Neurotoxins and the dementia of AIDS. *AIDS.Res.Rev.* 2: 157–170.

Giulian, D., Vaca, K. and Corpuz, M. 1993. Brain glia release factors with opposing actions upon neuronal survival. *J.Neurosci.* 13(1): 29–37.

Graeber, M.B. and Kreutzberg, G.W. 1988. Delayed astrocyte reaction following facial nerve axotomy. *J.Neurocytol.* 17: 209–220.

Graeber, M.B., Streit, W.J. and Kreutzberg, G.W. 1989. Formation of microglia-derived brain macrophages is blocked by adriamycin. *Acta.Neuropathol.* 78: 348–358.

Grafstein, B. The retina as a regenerating organ. 1986. In: The Retina, A model for cell biology studies Part II. (Adler, R and Farber, D. eds.) pp 275–335. New York. Academic Press.

Hatten, M.E., Lynch, M., Rydel, R.E., Sanchez, J., Joseph-Siverstein, J., Noscatelli, D. and Rifkin, D.B. 1988. *In vitro* neurite extension by granule neurons is dependent upon astroglial-derived fibroblast growth factor. *Dev.Biol.* 125: 280–289.

Hedgecock, E.M., Culotti, J.G. and Hall, D.H. 1990. The UNC-5, UNC-6 and UNC-40 genes guide circumferential migrations of pioneer axons and mesodermal cells on the epidermis in C.elegans. *Neuron* 4: 61–85.

Heywood, C.A. and Cowey, A. 1987. Effects on visual search of lesions of the superior colliculus in infant or adult rats. *Exp.Brain Res.* 65: 465–470.

Hinds, J.W. and Hinds, P.L. 1974. Early ganglion cell differentiation in the mouse retina: An electron microscopic analysis utilising serial sections. *Dev.Biol.* 37: 381–416.

Johns, P.R. and Easter, S.S. 1977. Growth of the adult goldfish eye. II. Increase in retinal number. *J.Comp.Neurol.* 176: 331–342.

Johnson, E.M. and Deckwerth, T.L. 1993. Molecular mechanisms of developmental neuronal death. *Ann. Rev. Neurosci.* 16: 31–46.

Johnson, J.E., Barde, Y.-A., Schwab, M.E. and Thoenen, H. 1986. Brain-derived neurotrophic factor supports the survival of cultured rat retinal ganglion cells. *J.Neurosci.* 6: 3031–3038.

Keirstead, S.A., Rasminsky, M., Fukuda, Y., Carter, D.A., Aguayo, A.J. and Vidal-Sanz, M. 1989. Electrophysiologic responses in hamster superior colliculus evoked by regenerating retinal axons. *Science.* 246. 255–258.

Kreutzberg, G.W. 1968. Über perineuronale Mikrogliazellen (autoradiographische Untersuchung). *Acta. Neuropathol.* 7: 149–161.

Kohno, T., Inomata, H. and Walz, W. 1993. Electrophysiological behaviour of microglia. *Glia.* 7: 93–101.

Kuhn, T.B., Schmidt, M.F., Kater, S.B. 1995. Laminin and fibronectin guide posts signal sustained but opposite effects to passing growth cones. *Neuron.* 14: 275–285.

Legg, C.R. and Cowey, A. 1977. The role of the ventral geniculate nucleus and posterior thalamus in visual intensity discrimination in rats. *Brain Res.* 123: 261–263.

Leventhal, A.G., Schall, J.D. and Ault, S.J. 1988. Extrinsic determinants of retinal ganglion cell structure in the cat. *J.Neurosci.* 8(6): 2028–2038.

Liesi, P. 1985. Laminin immunoreactive glia distinguish regenerative adult CNS systems from non-regenerative ones. *EMBO J.* 4: 2505–2511.

Linden, R.L.A. and Barrabas, P.C. 1986. Mononuclear phagocytes in the retina of developing rats. *Histochemistry.* 85: 335–339.

Malipiero, U.V., Frei, K. and Fontana, A. 1990. Production of hemopoietic colony-stimulating factors by astrocytes. *J.Immunol.* 144: 3816–3821.

Mallat, M., Houlgatte, R., Brachet, P. and Prochiantz, A. 1989. Lipopolysaccharide-stimulated rat brain macrophages release NGF *in vitro. Dev.Biol.* 133. 309–311.

Mansour-Rabaey, S., Bray, G.M. and Aguayo, A.J. 1992. *In vivo* effects of brain-derived neurotrophic factor (BDNF) and injury on the survival of axotomised retinal ganglion cells in adult rats. *Mol. Cell Biol.* 3: 333a.

Maslim, J., Webster, M. and Stone, J. 1986. Stages in the structural differentiation of retinal ganglion cells. *J.Comp.Neurol.* 254: 382–402.

Matsuda, S., Saito, H. and Nishiyama, N. 1990. Effect of basic fibroblast growth factor on neurons cultured from various regions of the rat brain. *Brain Res.* 520: 310–316.

Merril, J.E. 1992. Tumor necrosis factor alpha, interleukin 1 and related cytokines in brain development: normal and pathological. *Dev.Neurosci.* 14: 1–10.

Mey, J. and Thanos, S. 1993. Intravitreal injections of neurotrophic factors support the survival of axotomised retinal ganglion cells in adult rats *in vivo. Brain Res.* 602: 304–317.

Meyer, R.L. 1978. Evidence from thymidine labelling for continuing growth of retina and tectum in juvenile goldfish. *Exp.Neurol.* 59: 99–111.

Milligan, C.E., Cunningham, T.J., and Levitt,P. (1991). Differential immunochemical markers reveal the normal distribution of brain macrophages and microglia in the developing rat brain. *J.Comp.Neurol.* 314: 125–135.

Morrison, R.S., Sharma, A., De Vellis, J. and Bradshaw, R.A. 1986. Basic fibroblast growth factor supports the survival of cerebral cortical neurons in primary culture. *Proc.Natl.Acad.Sci.* 83: 7537–7541.

Nicotera, P., Bellomo, G. and Orrenius, S. 1992. Calcium-mediated mechanisms in chemically induced cell death. *Ann. Rev.Pharmacol.Toxicol.* 32: 449–470.

O'Leary, D.D.M. 1992. Development of connectional diversity and specificity in the mammalian brain by the pruning of collateral projections. *Curr.Opin.Neurobiol.* 2: 70–77.

Oppenheim, R.W. 1985. Naturally occurring cell death during neuronal development. *TINS.* 11: 487–493.

Oppenheim, R.W. 1991 Cell death during development of the nervous system. 1993. *Ann. Rev. Neurosci.* 14: 453–501.

Pearson, H.E., Payne, B.R. and Cunnigham, T.J. 1983. Microglial invasion and activation in response to naturally occurring neuronal degeneration in the ganglion cell layer of the postnatal cat retina. *Dev.Brain.Res.* 76: 249–255.

Perry, V.H. and Gordon, S. 1988. Macrophages and microglia in the nervous system. *TINS.* 11: 273–277.

Piani, D., Frei, K., Do, K., Cuenod, M. and Fontana, A. 1991. Murine brain macrophages induce NMDA receptor mediated neurotoxicity *in vitro* by secreting glutamate. *Neurosci.Lett.* 133: 159–162.

Popovich, P.G., Wie, P. and Stokes, B.T. 1997 Cellular inflammatory response after spinal cord injury in Sprague-Dawley and Lewis rats. *J.Comp.Neurol.* 377: 443–464.

Porciatti, V., Pizzorusso, T., Cenni, M.C. and Maffei, L. 1996. The visual response of retinal ganglion cells is not altered by optic nerve transection in transgenic mice overexpressing Bcl-2. *Proc. Natl. Acad. Sci.* 93: 14955–14959.

Rabacchi, S.A., Ensini, M. Gravina, A., Fagiolini, M., Bonfanti, L., and Maffei, L. 1992. Retinal ganglion cell degeneration after optic nerve lesion and nerve growth factor protective effect in the developing rat. *Soc.Neurosci.Abstr.* 18: 28.3.

Rabacchi, S.A., Bonfanti, L., Liu, X.-H. and Maffei, L.1994. Apoptotic cell death induced by optic nerve lesion in the neonatal rat. *J.Neurosci.* 14(9): 5292–5301.

Reier, P.J., Stenasaas, L.J. and Guth, L. 1983. The astrocytic scar as an impediment to regeneration in the central nervous system. In: Spinal cord regeneration (Kao, C.C., Bunge, R.P. and Reier, P.J. eds), pp 163–195. New York: Raven Press.

Rio-Hortega, P. 1932. Microglia. In: Cytology and cellular pathology of the nervous system (Penfield, W. ed), pp 481–584. New York: Hocker Press.

Robinson, S.R., Dreher, B. and McCall, M.J. 1989. Nonuniform retinal expansion during the formation of the rabbit's visual streak: Implications for the ontogeny of mammalian retinal topography. *Vis. Neurosci.* 2: 201–219.

Ruckenstein, A., Rydel, R.E. and Greene, L.A. 1991. Multiple agents rescue PC12 cells from serum-free cell death by translation- and transcription-independent mechanisms. *J.Neurosci.* 11: 2552–2563.

Sandrock, W.W. and Matthew, W.D. 1987. An *in vitro* neurite promoting antigen functions in axonal regeneration *in vivo*. *Science.* 237: 1605–1608.

Sautter, J. and Sabel, B.A. 1993. Recovery of brightness discrimination in adult rats despite progressive loss of retrogradely labelled retinal ganglion cells after controlled optic nerve crush. *Eur.J.Neurosci.* 5: 1156–1171.

Sauve, Y., Sawai, H. and Rasminsky, M. 1995. Functional synaptic connections made by regenerated retinal ganglion cell axons in the superior colliculus of adult hamsters. *J.Neurosci.* 15(1): 665–675.

Schmidt, C.E., Dai, J., Lauffenburger, D.A., Sheetz, M.P. and Horwitz, A.F. 1995. Integrin-cytoskeletal interactions in neuronal growth cones. *J.Neurosci.* 15: 3400–3407.

Schnell, L., Schneider, R., Kolbeck, R., Barde, Y. -A. and Schwab, M.E. 1994. Neurotrophin-3 enhances sprouting of corticospinal tract during development and after adult spinal cord lesion. *Nature.* 367: 170–173.

Schwab, M.E. 1990. Myelin-associated inhibitors of neurite growth. *Exp.Neurol.* 109: 2–5.

Schwab, M. and Selkoe, D. 1995. Disease transplantation and regeneration. *Curr.Opin.Neurobiol.* 5: 613–615.

Sidman, R.L. 1961. Histogenesis of mouse retina studied with thymidine H3. In: Structure of the eye. (Smelser, G.K. ed) pp 487–506. New York: Academic Press.

Skene, J.H.P. and Willard, M. 1981. Axonally transported proteins associated with axon growth in rabbit central and peripheral nervous systems. *J.Cell.Biol.* 89: 96–103.

Srimal, S. and Nathan, C. 1990. Purification of macrophage deactivating factor. *J.Exp.Med.* 171: 1347–1361.

Streilein, J.W., Wilbanks, G.A. and Cousins, S.W. 1992. Immunoregulatory mechanisms of the eye. *J. Neuroimuunol.* 39: 185–200.

Streit, W.J. and Kreutzberg, G.W. 1987. Lectin binding by resting and reactive microglia. *J.Neurocytol.* 16: 249–260.

Suzumura, A., Mezitis, S.G.E., Gonatas, N.K. and Silberberg, D.H. 1987. MHC antigen expression on bulk isolated macrophage-microglia from newborn mouse brain: induction of Ia antigen expression by gamma-interferon. *J.Neuroimmunol.* 15: 263–278.

Thanos, S., Bähr, M., Barde, Y.-A. and Vanselow, J. 1989. Survival and axonal elongation of adult rat retinal ganglion cells. *Eur.J.Neurosci.* 1: 19–26.

Thanos, S. 1991a. Blockade of proteolytic activity retards retrograde degeneration of axotomised retinal ganglion cells and enhances axonal regeneration in organ cultures. In: The changing visual system (Bagnoli, P. and Hodos, W.eds) pp 77–93. New York, Plenum Press.

Thanos, S.1991b. The relationship of microglial cells to dying neurons during natural neuronal cell death and axotomy-induced degeneration of the rat retina. *Eur.J.Neurosci.* 3: 1189–1207.

Thanos, S. 1992. Adult retinofugal axons regenerating through peripheral nerve grafts can restore the light-induced pupilloconstriction reflex. *Eur. J. Neurosci.* 4: 691–699.

Thanos, S., Mey, J. and Wild, M. 1993. Treatment of the adult retina with microglia-suppressing factors retards axotomy-induced neuronal degradation and enhances axonal regeneration *in vivo* and *in vitro*. *J.Neurosci.* 13(2): 555–565.

Thanos, S. and Mey. J. 1995. Target-specific stabilisation and classification of regenerating ganglion cells in the rat retina. *J. Neurosci.* 15: 1057–1079.

Thanos, S., Naskar, R., Mey, J. and Schaeffel, F. 1996. Timing of fiber arrival and denervation of postsynaptic neurons is required for restoration of visual perception by regenerating axons. *J.Brain Res.* 37(2): 255–268.

Thanos, S., Naskar, R. and Heiduschka, P. 1997. Regenerating ganglion cell axons in the adult rat establish retino-fugal topography and restore visual function. *Exp. Brain Res.* 114:483–491.

Thery, C., Chamak, B. and Mallat, M. 1991. Free radical killing of neurons. *Eur.J.Neurosci.* 3: 1155–1164.

Toyota, B., Carbonetto, S. and David, S. 1990. A dual laminin-collagen receptor acts in peripheral nerve regeneration. *Proc. Natl. Acad. Sci.* 87: 1319–1322.

Vanselow, J. 1990. Untersuchungen zur Regenerationsfähigkeit adulter retinaler Ganglienzellen von Ratte und Huhn in Organkultur. Doctoral Thesis, submitted to the University of Tübingen, Germany.

Vanselow, J., Schwab, M.E. and Thanos, S. 1990. Responses of regenerating rat retinal ganglion cell axons to contacts with central nervous myelin *in vitro*. *Eur. J. Neurosci.* 2: 121–125.

Vidal-Sanz, M., Bray, G.M., Villegas-Perez, M.P. and Aguayo, A.J. 1987. Axonal regeneration and synapse formation in the superior colliculus by retinal ganglion cells in the adult rat. *J.Neurosci.* 7: 2894–2907.

Walsh, C., Pllley, E.H., Hickey, T.L. and Guillery, R.W. 1983. Generation of cat retinal ganglion cells in relation to central pathways. *Nature.* 302: 611–614.

Webster, M.J. and Rowe, M.H. 1991. Distribution and developmental timing in the albino rat retina. *J.Comp.Neurol.* 307: 460–474.

Wizenmann, A., Thies, E., Klostermann, S., Bonhoeffer, F. and Bähr, M. 1993. Appearance of target specific guidance information for regenerating axons after CNS lesions. *Neuron.* 11: 975–983.

17

THE ROLE OF ACTIVITY-DEPENDENT MECHANISMS IN PATTERN FORMATION IN THE RETINOGENICULATE PATHWAY

Catherine A. Leamey,* Karina S. Cramer, and Mriganka Sur

Department of Brain and Cognitive Sciences
Massachussetts Institute of Technology
Cambridge, Massachusetts 02139

1. INTRODUCTION

One of the most striking features of mammalian sensory systems is the extremely precise way that peripheral receptors are mapped onto the brain. A fundamental task of developmental neurobiology is to understand how these highly specific connection patterns arise. The development of the nervous system can be conveniently divided into 2 stages. The first one, which includes neurogenesis, differentiation, migration and axonal guidance, is usually regarded as being activity-independent. The second stage, which is primarily concerned with the fine-tuning of initial connections to produce the highly specific connection patterns characteristic of the adult, is often regarded as being activity-dependent. The visual system of some mammalian species has proven very useful for examination of the role which activity plays in the refinement of connections. The focus of the work described here is upon the developmental mechanisms through which the characteristic organization of the retinogeniculate pathway in the ferret is produced. This article will review this work in the context of other relevant literature in the field.

2. DEVELOPMENT OF THE RETINOGENICULATE PATHWAY

2.1. Afferent Arrival and Segregation of Inputs

In the ferret, axons of the retinal ganglion cells have grown out from the optic disc and have reached the optic chiasm by embryonic day (E)24. The first fibers reach their primary target, the lateral geniculate nucleus (LGN), by E27 (Johnson and Casagrande,

---

* Address correspondence to: Dr. Cathy Leamey, Dept. of Brain and Cognitive Sciences, E25-235, MIT, Cambridge, MA, 02139, USA. Ph: 617 253 8785; Fax: 617 253 9829; E-mail: cathy@mit.edu

Development and Organization of the Retina, edited by Chalupa and Finlay.
Plenum Press, New York, 1998.

1993). During the following weeks, the projection becomes more extensive and at birth (P0; gestation in ferrets is 41–42 days) the terminals of the fibers from the 2 eyes intermingle with each other in the LGN. By P7, however, the projection from each eye has become restricted to a single lamina within the LGN (Linden et al., 1981). Each eye's projection zone within the LGN is made up of individual axon arbors from that eye; single retinogeniculate axons are simple and unbranched at birth in ferrets, but form arbors that span approximately one eye-specific layer by P7 (Hahm et al., 1997). In the cat, retinal axons from both eyes reach the LGN by E35 (gestation in cats is 64 days). As in the ferret, the terminals from the two eyes initially overlap. By E47 the segregation of the two inputs has commenced, and clear ipsilateral and contralateral layers are present by E54 (Shatz, 1983). During the period of overlap, individual retinogeniculate axons are again relatively simple and restricted in extent. The appearance of the ipsi- and contralateral layers is characterized by the elaboration of terminal arbors in the appropriate LGN territory, and the retraction of a small number of minor side branches (Sretavan and Shatz, 1986). The fact that eye-specific layers form postnatally in the ferret makes the system more accessible for study in this species. In addition, in the ferret, during the 3rd and 4th postnatal weeks retinogeniculate projections undergo a further period of refinement into sublaminae. Each eye-specific layer becomes divided into an inner region which receives inputs from ON-centre retinal ganglion cells and an outer layer which receives inputs from OFF-centre retinal ganglion cells (Stryker and Zahs, 1983). Sublamination reflects the refinement of axon arbors which initially extend throughout both sublaminae, and are subsequently restricted to span only one sublamina (Hahm et al., 1991). Both the segregation of eye-specific layers and the subsequent sublamination of these layers involve the removal of inappropriate terminations and the elaboration of appropriate terminations, although this process is less pronounced for the eye-specific layers than for the ON/OFF sublaminae. The eye-specific layers and ON/OFF sublaminae also become apparent in the cellular organization of the LGN, where they are separated by narrow, relatively cell-sparse zones.

2.2. The Role of Activity in Retinogeniculate Development

2.2.1. Spontaneous Activity in the Developing Retina. Whilst lamination and sublamination occur well before eye-opening, there is considerable evidence that activity plays an important role in the fine-tuning of connections in the retinogeniculate pathway. It has been shown that waves of spontaneous activity sweep across the retina during development (Galli and Maffei, 1988; Meister et al., 1991; Wong et al., 1993). It is believed that these waves produce a form of retinotopic activity, long before photoreceptors are active. The waves are characterised by synchronised oscillations of intracellular calcium concentration among neighbouring subpopulations of retinal ganglion cells and a type of amacrine cell (Wong et al., 1995). Importantly, in the mouse, this pattern of activity has been shown to be capable of driving LGN cells (Mooney et al., 1996). The spontaneous retinal waves gradually subside with maturation, and have essentially disappeared by the time of eye-opening, which occurs at about one month of age (Wong et al., 1993).

2.2.2. Formation of Eye-Specific Layers. The possible role of spontaneous retinal activity in the formation of eye-specific layers in the LGN has been assessed *in vivo* by the use of pharmacological agents. For example, the infusion of the sodium channel blocker tetrodotoxin (TTX) in the vicinity of the optic chiasm has been reported to prevent the formation of eye-specific layers in the prenatal kitten (Shatz and Stryker, 1988). However, since this protocol non-specifically blocks both pre- and postsynaptic activity, it is difficult to determine whether it is activity in the afferent or target structures, or the two in combination, which is necessary for the segregation of the inputs from the two eyes. It is also possible that this protocol has other non-specific effects within the developing central nervous system. Several lines of evidence suggest that the segregation of eye-specific layers in the LGN does not rely a great deal on either afferent or target activity alone. A brief report (Cook et al., 1996) indicates that specifically blocking retinal activity with intravitreal application of TTX in ferrets during the first postnatal week has only a minor effect on eye-specific lamination; laminae still segregate in relatively normal fashion. Furthermore, blockade of the n-methyl-D-aspartate (NMDA) subtype of glutamate receptor during this period also has no effect on eye-specific lamination (Smetters et al., 1994). Finally, systemic inhibition of nitric oxide synthase (NOS) in the first postnatal week fails to prevent laminae from segregating (Cramer et al., 1996; see also below).

2.2.3. Formation of ON/OFF Sublaminae. In contrast to eye-specific segregation, there is strong evidence for the role of activity in the segregation of ON/OFF sublaminae. The blockade of retinal afferent activity specifically via intraocular injections of TTX disrupts ON/OFF sublamination in the ferret, demonstrating the importance of retinally driven activity in the this process (Cramer and Sur, 1997). However, activation of postsynaptic structures is also important, as the infusion of specific antagonists to NMDA receptors also prevents the formation of ON/OFF sublaminae (Hahm et al., 1991). This evidence for the involvement of NMDA-mediated activity in the refinement of connections suggests that a form of Hebbian synaptic plasticity, which permits the postsynaptic detection of temporally synchronous activity and the subsequent stabilization of the two inputs, is occurring. A great deal of attention has been focused on this phenomenon as a potential basis for learning and memory in the hippocampus, and more recently as a mechanism for developmental synaptic plasticity in other regions of the central nervous system.

3. THE ROLE OF NMDA RECEPTORS

3.1. Lessons from the Cortex

Hebbian mechanisms are believed to be involved in long term potentiation (LTP), and long term depression (LTD), which have been extensively studied in the CA1 region of the hippocampus (see Madison et al., 1991 for review). Here, the NMDA receptors act as detectors of correlated activity, allowing the influx of calcium into the postsynaptic cell following the depolarization-induced removal of the magnesium block (Collingridge and Bliss, 1987). There is considerable evidence that presynaptic mechanisms are also involved in the maintenance of LTP (Bekkers and Stevens, 1990; Malinow and Tsien, 1990; Malgoroli et al., 1995). It is likely that a retrograde messenger, secreted by the postsynaptic cell, may be involved in conveying the information necessary to effect changes in the presynaptic structure. Possible candidates for this include nitric oxide (NO) and the neurotrophins (these possibilities are discussed in more detail below). NMDA receptor dependent LTP and LTD have also been shown to be present in layer III of the visual cortex following stimulation of the white matter (Kirkwood and Bear, 1994a,b). In adult animals, LTP can only be induced by this protocol if an antagonist to the inhibitory neurotransmitter gamma-amino-butyric-acid (GABA) is present (Bear et al., 1992; Kirkwood and Bear, 1994a). In contrast, during a critical developmental period, LTP can be induced in layer III by white matter stimulation in the absence of a GABAergic antagonist. It has also been shown that LTP can be induced in layer III following stimulation of layer IV (in the absence of GABAergic receptor blockade), but that this ability does not decline with development (Kirkwood et al., 1993). These results suggest that the changes in the induction of LTP which occur with development do not occur at the level of the layer III cells, but reflect changes occurring in the layer IV cells. These changes in the layer IV cells are activity-dependent, as they are prevented by dark rearing (Kirkwood et al., 1995). An overall decline in the duration of the NMDA mediated response, which is similarly prevented by dark rearing, has also been demonstrated (Carmignoto and Vicini, 1992). In the somatosensory cortex, too, stimulation of thalamocortical afferents induces LTP in layer IV cells during a critical period of thalamocortical development (Crair and Malenka, 1995). Overall there appears to be a correlation between the time window during which LTP can be induced and developmental critical periods.

3.2. NMDA-Mediated Activity and the LGN

Neurotransmission between the retina and the LGN is glutamatergic and involves both NMDA and non-NMDA receptors (Sillito et al., 1990; Kwon et al., 1990). During the first postnatal month in the ferret, NMDA-mediated excitatory postsynaptic currents are characterized by a very slow decay time (Ramoa and McCormick, 1994a,b). Following eye-opening, which occurs at around 1 month of age, there is a marked decline in the duration of the NMDA-mediated component of the response. A form of NMDA-mediated LTP has also been reported in the developing retinogeniculate pathway during this early developmental period (Mooney et al., 1993). A recent study has demonstrated that the developmental decline in the duration of the NMDA-mediated response is prevented by the intraocular application of the sodium channel blocker TTX (Ramoa and Prusky, 1997). The same study also showed that in normal animals, there is a decrease in the degree of sensitivity to ifenprodil, which selectively blocks heteromeric NMDA receptors containing the 2B subunit. In animals where retinal activity was blocked with TTX, sensitivity to ifenprodil was retained. Together these results demonstrate that there is a developmental change in the subunit composition of NMDA receptors in the LGN, and that activity is necessary to effect this change from immature to mature forms (Ramoa and Prusky, 1997). Other work has shown that the infusion of NMDA-receptor antagonists *in vivo* prevents the segregation of ON/OFF sublaminae by permitting the retention of retinal arbors that are unusually large or inappropriately placed within the LGN (Hahm et al., 1991). The infusion of NMDA receptor antagonists during the 3rd postnatal week has also been shown to have pronounced effects on the dendritic morphology of LGN cells, resulting in a dramatic increase in the degree of dendritic branching and spine density (Rocha and Sur, 1995). Impressively, the increase in dendritic spine density was found to occur within a few hours of drug application (Rocha and Sur, 1995). This rapid response indicates that an active, local cellular mechanism is involved in regulating the addition of dendritic spines, and that spine formation is negatively regulated by NMDA-mediated afferent activity. Taken together, the perturbations observed in the organization of both pre- and postsynaptic elements following the blockade of NMDA receptors provide convincing evidence that NMDA-mediated activity is necessary for the production of normal pre- and postsynaptic structures.

3.3. Nitric Oxide as a Retrograde Messenger

3.3.1. Evidence for the Role of Nitric Oxide in LTP. How do NMDA receptors mediate changes in pre- and postsynaptic organisation? A strong contender for this role is NO.

In the hippocampus, it has been shown that calcium influx through NMDA receptors stimulates the production of NO by NO synthase (NOS). NO may then diffuse retrogradely to the presynaptic cell (Schuman and Madison, 1994a; Arancio et al., 1996), where it has been hypothesised to cause changes in the level of neurotransmitter release (Schuman and Madison, 1994b), possibly through activation of cGMP-dependent protein kinase (Arancio et al., 1995). The importance of NOS for LTP induction in stratum radiatum of CA1 has been demonstrated using mice genetically deficient for both the endothelial and neuronal forms of NOS (Son et al., 1996). In another study, an adenovirus vector which inserts a truncated (catalytically inactive) form of the endothelial NOS gene was used to study the effect of acutely reducing the expression of endothelial NOS on LTP induction (Kantor et al., 1996). It was found that endothelial NOS is required for LTP induction in stratum radiatum of CA1. Together these recent findings provide convincing evidence that NO acts as a neuronal messenger in hippocampal synaptic plasticity (see Holscher, 1997 for review).

3.3.2. Nitric Oxide and the LGN. In the LGN, NADPH-diaphorase, which is co-localized with NOS, shows a developmental regulation which peaks during the period of sublamination (Cramer et al., 1995). Importantly, application of the NOS inhibitor N^G-nitro-L-arginine, either systemically or focally via osmotic minipumps, during the 3rd and 4th postnatal weeks disrupts the formation of ON/OFF sublaminae, whereas the application of an inactive isomer had no effect. Arbors of individual axons are inappropriately positioned within the LGN (Cramer et al., 1996), just as they are after NMDA-receptor blockade. These effects are independent of the hypertensive effect of the NOS inhibitor. In contrast, NOS blockade during the first postnatal week does not appear to influence the formation of eye-specific layers (Cramer et al., 1996). This accords with work which has shown that the appearance of eye-specific layers in the ferret is not dependent on activation of NMDA receptors (Smetters et al., 1994) and argues against an NMDA-receptor-independent role for NO. It is uncertain what molecules mediate the effects of NO during LTP and whether these molecules may effect developmental changes downstream of NO. Candidate molecules include guanylyl cyclase (East and Garthwaite, 1991), adenosine diphosphatase ribosyl transferase (ADPRT; Schuman et al., 1994) or both of these (Abe et al., 1994). It is possible that ADPRT acts through the activation of growth associated protein 43 (see Cramer et al., 1996; Cramer and Sur, 1996 for reviews). The kinetics of NO diffusion argue for a substrate specific to recently active terminals.

3.4. The Neurotrophins

3.4.1. Neurotrophins and Synaptic Plasticity. Neurotrophins have long been known to be important for the survival and differentiation of neurons. Recently, a great deal of evidence has emerged that they may also play important roles in synaptic plasticity. As potential mediators of synaptic plasticity, the neurotrophins fulfill two important criteria: their production is regulated by neuronal activity, and they in turn have profound effects on the signalling properties of neurons (for review see Lo, 1995). For example, mRNA for brain-derived neurotrophic factor (BDNF) in the hippocampus increased 6-fold in response to epileptiform activity (Ernfors et al., 1992), and the same conditions which invoke LTP also increase mRNA for BDNF (Patterson et al., 1992). Acute application of BDNF or neurotrophin-3 (NT-3) to the developing neuromuscular synapse increases the frequency of spontaneous synaptic currents without affecting their amplitude (Lohof et al., 1993). In the hippocampus, Kang and Schuman (1995) found that BDNF and NT-3 potentiate evoked glutamatergic transmission by 200–300%. These effects were apparent within one hour of neurotrophin application. It has also been shown that LTP is impaired in mice deficient for the BDNF gene (Korte et al., 1995), and that the addition of recombinant BDNF in these mice permits the induction of LTP (Patterson et al., 1996). The effects of the neurotrophins on synaptic transmission appear to be mediated by tyrosine kinase (Trk) receptors as they are blocked by the addition of a specific Trk receptor antagonist. In addition to their effects on transmission, neurotrophins have been shown to induce changes in the degree of both axonal (Cohen-Cory and Fraser, 1995) and dendritic (McAllister et al., 1995) branching.

3.4.2. Possible Effector Pathways in Neurotrophin Signalling. While much is still to be learnt about how neurotrophins exert their influence on synaptic plasticity, it is believed that binding of a neurotrophin to its specific Trk receptor results in phosphorylation of the Trk receptor. This results in the recruitment of a number of adapter proteins which act through the Ras signalling pathway and lead to the activation of mitogen activated protein kinase (MAPK). MAPK then translocates to the nucleus where it phosphorylates transcription factors. It also causes activation of cAMP response element binding protein (CREB) kinase which leads to the production of immediate early genes (see Green and Kaplan, 1995; Segal and Greenberg, 1996 for reviews). A mechanism has been proposed through which this may then lead to the induction of synaptic growth (Kornhauser and Greenberg, 1997). Also, a recent study has shown that activation of the Trk/MAPK pathway by neurotrophins stimulates the phosphorylation of synapsin I (Jovanic et al., 1996). Synapsin I is believed to tether synaptic vesicles to the actin cytoskeleton in a phosphorylation dependent manner and thus regulate the number of synaptic vesicles available for release (Greengard et al., 1993; Pieribone et al., 1995). Synapsin-actin interactions have also been reported to play a role in the effects of synapsins on synapse formation (Han et al., 1991; Lu et al., 1992; Ferreira et al., 1994). Other studies have found that neurotrophins upregulate NOS expression (Holtzman et al., 1994; Baader et al., 1997), and that the addition of NO enhances the neuritogenic effects of neurotrophic factors (Hindley et al., 1997), suggesting that the reported roles of neurotrophins and nitric oxide (see above) in synaptic plasticity may not necessarily be entirely independent.

3.4.3. The Role of Neurotrophins in Visual System Development. The presence of both the neurotrophins (Schoups et al., 1995) and their specific Trk receptors (Allendoerfer et al., 1994; Cabelli et al. 1996) has been demonstrated in the developing visual system. Furthermore, the expression of neurotrophins is regulated by visual activity (Schoups et al., 1995). The infusion of either BDNF or neurotrophin 4 (NT-4) has been shown to disrupt the formation of ocular dominance columns in the visual cortex (Cabelli et al., 1995). The interpretation of these results is, however, confounded as it is uncertain whether the disruption of ocular dominance columns by the application of neurotrophins was caused by generalised sprouting of geniculocortical fibres, or whether the presence of an excess of neurotrophins prevented the normal developmental processes necessary for ocular dominance column formation from occurring. More convincing evidence for the importance of neurotrophins during developmental synaptic remodelling comes from a recent study where a specific antagonist to the Trk-B receptor was infused into the visual cortex to block endogenous BDNF and NT-4, and this was also found to disrupt the formation of ocular dominance columns (Cabelli et al., 1997). The application of NT-4 has also been found to block the effects of monocular deprivation of LGN neurons (Riddle et al., 1995), and the infusion of BDNF prevents the usual activity-dependent synaptic modifications following monocular deprivation (Galuske et al., 1996). Neurotrophins have also been shown to have differing effects on the branching patterns of dendrites of neurons from different cortical layers (McAllister et al., 1995). BDNF caused the greatest increase in the branching of layer IV neurons, whereas layers V and VI responded most to NT-4. Interestingly, blockade of Trk receptor function demonstrated that BDNF and NT-3 have opposing functions in the regulation of dendritic growth: NT-3 inhibits the growth of layer IV neurons, whereas BDNF inhibits the growth of the dendrites of neurons from layers V and VI (McAllister et al., 1997). The antagonistic actions of BDNF and NT-3 therefore provide a mechanism whereby the growth and retraction of dendrites can be dynamically regulated by an extracellular signalling mechanism. Further, the effects of neurotrophins on dendritic growth require activity (McAllister et al., 1996). This then provides a pathway through which electrical activity may be transduced into a signal which regulates the dendritic growth. It is likely that neurotrophins also play an important role during the activity-dependent refinement of connections in the retinogeniculate pathway. Experiments are currently underway to investigate this possibility.

4. CONCLUSIONS

The retinogeniculate pathway of the ferret is characterized by the presence of both eye-specific and ON/OFF sublaminae which arise in the LGN during the 1st and 3rd–4th postnatal weeks respectively. Consequently, this system provides an excellent model for the investigation of the mechanisms which are implemented by the developing nervous system to produce mature connection patterns. The formation of sublaminae is an active process involving the elaboration of appropriately placed terminals and retraction of inappropriately placed terminals. Both NMDA-receptor mediated activity and nitric oxide production are essential to this process. Recent studies have demonstrated that neurotrophins are important in mediating both structural and physiological changes in the hippocampus and the visual cortex, and they are likely to play a similar role in the retinogeniculate pathway.

REFERENCES

Abe, K., A. Mizutani and H. Saito (1994) Possible role of nitic oxide in long-term potentiation in the dentate gyrus. In: Nitric Oxide: roles in neuronal communication and neurotoxicity. H. Takagi, N. Toda and R.D. Hawkins (eds). Japan Scientific Series. Tokyo. pp149–159.

Allendoerfer, K.L., R.J Cabelli, E. Escandon, D.R. Kaplan, K. Nikolics and C.J. Shatz (1994) Regulation of neurotrophin receptors during the maturation of the mammalian visual system. J. Neurosci., 14: 1795–1811.

Arancio, O., E.R. Kandel and R.D. Hawkins (1995) Activity-dependent long-term enhancement of transmitter release by presynaptic 3′,5′-cyclic cGMP in cultured hippocampal neurons. Nature, 376: 74–80.

Arancio, O., M. Kiebler, C. Justin-Lee, V. Lev-Ram, R.Y. Tsien, E.R. Kandel and R.D. Hawkins (1996) Nitric oxide acts directly in the presynaptic neuron to produce long-term potentiation. Cell, 87: 1025–1035.

Baader, S.L., S. Bucher and K. Schilling (1997) The developmental expression of neuronal nitric oxide synthase in cerebellar granule cells is sensitive to GABA and neurotrophins. Developmental Neuroscience, 19: 283–290.

Bear, M.F., W.A. Press and B.W. Connors (1992) Long-term potentiation in slices of kitten visual cortex and the effects of NMDA receptor blockade.

Bekkers, J.M. and C.F. Stevens (1990) Presynaptic mechanism for long-term potentiation in the hippocampus. Nature, 346: 724–729.

Cabelli, R.J., A. Hohn and C.J. Shatz (1995) Inhibition of ocular dominance column formation by infusion of NT4/5 or BDNF. Science, 267: 1662–1666

Cabelli, R.J., K.L. Allendoerfer, M.J. Radeke, A.A. Welcher, S.C. Feinstein and C.J. Shatz. (1996) Changing patterns of expression and subcellular localization of TrkB in the developing visal system. J. Neurosci., 16: 7965–7980.

Cabelli, R.J., B.L. Shelton, R.A. Segal and C.J. Shatz (1997) Blockade of endogenous ligands of TrkB inhibits formation of ocular dominance columns. Neuron, 19: 63–76.

Carmignoto, G and S. Vicini (1992) Activity-dependent decrease in NMDA receptor responses during development of the visual cortex. Science, 258: 1007–1011.

Cohen-Cory, S. and S.E. Fraser (1995) Effects of brain-derived neurotrophic factor on optic axon remodelling in vivo. Nature, 378: 192–196.

Collingridge, G.L. and T.V.P Bliss (1987) NMDA receptors - their role in long-term potentiation. TINS, 10: 288–293.

Cook, P.M., G. Prusky and A.S. Ramoa (1996) Role of spontaneous activity in reorganization of retinogeniculate connections. Soc. Neurosci. Abst., 22: 761.

Crair, M.C. and R.C. Malenka (1995) A critical period for long-term potentiation at thalamocortical synapses. Nature, 365: 325–328.

Cramer, K.S. and M. Sur (1997) Blockade of afferent impulse activity disrupts on/off sublamination in the ferret lateral geniculate nucleus. Dev. Brain Res., 98: 287–290.

Cramer, K.S. and M. Sur (1996) The role of NMDA receptors and nitric oxide in retinogeniculate development. Progress in Brain Research, 108: 235–244.

Cramer, K.S., C.I. Moore and M.Sur (1995) Transient expression of NADPH-diaphorase in the lateral geniculate nucleus of the ferret during early postnatal development. J. Comp. Neurol., 353: 306–316.

Cramer, K.S., A. Angelucci, J-O. Hahm, M. Bogdanov and M. Sur (1996) A role for nitric oxide in the development of the ferret retinogeniculate projection. J. Neurosci., 16: 7995–8004.

East, S.J. and J. Garthwaite (1991) NMDA receptor activation in rat hipocampus induces cyclic GMP formation through the L-arginine-nitiric oxide pathway. Neurosci. Lett., 123: 17–19.

Ernfos, P., J. Bengzon, Z. Kokaia, H. Persson and O. Lindvall (1992) Increased levels of messenger RNAs for neurotrophic factors in the brain during kindling eliptogenesis. Neuron, 7: 165–176.

Ferreira, A., K.S. Kosikl, P. Greengard and H.-Q. Han (1994) Aberrant neurites and synaptic vesicle protein deficiency in synapsin II depleted neurons. Science, 264: 977–979.

Galli, L. and L. Maffei (1988) Spontaneous impulse activity of rat retinal ganglion cells in prenatal life. Science, 242: 90–91.

Galuske, R.A., D.-S. Kim, E. Castren, H. Thoenen and W. Singer, (1996) Brain-derived neurotrophic factor reverses experience-dependent synaptic modifications in kitten visual cortex. Eur. J. Neurosci., 8: 1554–1559.

Green, L.A. and D.R. Kaplan (1995) Early events in neurotrophin signalling via Trk and p75 receptors. Curr. Op. Neurobiol., 5: 579–587.

Greengard, P., F. Valtorta, A.J. Czernik and F. Benfenati (1993) Synaptic vesicle phosphoproteins and regulation of synaptic function. Science, 259: 780–785.

Hahm, J.-O., R.B. Langdon and M. Sur (1991) Disruption of retinogeniculate afferent segregation by antagonists to NMDA receptors. Nature, 351: 568–570.

Hahm, J., K. Cramer and M. Sur (1997) Pattern formation by retinal afferents in the ferret lateral geniculate nucleus: Developmental segregation and the role of NMDA receptors. J. Comp. Neurol., *submitted.*

Han, H.-Q., R.A. Nichols, M.R. Rubin, M. Bahler and P. Greengard (1991) Induction of formation of presynaptic terminals in neuroblastoma cells by synapsin IIb. Nature, 349: 697–800.

Hindley, S., B.H. Juurlink, J.W. Gysbers, P.J. Middlemas, M.A. Herman and M.P. Rathbone. Nitric oxide donors enhance neurotrophin-induced neurite outgrowth through a cGMP-dependent mechanism. J. Neurosci. Res., 47: 427–439.

Holscher, C. (1997) Nitric oxide, the enigmatic neuronal messenger: its role in synaptic plasticity. TINS, 20: 298–303.

Holtzman, D.M., J. Kilbridge, D.S. Bredt, S.M. Black, Y, Li, D.O. Clary, L.F. Reichardt and W.C. Mobley (1994) NOS induction by NGF in basal forebrain cholinergic neurones: evidence for regulation of brain NOS by a neurotrophin. Neurobiology of Disease, 1: 51–60.

Johnson, J.K. and V.A. Casagrande (1993) Prenatal development of axon outgrowth and connectivity in the ferret visual system. Vis. Neurosci., 10: 117–130.

Jovanovic, J., F. Benfenati, Y.L. Siow, T.S. Sihra, J.S. Sanghera, S.L. Pelech, P. Greengard and A.J. Czernik (1996) Neurotrophins stimulate phosphorylation of synapsin I by MAP kinase and regulate synapsin I-actin interactions. Proc. Natl Acad. Sci., 93: 3679–3683.

Kang, H. and E.M. Schuman (1995) Long-lasting neurotrophin-induced enhancement of synaptic transmission in the adult hippocampus. Science, 267: 1658–1662.

Kantor, D.B., M. Lanzrein, S.J. Stary, G.M. Sandoval, W.B. Smith, B.M. Sullivan, N. Davidson and E.M. Schuman (1996) A role for endothelial NO synthase in LTP revealed by adenovirus-mediated inhibition and rescue. Science, 274: 1744–1748.

Kirkwood, A. and M.F. Bear, (1992) Long-term potentiation in slices of kitten visual cortex and the effects of NMDA receptor blockade. J. Neurophysiol., 67: 841–851.

Kirkwood, A. and M.F. Bear (1994a) Hebbian synapses in visual cortex. J. Neurosci., 14: 1634–1645.

Kirkwood, A. and M.F. Bear (1994b) Homosynaptic long-term depression in the visual cortex. J. Neurosci., 14: 3404–3412.

Kirkwood, A., S.M. Dudek, J.T. Gold, C.C. Aizenman and M.F. Bear (1993) Common forms of synaptic plasticity in the hippocampus and neocortex in vitro. Science, 260: 1518–1521.

Kirkwood, A., H.-K. Lee and M.F. Bear (1995) Co-regulation of long-term potentiation and experience-dependent synaptic plasticity in visual cortex by age and experience. Nature, 375: 328–331.

Kornhauser, J. M. and M.E. Greenberg (1997) A kinase to remember: dual roles for MAP kinase in long-term memory. Neuron, 18: 839–842.

Korte, M. P. Carroll, E. Wolff, G. Brem, H. Thoenen and T. Bonhoeffer (1995) Hippocampal long-term potentiation is impaired in mice lacking brain-derived neurotrophic factor. Proc. Nat'l Acad. Sci. USA, 92: 8856–8860.

Kwon, Y.H., M. Esguerra and M. Sur (1991) NMDA and non-NMDA receptors mediate visual responses of neurons in the cat's lateral geniculate nucleus. J. Neurophysiol., 66: 414–428.

Linden, D.J., R.W. Guillery and J. Cucchiaro (1981) The dorsal lateral geniculate nucleus of the normal ferret and its postnatal development. J. Comp. Neurol., 203: 189–211.

Lo, D.C. (1995) Neurotrophic factors and synaptic plasticity. Neuron, 15: 979–981.

Lohof, A.M., N.Y. Yip and M.-m. Poo (1993) Potentiation of developing neuromuscular synapses by the neurotrophins NT-3 and BDNF. Nature, 363: 350–353.

Lu, B., P. Greengard and M.-M. Poo (1992) Exogenous synapsin I promotes functional maturation of developing neuromuscular synapses. Neuron, 8: 521–529.

Madison, D. V., R.C. Malenka and R.A. Nicoll (1991) Mechanisms underlying long-term potentiation of synaptic transmission. Ann. Rev. Neurosci., 14: 379–397.

McAllister, A.K., D.C. Lo and L.C. Katz (1995) Neurotrophins regulate dendritic growth in developing visual cortex. Neuron, 15: 791–803.

McAllister, A.K., L.C. Katz and D.C. Lo (1996) Neurotrophin regulation of cortical dendritic growth requires activity. Neuron, 17: 1057–1064.

McAllister, A.K., L.C. Katz and D.C. Lo (1997) Opposing roles for endogenous BDNF and NT-3 in regulating cortical dendritic growth. Neuron, 18: 767–778.

Malgoroli, A., A.E. Ting, B. Wendland, A. Bergamaschi, A. Villa, R.W. Tsien, R.H. Scheller (1995) Presynaptic component of long-term potentiation visualised at individual hippocampal synapses. Science, 268: 1624–1628.

Malinow, R. and R.W. Tsien (1990) Presynaptic enhancement shown by whole-cell recordings of long-term potentiation in hippocampal slices. Nature, 346: 177–180.

Meister, M., R.O.L. Wong, D.A. Baylor and C.J. Shatz (1991) Synchronous bursts of action potentials in ganglion cells of the developing mammalian retina. Science, 252: 939–943.

Mooney, R., D.V. Madison and Shatz (1993) Enhancement of transmission at the developing retinogeniculate synapse. Neuron, 10: 815–825.

Mooney, R., A.A. Penn, R. Gallego and C.J. Shatz (1996) Thalamic relay of spontaneous retinal activity prior to vision. Neuron, 17: 863–874.

Patterson, S.L., L.M. Grover, P.A. Schwartzkroin and M. Bothwell (1992) Neurotrophin expression in rat hippocampal slices: a stimulus paradigm inducing LTP in CA1 evokes increases in BDNF and NT-3 mRNAs. Neuron, 9: 1081–1088.

Patterson, S.L., T. Abel, T.A.S. Deuel, K.C. Martin, J.C. Rose and E.R. Kandel (1996) Recombinant BDNF rescues deficits in basal synaptic transmission and hippocampal LTP in BDNF knockout mice. Neuron, 16: 1137–1145.

Pieribone, V.A., O. Shupliakov, L. Brodin, S. Hilfoker-Rothenfluh, A.J. Czernik and P. Greengard (1995) Distinct pools of synaptic vesicles in neurotransmitter release. Nature, 375, 493–497.

Ramoa, A.S. and D.A. McCormick (1994a) Developmental changes in electrophysiological properties of LGNd neurons during reorganization of retinogeniculate connections. J. Neurosci., 14: 2089–2097.

Ramoa, A.S. amd D.A. McCormick (1994b) Enhanced activation of NMDA receptor responses at the immature retinogeniculate synapse. J. Neurosci., 14: 2098–2105.

Ramoa, A.S. and G. Prusky (1997) Retinal activity regulates developmental switches in functional properties and ifenprodil sensitivity of NMDA receptors in the lateral geniculate nucleus. Dev. Brain Res., 101: 165–175.

Riddle, D.R., D.C. Lo and L.C. Katz (1995) NT-4-mediated rescue of lateral geniculate neurons from effects of monocualr deprivation. Nature, 378: 189–191.

Rocha, M. and M. Sur (1995) Rapid acquisition of dendritic spines by visual thalamic neurons after blockade of N-methyl-D-aspartate receptors. Proc. Natl. Acad. Sci., USA, 92: 8026–8030.

Schoups, A.A., R.C. elliot, W.J. Friedman and I.B. Black, 1995. NGF and BDNF are differentially modified by visual experience in the developing geniculocortical pathway. Dev. Brain Res., 86: 326–334.

Schuman, E.M. and D.V. Madison (1994a) Locally distributed synaptic potentiation in the hippocampus. Science, 263: 532–536.

Schuman, E.M. and D.V. Madison (1994b) Nitric oxide and synpatic function. Ann. Rev. Neurosci., 17: 153–183.

Schuman, E.M., M.K. Meffert, H. Schulman and D.V. Madison (1994) An ADP-ribosyltransferase as a potential target for nitric oxide action in hippocampal long-term potentiation. Proc. Natl. Acad. Sci., USA, 91: 11958–11962.

Segal, R.A. and M.E. Greenberg (1996) Intracellular signalling pathways activated by neurotrophic factors. Ann. Rev. Neurosci., 19: 463–489.

Shatz, C.J. (1983) The prenatal development of the cat's retinogeniculate pathway. J. Neurosci., 3: 482–499.

Shatz, C.J. and M.P. Stryker (1988) Prenatal tetrodotoxin infusion blocks segregation of retinogeniculate afferents. Science, 242: 87–89.

Sillito, A.M., P.C. Murphy and T.E. Salt (1990) The contribution of the non-*N*-methyl-D-aspartate group of excitatory amino acid receptors to retinogeniculate transmission in the cat. Neuroscience, 34: 273–280.

Smetters, D.K., J. Hahm and M. Sur (1994) An N-methyl-D-aspartate receptor antagonist does not prevent eye-specific segregation in the ferret retinogeniculate pathway. Brain Res., 658: 168–178.

Son, H., R.D. Hawkins, K. Martin, M. Kiebler, P.L. Huang, M.C. Fishman and E.R. Kandel (1996) Long-term potentiation is reduced in mice that are doubly mutant in endothelial and neuronal nitric oxide synthase. Cell, 87: 1015–1023.

Stretavan, D.W. and C.J. Shatz (1986) Prenatal development of retinal ganglion cell axons: segregation into eye-specific layers within the cat's lateral geniculate nucleus. J. Neurosci., 6: 234–251.

Styker, M.P. and K.R. Zahs (1983) ON and OFF sublaminae in the lateral geniculate nucleus of the ferret. J. Neurosci., 3: 1943–1951.

Wong, R.O.L., M. Meister and C.J. Shatz (1993) Transient period of correlated bursting activity during development of the mammalian retina. Neuron, 11: 923–938.

Wong, R.O.L., A. Chernjavsky, S.J. Smith and C.J. Shatz (1995) Early functional neural networks in the developing retina. Nature, 374: 716–718.

18

FUNCTIONAL CONSEQUENCES OF ELIMINATING PRENATAL BINOCULAR INTERACTIONS

S. Bisti,[1,4] S. Deplano,[2] and C. Gargini[3,4]

[1]Department STB, School of Medicine
University of L'Aquila, Italy
[2]Institute of Comparative Anatomy
University of Genova, Italy
[3]Institute of Biol. Disc.
University of Pisa, Italy
[4]Institute of Neurophysiol. CNR
Pisa, Italy

The pattern of neuronal connections in the visual system of mammals with highly developed binocular vision is a remarkable example of complexity and precision. Inputs from each retina are segregated in the geniculostriate system (Gerey et al., 1991), so that retinal ganglion cell axons terminate in separate, eye-specific layers within the dorsal lateral geniculate nucleus (dlgn). In turn, the axons of dlgn neurons project to the primary visual cortex in alternating clusters, providing the anatomical basis for ocular dominance columns (LeVay et al., 1975). However, during the early phases of development the projections from each eye are completely intermingled in the dlgn and the superior colliculus of fetal monkeys and cats. In these species, retinal axons segregate in the second half of gestation into the eye specific domains characteristic of the mature animal (Rakic, 1976, 1977; Shatz, 1983; Williams and Chalupa, 1982; White and Chalupa, 1991).

Little is understood about the factors responsible for the extensive rearrangement of early visual connections. However, it is known that this process is dependent upon binocular interaction since removal of one eye in fetal monkeys and cats causes a marked reorganization of the visual pathways (Rakic, 1981; Chalupa and Williams, 1984). After monocular enucleation, the remaining eye projects to all laminae in both the ipsilateral and contralateral dlgn, and these cells innervate layer IV of the visual cortex in continuous rather than periodic patterns. In addition, prenatal monocular enucleation attenuates naturally occurring ganglion cell loss in the remaining retina and prevents the formation of

Development and Organization of the Retina, edited by Chalupa and Finlay.
Plenum Press, New York, 1998.

319

distinct eye-specific laminae within the dlgn (Rakic and Riley, 1983; Chalupa et al., 1984; Chalupa and Williams, 1984).

The substantial neuroanatomical reorganization of retino-geniculate and geniculo-cortical projections evident after *in utero* eye removal, raises the question of the functional consequences of this manipulation, both at the single cell level and in terms of the behavioral capacity of the remaining eye.

GANGLION CELL PLASTICITY

It is well known that during normal development of the mammalian retina there is a massive overproduction of retinal ganglion cells followed by a period of ganglion cell loss. A substantial degree of this cell loss occurs during the period when ganglion cell projections become segregated in the dlgn (e.g., White and Chalupa, 1991). Removal of one eye during early development reduces neuronal death in the remaining eye so that the number of surviving ganglion cells is about 20–30% greater than normal in cat and monkey (Chalupa et al., 1984; Rakic and Riley, 1983). What is unknown is whether the rate of survival is the same for the three main classes of retinal ganglion cells (alpha-Y, beta-X, gamma-W). An accurate analyses of the alpha ganglion cell population in cats, after prenatal monocular enucleation, has shown that the density of these cells was about 25% greater than normal (Kirby and Chalupa, 1986). Moreover, the dendritic arbors and somas were reduced in size, possibly due to the dendrodendritic interactions among developing cells. Interestingly, morphological analyses of retinogeniculate alpha cell axons showed greatly expanded arbors, some spanning almost the entire dorsoventral width of the nucleus. In contrast, all beta-cell arbors appeared normal in size and were confined to the portion of the A/A1 layers that would normally have been innervated by the remaining eye (Garraghty et al., 1988; Sur, 1988).

SINGLE UNIT RECORDINGS

In normal adult animals, the dlgn is organized into two distinct dorsal layers (A and A1) and a ventral C complex. In the dlgn of prenatally enucleated cats, only a single dorsal layer and the ventral layer form. Chalupa and Williams (1984) have shown that the entire dlgn is innervated by the remaining eye, and that the input from that eye is organized in a retinotopic manner. Single unit recordings in the "fused" A/A1 layer of the dlgn (White et al., 1989) provided evidence that the response properties of geniculate neurons do not differ appreciably from of control animals. In particular, receptive field size, spatial resolution and topographical organization were similar in controls and enucleates. These results would seem to show that the functional organization of the dlgn is largely preserved in enucleates, even though the remaining eye innervates roughly twice as much territory as normal.

Interestingly, the main difference between normals and enucleates is not in the response properties (see Fig. 1), but in the relative probability of recording from X or Y cells. The probability of encountering cells with Y response properties is reduced in enucleates despite the expanded axonal arborization of Y-cell retinogeniculate arbors (see Fig. 2).

At the level of the primary visual cortex, all neurons in enucleates respond to stimulation of the remaining eye (Shook et al., 1985). Furthermore, retinotopy and orientation columns appear normally organized. However, receptive fields are significantly smaller

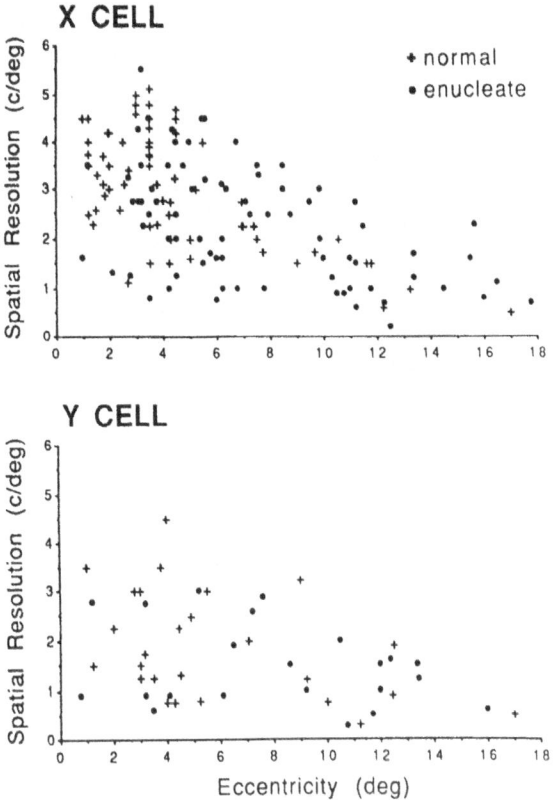

Figure 1. Spatial resolution of the first harmonic responses as a function of eccentricity for X and Y cells in normal and enucleated animals. For each group of cells, spatial resolution declines as eccentricity increases. At any given eccentricity, there is considerable overlap between normal and enucleated animals in spatial resolutions of both X cells (top) and Y cells (bottom). From White et al. (1989).

than in controls, and this appears to be due to a selective loss of cells with large receptive fields, rather than a mean reduction in the receptive-field size (see Fig. 3).

These results could be accounted for by a reduced Y-input from the dlgn. The reduced probability of recording from Y cells in the dlgn suggests that the Y system, notwithstanding its potentiation in terms of survival and territory of innervation, is no longer able to drive neuronal activity (White et al., 1989).

On the basis of all these results, one may suppose that binocular competition plays a key role during the development of the Y-system, but has little effect on the X-system which seems to continue its development normally, at least in term of axonal arborization, receptive-field properties and resolution limits.

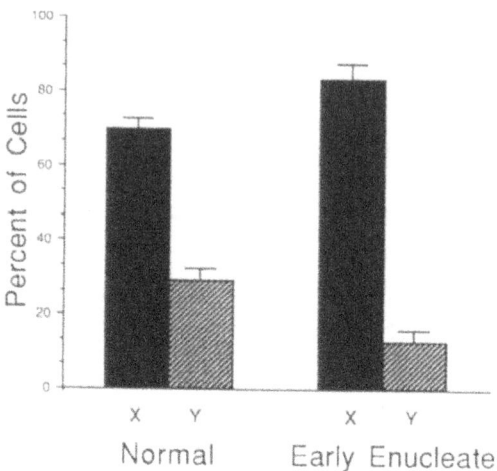

Figure 2. Histogram of the average percentage of neurons in normal animals and early enucleates that were identified as X and Y cells. Relative to normal animals, the early enucleates have a high percentage of X cells and a low percentage of Y cells. For normals, n = 6 animals; for early enucleates, n = 3. From White et al. (1989).

VISUAL EVOKED POTENTIALS (VEPs)

To test the hypothesis that prenatal binocular competition is an essential factor for the normal development of the Y-system, a series of electrophysiological and behavioral experiments were carried out on prenatal enucleated cats (Bisti and Trimarchi, 1993; Bisti et al., 1995). For this purpose, two aspects of the visual performance of the remaining eye were tested: a) contrast sensitivity function and b) visual acuity. Evidence was provided to support the notion that the behavioral changes detected in the visual processing of the remaining eye reflect an abnormality of the Y-system in the prenatal enucleates.

The activity of visual cortical neurons in areas 17 and 18 was sampled by recording Visual Evoked Potentials (VEPs) in response to sinusoidal gratings modulated sinusoidally in contrast. A series of spatial frequencies and contrast values was tested. According to previous data, a difference between control animals and early enucleates might be expected at low frequencies which are mainly associated with large receptive-fields. Figure 4 shows an example of VEP responses at two different spatial frequencies in one control, one prenatally enucleated, and one neonatally enucleated cat. Note that only the responses recorded in the prenatally enucleated animal show a strong attenuation in amplitude at low spatial frequencies. By contrast, there was no effect in the postnatal enucleate.

It can also be seen that from 0.2 to 0.7 (c/deg) the phase of the response shifted more steeply for the control and neonatally enucleated animals than for the prenatal enucleate. This could reflect a change in the spectrum of conducting fibres in the prenatal enucleate. Y cells respond preferentially to low spatial frequencies and their axons conduct impulses at a high velocity within the optic nerve. Thus, information about large visual stimuli is

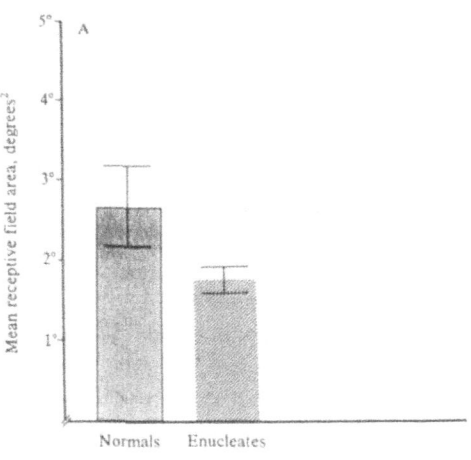

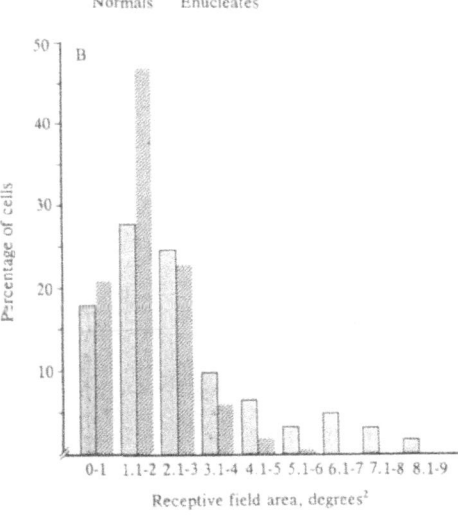

Figure 3. (A) The mean receptive field area of neurons in area 17 of normal (black bars) and prenatally enucleated (shaded bars) cats. Only cells with receptive fields within 5 degrees of the area centralis representation were included in this analysis, which is based on a sample of 60 cells from normal cats and 145 neurons from animals that had an eye removed in utero. The error bar denotes the 95% confidence interval. (B) Breakdown of the data presented in A showing the percentage of cells with a given receptive field area. Note that the prenatal enucleates have a higher percentage of cells with receptive field areas between 1.1 and 2.0 degrees^2 with a concomitant decrease of cells with field areas greater than 4.1 degrees^2. From Shook et al. (1985).

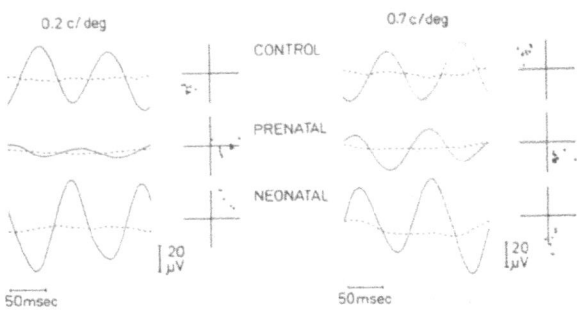

Figure 4. Examples of VEP records from three cats (normal, prenatally, and neonatally enucleated cats) averaged over one stimulus period. The stimulus was a vertical sinusoidal grating reversed in contrast sinusoidally at 5 Hz. Stimulus frequency, 0.2 c/deg and 0.7 c/deg; contrast was 20%. The dotted lines through each record show records obtained by averaging at 1.1 times the stimulus frequency to indicate intrinsic noise levels. The polar graphs (at right) plot amplitude and phase for partial averages of the total record (second harmonic). The distance from the origin of each point (norm) gives the response amplitude. The two-dimensional scatter of this plot gives an estimate of the variance of amplitude and phase, reported as error bars in the following curves. From Bisti et al. (1995).

the first to be transferred to the primary visual cortex. At higher spatial frequencies, the X system, which conveys impulses at a moderate velocity, becomes activated so that the latency of the response would increase (see Fig. 5, filled symbols in A and B).

In the prenatal enucleates this phase lag is reduced (Fig. 5A, open symbols), possibly due to a malfuctioning of the Y fibres. It was possible to validate this hypothesis by recording VEPs in the primary visual area of an animal which had the Y input surgically removed. A unilateral ventral section of the optic tract selectively eliminates the Y-ganglion cell input to the ipsilateral visual cortex (Reese et al., 1991). Data from a deafferentated (open symbols) and a normal (filled symbols) visual area is shown in Fig. 5B. Note that in both the prenatal enucleates and in animals with a Y-cell deafferented cortex there was a similar shift in the phase response as a function of the spatial frequency. This observation is consistent with the idea that a change in the spectrum of conducting fibers occurs under both conditions.

When the amplitude of the response is reported as a function of the spatial frequency in prenatally and neonatally enucleated animals and the spatial frequency-response curves are compared (Fig. 6), it appears that prenatal enucleates have a strong attenuation of responses at low spatial frequencies, a slight increase in sensitivity at intermediate frequencies, but no appreciable change in the resolution limit.

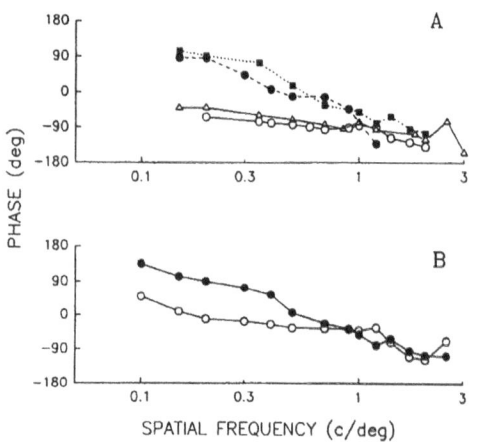

Figure 5. Phase of the second-harmonic modulation as a function of spatial frequency. (A) Prenatally enucleated cats (open symbols, solid line), neonatally enucleated (solid circles, broken line), normal cat (solid squares, dotted line). (B) Cat with a ventral lesion of one optic tract (solid symbols, normal hemicortex; open symbols, deafferented hemicortex). Temporal frequency was 5 Hz; contrast was 17% in A and 30% in B. From Bisti et al. (1995).

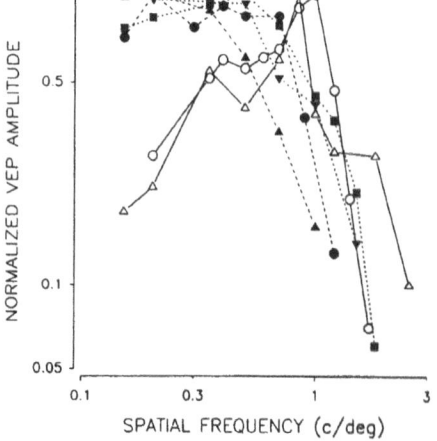

Figure 6. Normalized VEP amplitude reported as a function of spatial frequency for two prenatally enucleated cats (open symbols, continuous lines), two neonatally enucleated cats (upright solid triangles and solid circles, broken lines) and two normal cats (solid squares and solid inverted triangles, dotted lines). Temporal frequency was 5 Hz; contrast was 17%. From Bisti et al. (1995).

BEHAVIORAL EXPERIMENTS

To assess the behavioral effects of terminating prenatal binocular interactions, contrast sensitivity functions, visual acuities and visual fields were measured in prenatal enucleates and in control animals (Bisti and Trimarchi, 1993). Figure 7 shows an example of contrast thresholds at two different frequencies measured in two cats (control and monocularly enucleated). It may be seen that at the low spatial frequency (0.25 c/deg) the contrast threshold in the control animal (open symbols) is much lower than in the enucleate. However, at a higher spatial frequency (0.7 c/deg), the threshold in the enucleate was five times lower than in the control. It is also interesting to point out that it was not

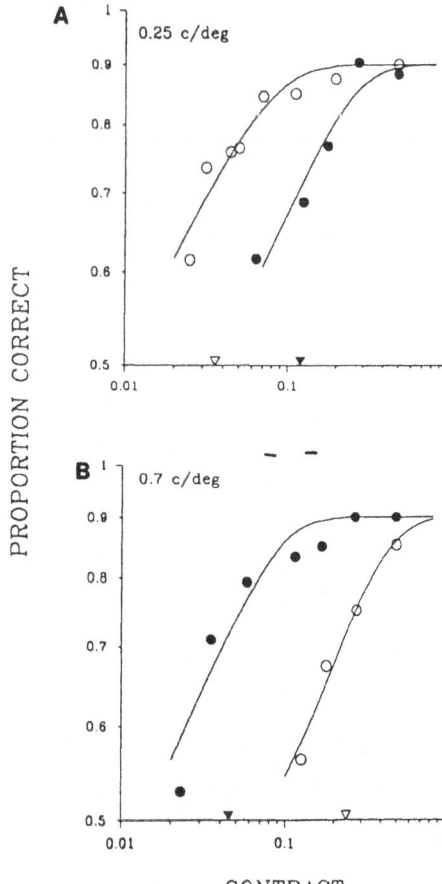

Figure 7. Learning curves of a normal (open symbols) and a prenatal enucleated cat (filled symbols) for the monocular forced choice discrimination between a sinusoidal grating and an equiluminant uniform field as a function of grating contrast for two spatial frequencies (A, 0.25 c/deg; B, 0.7 c/deg). Each point is the percentage of correct responses for at least 100 trials for each value of contrast around threshold. The arrowheads on the abscissa indicate the extrapolated threshold as the value of contrast at which the cat performance was at 70% correct. Data are fitted with the Weibull equation. From Bisti and Trimarchi (1993).

possible to determine a threshold in enucleated animals below 0.25 c/deg. It is also the case, that when normal and enucleated animals were tested specifically for the parameter visual acuity they showed comparable results.

Thus, the remaining eye of a prenatally enucleated animal develops normal visual acuity and normal visual fields (Bisti and Trimarchi, 1993), but reduced sensitivity at low spatial frequencies and increased sensitivity at middle spatial frequencies. In agreement with the results obtained in cats, behavioral experiments in monkeys, which had one eye removed during prenatal life, showed that the vernier hyperacuity does not differ significantly in comparison with that of a normal control (MacAvoy et al., 1987).

Taken all together, these data support the idea that prenatal binocular competition plays a key role in the development of one class of retinal ganglion cells, the alpha cells. After fetal removal of one eye these neurons seem to become unable to drive target cells, notwithstanding the reduced ganglion cell death and expanded axonal arborization, possibly because their synaptic efficency is reduced. Beta cells, however, seem to follow a developmental plan which is not dependent upon bionocular interactions. Functional properties of dlgn neurons do not differ from those recorded in normal animals and visual acuity is unchanged. The visual performance of the remaining eye in an early enucleate corresponds to that of an animal that had the Y system. Thus, the reduced sensitivity at low spatial frequencies in prenatal enucleates might be directly correlated with the functional loss of the Y fibres. The lower threshold at intermediate spatial frequencies might reflect the increased percentage of neurons with medium size receptive fields (see Fig. 3B), which may be simply due to an increased number of cortical neurons driven by an X input (as pointed out by White et al. 1989). Finally, visual acuity is unchanged in prenatal enucleates since it is defined by the X system. It is interesting to note that interrupting binocular competition at birth has no effect on the visual performance of the remaining eye, even though the retino-geniculate projection of the Y-ganglion cells is markedly expanded (Sur, 1988).

Since the pioneer experiments of Hubel and Wiesel (1963, 1965) on monocular deprivation, it remains to be resolved if the functional impairment of the deprived eye could be compensated and enhanced visual capability of the eye placed at a competitive advantage during development. Conflicting results have been provided by authors working on human subjects. Freeman and Bradley (1980) have claimed that vernier acuity was enhanced in the non-deprived eye of human amblyopes, but this result was challenged by Johnson et al. (1982) in their study on identical twins, one of which was monocularly deprived since birth. In view of the experimental evidence obtained in early enucleated animals, it seems reasonable to rule out the possibility that an eye might enhance its visual capability even in extreme conditions of developmental advantage.

REFERENCES

Bisti S. and Trimarchi C. (1993) Visual performance in behaving cats after prenatal unilateral enucleation. Proc. Natl. Acad. Sci. USA, 90, 11142–11146.

Bisti S., Trimarchi C. and Turlejski K. (1995) Prenatal monocular enucleation induces a selective loss of low-spatial-frequency cortical responses to the remaining eye. Proc. Natl. Acad. Sci. USA, 92, 3908–3912.

Chalupa L.M. and Williams R.W. (1984) Organization of the cat's lateral geniculate nucleus following interruption of prenatal binocular competition. Hum. Neurobiol. 3, 103–107.

Chalupa L.M., Williams R.W. and Henderson Z. (1984) Binocular interaction in fetal catregulates the size of the ganglion cell population. Neuroscience, 12, 1139–1146.

Freeman R.D. and Bradley A. (1980) Monocular deprived humans: non deprived eye has supernormal vernier acuity. J. Neurophysiol. 43, 1645–1653.

Garey L.F., Dreher B. and Robinson S. (1991) The organization of the visual thalamus Vision and Visual Dysfunction, eds. Dreher B. & Robinson S.R. (Macmillan, London) Vol. 3, pp. 176–234.

Garraghty P.E., Shatz C.J., Sretavan D.W. and Sur M. (1988) Axon arbors of X and Y retinal ganglion cells are differentially affected by prenatal disruption of binocular input. Proc. Natl. Acad. Sci. USA, 85, 7361–7365.

Johnson C.A., Post R.B., Chalupa L.M. and Lee T.J. (1982) Monocular deprivation in humans : a study of identical twins. Invest. Ophthal. & Vis. Sci. 23, 135–138.

Kirby M.A. and Chalupa L.M. (1986) Retinal crowding alters the morphology of alpha ganglion cells. J. Comp. Neurol. 251, 532–541.

LeVay S., Hubel D.H. and Wiesel T.N. (1975) The pattern of ocular dominannce columns in macaque visual cortex revealed by a reduced silver stain. J. Comp. Neurol., 159, 559–575.

MacAvoy M.G., Bruce C.J. and Rakic P.. (1987) Effect of prenatal monocular enucleation on vernier hyperacuity in rhesus monkeys. Soc. Neurosci. Abs., 13, 1244.

Rakic P. (1976) Prenatal genesis of connections subserving ocular dominance in the rhesus monkey. Nature, 261, 467–471.

Rakic P. (1977) development of the visual system in the rhesus monkey. Phil. Trans. Roy. Soc. Lond. B. 278, 245–260.

Rakic P. (1981) Development of visual centers in the primate brain depends on binocular competition before birth. Science, 214, 928–931.

Rakic P. and Riley K.P. (1983) Regulation of axon number in the primate optic nerve by prenatal binocular competition. Nature, 305, 135–137.

Reese B.E., Guillery R.W., Marzi C.A. and Tassinari G. (1991) J. Comp. Neurol., 306, 539–553.

Shook B.L., Maffei L, and Chalupa L.M. (1985) Functional organization of the cat's visual cortex after prenatal interruption of binocular interactions. Proc. Natl. Acad. Sci. USA, 82, 3901–3905.

Shatz C.H. (1983) The prenatal development of the cat's retino geniculate pathways . J. Neurosci. 3, 482–499.

Sur M. (1988) Development and plasicity of retinal X and Y axon terminations in the cat's lateral geniculate nucleus. Braib Behav. Evol., 31, 243–251.

White C.A., Chalupa L.M., Maffei L., Kirby M.A. and Lia B. (1989) Response properties in the dorsal lateral geniculate nucleus of the adult cat after interruption of prenatal binocular interactions. J. Neurophysiol. 62, 1039–1051.

White C.A. and Chalupa L.M. (1991) Development of the mammalian retinofugal pathways in Vision and Visual Dysfunction, eds. Dreher B. & Robinson S.R. (Macmillan, London) Vol. 3, pp. 129–143.

Wiesel T.N. and Hubel D.H. (1963) Single-cell responsesin striate cortex of kittens deprived of vision in one eye. J. Neurophysiol. 26, 1003–1018.

Wiesel T.N. and Hubel D.H. (1965) Comparison of the effects of unilateral and bilateral eye closure on cortical unit responses in kittens. J. Neurophysiol. 28, 1029–1040.

Williams R.W. and Chalupa L.M. (1982) Prenatal development of retino collicular projections in the cat; an anterograde tracer transport study. J. Neurosci. 2, 604–622.

19

ROLE OF SEROTONIN IN THE DEVELOPMENT OF THE RETINOTECTAL PROJECTION

Robert W. Rhoades,* Tracy Crnko Hoppenjans, and Richard D. Mooney

Department of Anatomy and Neurobiology
Medical College of Ohio
P. O. Box 10008
Toledo, Ohio 43699

1. ABSTRACT

Serotonin (5-HT)-containing axons and 5-HT receptors are present in the hamster superior colliculus (SC) before birth and are coincident with the development of retinal ganglion cell axon terminals in this nucleus (retinotectal projection). Neural activity in the SC is suppressed by 5-HT presynaptically at receptors located on retinotectal axons. Elevation of 5-HT levels in the SC postnatally, by subcutaneous injection of 5,7-dihydroxytryptamine or by direct release of 5-HT within the SC, prevents refinement of the retinotectal terminal field and maintains it in an immature state. Retinal ganglion cell number and density remain normal in such animals. The persistence of a poorly differentiated retinotectal projection following 5-HT elevation may be a consequence of reduced retinotectal activity.

2. INTRODUCTION

A growing body of literature (see Lipton and Kater, 1989; Kater and Mills, 1991; and Killackey et al., 1995 for reviews) provides support for the proposal that molecules used for signaling between neurons in maturity may also modulate neuronal development. A large number of experiments have suggested such a role for serotonin (5-HT, e.g., Lauder, 1983; Gromova et al., 1983; Forda and Kelly, 1985; Chubakov et al., 1986; Lipton and Kater, 1989; Blue et al., 1991; Rhoades et al., 1993; Bennett-Clarke et al., 1994). Most investigators who have demonstrated influences of 5-HT upon developing neurons or axons have suggested that this amine's effects result from direct actions upon these

* Address correspondence to: Robert W. Rhoades (address above, ph. 419-381-4197, fax. 419-381-3008, e-mail: rrhoades@gemini.mco.edu)

Development and Organization of the Retina, edited by Chalupa and Finlay.
Plenum Press, New York, 1998.

elements which may be mediated by its ability to alter intracellular Ca^{2+} or activate second messenger systems (see below for citations). However, 5-HT also strongly modulates the activity of neurons in many structures (see Andrade and Chaput, 1991, and Zifa and Fillion, 1992, for reviews) and there is evidence that it may specifically antagonize activation of the NMDA receptor (Murase et al., 1990; Holohean et al., 1992), an element shown to be involved in the activity-dependent refinement of axonal projections (e.g., Cline et al. 1987; Cline and Constantine-Paton, 1990; Simon et al., 1992). The ability of 5-HT to alter neuronal activity, including that mediated by NMDA receptors, raises the possibility that it may influence development of axonal projections, particularly the refinement of axon arbors, *indirectly* by modulation of activity-dependent processes. This chapter summarizes the results of several studies that have evaluated the ability of serotonin to modulate retinotectal activity in the hamster and to influence retinotectal development in this species.

3. ORGANIZATION AND ACTIONS OF 5-HT IN THE MATURE AND DEVELOPING SUPERIOR COLLICULUS (SC)

In all adult mammals examined to date, 5-HT fibers heavily innervate the retinorecipient SC laminae (e.g., Harvey and Macdonald, 1987; Villar et al. 1988; Rhoades et al., 1990). The possibility that 5-HT might strongly influence retinotectal development is attractive because serotoninergic fibers are present in SC when retinal axons are beginning to arrive (Lund and Bunt, 1976; Lidov and Molliver, 1982; Godement et al., 1984; Edwards et al., 1986; Aitken and Tork, 1988). Serotonin-immunoreactive fibers are visible in the hamster's SC by embryonic day 14, and at birth, they are present in all SC laminae (Rhoades et al., 1990). Collateralization of retinal axons in the hamster's SC begins on the first postnatal day (Jhaveri and Schneider, 1994).

Considerable effort has been directed toward localization of 5-HT receptor subtypes in the mature SC, but relatively little is known regarding their organization in developing animals. Studies in adult rodents have demonstrated binding of ligands to both 5-HT_{1A} (Marcinkiewicz et al., 1984; Welner et al., 1989) and 5-HT_{1D} or 5-HT_{1B} receptors (Waeber and Palacios, 1990; Mooney et al., 1994). Nakada et al. (1984) have reported that 5-HT_2 receptors are present in the SC, but the data of Blue et al. (1988) indicate that their density may be very low. While 5-HT_{1A} and 5-HT_2 receptors are generally thought to be located on somata and dendrites (e.g., Zifa and Filion, 1992), there is considerable evidence that 5-HT_{1B} receptors are distributed on the terminal arbors of retinal ganglion cell axons in adult rodents (Segu et al., 1986; Boulenguez et al., 1993; Mooney et al., 1994; Boschert et al., 1994) and that these receptors are present very early in development (Figure 1).

Over 20 years ago, Straschill and Perwein (1971) showed that iontophoresis of 5-HT depressed the visual responses of about one-half of the neurons tested in the cat's SC, but that it increased the excitability of other cells. Our own *in vivo* and *in vitro* experiments have indicated that 5-HT inhibits responses evoked by visual- or optic chiasm (OX)-stimulation in the vast majority of visual SC neurons. This effect is almost certainly mediated by a presynaptic effect of this amine at 5-HT_{1B} receptors located on retinotectal axons (Huang et al., 1993; Mooney et al., 1994).

Iontophoresis of 5-HT *in vivo* produced a suppression of visual responses of 40% or greater in 78.1% (N=50) of 64 neurons evaluated and did not augment the visual responses of any of the cells tested. The average response suppression was 75.3 ± 21.2%. Iontophoresis of 5-HT had significantly different effects upon activation of SC cells by OX and visual cortical (CTX) stimulation (Figure 2). Application of 5-HT suppressed the OX-evoked

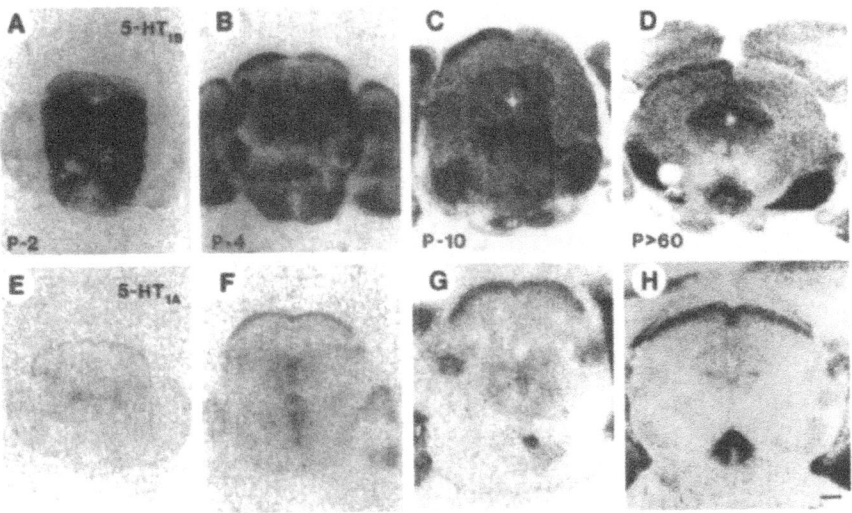

Figure 1. Demonstration of 5-HT$_{1B}$ (A–D) and 5-HT$_{1A}$ (E–H) receptors in the SC of hamsters killed between P-2 and adulthood (P > 60). All animals had the left eye removed 2 days prior to sacrifice. Note the presence of 5-HT$_{1B}$ and 5-HT$_{1A}$ receptors on P-4 and the fact that eye removal reduces the density of 5-HT$_{1B}$ receptors, but not 5-HT$_{1A}$ receptors, in the superficial layers of the contralateral SC at all ages. The calibration is 250 μm.

responses of 96.9% (N=31) of the 32 SC cells tested by at least 40% and the average response suppression for all 32 neurons tested was 87.1 ± 22.5%. Application of 5-HT suppressed the responses of only 35.7% (N=10) of the 28 cells tested with CTX-stimulation by at least 40%. The average response suppression for all 28 cells was 35.3 ± 38.8%. The effects of 5-HT upon the glutamate-evoked responses of SC cells that were synaptically "isolated" by concurrent application of Mg^{2+} were also evaluated. Application of 5-HT produced a response suppression ≥40% in 29.7% (N=19) of the 64 neurons tested under these conditions. The average response suppression for all of the cells evaluated was 28.4 ± 35.7%. This effect of 5-HT was weaker than that upon visually evoked responses. These results demonstrate that 5-HT markedly depresses the visual responses of most superficial layer SC neurons, and they suggest further that much of this effect may be mediated by presynaptic inhibition of retinotectal transmission.

The presynaptic effects of 5-HT upon retinotectal transmission suggested by the results of the *in vivo* recording experiments were confirmed by *in vitro* intracellular recording studies. The effects of 5-HT on epsps evoked by electrical stimulation of the optic tract (OT) were evaluated for 67 SC neurons in adult hamsters (Figure 3). Application of 5-HT produced at least a 50% reduction in OT-evoked epsps in 85% of these neurons. The average epsp amplitude was 7.8 ± 2.1 mV under control conditions and 2.7 ± 1.9 mV in the presence of 5-HT (p < 0.01). For most of these neurons, application of 5-HT had little effect on either membrane potential or input resistance. The average percent change in membrane potential for cells tested with 5-HT was 0.5 ± 6.0% and the average percent change in input resistance was 0.6 ± 22.9%. For 4 of 6 cells tested, application of 5-HT had no significant effects upon the responses evoked by application of glutamate either under normal bathing conditions or when the medium included low Ca^{2+} and high Mg^{2+}. Pharmacologic experiments indicated that the effects of 5-HT upon retinotectal transmission were mimicked by the 5-HT$_{1B}$ agonists, TFMPP and CGS12066B, and antagonized

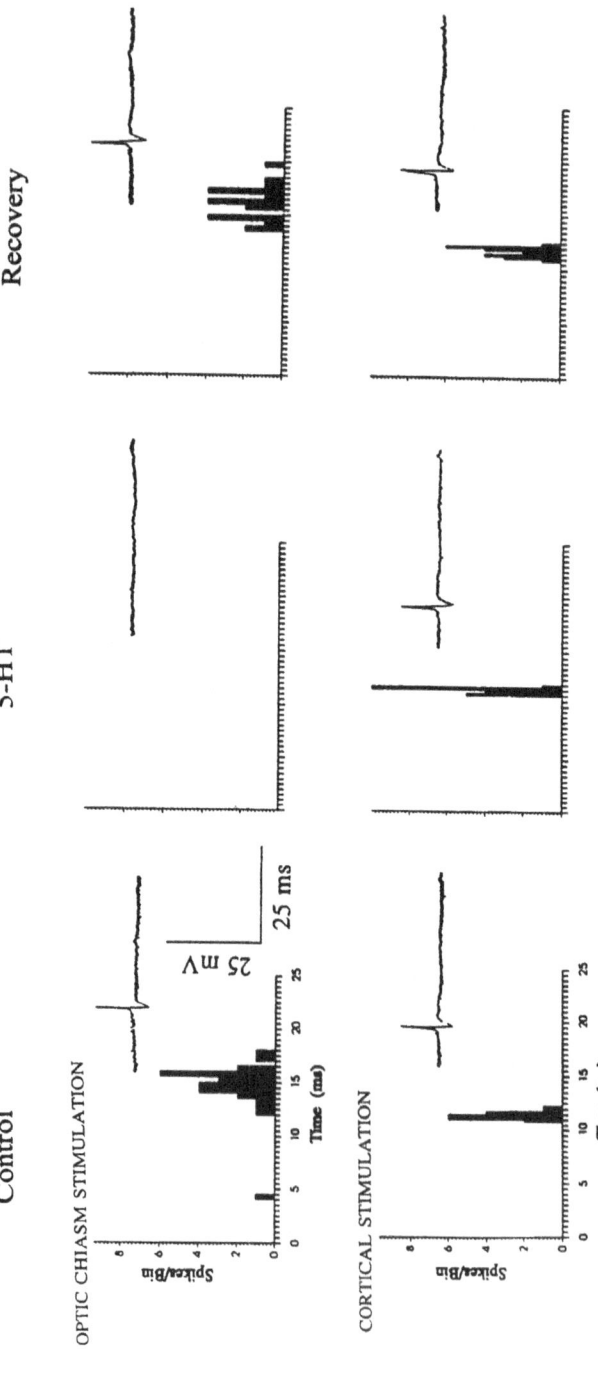

Figure 2. Peristimulus time histograms of a neuron in which 5-HT suppressed responses to OX- but not to CTX-stimulation. Electrical pulses were delivered in alternation to OX and CTX, and 5-HT was applied by ionophoresis (100 nA current for 110 sec). Recovery PSTHs were obtained 100 and 130 sec after termination of 5-HT ejecting current for CTX- and OX-stimulation, respectively. Insets show individual oscilloscope traces during each period.

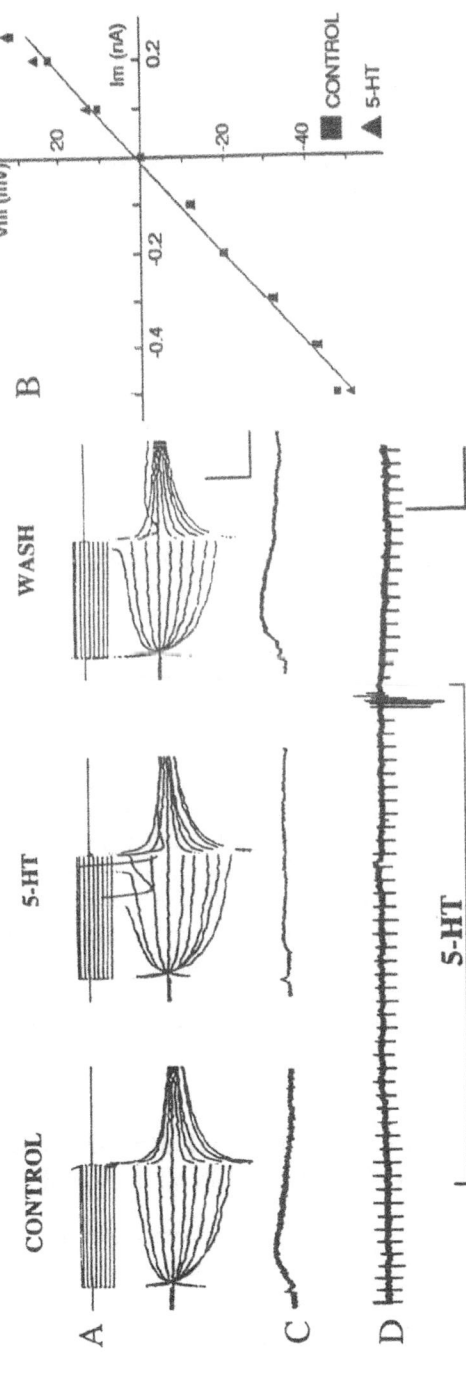

Figure 3. Serotonin reduced epsp amplitude but has relatively little effect on membrane potential or input resistance in an SC neuron recorded *in vitro*. **A** shows superimposed oscilloscope traces of current pulses (upper records) and membrane potential (lower records); these current-voltage data are plotted in **B** (regression line, the input resistance, is 107.3 MΩ for control and 108.1 MΩ for 5-HT [line omitted for clarity]). **C** shows individual epsps to electrical OT-stimulation (early transient is stimulus artifact). **D** shows the chart recording of membrane potential throughout entire test (brief vertical excursions are OT-stimulus artifacts and epsps and injected current pulses). Calibrations are 20 mV by 5 ms for **A** and **C** and 20 mV by 1 minute for **D**.

by the 5-HT$_{1B}$ antagonists, (-)-pindolol and methiothepin. The effects of 5-HT on the OT-evoked epsp were not antagonized by spiperone, NAN-190, or MDL72222, respectively 5-HT$_{1A \text{ and } 2}$, 5-HT$_{1A}$, and 5-HT$_3$ antagonists. These findings are consistent with earlier *in vitro* results which suggested a presynaptic effect of 5-HT upon SC cells (Kawai and Yamamoto, 1969). Unpublished studies from our laboratory indicate that 5-HT presynaptically inhibits retinotectal transmission in the same manner in developing hamsters.

4. EFFECTS OF ALTERED 5-HT LEVELS ON RETINOTECTAL DEVELOPMENT

Given the early arrival of 5-HT-immunoreactive fibers in SC and the remarkable change in their synaptic organization during early postnatal development, these axons and the amine they contain seemed a likely candidate for modulation of retinotectal development. Over 20 years ago, Sachs and Jonsson (1975) showed that neonatal injection of 5,7-dihydroxytryptamine (5,7-DHT) decreased the density of the 5-HT innervation of the cerebral and cerebellar cortices, but increased the density of these fibers in the brainstem including the SC. We used this approach to determine whether sprouting of 5-HT-containing fibers and the resultant increase in the concentration of this amine in the developing SC altered retinotectal projections. Anterograde tracing with HRP was used to compare the organization of retinotectal projections in normal adult hamsters and in animals that sustained subcutaneous injections of 5,7-DHT on the day of birth. Analysis of tissue from the retinorecipient laminae of the SC by high-pressure liquid chromatography (HPLC) indicated that 5,7-DHT treatment increased the amount of 5-HT in the adult SC by 47%. The increased 5-HT innervation of SC was associated with a marked change in the distribution of the uncrossed retinotectal projection. In normal adult hamsters, fibers from the ipsilateral eye form dense clusters in the lowermost *stratum griseum superficiale* (SGS) and *stratum opticum* (SO, Figure 4). A small number of uncrossed fibers are also visible in the more caudal portions of these layers. In adult animals that sustained neonatal 5,7-DHT injections, uncrossed retinotectal fibers formed a nearly continuous band in rostral SO and lower SGS and numerous labelled fibers were present in the caudal SC, primarily in the SO (Figure 5). Neonatal treatment with 5,7-DHT also produced alterations in the crossed retinotectal pathway and in the crossed and uncrossed retinogeniculate projections. These results suggested that altering the 5-HT input to the brainstem may strongly influence the development of retinofugal projections. It must also be acknowledged, however, that these changes may have resulted from direct effects of the neurotoxin on retinotectal axons or from changes in the concentration of 5-HT in the retina or visual cortex.

5. ELEVATING 5-HT DOES NOT CHANGE RETINAL GANGLION CELL NUMBER OR AXONAL DECUSSATION PATTERNS

We carried out electron microscopic and retrograde tracing experiments to determine whether the changes observed in the retinotectal projections of 5,7-DHT-treated hamsters were associated with alterations in the number, size, or distribution of retinal ganglion cells in these animals. Nissl staining of retinas from normal adult and 5,7-DHT treated hamsters revealed no differences in the number or average diameter of cells in the retinal ganglion cell layer. Retrograde labelling with HRP also demonstrated no effect of 5,7-DHT treatment on the number or distribution of ipsilaterally or contralaterally projecting ganglion cells (Figure 6). Neonatal 5,7-DHT administration also had no effect on the

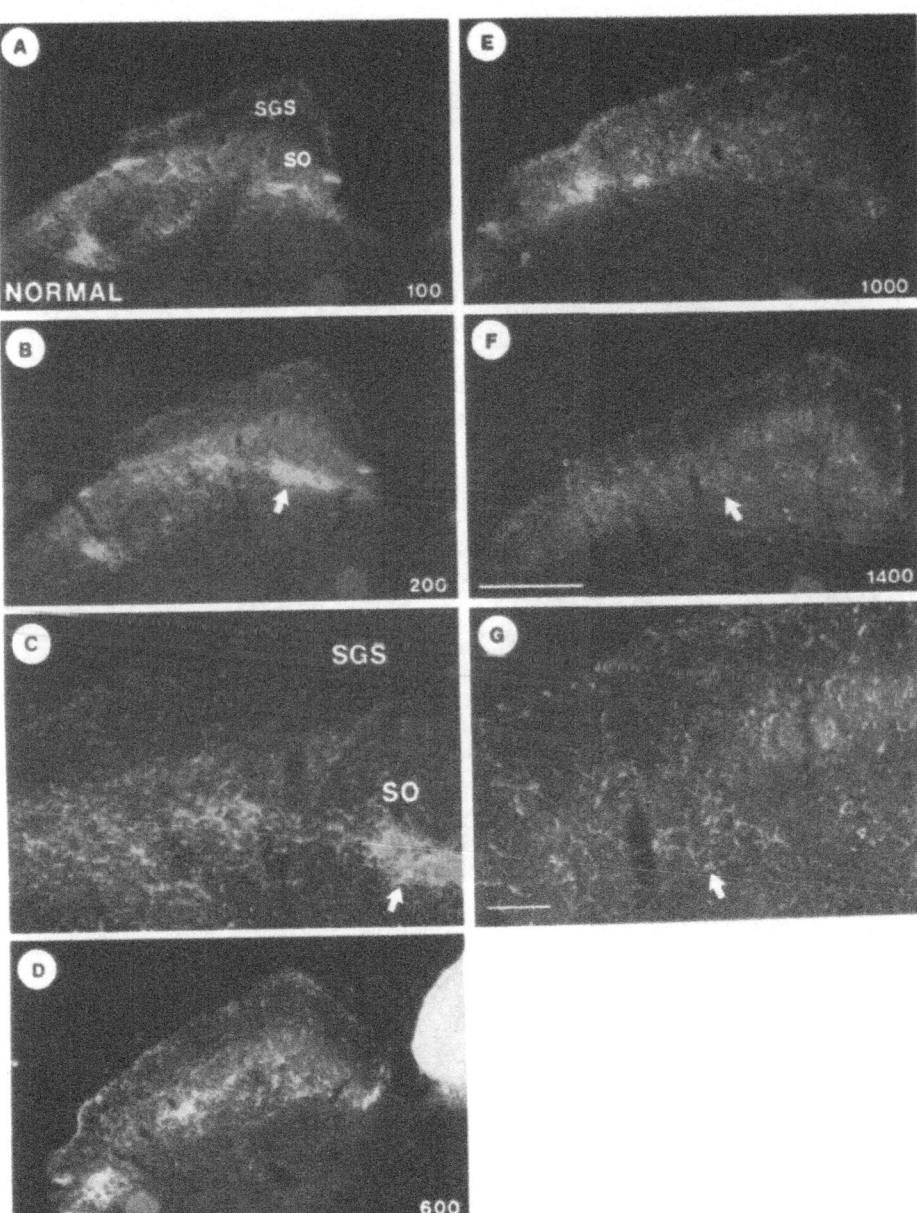

Figure 4. A–G show polarized darkfield photomicrographs of the HRP-labelled ipsilateral retinotectal projection in a normal adult hamster. The numbers at the lower right of panels A, B, D, E, and F indicate distance in μm from the respective section to the rostral border of SC. Panels C and G are higher power photomicrographs of the sections shown in B and F, respectively (arrows denote corresponding points in each pair). Note discrete patches of labeling and additional isolated fibers in the rostral sections and scattered fibers in caudal sections. The calibration bar in F is 500 μm and also applies to panels A, B, D, and E; the calibration bar in G is 100 μm and also applies to C.

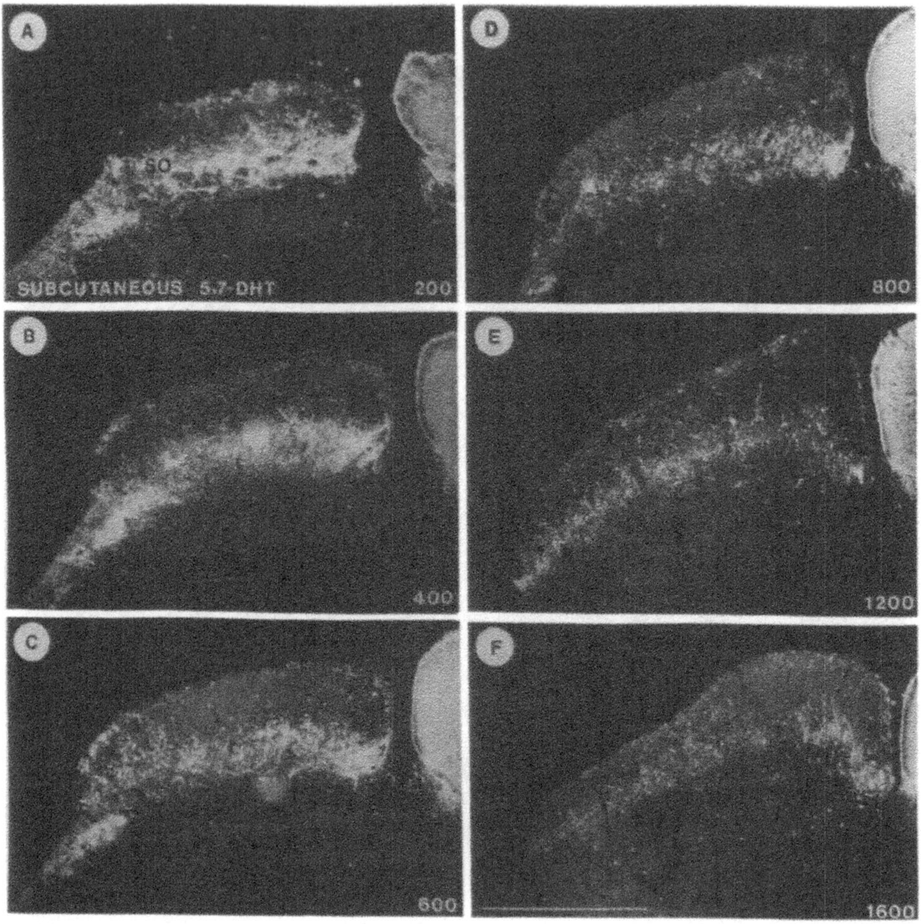

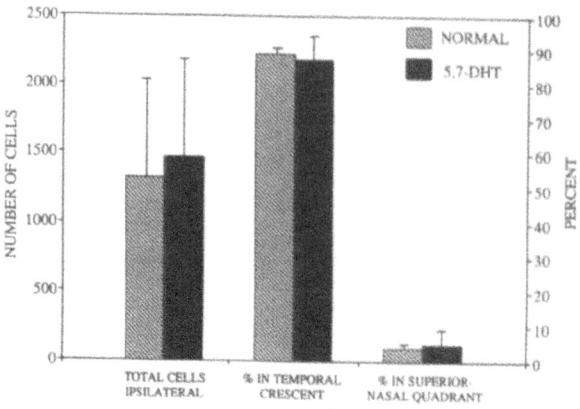

Figure 5. A-F show polarized darkfield photomicrographs of the uncrossed retinotectal projection in an adult hamster that received a subcutaneous injection of 5,7-DHT at birth. Conventions are the same as in Fig. 4. The calibration in F is calibration bar in F is 500 µm and applies to all panels.

Figure 6. Bar graphs showing the average (mean ± standard deviation) of the total number of ipsilaterally projecting retinotectal ganglion cells, the number of labelled cells in the temporal crescent of the retina, and the number of such cells in the superior nasal retinal quadrant following injections of HRP into the ipsilateral SC in normal adult hamsters and in animals treated at birth with 5,7-DHT. There were no statistically significant between-group differences for any of these measures.

distribution of soma diameters for HRP-labelled retinal ganglion cells. Electron micro-scopic analysis demonstrated no significant difference between the numbers of optic nerve fibers in the normal and 5,7-DHT-treated hamsters. All of these results are consistent with the conclusion that the effect of 5,7-DHT upon the retinotectal projection may be primar-ily a function of this toxin or the increase in 5-HT it induces on the terminal arbors of reti-notectal axons rather than their parent cells.

6. DIRECT APPLICATION OF 5-HT TO THE DEVELOPING SC ALTERS RETINOTECTAL DEVELOPMENT

The results of the experiments described in the preceding sections suggested that the effect of increased SC 5-HT on retinotectal development was due to increased presynaptic inhibition of transmission at the retinotectal synapse (Mooney et al., 1994). However, the abnormal projection pattern may have actually been due to: (i) direct effects of either 5-HT or 5,7-DHT upon developing retinal axons, (ii) alterations in SC architecture associ-ated with sprouting of 5-HT axons that were not dependent on the transmitter they con-tained, or (iii) an indirect effect of 5,7-DHT on synaptic transmission in the retina (Thier and Wässle, 1984; Brunken and Daw, 1987). To control for these possibilities, we carried out an additional experiment to further test the possibility that increasing the concentra-tion of 5-HT in the developing SC was responsible for the alterations in the uncrossed reti-notectal projection. In this trial, the level of 5-HT in the SGS and SO was raised by placing slow-release polymer (ELVAX) chips impregnated with 5-HT over the developing SC and assessing the organization of the uncrossed retinotectal projection via anterograde tracing with HRP.

Serotonin-impregnated ELVAX implants increased 5-HT concentrations in SC through P-12 (11 days after implantation, Figure 7). Analysis of data from hamsters im-planted on P-1 with 5-HT-containing or control chips and sacrificed on P-4 through P-21 indicated that the elevation of 5-HT concentrations in SC was statistically significant (p < 0.001) and was inversely correlated with age for the 5-HT-implanted hamsters (r = −0.54; p < 0.003). The difference between control and experimental groups was due to samples taken on or before P-12 (p < 0.0005); no significant difference was detected in groups of

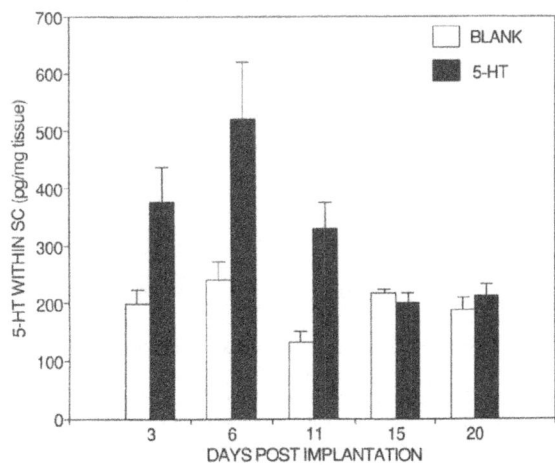

Figure 7. Quantity of 5-HT, normalized by tissue weight, in superior colliculi (SC) underlying 5-HT-impregnated and blank ELVAX implants. Implants were inserted on P-1 and remained for the number of days indicated. Each bar represents the mean and s.e.m. for 3 or more animals.

animals sacrificed ≥ P-16 (p > 0.7). It is important to note that the increases demonstrated by this analysis probably underestimate the rise in 5-HT in the superficial laminae. The dissections of the SC used in this analysis also included at least some portion of the deep laminae, and it is reasonable to suggest that concentrations of 5-HT in these layers were lower than those closer to the pial surface.

In order to test whether 5-HT released from implants was sufficient to alter neuronal activity in the superficial layers of the SC, recordings of single or multiple unit activity were made acutely from adult hamsters during placement of control or 5-HT-containing implants on the SC. Figure 8 shows peristimulus time histograms (PSTHs) of visual responses recorded just after a 5-HT implant was positioned on the SC. A diminution of the responses was apparent by about 10 minutes after data collection started. By 1 hour, the responses were much weaker; and about 10 minutes later, just after removal of the implant, visually evoked activity had nearly disappeared. The responses fully recovered over the succeeding 30 minutes. Control implants, which contained no 5-HT, had no significant effects on neuronal responsivity. These tests thus indicate that when applied in ELVAX chips, 5-HT suppresses visually elicited SC activity. This effect is similar to, but much slower in onset than, that observed during iontophoresis or micropressure ejection of 5-HT onto SC neurons (Huang et al., 1993).

Application of 5-HT directly to the developing SC produced alterations in the uncrossed retinotectal projection similar to those observed in the hamsters that received neonatal 5,7-DHT injections (Figure 9). In hamsters that received control implants, the organization of the uncrossed retinotectal pathway was essentially the same as that reported previously for normal adult hamsters by Frost et al. (1979), Rhoades et al. (1982, 1993), and Woo et al. (1985). There are dense patches of labelled axons in the rostral one-third of the SC (Fig. 4A and C) and sparsely scattered labelled fibers in the SGS and SO throughout the rostrocaudal extent of the nucleus.

The organization of the uncrossed retinotectal projection in animals that had 5-HT-impregnated implants was qualitatively different from that in the control animals. The

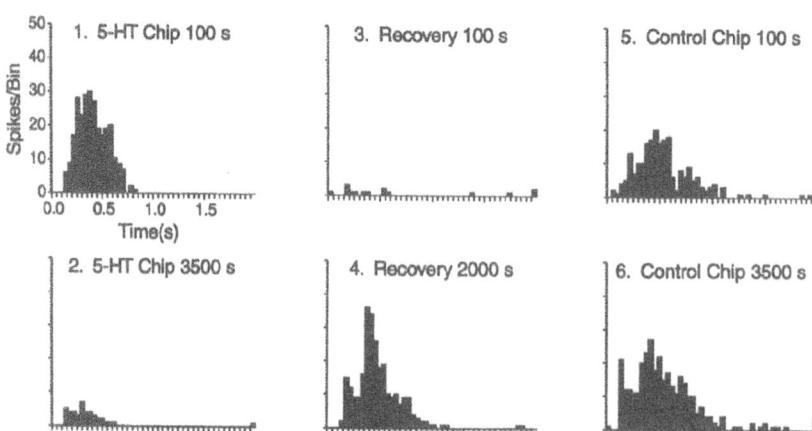

Figure 8. Suppression and recovery of visual activity by 5-HT released from an ELVAX implant. Panels 1–6 show PSTHs of visually evoked responses from a neuron 210 μm below the pial surface of the SC. The PSTH in each panel was accumulated over 100 s, beginning at the times indicated, which are referenced to implant placement or removal. Data for PSTHs 1 and 2 were obtained while the implant remained on the SC, and PSTHs 3 and 4 were collected after its removal. Following placement of a blank, but otherwise identical, implant upon the SC, data shown in PSTHs 5 and 6 were obtained.

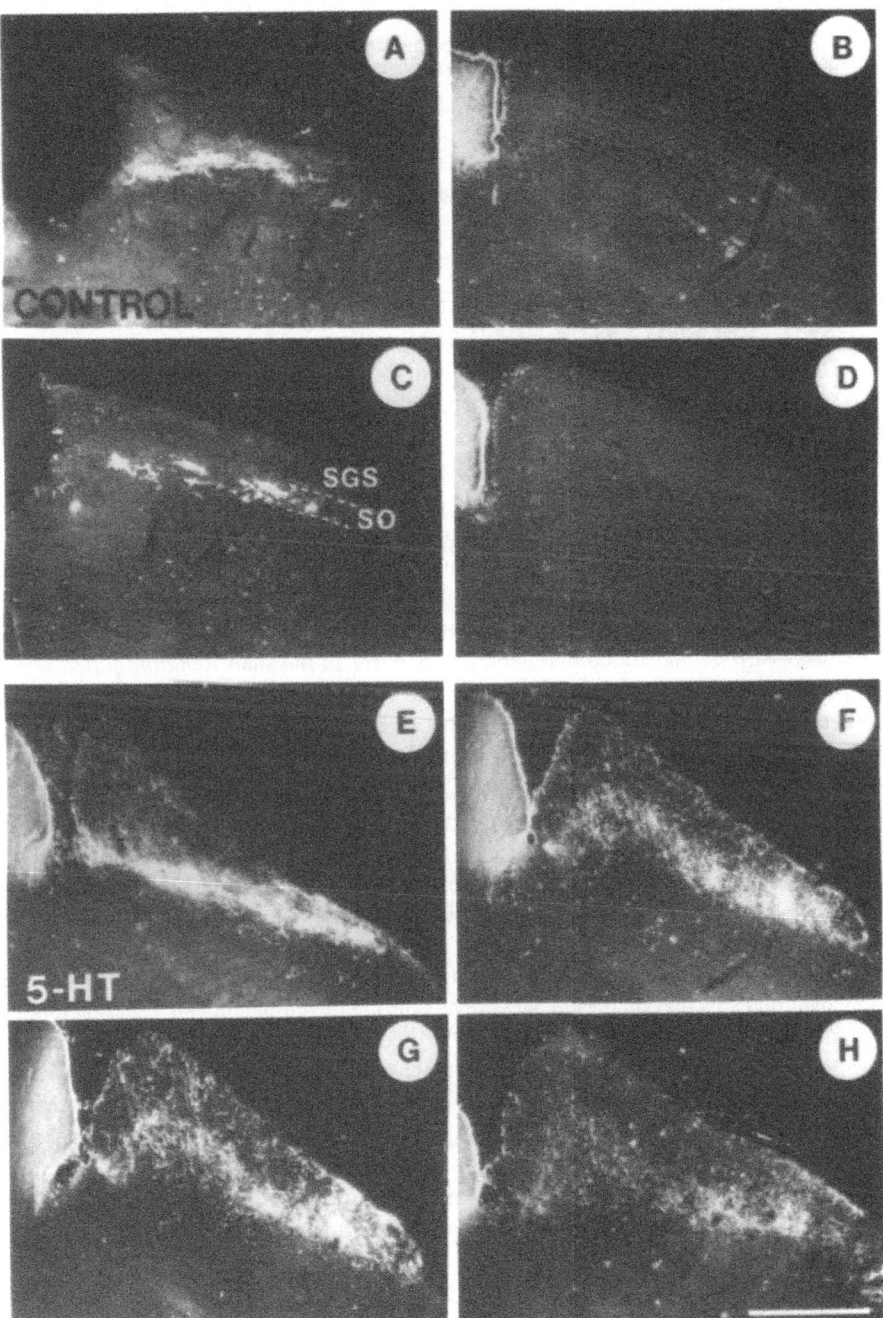

Figure 9. Polarized darkfield photomicrographs showing HRP-labelled uncrossed retinotectal terminals in the SC of 2 animals with control implants (A–D) and 2 others with 5-HT-impregnated implants (E–H). Left hand sections of each pair (A, C, E, G) were taken from the rostral 200 μm of the SC, and right hand sections (B, D, F, H) were from the caudal 300 μm. Note that terminal labelling in both 5-HT-implanted animals is continuous in the SO and lower SGS throughout the mediolateral and rostrocaudal extents of the SC. Borders of the SGS and SO are outlined in C. The calibration line in H is 1 mm and applies to all panels.

patches characteristic of the uncrossed retinal projection to the rostral one-third of the SC in normal hamsters were much less distinct in the hamsters that received the 5-HT-impregnated implants. Instead, there was a relatively dense, but diffuse, innervation of the entire SO. Such labelling was very dense in the rostral one-third of the SC, but relatively dense labelling, restricted primarily to the SO, was present at all levels.

Quantitative analyses of the density of the uncrossed retinotectal projection confirmed that the animals which received 5-HT-impregnated implants differed significantly from those that received vehicle-impregnated ELVAX chips and differed also from normal animals.

7. SUMMARY AND CONCLUSIONS

The results of the experiments described above demonstrate that increasing 5-HT levels in the developing SC results in an abnormally widespread distribution of the uncrossed retinotectal projection. There could be two reasonable explanations for these observations. First, it is possible that 5-HT has direct positive effects on axonal growth. *In vitro* studies have shown that increasing the concentration of 5-HT enhances neuropil formation and synaptic development in the growing cortex (e.g. Chubakov et al., 1986). However, Kater and his colleagues (Haydon et al., 1984; 1987; McCobb et al., 1988a,b) have shown that addition of 5-HT to the culture medium surrounding growing neurites from *Helisoma* neurons caused an immediate cessation of their elongation. Baker et al. (1993) have also demonstrated that depletion of 5-HT in the snail *Achatina fulica* results in axonal sprouting by buccal ganglion neurons. The second potential explanation for the effects of 5-HT on retinal development is that it modulates activity-dependent processes known to be involved in the refinement of this and other pathways (Fawcett et al., 1984; Thompson and Holt, 1989; Simon et al., 1992). It will be difficult to resolve the two possibilities put forward in the preceding paragraphs *in vivo*. However, it may be possible to determine whether or not 5-HT has direct activity-independent effects on axon-outgrowth of retinal ganglion cells maintained in dissociated or even single cell cultures where activity-dependent interneuronal interactions should have little opportunity to influence axonal outgrowth or arborization.

The present results are consistent with those of another recently published study which also indicates that increased 5-HT levels during development can affect the organization of axonal projections. Cases et al. (1996) reported that 5-HT levels in mice lacking the gene for monoamine oxidase A were as much as 900% higher than those in wild type animals, and that thalamic axons in these mice did not form a normal pattern corresponding to the mystacial vibrissae in the primary somatosensory cortex. The conclusion that elevated 5-HT levels were responsible for this abnormality was supported by an additional experiment in which endogenous levels of this amine were reduced by administration of parachloroamphetamine, and a vibrissae-related pattern appeared in the cortices of transgenic mice.

REFERENCES

Aitken, A.R., and I. Tork (1988) Early development of serotonin-containing neurons and pathways as seen in wholemount preparations of the fetal rat brain. J. Comp. Neurol. 274:32–47.

Andrade, R., and Y. Chaput (1991) The electrophysiology of serotonin receptor subtypes. In Venter, J.C. and Harrison, L.C. (eds): Serotonin Receptor Subtypes: Basic and Clinical Aspects. New York, Wiley-Liss, Inc., pp. 103–124.

Baker, M.W., M.M. Vohra, and R.P. Croll, (1993) Serotonin depletors, 5,7-dihydroxytryptamine and p-chlorophenylanine, cause sprouting in the CNS of the adult snail. Brain Res. 623:311–315.

Bennett-Clarke, C.A., M.J. Leslie, R.D. Lane, and R.W. Rhoades (1994) Effect of serotonin depletion on vibrissae-related patterns of thalamic afferents in the rat's somatosensory cortex. J. Neurosci. 14:7594–7606.

Blue, M.E., J.W. McDonald, and M.V. Johnston (1991) Ontogeny of excitatory amino acid receptors in barrel field cortex of immature rat. Soc. Neurosci. Absts. 17:625.

Blue, M.E., K.A. Yagaloff, L.A. Mamounas, P.R. Hartiag, and M.E. Molliver, (1988) Correspondence between 5-HT_2 receptors and serotonergic axons in rat neocortex. Brain Res. 453:315–328.

Boschert, U., D.A. Amara, L. Segu, and R. Hen, (1994) The mouse 5-hydroxytryptamine$_{1B}$ receptor is localized predominantly on axon terminals. Neuroscience 58:167–182.

Boulenguez, P., J. Abdelkefi, R. Pinard, A. Christolomme, and L. Segu, (1993) Effects of retinal deafferentation on serotonin receptor types in the superficial grey layer of the superior colliculus of the rat. J. Chem. Neuroanat. 6:167–175.

Brunken, W.J., and N.G. Daw (1987) The actions of serotonergic agonists and antagonists on the activity of brisk ganglion cells in the rabbit retina. J. Neurosci. 7:4054–4065.

Cases, O., T. Vitalls, I. Self, E. DeMaeyer, C. Sotelo, and P. Gaspar (1996) Lack of barrels in the somatosensory cortex of monoamine oxidase A-deficient mice: role of a serotonin excess during the critical period. Neuron 16:297–307.

Chubakov, A.R., E.A. Gromova, G.V. Konovalov, E.F. Sarkisova, and E.I. Chumasov (1986) The effects of serotonin on the morphofunctional development of rat cerebral neocortex in tissue culture. Brain Res. 369:285–297.

Cline, H.T., and M. Constantine-Paton (1990) NMDA receptor agonist and antagonists alter retinal ganglion cell arbor structure in the developing frog retinotectal projection. J. Neurosci., 10:1197–1216.

Cline, H.T., E. A. Debski, and M. Constantine-Paton (1991) N-methyl-D-aspartate receptor antagonist desegregates eye-specific stripes. Proc. Natl. Acad. Sci. USA 84:4342–4345.

Edwards, M.A., G.E. Schneider, and V.S. Caviness, Jr. (1986) Development of the crossed retinocollicular projection in the mouse. J. Comp. Neurol. 248:410–421.

Fawcett, J.W., D.D.M. O'Leary, and W.M. Cowan (1984) Activity and the control of ganglion cell death in the rat retina. Proc. Natl. Acad. Sci. USA 81:5589–5593.

Forda, S., and J.S. Kelly (1985) The possible modulation of the development of rat dorsal root ganglion cells by the presence of 5-HT-containing neurones of the brainstem in dissociated cell culture. Dev. Brain Res. 22:55–65.

Frost, D.O., K.F. So, and G.E. Schneider (1979) Postnatal development of retinal projections in Syrian hamsters: a study using autoradiographic and anterograde degeneration techniques. Neuroscience 4:1649–1677.

Godement, P., J. Salaun, and M. Imbert (1984) Prenatal and postnatal development of retinogeniculate and retinocollicular projections in the mouse. J. Comp. Neurol. 230:552–575.

Gromova, H.A., A.R. Chubakov, E.I. Chumasov, and H.V. Knonvalov (1983) Serotonin as a stimulator of hippocampal cell differentiation in tissue culture. Int. J. Dev. Neurosci. 1:339–349.

Harvey, A.R., and A.M. MacDonald (1987) The host serotonin projection to tectal grafts in young rats: an immunohistochemical study. Exp. Neurol. 95:688–696.

Haydon, P.G., D.P. McCobb, and S.R. Kater (1984) Serotonin selectively inhibits growth cone motility and synaptogenesis of specific identified neurons. Science 226:561–564.

Haydon, P.G., D.P. McCobb, and S.B. Kater (1987) The regulation of neurite outgrowth cone motility, and electrical synaptogenesis by serotonin. J. Neurobiol. 18:197–215.

Holohean, A.M., J.C. Hackman, S.B. Shope, and R.A. Davidoff (1992) Activation of 5-$HT_{1C/2}$ receptors depresses polysynaptic reflexes and excitatory amino acid-induced motoneuron responses in frog spinal cord. Brain Res. 579:8–16.

Huang, X., R.D. Mooney, and R.W. Rhoades (1993) Effects of serotonin (5-HT) upon retinotectal, corticotectal, and glutamate-induced activity in the superior colliculus of the hamster. J. Neurophysiol. 70:723–732.

Jhaveri, S., and G.E. Schneider (1994) Pathway navigation and target innervation in the visual system: normal axon growth and some effects of brain injury. In D.M. Albert and F.A. Jakobiec (eds) Principles and Practice of Ophthalmology: Basic Sciences. Philadelphia, W.B. Saunders Co., pp. 511–521.

Kater, S.B., and L.R. Mills (1991) Regulation of growth cone behavior by calcium. J. Neurosci. 11:891–899.

Kawai, N., and C. Yamamoto (1969) Effects of 5-hydroxytryptamine, LSD, and related compounds on electrical activities evoked in vitro in thin sections from the superior colliculus. Int. J. Neuropharmacol. 8:437–449.

Killackey, H.P., R.W. Rhoades, and C.A. Bennett-Clarke (1995) The formation of a cortical somatotopic map. T.I.N.S., 18:402–407.

Lauder, J.M. (1983) Hormonal and humoral influences on brain development. Psychoneuroendocrinology 8:121–155.

Lidov, H.G.W., and M.E. Molliver (1982) an immunohistochemical study of serotonin neuron development in the rat: ascending pathways and terminal fields. Brain Res. 8:389–430.

Lipton, S.A., and S.B. Kater (1989) Neurotransmitter regulation of neuronal outgrowth, plasticity and survival. T.I.N.S. 12:265–270.

Lund, R.D., and A.H. Bunt (1976) Prenatal development of central optic pathways in albino rats. J. Comp. Neurol. 165:247–264.

Marcinkiewicz, M., D. Vergé, H. Gozlan, L. Pichat, and M. Hamon (1984) Autoradiographic evidence for the heterogeneity of 5-HT, sites in the rat brain. Brain Res. 291:159–163.

McCobb, D.P., C.S. Cohan, J.A. Connor, and S.B. Kater (1988a) Interactive effect of serotonin and acetylcholine on neurite elongation. Neuron 1:377–385.

McCobb, D.P., P.G. Haydon, and S.B. Kater (1988b) Dopamine and serotonin inhibition of neurite elongation of different identified neurons. J. Neurosci. Res. 19:19–26.

Mooney, R.D., M.-Y. Shi, and R.W. Rhoades (1994) Modulation of retinotectal transmission by presynaptic 5-HT_{1B} receptors in the superior colliculus of the adult hamster. J. Neurophysiol. 72: 30–13.

Murase, K., M. Randic, T. Shirasaki, T. Nakagawa, and N. Akaike (1990) Serotonin suppresses N-methyl-D-aspartate responses in acutely isolated spinal dorsal horn neurons of the rat. Brain Res. 525:84–91.

Nakada, M.T., C.M. Wieczorek, and T.C. Rainbow (1984) Localization and characterization by quantitative autoradiography of [^{125}I] LSD binding sites in rat brain. Neurosci. Lett. 49:13–18.

Rhoades, R.W., C.A. Bennett-Clarke, N.L. Chiaia, F.A. White, G.J. MacDonald, J.H. Haring, and M.F. Jacquin (1990) Development and lesion induced reorganization of the cortical representation of the rat's body surface as revealed by immunocytochemistry for serotonin. J. Comp. Neurol. 293:190–207.

Rhoades, R.W., C.A. Bennett-Clarke, R.D. Lane, M.J. Leslie, and R.D. Mooney (1993) Increased serotoninergic innervation of the hamster's superior colliculus alters retinotectal projections. J. Comp. Neurol 334:397–409.

Rhoades, R.W., D.C. Kuo, and J.D. Polcer (1982) Effects of neonatal cortical lesions upon retinocollicular projections in the hamster. Neuroscience 7:2441–2458.

Sachs, C., and G. Jonsson (1975) 5,7-dihydroxytryptamine induced changes in the postnatal development of central 5- hydroxytryptamine neurons. Med. Biol. Eng. Comput. 53:156–164.

Segu, L. J. Abdelkefi, G. Dusticier, and J. Lanoir (1986) High-affinity serotonin binding sites: autoradiographic evidence for their location on retinal afferents in the rat superior colliculus. Brain Res. 384:205–217.

Simon, D.K., G.T. Prusky, D.D.M. O'Leary, and M. Constantine-Paton (1992) N-methyl-D-aspartate receptor antagonists disrupt the formation of mammalian neural map. Proc. Natl. Acad. Sci. 89:10593–10597.

Starchill, M., and J. Perwein (1971) Effect of iontophoretically applied biogenic amines and of cholinomimetic substances upon the activity of neurons in the superior colliculus and mesencephalic reticular formatin of the cat. Pflugers Arch. 324:43–55.

Thier, P., and H. Wässle (1984) Indoleamine-mediated reciprocal modulation of on-centre and off-centre ganglion cell activity in the retina of the cat. J. Physiol. Lond. 351:613–630.

Thompson, I., and C. Holt (1989) Effects of intraocular tetrodotoxin on the development of the retinocollicular pathway in the Syrian hamster. J. Comp. Neurol. 282:371–388.

Villar, M.J., M.L. Vitale, T. Hökfelt, and A.A.J. Verhofstad (1988) Dorsal raphe serotoninergic branching neurons projecting both to the lateral geniculate body and superior colliculus: a combined retrograde tracing-immunohistochemical study in the rat. J. Comp. Neurol. 277:126–140.

Waeber, C., and J.M. Placios (1990) 5-HT$_1$ receptor binding sites in the guinea pig superior colliculus are predominantly of the 5-HT_{1D} class and are presynaptically located on primary retinal afferents. Brain Res. 528:207–211.

Welner, S.A., C. De Montigny, J. Desroches, P. Desjardins, and B.E. Suranyi-Cadotte (1989) Autoradiographic quantification of serotonin $_{1A}$ receptors in rat brain following antidepressant drug treatment. Synapse 4:347–352.

Woo, H.H., L.S. Jen, and K.F. So (1985) The postnatal development of retinocollicular projections in normal hamsters and in hamsters following neonatal monocular enucleation: a horseradish peroxidase tracing study. Dev. Brain Res. 20:1–13.

Zifa, E., and G. Fillion (1992) 5-hydroxytryptamine receptors. Pharmacol. Rev. 44:401–458.

INDEX

The manufacturer's authorised representative in the EU is Springer
Nature Customer Service Centre GmbH, Europaplatz 3, 69115 Heidelberg,
Germany. If you have any concerns regarding our products, please
contact ProductSafety@springernature.com

Printed and bound by CPI Group (UK) Ltd, Croydon, CR0 4YY
23/04/2026
02095593-0003